1977-89 REPAIR MANUAL
2-60 HORSEPOWER, 1 AND 2 CYLINDER

Managing Partners	Dean F. Morgantini
	Barry L. Beck
Executive Editor	Kevin M. G. Maher, A.S.E.
Production Managers	Melinda Possinger
	Ronald Webb
Authors	Joan and Clarence Coles

Manufactured in USA
© 1989 Seloc Publishing, Inc.
104 Willowbrook Lane
West Chester, PA 19382
ISBN 13: 978-0-89330-015-9
ISBN 10: 0-89330-015-2
4567890123 5432109876

www.selocmarine.com
1-866-SELOC55

ALL RIGHTS RESERVED

No part of this publication may be reproduced, transmitted or stored in any form or by any means, electronic or mechanical, including photocopy, recording, or by information storage or retrieval system, without prior written permission from the publisher.

The materials contained in this manual are the intellectual property of Seloc Publishing, Inc., a Pennsylvania corporation, and are protected under the laws of the United States of America at Title 17 of the United States Code. Any efforts to reproduce any of the content of this manual, in any form, without the express written permission of Seloc Publishing, Inc. is punishable by a fine of up to $250,000 and 5 years in jail, plus the recovery of all proceeds including attorneys fees.

MARINE TECHNICIAN TRAINING

INDUSTRY SUPPORTED PROGRAMS
OUTBOARD, STERNDRIVE & PERSONAL WATERCRAFT

- Dyno Testing • Boat & Trailer Rigging • Electrical & Fuel System Diagnostics
- Powerhead, Lower Unit & Drive Rebuilds • Powertrim & Tilt Rebuilds
- Instrument & Accessories Installation

TRAIN IN SUNNY FLORIDA!

For information regarding housing, financial aid and employment opportunities in the marine industry, contact us today:

CALL TOLL FREE
1-800-528-7995

An Accredited Institution

SM

Name
Address
City   State   Zip
Phone

MARINE MECHANICS INSTITUTE
A Division of CTI
MEMBER NMMA
9751 Delegates Drive • Orlando, Florida 32837
2844 W. Deer Valley Rd. • Phoenix, AZ 85027

FINANCIAL ASSISTANCE AVAILABLE FOR THOSE WHO QUALIFY!

TABLE OF CONTENTS

1 SAFETY

INTRODUCTION	1-1
CLEANING, WAXING, & POLISHING	1-1
CONTROLLING CORROSION	1-2
PROPELLERS	1-2
LOADING	1-7
HORSEPOWER	1-8
FLOTATION	1-8
ANCHORS	1-10
BOATING ACCIDENT REPORTS	1-10

2 TUNING

INTRODUCTION	2-1
TUNE-UP SEQUENCE	2-1
COMPRESSION CHECK	2-3
SPARK PLUG INSPECTION	2-3
IGNITION SYSTEM	2-3
TIMING AND SYNCHRONIZING	2-4
ELECTRICAL POWER SUPPLY	2-5
CARBURETOR ADJUSTMENT	2-6
FUEL PUMPS	2-7
CRANKING MOTOR TEST	2-8
INTERNAL WIRING HARNESS	2-8
WATER PUMP CHECK	2-9
PROPELLER	2-9
LOWER UNIT	2-10
BOAT TESTING	2-10

3 MAINTENANCE

INTRODUCTION	3-1
OUTBOARD SERIAL NUMBERS	3-2
LUBRICATION - COMPLETE UNIT	3-3
EMERGENCY TETHER	3-3
PRE-SEASON PREPARATION	3-3
FIBERGLASS HULLS	3-6
BELOW WATERLINE SERVICE	3-6
SUBMERGED ENGINE SERVICE	3-7
PROPELLER SERVICE	3-8
INSIDE THE BOAT	3-9
LOWER UNIT	3-15
WINTER STORAGE	3-16

4 FUEL

INTRODUCTION	4-1
GENERAL CARBURETION INFORMATION	4-1
FUEL SYSTEM	4-3
Leaded Gasoline & Gasohol	4-3
Removing Fuel From System	4-4
TROUBLESHOOTING	
"Sour" Fuel	4-5
Fuel Pump Test	4-5
Fuel Line Test	4-7
Rough Engine Idle	4-7
Excessive Fuel Consumption	4-8
Engine Surge	4-8
CARBURETOR IDENTIFICATION	4-9
CARBURETOR "A"	
Removal & Disassembling	4-10
Cleaning & Inspecting	4-14
Assembling & Installation	4-15
Adjustments	4-19
CARBURETOR "B"	
Removal & Disassembling	4-20
Cleaning & Inspecting	4-23
Assembling & Installation	4-23
Adjustments	4-25
CARBURETOR "C"	
Removal & Disassembling	4-27
Cleaning & Inspecting	4-30
Assembling & Installation	4-31
Adjustments	4-34
CARBURETOR "D"	
Removal & Disassembling	4-35
Cleaning & Inspecting	4-37
Assembling & Installation	4-39
Adjustments	4-42
CARBURETOR "E"	
Removal & Disassembling	4-43
Cleaning & Inspecting	4-45
Assembling & Installation	4-47
Adjustments	4-49

4 FUEL (Continued)

CARBURETOR "F"
- Removal & Disassembling — 4-50
- Cleaning & Inspecting — 4-53
- Assembling & Installation — 4-56
- Adjustments — 4-61

CARBURETOR "G"
- Removal & Disassembling — 4-61
- Cleaning & Inspecting — 4-65
- Assembling & Installation — 4-66
- Adjustments — 4-70

CARBURETOR "H"
- Removal & Disassembling — 4-71
- Cleaning & Inspecting — 4-73
- Assembling & Installation — 4-75
- Adjustments — 4-77

CARBURETOR "J"
- Removal & Disassembling — 4-79
- Cleaning & Inspecting — 4-82
- Assembling & Installation — 4-84
- Adjustments — 4-88

CARBURETOR "K"
- Removal & Disassembling — 4-89
- Cleaning & Inspecting — 4-93
- Assembling & Installation — 4-94
- Adjustments — 4-99

CARBURETOR "L"
- Removal & Disassembling — 4-100
- Cleaning & Inspecting — 4-102
- Assembling & Installation — 4-104
- Adjustments — 4-107

FUEL PUMP
- Theory of Operation — 4-108
- Pump Pressure Check — 4-109
- Pump Removal — 4-110
- Cleaning & Inspecting — 4-110
- Assembling — 4-111

INTRODUCTION — 4-113
AUTO BLEND
- Description — 4-113
- Operation — 4-115
- Troubleshooting — 4-115
- System Storage — 4-117
- Preparation for Use — 4-118

5 IGNITION

INTRODUCTION AND CHAPTER COVERAGE — 5-1

SPARK PLUG EVALUATION — 5-2
POLARITY CHECK — 5-4

TYPE I IGNITION SYSTEM - MAGNETO WITH BREAKER POINT SET — 5-4
- Description & Operation — 5-4
- Troubleshooting — 5-6
- Disassembling — 5-9
- Testing Components — 5-12
- Assembling & Installation — 5-14
- Breaker Point Adjustment — 5-15

TYPE II IGNITION SYSTEM - CDI (CAPACITOR DISCHARGE) — 5-17
- Description & Operation — 5-17
- Troubleshooting — 5-18
- Testing One-Cylinder Components — 5-20
- Testing Two-Cylinder Components — 5-23

FLYWHEEL AND STATOR PLATE SERVICE — 5-31
- "Pulling" the Flywheel — 5-31
- Stator Plate Removal — 5-33
- Cleaning & Inspecting — 5-34
- Assembling & Installation — 5-35

6 TIMING AND SYNCHRONIZING

INTRODUCTION AND PREPARATION — 6-1
- Tachometer Connections — 6-2

MODELS W/BREAKER POINT TYPE IGNITION — 6-4
MODEL 4HP & 5HP — 6-5
MODEL 8C — 6-7
MODEL 9.9HP, 15HP, & ALL OTHER 8HP AFTER 1979 — 6-8
MODEL 20HP, 25HP, & 30HP — 6-11
MODEL 40HP /SINGLE CARB — 6-14
MODEL 40HP /DUAL CARB — 6-17
MODEL 48HP, 55HP, & 60HP — 6-21

7 POWERHEAD

INTRODUCTION AND CHAPTER ORGANIZATION — 7-1
TWO-STROKE POWERHEAD DESCRIPTION & OPERATION — 7-2

SERVICING POWERHEAD "A"
MODEL 2HP — 7-3
- Removal & Disassembling — 7-3
- Cleaning & Inspecting — 7-67
- Assembling & Installation — 7-8

SERVICING POWERHEAD "B"
   SINGLE-CYLINDER AIR-COOLED
   MODEL 3.5HP AND 5HP   7-13
      Removal & Disassembling   7-13
      Cleaning & Inspecting   7-67
      Assembling & Installation   7-18

SERVICING POWERHEAD "C"
   SINGLE-CYLINDER
   WATER-COOLED
   MODEL 4HP & 5HP   7-22
      Removal & Disassembling   7-22
      Cleaning & Inspecting   7-67
      Assembling & Installation   7-26

SERVICING POWERHEAD "D"
   TWO-CYLINDER,
   WATER-COOLED
   MODEL 8HP, 9.9HP, 15HP, 20HP,
   25HP, 28HP, 30HP, 40HP, 48HP,
   55HP, AND 60HP   7-33
      Powerhead Preparation   7-33
      Removal & Disassembling   7-36
      Cleaning & Inspecting   7-67
      Assembling & Installation   7-42
      Closing Tasks   7-66

CLEANING & INSPECTING   7-67
   Reed Block Service   7-67
   Exhaust Cover   7-68
   Crankshaft Service   7-69
   Connecting Rod Service   7-72
   Piston Service   7-73
   Ring End Gap Clearance   7-74
   Piston Ring Side Clearance   7-76
   Oversize Pistons & Rings   7-76
   Cylinder Block Service   7-77
      Honing Cylinder Walls   7-79
      Block & Cylinder Head
      Warpage   7-80

8 ELECTRICAL

INTRODUCTION   8-1
BATTERIES   8-1
   Jumper Cables   8-5
   Storage   8-6
THERMOMELT STICKS   8-6
TACHOMETER   8-6
ELECTRICAL SYSTEM
   GENERAL INFORMATION   8-7
CHARGING CIRCUIT SERVICE   8-9
CRANKING MOTOR
   CIRCUIT SERVICE   8-9
   Motor Troubleshooting   8-10
   Relay Testing   8-11

CRANKING MOTOR SERVICE   8-13
   Description   8-13
   Removal   8-14
   Disassembling   8-14
   Cleaning & Inspecting   8-17
   Testing Cranking Motor Parts   8-17
   Assembling   8-19
   Installation   8-21
TESTING OTHER
   ELECTRICAL PARTS   8-21

9 LOWER UNIT

DESCRIPTION   9-1
CHAPTER COVERAGE   9-1
TROUBLESHOOTING   9-2
PROPELLER REMOVAL   9-3
PROPELLER INSTALLATION   9-5

LOWER UNIT TYPE "A"
   NO REVERSE GEAR   9-6
   Removal   9-6
   Disassembling   9-7
   Cleaning & Inspecting   9-11
   Assembling   9-13
   Installation   9-18

LOWER UNIT TYPE "B"
   FORWARD, NEUTRAL & REVERSE,
   ONE-PIECE GEARCASE   9-20
   Description   9-20
   Removal   9-21
   Disassembling   9-23
      Water Pump   9-23
   Cleaning & Inspecting   9-33
   Assembling   9-43
      Pinion Gear Depth   9-49
      Backlash Measurement   9-54
      Water Pump   9-58
   Installation   9-60
   Propeller Installation   9-63
   Trim Tab Adjustment   9-64
   Closing Tasks   9-64
   Lower Unit Capacities   9-64

LOWER UNIT TYPE "C"
   FORWARD, NEUTRAL & REVERSE,
   TWO-PIECE GEARCASE   9-65
   Description   9-65
   Removal   9-65
   Disassembling   9-66
      Water Pump   9-67
   Cleaning & Inspecting   9-72
   Assembling   9-75
      Pinion Gear Depth   9-78
      Backlash Measurement   9-78
      Water Pump   9-79

9 LOWER UNIT (Continued)

LOWER UNIT TYPE "C" (Cont.)
Installation	9-80
Propeller Installation	9-82
Trim Tab Adjustment	9-82
Closing Tasks	9-82

10 REMOTE CONTROLS

INTRODUCTION	10-1
DISASSEMBLING	10-2
CLEANING & INSPECTING	10-9
ASSEMBLING	10-11

11 HAND REWIND STARTER

INTRODUCTION AND CHAPTER COVERAGE	11-1
STARTER IDENTIFICATION	11-1

SERVICING TYPE "A"
Removal & Disassembling	11-1
Cleaning & Inspecting	11-4
Assembling & Installation	11-5

SERVICING TYPE "B"
Removal & Disassembling	11-8
Cleaning & Inspecting	11-9
Assembling & Installation	11-10

SERVICING TYPE "C"
Removal & Disassembling	11-11
Cleaning & Inspecting	11-15
Assembling & Installation	11-15

SERVICING TYPE "D"
Removal & Disassembling	11-19
Cleaning & Inspecting	11-20
Assembling & Installation	11-20

SERVICING TYPE "E"
Removal & Disassembling	11-22
Cleaning & Inspecting	11-24
Assembling & Installation	11-24

SERVICING TYPE "F"
Removal & Disassembling	11-25
Cleaning & Inspecting	11-27
Assembling & Installation	11-28

APPENDIX

METRIC CONVERSION CHART	A-1
ENGINE SPECIFICATIONS AND TUNE-UP ADJUSTMENTS	A-2
WIRING DIAGRAMS	
Model 2hp	A-10
Model 3.5hp & 5hp (Air-Cooled)	A-10
Model 4hp & 5hp (Water-Cooled)	A-11
Model 8hp, 15hp, 20hp, & 28hp (Breaker Point Ignition System)	A-12
Model 8hp, 15hp, & 20hp (Electric Start System)	A-13
Model 8hp, 9.9hp, & 15hp (CDI Ignition System)	A-14
Model 20hp, 25hp, & 30hp (CDI Ignition System)	A-15
Model 28hp (Electric Start System)	A-16
Model 48hp, 55hp, & 60hp	A-18

1
SAFETY

1-1 INTRODUCTION

In order to protect the investment for the boat and outboard, they must be cared for properly while being used and when out of the water. Always store the boat with the bow higher than the stern and be sure to remove the transom drain plug and the inner hull drain plugs. If any type of cover is used to protect the boat, be sure to allow for some movement of air through the hull. Proper ventilation will assure evaporation of any condensation that may form due to changes in temperature and humidity.

1-2 CLEANING, WAXING, AND POLISHING

Any boat should be washed with clear water after each use to remove surface dirt and any salt deposits from use in salt water. Regular rinsing will extend the time between waxing and polishing. It will also give you "pride of ownership", by having a sharp looking piece of equipment. Elbow grease, a mild detergent, and a brush will be required to remove stubborn dirt, oil, and other unsightly deposits.

Stay away from harsh abrasives or strong chemical cleaners. A white buffing com-

A "fun day" on the water is always enhanced by prudent activities, including boat operation and attention to safety.

1-2 SAFETY

Zinc installation also used as the trim tab. The tab assists the helmsperson to maintain a true course without "fighting" the wheel.

pound can be used to restore the original gloss to a scratched, dull, or faded area. The finish of your boat should be thoroughly cleaned, buffed, and polished at least once each season. Take care when buffing or polishing with a marine cleaner not to overheat the surface you are working, because you will burn it.

1-3 CONTROLLING CORROSION

Since man first started out on the water, corrosion on his craft has been his enemy. The first form was merely rot in the wood and then it was rust, followed by other forms of destructive corrosion in the more modern materials. One defense against corrosion is to use similar metals throughout the boat. Even though this is difficult to do in designing a new boat, particularly the undersides, similar metals should be used whenever and wherever possible.

An anode installed within the water jacket of the Model 40hp powerhead provides added protection against corrosion.

A second defense against corrosion is to insulate dissimilar metals. This can be done by using an exterior coating of Sea Skin or by insulating them with plastic or rubber gaskets.

Using Zinc

The proper amount of zinc attached to a boat is extremely important. The use of too much zinc can cause wood burning by placing the metals close together and they become "hot". On the other hand, using too small a zinc plate will cause more rapid deterioration of the metal you are trying to protect. If in doubt, consider the fact that it is far better to replace the zincs than to replace planking or other expensive metal parts from having an excess of zinc.

When installing zinc plates, there are two routes available. One is to install many different zincs on all metal parts and thus run the risk of wood burning. Another route, is to use one large zinc on the transom of the boat and then connect this zinc to every underwater metal part through internal bonding. Of the two choices, the one zinc on the transom is the better way to go.

The small outboard units covered in this manual all have a zinc attached somewhere. Some larger horsepower late model powerheads also have a zinc installed deep inside the powerhead. Therefore, the zinc remains with the engine at all times.

1-4 PROPELLERS

As you know, the propeller is actually what moves the boat through the water. This is how it is done. The propeller oper-

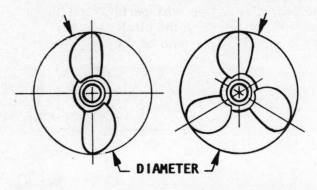

Diameter and pitch are the two basic dimensions of a propeller. The diameter is measured across the circumference of a circle scribed by the propeller blades, as shown.

ates in water in much the same manner as a wood screw does in wood. The propeller "bites" into the water as it rotates. Water passes between the blades and out to the rear in the shape of a cone. The propeller "biting" through the water in much the same manner as a wood auger is what propels the boat.

All units covered in this manual **EXCEPT** the 3.5hp and some 8 and 15hp models are equipped, from the factory, with a through the propeller exhaust. With these units, exhaust gas is forced out through the propeller.

Diameter and Pitch

Only two dimensions of the propeller are of real interest to the boat owner: The diameter and the pitch. These two dimensions are stamped on the propeller hub and always appear in the same order: the diameter first and then the pitch. Propellers furnished with the outboard by the manufacturer for the units covered in this manual have a letter designation following the pitch size. This letter indicates the propeller type. For instance, the numbers and letter 9-7/8 x 10-1/2 - F stamped on the back of one blade indicates the propeller diameter to be 9-7/8", with a pitch of 10-1/2" and it is a Type "F".

The diameter is the measured distance from the tip of one blade to the tip of the other as shown in the accompanying illustration.

The pitch of a propeller is the angle at which the blades are attached to the hub. This figure is expressed in inches of water travel for each revolution of the propeller. In our example of a 9-7/8 x 10-1/2 propeller, the propeller should travel 10-1/2 inches through the water each time it revolves. If the propeller action was perfect and there was no slippage, then the pitch multiplied by the propeller rpms would be the boat speed.

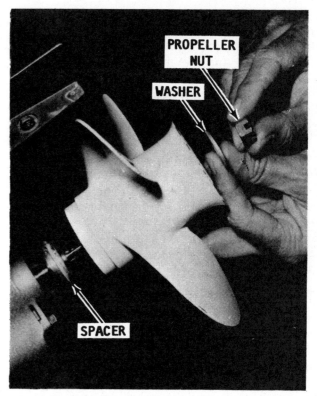

Typical attaching hardware for a propeller.

Most outboard manufacturers equip their units with a standard propeller having a diameter and pitch they consider to be best suited to the engine and the boat. Such a propeller allows the engine to run as near to the rated rpm and horsepower (at full throttle) as possible for the boat design.

The blade area of the propeller determines its load-carrying capacity. A two-blade propeller is used for high-speed running under very light loads.

A four-blade propeller is installed in boats intended to operate at low speeds under very heavy loads such as tugs, barges, or large houseboats. The three-blade propeller is the happy medium covering the wide range between the high performance units and the load carrying workhorses.

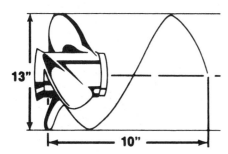 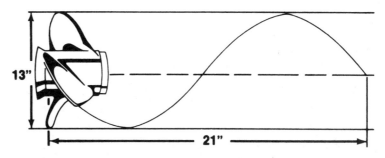

Diagram to explain the pitch dimension of a propeller. The pitch is the theoretical distance a propeller would travel through water if there were no friction.

Propeller Selection

There is no standard propeller that will do the proper job in very many cases. The list of sizes and weights of boats is almost endless. This fact coupled with the many boat-engine combinations makes the propeller selection for a specific purpose a difficult task. Actually, in many cases the propeller may be changed after a few test runs. Proper selection is aided through the use of charts set up for various engines and boats. These charts should be studied and understood when buying a propeller. However, bear in mind, the charts are based on average boats with average loads, therefore, it may be necessary to make a change in size or pitch, in order to obtain the desired results for the hull design or load condition.

Propellers are available with a wide range of pitch. Remember, a low pitch takes a smaller bite of the water than the high pitch propeller. This means the low pitch propeller will travel less distance through the water per revolution. The low pitch will require less horsepower and will allow the engine to run faster.

All engine manufacturers design their units to operate with full throttle at, or slightly above, the rated rpm. If the powerhead is operated at the rated rpm, several positive advantages will be gained.

1- Spark plug life will be increased.
2- Better fuel economy will be realized.
3- Easier steering qualities.
4- Best performance received from the boat and power unit.

Therefore, take time to make the proper propeller selection for the rated rpm of the engine at full throttle with what might be considered an "average" load. The boat will then be correctly balanced between engine and propeller throughout the entire speed range.

A reliable tachometer must be used to measure powerhead speed at full throttle to ensure the engine will achieve full horsepower and operate efficiently and safely. To test for the correct propeller, make a test run in a body of smooth water with the lower unit in forward gear at full throttle. If the reading is above the manufacturer's recommended operating range, try propellers of greater pitch, until one is found allowing the powerhead to operate continually within the recommended full throttle range.

If the engine is unable to deliver top performance and the powerhead is properly tuned, then the propeller may not be to blame. Operating conditions have a marked effect on performance. For instance, an engine will lose rpm when run in very cold water. It will also lose rpm when run in salt water as compared with fresh water. A hot, low-barometer day will also cause the engine to lose power.

Cavitation

Cavitation is the forming of voids in the water just ahead of the propeller blades. Marine propulsion designers are constantly fighting the battle against the formation of

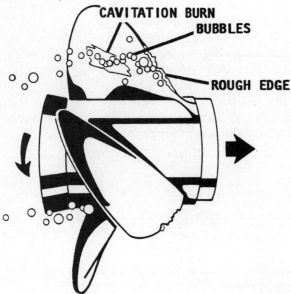

Cavitation (air bubbles) formed at the propeller. Manufacturers are constantly fighting this problem, as explained in the text.

Example of a damaged propeller. This unit should have been replaced long before this amount of damage was sustained.

these voids due to excessive blade tip speed and engine wear. The voids may be filled with air or water vapor, or they may actually be a partial vacuum. Cavitation may be caused by installing a piece of equipment too close to the lower unit, such as the knot indicator pickup, depth sounder, or bait tank pickup.

Vibration

The propeller should be checked regularly to ensure all blades are in good condition. If any of the blades become bent or nicked, this condition will set up vibrations in the drive unit and the motor. If the vibration becomes very serious it will cause a loss of power, efficiency, and boat performance. If the vibration is allowed to continue over a period of time it can have a damaging effect on many of the operating parts.

Vibration in boats can never be completely eliminated, but it can be reduced by keeping all parts in good working condition and through proper maintenance and lubrication. Vibration can also be reduced in some cases by increasing the number of blades. For this reason, many racers use two-blade props and luxury cruisers have four- and five-blade props installed.

Shock Absorbers

The shock absorber in the propeller plays a very important role in protecting the shafting, gears, and engine against the shock of a blow, should the propeller strike an underwater object. The shock absorber allows the propeller to stop rotating at the instant of impact while the power train continues turning.

How much impact the propeller is able to withstand, before causing the shock absorber to slip, is calculated to be more than the force needed to propel the boat, but less than the amount that could damage any part of the power train. Under normal propulsion loads of moving the boat through the water, the hub will not slip. However, it will slip if the propeller strikes an object with a force that would be great enough to stop any part of the power train.

If the power train was to absorb an impact great enough to stop rotation, even

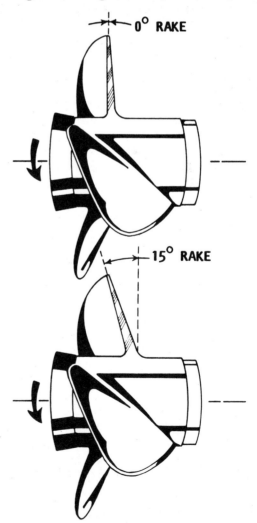

Illustration depicting the rake of a propeller, as explained in the text.

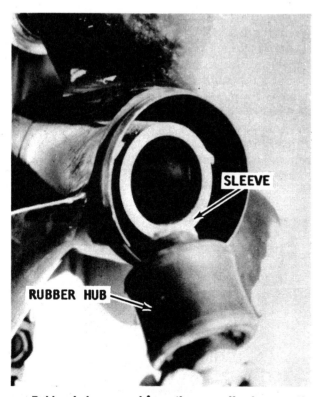

Rubber hub removed from the propeller because the hub was slipping in the propeller.

1-6 SAFETY

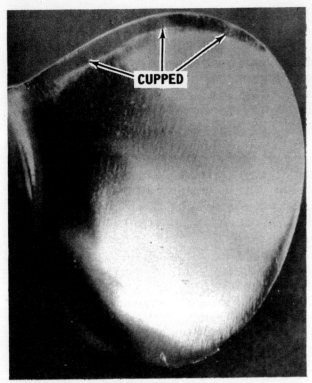

Propeller with a "cupped" leading edge. "Cupping" gives the propeller a better "hold" in the water.

for an instant, something would have to give, resulting in severe damage. If a propeller is subjected to repeated striking of underwater objects, it would eventually slip on its clutch hub under normal loads. If the propeller should start to slip, a new shock absorber/cushion hub would have to be installed.

Propeller Rake

If a propeller blade is examined on a cut extending directly through the center of the hub, and if the blade is set vertical to the propeller hub, as shown in the accompanying illustration, the propeller is said to have a zero degree ($0°$) rake. As the blade slants back, the rake increases. Standard propellers have a rake angle from $0°$ to $15°$.

A higher rake angle generally improves propeller performance in a cavitating or ventilating situation. On lighter, faster boats, higher rake often will increase performance by holding the bow of the boat higher.

Progressive Pitch

Progressive pitch is a blade design innovation that improves performance when forward and rotational speed is high and/or the propeller breaks the surface of the water.

Progressive pitch starts low at the leading edge and progressively increases to the trailing edge, as shown in the accompanying illustration. The average pitch over the entire blade is the number assigned to that propeller. In the illustration of the progressive pitch, the average pitch assigned to the propeller would be 21.

Cupping

If the propeller is cast with an edge curl inward on the trailing edge, the blade is said to have a cup. In most cases, cupped blades improve performance. The cup helps the blades to "HOLD" and not break loose, when operating in a cavitating or ventilating situation.

The cup has the effect of adding to the propeller pitch. Cupping usually will reduce full-throttle engine speed about 150 to 300 rpm below the same pitch propeller without a cup to the blade. A propeller repair shop is able to increase or decrease the cup on the blades. This change, as explained, will alter powerhead rpm to meet specific operating demands. Cups are rapidly becoming standard on propellers.

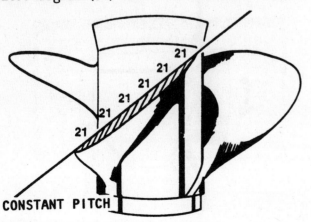
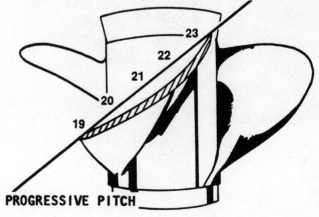

Comparison of a constant and progressive pitch propeller. Notice how the pitch of the progressive propeller, right, changes to give the blade more thrust and therefore, the boat more speed.

In order for a cup to be the most effective, the cup should be completely concave (hollowed) and finished with a sharp corner. If the cup has any convex rounding, the effectiveness of the cup will be reduced.

Rotation

Propellers are manufactured as right-hand rotation (RH), and as left-hand rotation (LH). The standard propeller for outboard units is RH rotation.

A right-hand propeller can easily be identified by observing it as shown in the accompanying illustration. Observe how the blade of the right-hand propeller slants from the lower left to upper right. The left-hand propeller slants in the opposite direction, from lower right to upper left.

When the RH propeller is observed rotating from astern the boat, it will be rotating clockwise when the outboard unit is in forward gear. The left-hand propeller will rotate counterclockwise.

1-5 LOADING

In order to receive maximum enjoyment, and performance, with safety, from the boat, an earnest effort should be made not to exceed the load capacity given by the manufacturer. A plate attached to the hull indicates the U.S. Coast Guard capacity information in pounds for persons and gear. If the plate states the maximum person capacity to be 750 pounds and the assumption is made each person weighs an average of 150 lbs., then the boat could carry five persons safely. If another 250 lbs. is added for motor and gear, and the maximum weight capacity for persons and gear is 1,000 lbs. or more, then the five persons and gear would be within the limit.

Try to load the boat evenly port and starboard. If more weight is placed on one side than on the other, the boat will list to the heavy side and make steering difficult. Performance will also be increased by placing heavy supplies aft of the center to keep the bow light for more efficient planing.

Clarification

Much confusion arises from the terms, certification, requirements, approval, regulations, etc. Perhaps the following may clarify a couple of these points.

1- The Coast Guard does not approve boats in the same manner as they "Approve" life jackets. The Coast Guard applies a formula to inform the public of what is safe for a particular craft.

2- If a boat has to meet a particular regulation, it must have a Coast Guard certification plate. The public has been led to believe this indicates approval of the Coast Guard. Not so.

3- The certification plate means a willingness of the manufacturer to meet the Coast Guard regulations for that particular craft. The manufacturer may recall a boat if it fails to meet the Coast Guard requirements.

4- The Coast Guard certification plate, see accompanying illustration, may or may not be metal. The plate is a regulation for the manufacturer. It is only a warning plate and the public does not have to adhere to the restrictions set forth on it. Again, the plate sets forth information as to the Coast Guard's opinion for safety on that particular boat.

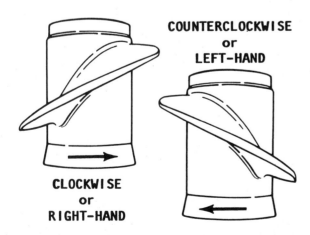

Right- and left-hand propellers showing how the angle of the blades is reversed. Right-hand propellers are by far the most popular for outboard units.

U.S. Coast Guard plate affixed to all new boats. When the blanks are filled in, the plate will indicate the Coast Guard's recommendations for persons, gear, and horsepower to ensure safe operation of the boat. These recommendations should not be exceeded, as explained in the text.

1-8 SAFETY

5- Coast Guard Approved equipment is equipment which has been approved by the Commandant of the U.S. Coast Guard and has been determined to be in compliance with Coast Guard specifications and regulations relating to the materials, construction, and performance of such equipment.

1-6 HORSEPOWER

The maximum horsepower engine for each individual boat should not be increased by any significant amount without checking requirements from the Coast Guard in the local area. The Coast Guard determines horsepower requirements based on the length, beam, and depth of the hull. **TAKE CARE NOT** to exceed the maximum horsepower listed on the plate or the warranty, and possibly the insurance, on the boat may become void.

1-7 FLOTATION

If the boat is less than 20 ft. overall, a Coast Guard or BIA (Boating Industry of America), now changed to NMMA (National Marine Manufacturers Association) requirement is that the boat must have buoyant material built into the hull (usually foam) to keep it from sinking if it should become swamped. Coast Guard requirements are mandatory but the NMMA is voluntary.

"Kept from sinking" is defined as the ability of the flotation material to keep the boat from sinking when filled with water and with passengers clinging to the hull. One restriction is that the total weight of the motor, passengers, and equipment aboard does not exceed the maximum load capacity listed on the plate.

Life Preservers —Personal Flotation Devices (PFDs)

The Coast Guard requires at least one Coast Guard approved life-saving device be carried on board all motorboats for each person on board. Devices approved are identified by a tag indicating Coast Guard approval. Such devices may be life preservers, buoyant vests, ring buoys, or buoyant cushions. Cushions used for seating are serviceable if air cannot be squeezed out of it. Once air is released when the cushion is squeezed, it is no longer fit as a flotation device. New foam cushions dipped in a rubberized material are almost indestructable.

Life preservers have been classified by the Coast Guard into five type categories. All PFDs presently acceptable on recreational boats fall into one of these five designations. All PFDs **MUST** be U.S. Coast Guard approved, in good and serviceable condition, and of an appropriate size for the persons who intend to wear them. Wearable PFDs **MUST** be readily accessible and throwable devices **MUST** be immediately available for use.

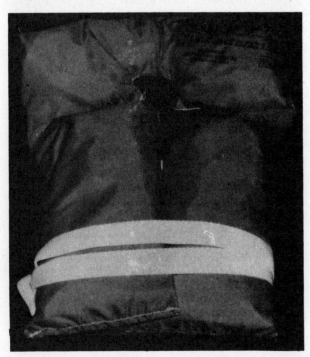

Type I PFD Coast Guard approved life jacket. This type flotation device provides the greatest amount of buoyancy. **NEVER** *use them for cushions or other purposes.*

A Type IV PFD cushion device intended to be thrown to a person in the water. If air can be squeezed out of the cushion, it is no longer fit for service as a PFD.

FLOTATION 1-9

Type I PFD has the greatest required buoyancy and is designed to turn most **UNCONSCIOUS** persons in the water from a face down position to a vertical or slightly backward position. The adult size device provides a minimum buoyancy of 22 pounds and the child size provides a minimum buoyancy of 11 pounds. The Type I PFD provides the greatest protection to its wearer and is most effective for all waters and conditions.

Type II PFD is designed to turn its wearer in a vertical or slightly backward position in the water. The turning action is not as pronounced as with a Type I. The device will not turn as many different type persons under the same conditions as the Type I. An adult size device provides a minimum buoyancy of 15½ pounds, the medium child size provides a minimum of 11 pounds, and the infant and small child sizes provide a minimum buoyancy of 7 pounds.

Type III PFD is designed to permit the wearer to place himself (herself) in a vertical or slightly backward position. The Type III device has the same buoyancy as the Type II PFD but it has little or no turning ability. Many of the Type III PFD are designed to be particularly useful when water skiing, sailing, hunting, fishing, or engaging in other water sports. Several of this type will also provide increased hypothermia protection.

Type IV PFD is designed to be thrown to a person in the water and grasped and held by the user until rescued. It is **NOT** designed to be worn. The most common Type IV PFD is a ring buoy or a buoyant cushion.

Type V PFD is any PFD approved for restricted use.

Coast Guard regulations state, in general terms: All boats less than 16 ft. overall, one Type I, II, III, or IV device shall be carried on board for each person in the boat. On boats over 26 ft., one Type I, II, or III device shall be carried on board for each person in the boat **plus** one Type IV device.

It is an accepted fact, most boating people own life preservers, but too few actually wear them. There is little or no excuse for not wearing one because the modern comfortable designs available today do not subtract from an individual's boating pleasure. Make a life jacket available to the crew and advise each member to wear it. If you are a crew member, ask the skipper to issue one, especially when boating in rough weather, cold water, or when running at high speed. Naturally, a life jacket should be a must for non-swimmers any time they are out on the water in a boat.

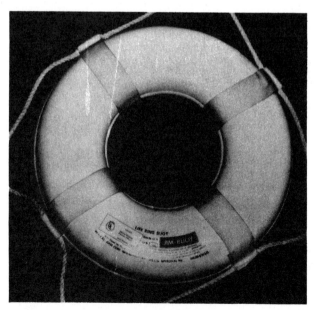

Type IV ring buoy also designed to be thrown to a person in the water. On ocean cruisers, this type device usually has a weighted pole with flag and light attached to the buoy.

Moisture-protected flares should be carried on board for use as a distress signal.

1-8 ANCHORS

One of the most important pieces of equipment in the boat next to the power plant is the ground tackle carried. The engine makes the boat go and the anchor and its line are what hold it in place when the boat is not secured to a dock or on the beach.

The anchor must be of suitable size, type, and weight to give the skipper "peace of mind" when the boat is at anchor. Under certain conditions, a second, smaller, lighter anchor may help to keep the boat in a favorable position during a non-emergency daytime situation.

In order for the anchor to hold properly, a piece of chain must be attached to the anchor and then the nylon anchor line attached to the chain. The amount of chain should equal or exceed the length of the boat. Such a piece of chain will ensure that the anchor stock will lay in an approximate horizontal position and permit the flutes to dig into the bottom and hold.

1-9 BOATING ACCIDENT REPORTS

In the United States, new federal and state regulations require an accident report to be filed with the nearest state boating authority within 48 hours, if a person is lost, disappears, or is injured. "Injured" is defined as requiring medical attention beyond "First Aid".

Accidents involving only property or equipment damage **MUST** be reported within 10 days if the damage is in excess of $200. Some states are more stringent and require reporting of accidents with property damage less than $200.

A **$500 PENALTY** may be assessed for failure to submit the report.

WORD OF ADVICE

Take time to make a copy of the report to keep for your records or for the insurance company. Once the report is filed, the Coast Guard will not give out a copy, even to the person who filed the report.

The report must give details of the accident and include:

1- The date, time, and exact location of the occurrence.

2- The name of each person who died, was lost, or injured.

3- The number and name of the vessel.

4- The names and addresses of the owner and operator.

If the operator cannot file the report for any reason, each person on board **MUST** notify the authorities, or determine that the report has been filed.

2
TUNING

2-1 INTRODUCTION

The efficiency, reliability, fuel economy and enjoyment available from engine performance are all directly dependent on having it tuned properly. The importance of performing service work in the sequence detailed in this chapter cannot be over emphasized. Before making any adjustments, check the specifications in the Appendix. **NEVER** rely on memory when making critical adjustments.

Before beginning to tune any engine, check to be sure the engine has satisfactory compression. An engine with worn or broken piston rings, burned pistons, or scored cylinder walls, cannot be made to perform properly no matter how much time and expense is spent on the tune-up. Poor compression must be corrected or the tune-up will not give the desired results.

A practical maintenance program that is followed throughout the year, is one of the best methods of ensuring the engine will give satisfactory performance at any time.

The extent of the engine tune-up is usually dependent on the time lapse since the last service. A complete tune-up of the entire engine would entail almost all of the work outlined in this manual. A logical sequence of steps will be presented in general terms. If additional information or detailed service work is required, the chapter containing the instructions will be referenced.

Each year higher compression ratios are built into modern outboard engines and the electrical systems become more complex, especially with electronic (capacitor discharge) units. Therefore, the need for reliable, authoritative, and detailed instructions becomes more critical. The information in this chapter and the referenced chapters fulfill that requirement.

2-2 TUNE-UP SEQUENCE

During a major tune-up, a definite sequence of service work should be followed to return the engine to the maximum performance desired. This type of work should not be confused with attempting to locate problem areas of "why" the engine is not performing satisfactorily. This work is classified as "trouble shooting". In many cases, these two areas will overlap, because many times a minor or major tune-up will correct the malfunction and return the system to normal operation.

The following list is a suggested sequence of tasks to perform during the tune-

This clean powerhead reflects the owner's pride in his unit. A safe "bet" might be to wager it is also properly tuned, synchronized, and well lubricated.

2-2 TUNING

up service work. The tasks are merely listed here. Generally procedures are given in subsequent sections of this chapter. For more detailed instructions, see the referenced chapter.

1- Perform a compression check of each cylinder. See Chapter 5.
2- Inspect the spark plugs to determine their condition. Test for adequate spark at the plug. See Chapter 5.
3- Start the engine in a body of water and check the water flow through the engine. See Chapter 9.
4- Check the gear oil in the lower unit. See Chapter 9.
5- Check the carburetor adjustments and the need for an overhaul. See Chapter 4.
6- Check the fuel pump for adequate performance and delivery. See Chapter 4.
7- Make a general inspection of the ignition system. See Chapter 5.
8- Test the starter motor and the solenoid, if so equipped. See Chapter 8.
9- Check the internal wiring.
10- Check the timing and synchronization. See Chapter 6.

2-3 COMPRESSION CHECK

A compression check is extremely important, because an engine with low or uneven compression between cylinders **CANNOT** be tuned to operate satisfactorily. Therefore, it is essential that any compression problem be corrected before proceeding with the tune-up procedure. See Chapter 5.

If the powerhead shows any indication of overheating, such as discolored or scorched paint, inspect the cylinders visually thru the transfer ports for possible scoring. It is possible for a cylinder with satisfactory compression to be scored slightly. Also, check the water pump. The overheating condition may be caused by a faulty water pump.

Checking Compression

Remove the spark plug wires. **ALWAYS** grasp the molded cap and pull it loose with a twisting motion to prevent damage to the connection. Remove the spark plugs and keep them in **ORDER** by cylinder for evaluation later. Ground the spark plug leads to the engine to render the ignition system inoperative while performing the compression check.

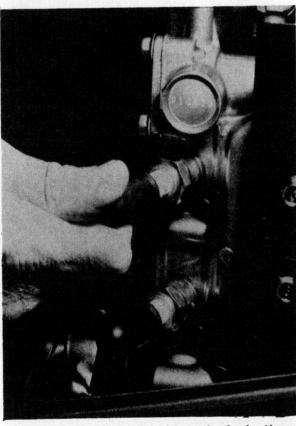

Removing the spark plug high tension lead. Always use a pulling and twisting motion on only the cap, not the wire, to prevent damage to the cap or the boot.

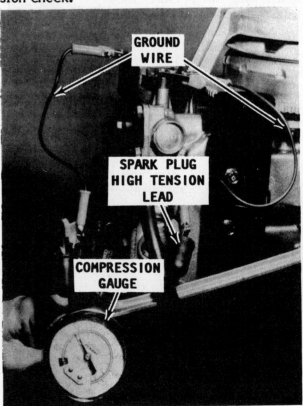

All spark plugs should be "grounded" while making compression tests. This action will prevent placing an extra load on the ignition coil.

Insert a compression gauge into the No. 1, top, spark plug opening. Crank the engine with the starter thru at least 4 complete strokes with the throttle at the wide-open position, to obtain the highest possible reading. Record the reading. Repeat the test and record the compression for each cylinder. A variation between cylinders is far more important than the actual readings. A variation of more than 15 psi (103 kPa), between cylinders indicates the lower compression cylinder is defective. The problem may be worn, broken, or sticking piston rings, scored pistons or worn cylinders.

Use of an engine cleaner will help to free stuck rings and to dissolve accumulated carbon. Follow the directions on the can.

2-4 SPARK PLUG INSPECTION

Inspect each spark plug for badly worn electrodes, glazed, broken, blistered, or lead fouled insulators. Replace all of the plugs, if one shows signs of excessive wear.

Make an evaluation of the cylinder performance by comparing the spark condition with those shown in Chapter 5. Check each spark plug to be sure they are all of the same manufacturer and have the same heat range rating.

Inspect the threads in the spark plug opening of the block, and clean the threads before installing the plug.

When purchasing new spark plugs, **ALWAYS** ask the marine dealer if there has been a spark plug change for the engine being serviced.

Crank the engine through several revolutions to blow out any material which might have become dislodged during cleaning.

ALWAYS use a new gasket and wipe the seats in the block clean. The gasket must be fully compressed on clean seats to complete the heat transfer process and to provide a gas tight seal in the cylinder.

Install the spark plug/s and tighten them to the torque value given in Chapter 7 for the powerhead being serviced. Overtightening the spark plug/s may cause the porcelain insulator to crack, undertightening the spark plug/s may cause them to unseat due to constant vibration.

Broken Reed

A broken reed is usually caused by metal fatigue over a long period of time. The failure may also be due to the reed flexing too far because the reed stop has not been adjusted properly or the stop has become distorted.

If the reed is broken, the loose piece **MUST** be located and removed, before the powerhead is returned to service. The piece of reed may have found its way into the crankcase, behind the bypass cover. If the broken piece cannot be located, the powerhead must be completely disassembled until it is located and removed.

An excellent check for a broken reed on an operating powerhead is to hold an ordinary business card in front of the carburetor. Under normal operating conditions, a very small amount of fine mist will be noticeable, but if fuel begins to appear rapidly on the card from the carburetor, one of the reeds is broken and causing the backflow through the carburetor onto the card.

A broken reed will cause the powerhead to operate roughly and with a "pop" back through the carburetor.

The reeds must **NEVER** be turned over in an attempt to correct a problem. Such action would cause the reed to flex in the opposite direction and the reed would break in a very short time.

2-5 IGNITION SYSTEM

Only two different ignition systems are used on the outboard engines covered in this manual. If the powerhead performance is less than expected, and the ignition is diagnosed as the problem area, refer to Chapter 5 for detailed service procedures. To properly time and synchronize the ignition system with the fuel system, see Chapter 6.

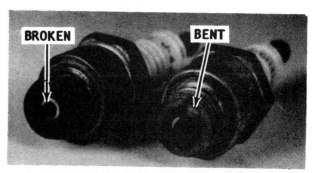

Damaged spark plugs. Notice the broken electrode on the left plug. The broken part MUST be found and removed before returning the powerhead to service.

2-4 TUNING

Breaker Points
Type I Ignition System Only

SOME GOOD WORDS

High primary voltage will darken and roughen the breaker points within a short period. This is not cause for alarm. Normally points in this condition would not operate satisfactorily in the conventional magneto, but they will give good service in this Type I system.

Therefore, **DO NOT** replace the points unless an obvious malfunction exists, or the contacts are loose or burned. Rough or discolored contact surfaces are **NOT** sufficient reason for replacement. The cam follower will usually have worn away by the time the points have become unsatisfactory for efficient service.

Check the resistance across the contacts. If the test indicates zero resistance, the points are serviceable. A slight resistance across the points will affect idle operation. A high resistance may cause the ignition system to malfunction and loss of spark. Therefore, if any resistance across the points is indicated, the point set should be replaced.

2-6 TIMING AND SYNCHRONIZING

Correct timing and synchronization are essential to efficient engine operation. An engine may be in apparent excellent mechanical condition, but perform poorly, unless the timing and synchronization have been adjusted precisely. To time and synchronize the powerhead, see the Table of Contents -- Chapter 6.

2-7 BATTERY CHECK

Inspect and service the battery, cables and connections. Check for signs of corrosion. Inspect the battery case for cracks or bulges, dirt, acid, and electrolyte leakage. Check the electrolyte level in each cell.

Fill each cell to the proper level with distilled water or water passed thru a demineralizer.

Clean the top of the battery. The top of a 12-volt battery should be kept especially clean of acid film and dirt, because of the high voltage between the battery terminals. For best results, first wash the battery with a diluted ammonia or baking soda solution to neutralize any acid present. Flush the solution off the battery with clean water. Keep the vent plugs tight to prevent the neutralizing solution or water from entering the cells.

Check to be sure the battery is fastened securely in position. The hold-down device should be tight enough to prevent any movement of the battery in the holder, but not so tight as to place a strain on the battery case.

If the battery posts or cable terminals are corroded, the cables should be cleaned separately with a baking soda solution and a wire brush. Apply a thin coating of Multi-

Type I ignition point set as seen through the access window in the flywheel. This ignition system is used on small single cylinder powerheads, and on some early model 2-cylinder powerheads.

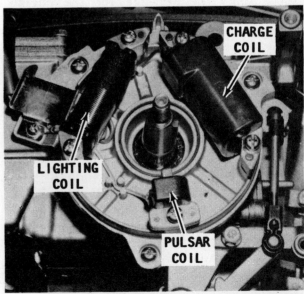

Type II CDI ignition system used on all late model powerheads covered in this manual, except the little 2hp unit.

purpose Lubricant to the posts and cable clamps before making the connections. The lubricant will help to prevent corrosion.

If the battery has remained under-charged, check for high resistance in the charging circuit. If the battery appears to be using too much water, the battery may be defective, or it may be too small for the job.

Jumper Cables

If booster batteries are used for starting an engine the jumper cables must be connected correctly and in the proper sequence to prevent damage to either battery, or diodes in the circuit.

ALWAYS connect a cable from the positive terminals of the dead battery to the positive terminal of the good battery **FIRST**. **NEXT**, connect one end of the other cable to the negative terminals of the good battery and the other end of the **ENGINE** for a good ground. By making the ground connection on the engine, if there is an arc when you make the connection it will not be near the battery. An arc near the battery could cause an explosion, destroying the battery and causing serious personal injury.

DISCONNECT the battery ground cable before replacing any part of the ignition or cranking system, or before connecting any type of meter to the ignition system.

If it is necessary to use a fast-charger on a dead battery, **ALWAYS** disconnect one of the boat cables from the battery first, to prevent burning out the diodes in the circuit.

NEVER use a fast charger as a booster to start the engine because the diodes will be **DAMAGED**.

2-8 FUEL AND FUEL TANKS

Take time to check the fuel tank and all of the fuel lines, fittings, couplings, valves, flexible tank fill and vent. Turn on the fuel supply valve at the tank. If the gas was not drained at the end of the previous season, make a careful inspection for gum formation. When gasoline is allowed to stand for long periods of time, particularly in the presence of copper, gummy deposits form. This gum can clog the filters, lines, and passageway in the carburetor. Chapter 4 has good information regarding "sour" fuel, unleaded fuel, and gasohol.

A check of the electrolyte in the battery should be a regular task on the maintenance schedule on any boat.

Common set of jumper cables for use with a second battery to crank and start the engine. EXTREME care should be exercised when using a second battery, as explained in the text.

2-6 TUNING

Check the fuel tank for the following:
1- Adequate air vent in the fuel cap.
2- Fuel line of sufficient size, should be 5/16" to 3/8" (8mm to 9.5mm).
3- Filter on the end of the pickup is too small or is clogged.
4- Fuel pickup tube is too small.

2-9 CARBURETOR ADJUSTMENTS

SPECIAL WORDS ON TACHOMETERS AND CONNECTIONS

The 8 to 30hp powerheads use a CDI system firing a twin lead ignition coil twice for each crankshaft revolution. If an induction tachometer is installed to measure powerhead speed, the tachometer will probably indicate **DOUBLE** the actual crankshaft rotation. Check the instructions with the tachometer to be used. Some tachometer manufacturers have allowed for the double reading and others have not.

Connections:

For all models with the Type I (with points), ignition system, connect one tachometer lead (may be White, Red, or Yellow, depending on the manufacturer), to the primary negative terminal of the coil (usually a small black lead), and the other tachometer lead, Black, to a suitable ground.

For all models with the Type II (CDI) ignition system, connect the two tachometer leads to the two green leads from the stator. Either tachometer lead may be connected to either green lead.

Carburetor Adjustments

Due to local conditions, it may be necessary to adjust the carburetor while the outboard unit is running in a test tank or with the boat in a body of water. For maximum performance, the idle rpm should be adjusted under actual operating conditions.

Remove the cowling and attach a tachometer to the powerhead as directed under "Connections" above.

Start the engine and allow it to warm to operating temperature.

REMEMBER, if the powerhead is equipped with a "kill" switch knob, the powerhead will **NOT** start without the emergency tether in place behind the "kill" switch knob.

CAUTION

Water must circulate through the lower unit to the powerhead anytime the powerhead is operating to prevent damage to the water pump in the lower unit. Just five seconds without water will damage the water pump impeller.

NEVER, AGAIN, NEVER operate the engine at high speed with a flush device attached. The engine, operating at high speed with such a device attached, would **RUNAWAY** from lack of a load on the propeller, causing extensive damage.

All Units Except Model 2hp

The idle mixture screw, also known as the pilot screw, regulates the air/fuel mixture. The setting for this screw varies for each unit. The setting is given in Chapter 4, for each model covered in this manual. Check the Table of Contents, Chapter 4 for the model being serviced. Actually, this setting is **NOT** an adjustment, it is a specification.

The idle speed is regulated by the throttle stop screw which "sets" the position of the throttle plate inside the carburetor throat. Idle speed recommendations are given in Chapter 4. Rotating the throttle stop **CLOCKWISE** increases powerhead speed. Rotating the screw **COUNTERCLOCKWISE** decreases powerhead speed.

The idle rpm is adjusted under actual operation in Chapter 4 as follows:

Two Green lighting coil leads will be found on all powerheads 8hp and above covered in this manual. These leads are used when connecting a tachometer, as explained in the text.

CARBURETOR ADJUSTMENTS 2-7

HP MODEL	YEAR	CARB	IDLE ADJUST. PAGE NO.
2	1977 & On	A	4-19
3.5	1978-81	B	4-26
4	1982 & On	C	4-34
5	1977-79	D	4-42
5	1979 & On	C	4-34
8A & 8	1977-79	E	4-49
8B, 8W & Mrthn 8	1979 & On	F/G	4-61 or 4-70
9.9C & 15C	1977 & On	F/G	4-61 or 4-70
8C	1979 & On	G	4-70
15W & 15	1977-79	H	4-77
20	1977 & On	J	4-88
25	1980 & On	J	4-88
28	1977-79	J	4-88
30	1980 & On	J	4-88
40	1978 & On	K	4-99
48	1977-79	L	4-107
55	1986 & On	L	4-107
60	1977-83	L	4-107

If the condition of the fuel is in doubt, drain, clean, and fill the tank with fresh fuel.

Check the line between the fuel pump and the carburetor while the powerhead is operating and the line between the fuel tank and the pump when the powerhead is not operating. A leak between the tank and the pump many times will not appear when the powerhead is operating, because the suction created by the pump drawing fuel will not allow the fuel to leak. Once the powerhead is shut down and the suction no longer exists, fuel may begin to leak.

Repairs and Adjustments

Chapter 4 contains detailed, comprehensive procedures to disassemble, clean, assemble, and adjust all carburetors used on the powerheads covered in this manual. See the Table of Contents for the appropriate section for the carburetor installed on the powerhead being serviced.

2-10 FUEL PUMPS

If the powerhead operates as if the load on the boat is being constantly increased and decreased, even though an attempt is being made to hold a constant powerhead speed, the problem can most likely be attributed to the fuel pump.

Many times, a defective fuel pump diaphragm is mistakenly diagnosed as a problem in the ignition system. The most common problem is a tiny pin-hole in the diaphragm or a bent check valve on the Model 20hp, 25hp, 28hp, or 30hp. Such a small hole will permit gas to enter the crankcase and wet foul the spark plug at idle-speed.

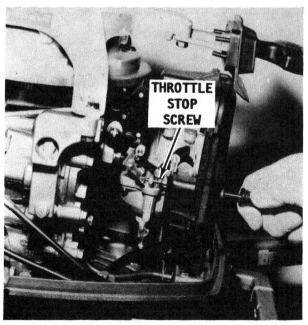

Location of the throttle stop screw on a Model 25hp powerhead.

Contents, at press time, of a typical Mariner carburetor rebuild kit. The items will vary with each model carburetor.

2-8 TUNING

During high-speed operation, gas quantity is limited, the plug is not foul and will therefore fire in a satisfactory manner.

If the fuel pump fails to perform properly, an insufficient fuel supply will be delivered to the carburetor. This lack of fuel will cause the engine to run lean, lose rpm or cause piston scoring.

Tune-up Task

Remove the fuel filter on the carburetor. Wash the parts in solvent and then dry them with compressed air. Install the clean element. A fuel pump pressure test should be made any time the engine fails to perform satisfactorily at high speed.

NEVER use liquid Neoprene on fuel line fittings. Always use Permatex when making fuel line connections. Permatex is available at almost all marine and hardware stores.

To service the fuel pump, see Chapter 4.

2-11 CRANKING MOTOR

Cranking Motor Test

Check to be sure the battery has a 70-ampere rating and is fully charged. Would you believe, many cranking motors are needlessly disassembled, when the battery is actually the culprit.

Lubricate the pinion gear and screw shaft with No. 10 oil.

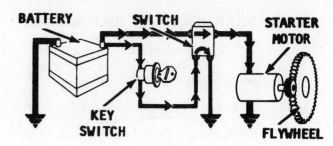

Functional diagram of a typical cranking circuit.

Connect one lead of a voltmeter to the positive terminal of the cranking motor. Connect the other meter lead to a good ground on the engine. Check the battery voltage under load by turning the ignition switch to the **START** position and observing the voltmeter reading.

If the reading is 9-1/2 volts or greater, and the cranking motor fails to operate, repair or replace the cranking motor. See Chapter 8.

2-12 INTERNAL WIRING HARNESS

Check the internal wiring harness if problems have been encountered with any of the electrical components. Check for frayed or chafed insulation and/or loose connections between wires and terminal connections.

Check the harness connector for signs of corrosion. Inspect the electrical "prongs" to be sure they are not bent or broken. If the harness shows any evidence of the foregoing problems, the problem must be corrected before proceeding with any harness testing.

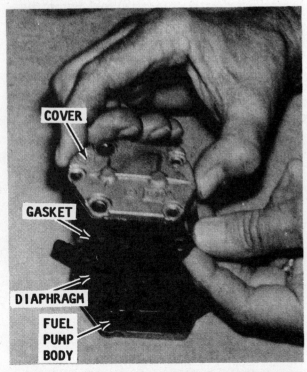

Arrangement of vacuum operated fuel pump parts. A tiny hole in the diaphragm can affect performance.

Harness connector, with the prongs properly cleaned, ready for continued service.

Verify the "prongs" of the harness connector are clean and free of corrosion. Convince yourself a good electrical connection is being made between the harness connector and the remote control harness.

2-13 WATER PUMP CHECK

FIRST, SOME GOOD WORDS

The water pump **MUST** be in very good condition for the engine to deliver satisfactory service. The pump performs an extremely important function by supplying enough water to properly cool the engine. Therefore, in most cases, it is advisable to replace the complete water pump assembly at least once a year, or anytime the lower unit is disassembled for service.

Sometimes during adjustment procedures, it is necessary to run the engine with a flush device attached to the lower unit. **NEVER** operate the engine over 1000 rpm with a flush device attached, because the engine may **"RUNAWAY"** due to the no-load condition on the propeller. A "runaway" engine could be severely damaged.

As the name implies, the flush device is primarily used to flush the engine after use in salt water or contaminated fresh water. Regular use of the flush device will prevent salt or silt deposits from accumulating in the water passage-way. During and immediately after flushing, keep the outboard unit in an upright position until all of the water has drained from the intermediate housing. This will prevent water from entering the powerhead by way of the intermediate housing and the exhaust ports, during the flush. It will also prevent residual water from being trapped in the intermediate housing and other passageways.

All powerheads covered in this manual have water exhaust ports which deliver a tattle-tale stream of water, if the water pump is functioning properly during engine operation. Water pressure at the cylinder block should be checked if an overheating condition is detected or suspected.

To test the water pump, the lower unit **MUST** be placed in a test tank or the boat moved into a body of water. The pump must now work to supply a volume to the engine. A tattle-tale stream of water should be visible from the pilot hole beneath the cover cowling.

Lack of adequate water supply from the water pump thru the engine will cause any number of powerhead failures, such as stuck rings, scored cylinder walls, burned pistons, etc.

For water pump service, see Chapter 9.

2-14 PROPELLER

Check the propeller blades for nicks, cracks, or bent condition. If the propeller is damaged, the local marine dealer can make repairs or send it out to a shop specializing in such work.

Remove the propeller and the thrust hub. Check the propeller shaft seal to be sure it is not leaking. Check the area just forward of the seal to be sure a fish line is not wrapped around the shaft.

Worn water pump impeller, unfit for further service.

Lower unit and propeller of a Model 30hp unit, cleaned, serviced, and ready for work.

Operation At Recommended RPM

Check with the local marine dealer, or a propeller shop for the recommended size and pitch for a particular size engine, boat, and intended operation. The correct propeller should be installed on the engine to enable operation at the upper end of the factory recommended rpm.

2-15 LOWER UNIT

NEVER remove the vent or filler plugs when the lower unit is hot. Expanded lubricant would be released through the plug hole. Check the lubricant level after the unit has been allowed to cool. Add only Quicksilver Case Lubricant. **NEVER** use regular automotive-type grease in the lower unit, because it expands and foams too much. Outboard lower units do not have provisions to accommodate such expansion.

If the lubricant appears milky brown, or if large amounts of lubricant must be added to bring the lubricant up to the full mark, a thorough check should be made to determine the cause of the loss.

Draining Lower Unit

Remove the **FILL** plug from the lower end of the gear housing on the port side and the **OIL LEVEL** plug just above the anti-cavitation plate.

Filling Lower Unit

Position the drive unit approximately vertical and without a list to either port or starboard. Insert the lubricant tube into the **OIL FILL** hole at the bottom plug hole, and inject lubricant until the excess begins to come out the **OIL LEVEL** hole. Install the **OIL LEVEL** plug first then replace the **OIL FILL** plug with **NEW** gaskets. Check to be sure the gaskets are properly positioned to prevent water from entering the housing.

For detailed lower unit service procedures, **AND** lower unit lubrication capacities, see Chapter 9.

2-16 BOAT TESTING

Operation of the outboard unit, mounted on a boat with some type of load, is the ultimate test. Failure of the power unit or the boat under actual movement through the water may be detected much more quickly than operating the power unit in a test tank.

Hook and Rocker

Before testing the boat, check the boat bottom carefully for marine growth or evidence of a "hook" or a "rocker" in the bottom. Either one of these conditions will greatly reduce performance.

Performance

Mount the motor on the boat. Install the remote control cables (if used), and check for proper adjustment.

Make an effort to test the boat with what might be considered an average gross load. The boat should ride on an even keel, without a list to port or starboard. Adjust the motor tilt angle, if necessary, to permit the boat to ride slightly higher than the stern. If heavy supplies are stowed aft of the center, the bow will be light and the boat will "plane" more efficiently. For this test the boat must be operated in a body of water.

If the motor is equipped with an adjustable trim tab, the tab should be adjusted to permit boat steerage in either direction with equal ease.

Check the engine rpm at full throttle. The rpm should be within the Specifications in the Appendix. If the rpm is not within specified range, a propeller change may be in order. A higher pitch propeller will decrease rpm, and a lower pitch propeller will increase rpm.

For maximum low speed engine performance, the idle mixture and the idle rpm should be readjusted under actual operating conditions.

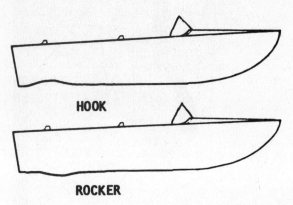

Boat performance will be drastically impaired, if the bottom is damaged by a dent (hook) or bulge (rocker).

3
MAINTENANCE

3-1 INTRODUCTION

GOOD WORDS

The authors estimate 75% of engine repair work can be directly or indirectly attributed to lack of proper care for the engine. This is especially true of care during the off-season period. There is no way on this green earth for a mechanical engine, particularly an outboard motor, to be left sitting idle for an extended period of time, say for six months, and then be ready for instant satisfactory service.

Imagine, if you will, leaving your automobile for six months, and then expecting to turn the key, have it roar to life, and be able to drive off in the same manner as a daily occurrence.

It is critical for an outboard engine to be run at least once a month, preferably, in the water, but if this is not possible, then a flush attachment MUST be connected to the lower unit.

CAUTION

Water must circulate through the lower unit to the powerhead anytime the powerhead is operating to prevent damage to the water pump in the lower unit. Just five seconds without water will damage the water pump impeller.

NEVER, AGAIN NEVER, operate the engine at high speed with a flush device attached. The engine, operating at high speed with such a device attached, would **RUNAWAY** from lack of load on the propeller, causing extensive damage.

At the same time, the shift mechanism should be operated through the full range several times and the steering operated from hard-over to hard-over.

Only through a regular maintenance program can the owner expect to receive long life and satisfactory performance at minimum cost.

The material presented in this chapter is divided into four general areas.

1- General information every boat owner should know.
2- Maintenance tasks that should be performed periodically to keep the boat operating at minimum cost.
3- Care necessary to maintain the appearance of the boat and to give the owner that "Pride of Ownership" look.
4- Winter storage practices to minimize damage during the off-season when the boat is not in use.

In nautical terms, the front of the boat is the **bow**; the rear is the **stern**; the right side, when facing forward, is the **starboard** side; and the left side is the **port** side. All directional references in this manual use this terminology. Therefore, the direction

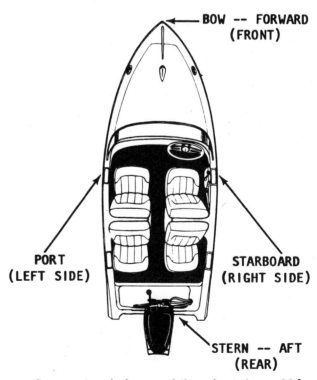

Common terminology used throughout the world for reference designation on boats of ALL sizes. "Port", "Starboard", "Forward", and "Aft", never change, even if standing on your head.

3-2 MAINTENANCE

from which an item is viewed is of no consequence, because **starboard** and **port** **NEVER** change no matter where the individual is located or his position -- even standing on his/her head.

3-2 OUTBOARD SERIAL NUMBERS

The outboard serial number and the engine serial number are the manufacturer's key to engine changes. These numbers identify the year of manufacture, the qualified horsepower rating, and the parts book identification. If any correspondence or parts are required, the engine serial number **MUST** be used or proper identification is not possible. The accompanying illustration will be very helpful in locating the engine identification tag for the various models.

The outboard number is stamped on the plate usually attached to the port side of the clamp bracket.

The powerhead serial number is usually stamped on the port side of the cylinder block.

ONE MORE WORD

As a theft prevention measure, a special label with the outboard serial number is bonded to the starboard side of the clamp bracket. Any attempt to remove this label will result in cracks across the serial number.

3-3 EMERGENCY TETHER

Late model outboard units covered in this manual are equipped with an emergency tether by the manufacturer. This tether

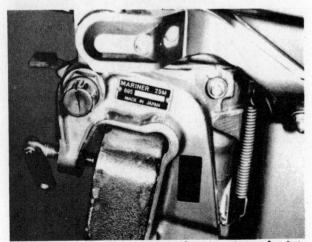

Serial numbers on the identification plate of a late model outboard unit covered in this manual. This is a standard location for the plate, although some units may have it elsewhere.

must be in place behind the "kill" switch or the powerhead cannot be started. If the powerhead is operating and the tether is removed, the unit will immediately shut down.

Explanation

The plastic tether acts as a spacer, moving the "kill" button out slightly and allowing an internal contact to be closed, permitting the ignition circuit to be completed and the powerhead to be started.

The "kill" button may now be depressed to shut the powerhead down with the tether in place. If the tether is pulled free from behind the "kill" button, the button pops inward and the ignition circuit is opened. With the "kill" switch in this position, no amount of cranking will result in powerhead startup.

The purpose of this tether is two fold.

As a Safety Feature

When this tether is used as intended by the manufacturer, the boat operater attaches the belt hook onto his/her clothing at any convenient location. Should the operator be thrown overboard or knocked forward away from the outboard unit, the tether will be pulled free of the "kill" button, and the powerhead will be shut down.

As a Security Feature

If the boat is moored and will be left unattended, the owner may take the entire emergency tether with him/her. Without the tether in place behind the "kill" button, the powerhead cannot be started. Any attempt to start the powerhead and steal the boat will be unsuccessful -- unless the thief is familiar with this particular security device.

Temporary Replacement

If the boat owner loses the emergency tether and is unable to obtain one immediately from the local Mariner dealer, an emergency substitute tether may be made using only a common knife and a couple pieces of plastic.

First, obtain a piece of plastic from the cover of a container of margarine, whipped topping, or similar product.

Next, using the pattern shown on the following page cut out about four shapes, as shown. Stack the four cutouts together,

PRE-SEASON PREP. 3-3

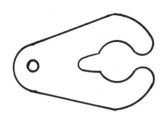

Pattern to be used to fabricate a "homemade" emergency tether, as explained in the text.

secure them with a paper clip, or similar object, and then insert them behind the "kill" switch.

SPECIAL WORDS

If the material described is not available, obtain some other pliable material and cut the shape indicated. The thickness of the substitute tether device should be approximately 1/8" (3mm). About 100 pages (50 sheets) of this manual is approximately the proper thickness.

REMEMBER, use this device only in an emergency situation and purchase the proper tether from the local Mariner dealer at the first opportunity. By substituting this "home made" tether, both the safety and the security features intended by the manufacturer have been lost.

3-4 LUBRICATION - COMPLETE UNIT

As with every type mechanical invention with moving parts, lubrication plays a prominent role in operation, enjoyment, and longevity of the unit.

If an outboard unit is operated in salt water the frequency of applying lubricant to fittings is usually cut in half for the same fitting if the unit is used in fresh water. The few minutes involved in moving around the outboard applying lubricant and at the same time making a visual inspection of its general condition will pay in rich rewards with years of continued service.

It is not uncommon to see outboard units well over 20-years of age moving a boat through the water as if the unit had recently been purchased from the current line of models. An inquiry with the proud owner will undoubtedly reveal his main credit for its performance to be regular periodic maintenance.

The accompanying chart can be used as a guide to periodic maintenance while the outboard is being used during the season.

In addition to the normal lubrication listed in the lubrication chart, the prudent owner will inspect and make checks on a regular basis as listed in the accompanying chart.

3-5 PRE-SEASON PREPARATION

Satisfactory performance and maximum enjoyment can be realized if a little time is spent in preparing the outboard unit for service at the beginning of the season. Assuming the unit has been properly stored, as outlined in Section 3-12, a minimum amount of work is required to prepare the unit for use.

The following steps outline an adequate and logical sequence of tasks to be performed before using the outboard the first time in a new season.

1- Lubricate the outboard according to the manufacturer's recommendations. Refer to the lubrication chart. Remove, clean, inspect, adjust, and install the spark plugs with new gaskets (if they require gaskets). Make a thorough check of the ignition system. This check should include: the points, coil, condenser, stator assembly, condition of the wiring, and the battery electrolyte level and charge.

2- If a built-in fuel tank is installed, take time to check the gasoline tank and all

Various lubrication points called out in the chart on the following page.

LUBRICATION POINT/FREQUENCY CHART

DESCRIPTION	LUBRICANT	FREQUENCY FRESH WATER	FREQUENCY SALT WATER
Throttle Linkage	Quicksilver All-Purpose Lubricant	Every 100 hrs. or 60 days	Every 50 hrs. or 30 days
Throttle Control Lever			
Throttle Grip Housing			
Throttle Link Journal			
Shift Lever Journal			
Shift Mechanism			
Steering Pivot Shaft			
Top or Bottom Cowling Clamp Lever Journal			
Choke Lever			
Swivel Bracket			
Clamp Bolt			
Tilt Mechanism			
Propeller Shaft			
Gear Oil	Quicksilver Gear Case Lubricant	Every 100 hrs. or 60 days	Every 50 hrs. or 30 days

PRE-SEASON PREP.   3-5

fuel lines, fittings, couplings, valves, including the flexible tank fill and vent. Turn on the fuel supply valve at the tank. If the fuel was not drained at the end of the previous season, make a careful inspection for gum formation. If a six-gallon fuel tank is used, take the same action. When gasoline is allowed to stand for long periods of time, particularly in the presence of copper, gummy deposits form. This gum can clog the filters, lines, and passageways in the carburetor. See Chapter 4, Fuel System Service.

3- Check the oil level in the lower unit by first removing the vent screw on the port side just above the anti-cavitation plate. Insert a short piece of wire into the hole and check the level. Fill the lower unit according to procedures outlined in Section 3-11.

GOOD WORDS

The manufacturer recommends the fuel filter be replaced at the start of each season or at least once a year. The manufacturer also recommends oil be added to the fuel tank at the ratio of 25:1 for the first ten hours of operation after the unit is brought out of storage. This ratio will **ENSURE** adequate lubrication of moving parts which have been drained of oil during the storage period. After the first ten hours of operation, the normal 100:1 oil/fuel mixture may be used. Use only outboard marine oil in the mixture, never automotive oils.

ALL UNITS

4- Close all water drains. Check and replace any defective water hoses. Check to be sure the connections do not leak.

Mariner recommended lubricants and additives will not only keep the unit within the limits of the warranty, but will be a major contributing factor to dependable performance and reduced maintenance costs.

After 60 seconds at 1500 rpm.

After 90 seconds at 1500 rpm.

After 30 seconds at 2000 rpm.

After 45 seconds at 2000 rpm.

After 60 seconds at 2000 rpm

Cautions throughout this manual point out the danger of operating the powerhead without water passing through the water pump. The above photographs are self evident.

3-6 MAINTENANCE

Replace any spring-type hose clamps, if they have lost their tension, or if they have distorted the water hose, with band-type clamps.

5- The engine can be run with the lower unit in water to flush it. If this is not practical, a flush attachment may be used. This unit is attached to the water pick-up in the lower unit. Attach a garden hose, turn on the water, allow the water to flow into the engine for awhile, and then run the engine.

CAUTION

Water must circulate through the lower unit to the powerhead anytime the powerhead is operating to prevent damage to the water pump in the lower unit. Just five seconds without water will damage the water pump impeller.

Check the exhaust outlet for water discharge. Check for leaks. Check operation of the thermostat.

6- Check the electrolyte level in the battery and the voltage for a full charge. Clean and inspect the battery terminals and cable connections. **TAKE TIME** to check the polarity, if a new battery is being installed. Cover the cable connections with grease or special protective compound as a prevention to corrosion formation. Check all electrical wiring and grounding circuits.

7- Check all electrical parts on the engine and lower portions of the hull to be sure they are not of a type that could cause ignition of an explosive atmosphere. Rubber caps help keep spark insulators clean and reduce the possibility of arcing. Electric cranking motors and high-tension wiring harnesses should be of a marine type that cannot cause an explosive mixture to ignite.

3-6 FIBERGLASS HULLS

Fiberglass-reinforced plastic hulls are tough, durable, and highly resistant to impact. However, like any other material they can be damaged. One of the advantages of this type of construction is the relative ease with which it may be repaired. Because of its break characteristics, and the simple techniques used in restoration, these hulls have gained popularity throughout the world. From the most congested urban marina, to isolated lakes in wilderness areas, to the severe cold of far off northern seas, and in sunny tropic remote rivers of primative islands or continents, fiberglass boats can be found performing their daily task with a minimum of maintenance.

A fiberglass hull has almost no internal stresses. Therefore, when the hull is broken or stove-in, it retains its true form. It will not dent to take an out-of-shape set. When the hull sustains a severe blow, the impact will be either absorbed by deflection of the laminated panel or the blow will result in a definite, localized break. In addition to hull damage, bulkheads, stringers, and other stiffening structures attached to the hull may also be affected and therefore, should be checked. Repairs are usually confined to the general area of the rupture.

3-7 BELOW WATERLINE SERVICE

A foul bottom can seriously affect boat performance. This is one reason why racers, large and small, both powerboat and sail, are constantly giving attention to the condition of the hull below the waterline.

In areas where marine growth is prevalent, a coating of vinyl, anti-fouling bottom paint should be applied. If growth has developed on the bottom, it can be removed with a solution of muriatic acid applied with a brush or swab and then rinsed with clear water. **ALWAYS** use rubber gloves when working with muriatic acid and **TAKE EXTRA CARE** to keep it away from your face and hands. The **FUMES ARE TOXIC.** Therefore, work in a well-ventilated area,

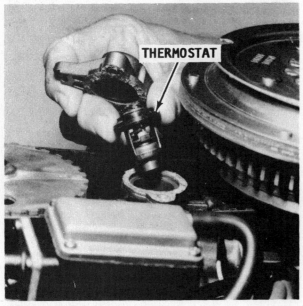

The thermostat is usually located in an accessible place for easy maintenance or replacement.

or if outside, keep your face on the windward side of the work.

Barnacles have a nasty habit of making their home on the bottom of boats which have not been treated with anti-fouling paint. Actually they will not harm the fiberglass hull, but can develop into a major nuisance.

If barnacles or other crustaceans have attached themselves to the hull, extra work will be required to bring the bottom back to a satisfactory condition. First, if practical, put the boat into a body of fresh water and allow it to remain for a few days. A large percentage of the growth can be removed in this manner. If this remedy is not possible, wash the bottom thoroughly with a high-pressure fresh water source and use a scraper. Small particles of hard shell may still hold fast. These can be removed with sandpaper.

3-8 SUBMERGED ENGINE SERVICE

A submerged engine is always the result of an unforeseen accident. Once the engine is recovered, special care and service procedures **MUST** be closely followed in order to return the unit to satisfactory performance.

NEVER, again we say **NEVER** allow an engine that has been submerged to stand more than a couple hours before following the procedures outlined in this section and making every effort to get it running. Such delay will result in serious internal damage. If all efforts fail and the engine cannot be started after the following procedures have been performed, the engine should be disassembled, cleaned, assembled, using new gaskets, seals, and O-rings, and then started as soon as possible.

Submerged engine treatment is divided into three unique problem areas: Submersion in salt water; submerged while powerhead was running; and a submerged unit in fresh water.

The most critical of these three circumstances is the engine submerged in salt water, with submersion while running, a close second.

Salt Water Submersion

NEVER attempt to start the engine after it has been recovered. This action will only result in additional parts being damaged and the cost of restoring the engine increased considerably. If the engine was submerged in salt water the complete unit **MUST** be disassembled, cleaned, and assembled with new gaskets, O-rings, and seals. The corrosive effect of salt water can only be eliminated by the complete job being properly performed.

Submerged While Running
Special Instructions

If the engine was running when it was submerged, the chances of internal engine damage is greatly increased. After the engine has been recovered, remove the spark plugs to prevent compression in the cylinders. Make an attempt to rotate the crankshaft with the rewind starter or the flywheel. On larger horsepower engines without a rewind starter, use a socket wrench on the flywheel nut to rotate the crankshaft. If the attempt fails, the chances of serious internal damage, such as: bent connecting rod, bent crankshaft, or damaged cylinder, is greatly increased. If the crankshaft cannot be rotated, the powerhead must be completely disassembled.

CRITICAL WORDS

Never attempt to start powerhead that has been submerged. If there is water in the cylinder, the piston will not be able to compress the liquid. The result will most likely be a bent connecting rod.

Easy removal of the exhaust cover will provide access to inspect the condition of the cylinder bores, the pistons, and the rings. Such inspection may reveal the cause of "strange" noises in the powerhead.

3-8 MAINTENANCE

Submerged Engine -- Fresh Water

SPECIAL WORD: As an aid to performing the restoration work, the following steps are numbered and should be followed in sequence. However, illustrations are not included with the procedural steps because the work involved is general in nature.

1- Recover the engine as quickly as possible.

2- Remove the cowling and the spark plug/s.

3- Remove the carburetor float bowl cover, or the bowl.

4- Flush the outside of the engine with fresh water to remove silt, mud, sand, weeds, and other debris. **DO NOT** attempt to start the engine if sand has entered the powerhead. Such action will only result in serious damage to powerhead components. Sand in the powerhead means the unit must be disassembled.

CRITICAL WORDS

Never attempt to start powerhead that has been submerged. If there is water in the cylinder, the piston will not be able to compress the liquid. The result will most likely be a bent connecting rod.

5- Remove as much water as possible from the powerhead. Most of the water can be eliminated by first holding the engine in a horizontal position with the spark plug holes **DOWN**, and then cranking the powerhead with the rewind starter or with a socket wrench on the flywheel nut. Rotate the crankshaft through at least 10 complete revolutions. If you are satisfied there is no water in the cylinders, proceed with Step 6 to remove moisture.

6- Alcohol will absorb moisture. Therefore, pour alcohol into the carburetor throat and again crank the powerhead.

7- Rotate the outboard in the horizontal position until the spark plug openings are facing **UPWARD**. Pour alcohol into the spark plug openings and again rotate the crankshaft.

8- Rotate the outboard in the horizontal position until the spark plug openings are again facing **DOWN**. Pour engine oil into the carburetor throat and, at the same time, rotate the crankshaft to distribute oil throughout the crankcase.

9- Rotate the outboard in the horizontal position until the spark plug holes are again facing **UPWARD**. Pour approximately one teaspoon of engine oil into each spark plug opening. Rotate the crankshaft to distribute the oil in the cylinders.

10- Install and connect the spark plugs.

11- Install the carburetor float bowl cover, or the bowl.

12- Obtain **FRESH** fuel and attempt to start the engine. If the powerhead will start, allow it to run for approximately an hour to eliminate any unwanted moisture remaining in the powerhead.

CAUTION

Water must circulate through the lower unit to the powerhead anytime the powerhead is operating to prevent damage to the water pump in the lower unit. Just five seconds without water will damage the water pump impeller.

13- If the powerhead fails to start, determine the cause, electrical or fuel, correct the problem, and again attempt to get it running. **NEVER** allow a powerhead to remain unstarted for more than a couple hours without following the procedures in this section and attempting to start it. If attempts to start the powerhead fail, the unit should be disassembled, cleaned, assembled, using new gaskets, seals, and O-rings, just as **SOON** as possible.

3-9 PROPELLER SERVICE

The propeller should be checked regular-

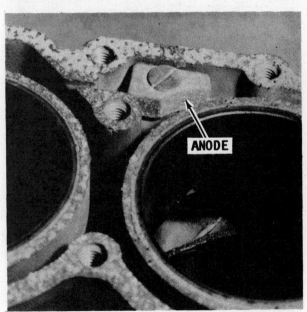

An anode installed within the water jacket of the Model 40hp powerhead provides added protection against corrosion.

ly to be sure all the blades are in good condition. If any of the blades become bent or nicked, this condition will set up vibrations in the motor. Remove and inspect the propeller. Use a file to trim nicks and burrs. **TAKE CARE** not to remove any more material than is absolutely necessary. For a complete check, take the propeller to your marine dealer where the proper equipment and knowledgeable mechanics are available to perform a proper job at modest cost.

Inspect the propeller shaft to be sure it is still true and not bent. If the shaft is not perfectly true, it should be replaced.

Install the thrust hub. Coat the propeller shaft splines with Perfect Seal No. 4, and the rest of the shaft with a good grade of anti-corrosion lubricant. Install the front spacer, the propeller, the washer and the propeller nut.

Position a block of wood between the propeller and the anti-cavitation tab to keep the propeller from turning. Tighten the propeller nut to the torque specifications given in Chapter 9. Adjust the nut to enable the cotter pin to be threaded through the propeller nut and shaft. Bend the two ends of the cotter pin in opposite directions around the nut. This action will prevent the nut from backing off the shaft.

Trim Tabs

Check the trim tab and the anodic heads. Replace them, if necessary. The trim tab must make a good ground inside the lower unit. Therefore, the trim tab and the cavity **MUST NOT** be painted. In addition to trimming the boat, the trim tab acts as a zinc electrode to prevent electrolysis from acting on more expensive parts. It is normal for the tab to show signs of erosion. The

Three different type anodes used on the outboard units covered in this manual.

tabs are inexpensive and should be replaced frequently.

Clean the exterior surface of the unit thoroughly. Inspect the finish for damage or corrosion. Clean any damaged or corroded areas, and then apply primer and matching paint.

Check the entire unit for loose, damaged, or missing parts.

3-10 INSIDE THE BOAT

The following points may be lubricated with Quicksilver All-Purpose lubricant:

a- Remote control cable ends next to the hand nut. **DO NOT** over-lubricate the cable.

Such extensive erosion of a trim tab compared with a new tab, suggests an electrolysis problem, or complete disregard for periodic maintenance.

Manufacturer's approved paint products to dress the outboard unit and give a special "pride of ownership" look. After the paint (left), is used, the leveler (right), can be applied for a smooth professional finish.

b- Steering arm pivot socket.
c- Exposed shaft of the cable passing through the cable guide tube.
d- Steering link rod to the steering cable.

3-11 LOWER UNIT

Draining Lower Unit

Remove the **FILL** plug from the lower end of the gear housing on the port side and the **VENT** plug just above the anti-cavitation plate.

CAUTION WORDS

Do not remove the plugs if the outboard unit has been operated recently, or if the unit has been sitting exposed to the hot sun. If one of the plugs should be removed when the lubricant is hot, the material will squirt out under considerable pressure.

Add only Quicksilver gear case lubricant. **NEVER** use regular automotive-type grease in the lower unit because it expands and foams too much. Lower units do not have provisions to accommodate such expansion.

If the lubricant appears milky brown, or if large amounts of lubricant must be added to bring the lubricant up to the full mark, a thorough check should be made to determine the cause of the loss.

Water in the Lower Unit

Water in the lower unit is usually caused by fish line becoming entangled around the propeller shaft behind the propeller and damaging the propeller seal. If the line is not removed, it will cut the propeller shaft seal and allow water to enter the lower unit. Fish line has also been known to cut a groove in the propeller shaft.

The propeller should be removed each time the boat is hauled from the water at the end of an outing and any material entangled behind the propeller removed before it can cause expensive damage. The small amount of time and effort involved in pulling the propeller is repaid many times by reduced maintenance and service work, including the replacement of expensive parts.

Filling Lower Unit

Position the drive unit approximately vertical and without a list to either port or starboard. Insert the lubricant tube into the **FILL/DRAIN** hole at the bottom plug hole, and inject lubricant until the excess begins to come out the **VENT** hole. Install the

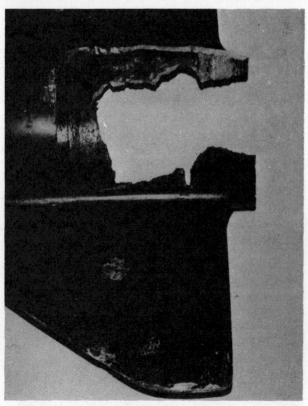

This lower unit was destroyed because the bearing carrier was "frozen" -- possibly due to inadequate lubrication during the last installation. The drastic action used for removal of the carrier did not damage other expensive parts, which were saved for further service.

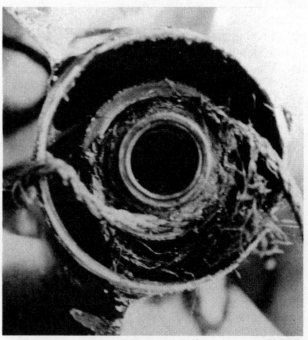

Excellent view of rope and fish line entangled behind the propeller. Entangled fish line can actually cut through the seals allowing water to enter and oil to escape from the lower unit.

VENT and **FILL** plugs with **NEW** gaskets. Check to be sure the gaskets are properly positioned to prevent water from entering the housing.

See Chapter 9 for lower unit capacities.

3-12 WINTER STORAGE

Taking extra time to store the boat properly at the end of each season, will increase the chances of satisfactory service at the next season. **REMEMBER**, idleness is the greatest enemy of an outboard motor. The unit should be run on a monthly basis. The boat steering and shifting mechanism should also be worked through complete cycles several times each month. The owner who spends a small amount of time involved in such maintenance will be rewarded by satisfactory performance, and greatly reduced maintenance expense for parts and labor.

Proper storage involves adequate protection of the unit from physical damage, rust, corrosion, and dirt.

The amount of lubricant in the lower unit should be checked on a daily basis during the operating season. A fish line wrapped around the propeller shaft could cut through a seal and oil be lost without warning.

The following steps provide an adequate maintenance program for storing the unit at the end of a season.

1- Empty all fuel from the carburetor.

For many years there has been the widespread belief simply shutting off the fuel at the tank and then running the powerhead until it stops is the proper procedure before storing the engine for any length of time. Right? **WRONG!**

First, it is **NOT** possible to remove all fuel in the carburetor by operating the powerhead until it stops. Considerable fuel is trapped in the float chamber and other passages and in the line leading to the carburetor. The **ONLY** guaranteed method of removing **ALL** fuel is to take the time to remove the carburetor, and drain the fuel.

Secondly, if the powerhead is operated with the fuel supply shut off until it stops, the fuel and oil mixture inside the powerhead is removed, leaving bearings, pistons, rings, and other parts without any protective lubricant.

2- Drain the fuel tank and the fuel lines. Pour approximately one quart (0.96 liters) of benzol (benzine) into the fuel tank, and then rinse the tank and pickup filter with the benzol. Drain the tank. Store the fuel tank in a cool dry area with the vent **OPEN** to allow air to circulate through the tank. **DO NOT** store the fuel tank on bare concrete. Place the tank to allow air to circulate around it.

3- Clean the carburetor fuel filter with benzol, see Chapter 4, Carburetor Repair Section.

4- Drain, and then fill the lower unit with Quicksilver gear case lubricant, as outlined in Section 3-11.

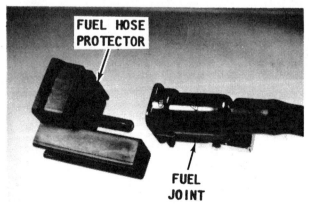

A Kleen-Klip fuel hose protector, available at marine dealers and used on the end of a disconnected fuel line, will keep the fitting free of almost any type contamination and most damage.

5- Lubricate the throttle and shift linkage. Lubricate the steering pivot shaft with multi-purpose water reststant lubricant or equivalent.

Clean the outboard unit thoroughly. Coat the powerhead with a commercial corrosion and rust preventative spray. Install the cowling, and then apply a thin film of fresh engine oil to all painted surfaces.

Remove the propeller. Apply Perfect Seal or a waterproof sealer to the propeller shaft splines, and then install the propeller back in position.

FINAL WORDS

Be sure all drain holes in the gear housing are open and free of obstruction. Check to be sure the **FLUSH** plug has been removed to allow all water to drain. Trapped water could freeze, expand, and cause expensive castings to crack.

ALWAYS store the outboard unit off the boat with the lower unit below the powerhead to prevent any water from being trapped inside.

BATTERY STORAGE

Remove the batteries from the boat and keep them charged during the storage period. Clean the batteries thoroughly of any dirt or corrosion, and then charge them to full specific gravity reading. After they are fully charged, store them in a clean cool dry place where they will not be damaged or knocked over.

NEVER store the battery with anything on top of it or cover the battery in such a manner as to prevent air from circulating around the fillercaps. All batteries, both new and old, will discharge during periods of storage, more so if they are hot than if they remain cool. Therefore, the electrolyte level and the specific gravity should be checked at regular intervals. A drop in the specific gravity reading is cause to charge them back to a full reading.

In cold climates, **EXERCISE CARE** in selecting the battery storage area. A fully-charged battery will freeze at about 60 degrees below zero. A discharged battery, almost dead, will have ice forming at about 19 degrees above zero.

ALWAYS remove the drain plug and position the boat with the bow higher than the stern. This will allow any rain water and melted snow to drain from the boat and prevent "trailer sinking". This term is used to describe a boat that has filled with rain water and ruined the interior, because the plug was not removed or the bow was not high enough to allow the water to drain properly.

A Model 25hp powerhead, serviced, adjusted, and ready to "work" for its owner.

4
FUEL

4-1 INTRODUCTION

The carburetion and ignition principles of two-cycle engine operation **MUST** be understood in order to perform a proper tune-up on an outboard motor.

If any doubts exist concerning an understanding of two-cycle engine operation, it would be best to study the Introduction section in the first portion of Chapter 7, Powerhead, before tackling any work on the fuel system.

The fuel system includes the fuel tank, fuel pump, fuel filters, carburetors, a squeeze bulb, and the associated parts to connect it all together. Regular maintenance of the fuel system to obtain maximum performance, is limited to changing the fuel filter at regular intervals and using fresh fuel.

If a sudden increase in gas consumption is noticed, or if the engine does not perform properly, a carburetor overhaul, including boil-out, or replacement of the fuel pump may be required.

4-2 GENERAL CARBURETION INFORMATION

The carburetor is merely a metering device for mixing fuel and air in the proper proportions for efficient engine operation. At idle speed, an outboard engine requires a mixture of about 8 parts air to 1 part fuel. At high speed or under heavy duty service, the mixture may change to as much as 12 parts air to 1 part fuel.

Float Systems

A small chamber in the carburetor serves as a fuel reservoir. A float valve admits fuel into the reservoir to replace the fuel consumed by the engine. If the carburetor has more than one reservoir, the fuel level in each reservoir (chamber) is controlled by identical float systems.

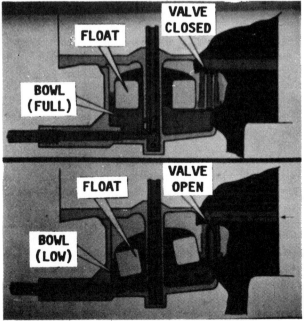

Fuel flow principle of a modern carburetor.

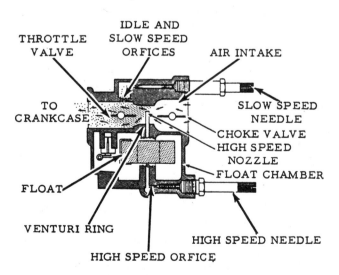

Fuel flow through the venturi, showing principle and related parts controlling intake and outflow.

Fuel level in each chamber is extremely critical and must be maintained accurately. Accuracy is obtained through proper adjustment of the float/s. This adjustment will provide a balanced metering of fuel to each cylinder at all speeds.

Following the fuel through its course, from the fuel tank to the combustion chamber of the cylinder, will provide an appreciation of exactly what is taking place. In order to start the engine, the fuel must be moved from the tank to the carburetor by a squeeze bulb installed in the fuel line. This action is necessary because the fuel pump does not have sufficient pressure to draw fuel from the tank during cranking before the engine starts.

The fuel for some small horsepower units is by gravity feed from a tank mounted at the rear of the powerhead. Even with the gravity feed method, a small fuel pump may be an integral part of the carburetor.

After the engine starts, the fuel passes through the pump to the carburetor. All systems have some type of filter installed somewhere in the line between the tank and the carburetor. Many units have a filter as an integral part of the carburetor.

At the carburetor, the fuel passes through the inlet passage to the needle and seat, and then into the float chamber (reservoir). A float in the chamber rides up and down on the surface of the fuel. After fuel enters the chamber and the level rises to a predetermined point, a tang on the float closes the inlet needle and the flow entering the chamber is cutoff. When fuel leaves the chamber as the engine operates, the fuel level drops and the float tang allows the inlet needle to move off its seat and fuel once again enters the chamber. In this manner a constant reservoir of fuel is maintained in the chamber to satisfy the demands of the engine at all speeds.

A fuel chamber vent hole is located near the top of the carburetor body to permit atmospheric pressure to act against the fuel in each chamber. This pressure assures an adequate fuel supply to the various operating systems of the powerhead.

Air/Fuel Mixture

A suction effect is created each time the piston moves upward in the cylinder. This suction draws air through the throat of the carburetor. A restriction in the throat, called a venturi, controls air velocity and has the effect of reducing air pressure at this point.

The difference in air pressures at the throat and in the fuel chamber, causes the fuel to be pushed out of metering jets extending down into the fuel chamber. When the fuel leaves the jets, it mixes with the air passing through the venturi. This fuel/air mixture should then be in the proper proportion for burning in the cylinder/s for maximum engine performance.

In order to obtain the proper air/fuel mixture for all engine speeds, some models have high and low speed jets. These jets have adjustable needle valves which are used to compensate for changing atmospheric conditions. In almost all cases, the high-speed circuit has fixed high-speed jets and are not adjustable.

A throttle valve controls the flow of air/fuel mixture drawn into the combustion

Typical fuel filter installation.

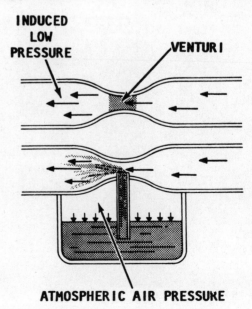

Air flow principle of a modern carburetor.

chambers. A cold powerhead requires a richer fuel mixture to start and during the brief period it is warming to normal operating temperature. A choke valve is placed ahead of the metering jets and venturi. As this valve begins to close, the volume of air intake is reduced, thus enriching the mixture entering the cylinder/s.

When this choke valve is fully closed, a very rich fuel mixture is drawn into the cylinder/s.

The throat of the carburetor is usually referred to as the "barrel". Carburetors with single, double, or four barrels have individual metering jets, needle valves, throttle and choke plates for each barrel. Single and two barrel carburetors are fed by a single float and chamber.

4-3 FUEL SYSTEM

The fuel system includes the fuel tank, fuel pump, fuel filters, carburetor, connecting lines with a squeeze bulb, and the associated parts to connect it all together. Regular maintenance of the fuel system to obtain maximum performance, is limited to changing the fuel filter at regular intervals and using fresh fuel. Even with the high price of fuel, removing gasoline that has been standing unused over a long period of time, is still the easiest and least expensive preventive maintenance possible. In most cases this old gas, even with some oil mixed with it, can be used without harmful effects in an automobile using regular gasoline.

If a sudden increase in gas consumption is noticed, or if the engine does not perform properly, a carburetor overhaul, including boil-out, or replacement of the fuel pump may be required.

LEADED GASOLINE AND GASOHOL

The manufacturer of the units covered in this manual recommends the powerheads be operated using either regular unleaded or regular leaded gasoline with a minimum octane rating of 84 or higher.

In the United States, the Environmental Protection Agency (EPA) has slated a proposed national phase-out of leaded fuel, "Regular" gasoline, by 1988. Lead in gasoline boosts the octane rating (energy). Therefore, if the lead is removed, it must be replaced with another agent. Unknown to the general public, many refineries are adding alcohol in an effort to hold the octane rating.

Alcohol in gasoline can have a deteriorating effect on certain fuel system parts. Seals can swell, pump check valves can swell, diaphragms distort, and other rubber or neoprene composition parts in the fuel system can be affected.

Since about 1981, the manufacturer has made every effort to use materials that will resist the alcohol being added to fuels.

Fuels containing alcohol will slowly absorb moisture from the air. Once the mois-

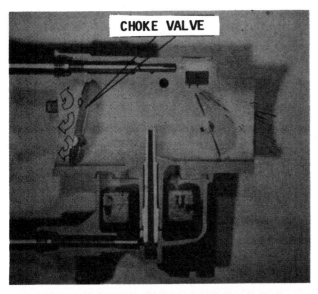

Choke valve location in the carburetor venturi. The choke valve is always located in front of the venturi to restrict air flow and assist cold powerhead startup.

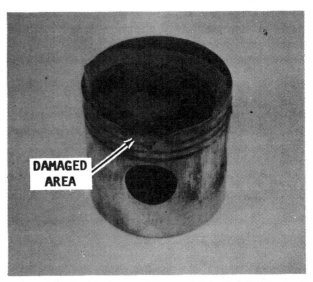

Damaged piston, possibly caused by insufficient oil mixed with the fuel; using too-low an octane fuel; or using "soured" fuel (fuel stood too long without a preservative added).

4-4 FUEL

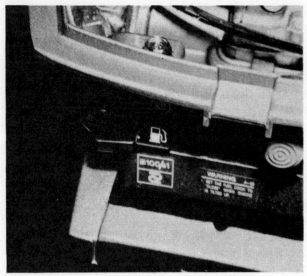

Fuel shutoff knob on a 4hp powerhead.

ture content in the fuel exceeds about 1%, it will separate from the fuel taking the alcohol with it. This water/alcohol mixture will settle to the bottom of the fuel tank. The engine will fail to operate. Therefore, storage of this type of gasoline for use in marine engines is not recommended for more than just a few days.

One temporary, but aggravating, solution to increase the octane of "Unleaded" fuel is to purchase some aviation fuel from the local airport. Add about 10 to 15 percent of the tank's capacity to the unleaded fuel.

REMOVING FUEL FROM THE SYSTEM

For many years there has been the widespread belief that simply shutting off the fuel at the tank and then running the engine until it stops is the proper procedure before storing the engine for any length of time. Right? **WRONG**.

Typical fuel line quick disconnect fitting.

It is **NOT** possible to remove all of the fuel in the carburetor by operating the engine until it stops. Some fuel is trapped in the float chamber and other passages and in the line leading to the carburetor. The **ONLY** guaranteed method of removing **ALL** of the fuel is to take the time to remove the carburetor, and drain the fuel.

If the engine is operated with the fuel supply shut off until it stops the fuel and oil mixture inside the engine is removed, leaving bearings, pistons, rings, and other parts with little protective lubricant, during long periods of storage.

Proper procedure involves: Shutting off the fuel supply at the tank; disconnecting the fuel line at the tank; operating the engine until it begins to run **ROUGH**; then stopping the engine, which will leave some fuel/oil mixture inside; and finally removing and draining the carburetor. By disconnecting the fuel supply, all **SMALL** passages are cleared of fuel even though some fuel is left in the carburetor. A light oil should be put in the combustion chamber as instructed in the Owner's Manual. On some model carburetors the high-speed jet plug can be removed to drain the fuel from the carburetor.

For short periods of storage, simply running the carburetor dry may help prevent severe gum and varnish from forming in the carburetor. This is especially true during hot weather.

4-4 TROUBLESHOOTING

The following paragraphs provide an orderly sequence of tests to pinpoint problems in the system. It is very rare for the carburetor by itself to cause failure of the engine to start.

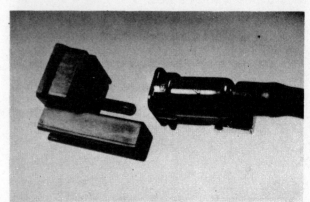

An excellent method of protecting fuel hose connectors against damage and contamination is an end cap by "Kleen Klip". More information on this device in Chapter 3.

FUEL PROBLEMS

Many times fuel system troubles are caused by a plugged fuel filter, a defective fuel pump, or by a leak in the line from the fuel tank to the fuel pump. A defective choke may also cause problems. **WOULD YOU BELIEVE,** a majority of starting troubles which are traced to the fuel system are the result of an empty fuel tank or aged "sour" fuel.

"SOUR" FUEL

Under average conditions (temperate climates), fuel will begin to breakdown in about four months. A gummy substance forms in the bottom of the fuel tank and in other areas. The filter screen between the tank and the carburetor and small passages in the carburetor will become clogged. The gasoline will begin to give off an odor similar to rotten eggs. Such a condition can cause the owner much frustration, time in cleaning components, and the expense of replacement or overhaul parts for the carburetor.

Even with the high price of fuel, removing gasoline that has been standing unused over a long period of time is still the easiest and least expensive preventative maintenance possible. In most cases, this old gas can be used without harmful effects in an automobile using regular gasoline.

A gasoline preservative additive Quicksilver Fuel Conditioner and Stabilizer for 2 and 4 Cycle Engines, will keep the fuel "fresh" for up to twelve months. If this particular product is not available in your area, other similar additives are produced under various trade names.

Choke Problems

When the engine is hot, the fuel system can cause starting problems. After a hot engine is shut down, the temperature inside the fuel bowl may rise to $200°F$ and cause the fuel to actually boil. All carburetors are vented to allow this pressure to escape to the atmosphere. However, some of the fuel may percolate over the high-speed nozzle.

If the choke should stick in the open position, the engine will be hard to start. If the choke should stick in the closed position, the engine will flood making it very difficult to start.

In order for this raw fuel to vaporize enough to burn, considerable air must be added to lean out the mixture. Therefore, the only remedy is to remove the spark plug/s; ground the leads; crank the powerhead through about ten revolutions; clean the plugs; install the plugs again; and start the engine.

If the needle valve and seat assembly is leaking, an excessive amount of fuel may enter the reed housing in the following manner: After the powerhead is shut down, the pressure left in the fuel line will force fuel past the leaking needle valve. This extra fuel will raise the level in the fuel bowl and cause fuel to overflow into the reed housing.

A continuous overflow of fuel into the reed housing may be due to a sticking inlet needle or to a defective float which would cause an extra high level of fuel in the bowl and overflow into the reed housing.

FUEL PUMP TEST

First, These Words

On very small horsepower units, a fuel pump may not be used because the fuel is

Adding a quality brand fuel stabilizer and conditioner can prevent fuel from "souring" up to twelve months.

4-6 FUEL

Testing the fuel pickup in the fuel tank AND operation of the squeeze bulb by observing fuel flow from the disconnected line at the fuel pump, discharged into a suitable container.

Working the squeeze bulb and observing the fuel flow from the line disconnected at the carburetor and discharged into a suitable container. This simple operation will verify fuel flow through the fuel pump. (The two photographs in this column were taken with an older powerhead than those covered in this manual. However, the procedure is the same.)

gravity fed from a fuel tank mounted at the rear of the powerhead. The following test does not apply to models with an integral fuel pump as part of the carburetor. The test applies only to units with the fuel pump as a separate piece of equipment.

CAUTION Gasoline will be flowing in the engine area during this test. Therefore, guard against fire by grounding the high-tension wire to prevent it from sparking.

A good safety procedure is to ground each spark plug lead. Disconnect the fuel line at the carburetor. Place a suitable container over the end of the fuel line to catch the fuel discharged. Open the valve at the fuel tank. With gravity flow system, fuel should flow freely from the end of the hose. With a remote fuel tank, squeeze the primer bulb and observe if there is satisfactory flow of fuel from the line.

If there is no fuel discharged from the line, the check valve in the squeeze bulb may be defective, or there may be a break or obstruction in the fuel line.

If there is a good fuel flow, then crank the powerhead. If the fuel pump is operating properly, a healthy stream of fuel should pulse out of the line.

Continue cranking the engine and catching the fuel for about 15 pulses to determine if the amount of fuel decreases with each pulse or maintains a constant amount. A

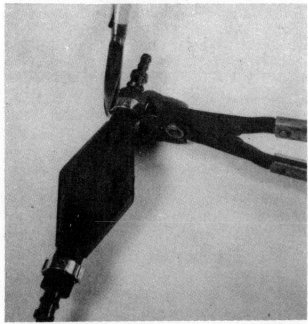

Using the proper tools to install a clamp around the squeeze bulb check valve.

TROUBLESHOOTING 4-7

decrease in the discharge indicates a restriction in the line. If the fuel line is plugged, the fuel stream may stop. If there is fuel in the fuel tank but no fuel flows out of the fuel line while the engine is being cranked, the problem may be in one of four areas:

1- The line from the fuel pump to the carburetor may be plugged as already mentioned.

2- The fuel pump may be defective.

3- The line from the fuel tank to the fuel pump may be plugged; the line may be leaking air; or the squeeze bulb may be defective.

4- If the engine does not start even though there is adequate fuel flow from the fuel line, the fuel filter in the carburetor inlet may be plugged or the fuel inlet needle valve and the seat may be gummed together and prevent adequate fuel flow.

FUEL LINE TEST

On most installations, the fuel line is provided with quick-disconnect fittings at the tank and at the engine. If there is reason to believe the problem is at the quick-disconnects, the hose ends should be replaced as an assembly. For a small additional expense, the entire fuel line can be replaced and thus eliminate this entire area as a problem source for many future seasons.

The primer squeeze bulb can be replaced in a short time. First, cut the hose line as close to the old bulb as possible. Slide a small clamp over the end of the fuel line from the tank. Next, install the **SMALL** end of the check valve assembly into this side of the fuel line. The check valve always goes towards the fuel tank. Place a large clamp over the end of the check valve assembly. Use Primer Bulb Adhesive when the connections are made. Tighten the clamps. Repeat the procedure with the other side of the bulb assembly and the line leading to the engine.

ROUGH ENGINE IDLE

If an engine does not idle smoothly, the most reasonable approach to the problem is

A replacement squeeze bulb kit includes parts necessary to return this section of the fuel line to service.

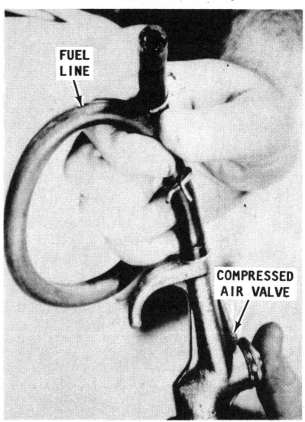

*Many times, restrictions such as foreign material may be cleared from the fuel line using compressed air. Use **CARE** to be sure the open end of the hose is pointing clear to avoid personal injury to the eyes.*

Fouled spark plug, possibly caused by the operator's habit of overchoking or a malfunction holding the choke closed. Either of these conditions delivered a too-rich fuel mixture to the cylinder.

to perform a tune-up to eliminate such areas as: defective points; faulty spark plugs; and timing out of adjustment.

Other problems that can prevent an engine from running smoothly include: An air leak in the intake manifold; uneven compression between the cylinders; and sticky or broken reeds.

Of course any problem in the carburetor affecting the air/fuel mixture will also prevent the engine from operating smoothly at idle speed. These problems usually include: Too high a fuel level in the bowl; a heavy float; leaking needle valve and seat; defective automatic choke; and improper adjustments for idle mixture or idle speed.

EXCESSIVE FUEL CONSUMPTION

Excessive fuel consumption can be the result of any one of three conditions, or a combination of all three.

1- Inefficient engine operation.
2- Faulty condition of the hull, including excessive marine growth.
3- Poor boating habits of the operator.

If the fuel consumption suddenly increases over what could be considered normal, then the cause can probably be attributed to the engine or boat and not the operator.

Marine growth on the hull can have a very marked effect on boat performance. This is why sail boats always try to have a haul-out as close to race time as possible. While you are checking the bottom, take note of the propeller condition. A bent blade or other damage will definitely cause poor boat performance.

If the hull and propeller are in good shape, then check the fuel system for possible leaks. Check the line between the fuel pump and the carburetor while the engine is running and the line between the fuel tank and the pump when the engine is not running. A leak between the tank and the pump many times will not appear when the engine is operating, because the suction created by the pump drawing fuel will not allow the fuel to leak. Once the engine is turned off and the suction no longer exists, fuel may begin to leak.

If a minor tune-up has been performed and the spark plugs, points, and timing are properly adjusted, then the problem most likely is in the carburetor and an overhaul is in order. Check the needle valve and seat for leaking. Use extra care when making any adjustments affecting the fuel consumption, such as the float level or automatic choke.

ENGINE SURGE

If the engine operates as if the load on the boat is being constantly increased and

Vacuum operated fuel pump installation on a 25hp or 30hp two-stroke powerhead.

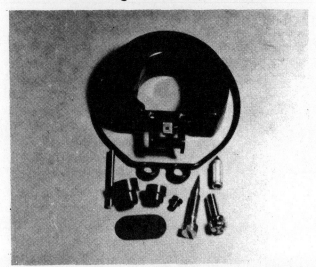

Major parts found in a typical carburetor repair kit.

TROUBLESHOOTING 4-9

Carburetor "C" cleaned, properly serviced with new parts, assembled, and ready for installation on the powerhead.

decreased, even though an attempt is being made to hold a constant engine speed, the problem can most likely be attributed to the fuel pump, or a restriction in the fuel line between the tank and the carburetor.

Operational description and service procedures for the fuel pump are given in Section 4-17.

4-5 CARBURETOR MODELS

Eleven, yes, eleven different carburetors have been used over the years on the outboard powerheads covered in this manual.

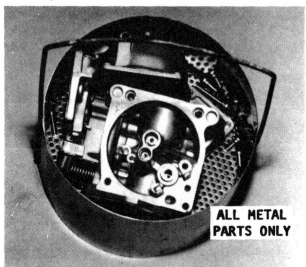

Metal carburetor parts in a basket ready to be immersed in a carburetor cleaning solution.

Complete, detailed, illustrated procedures for each carburetor are presented in separate sections of this chapter. To determine which carburetor is installed on the outboard unit being serviced, check the following table under powerhead size and year of production. Sometimes two different carburetors were installed on the same model. The letter designation in the third column is used throughout this chapter and in the Appendix.

HP MODEL	YEAR	CARB	COMMENTS
2	1977 & On	A	
3.5	1978-81	B	Air Cooled
4	1982 & On	C	
5	1977-79	D	Air Cooled
5	1979 & On	C	Water Cooled
8A & 8	1977-79	E	
8B, 8W & Mrthn 8	1979 & On	F/G	Both Used
9.9C & 15C	1977 & On	F/G	Both Used
8C	1979 & On	G	
15W & 15	1977-79	H	
20	1977 & On	J	
25	1980 & On	J	
28	1977-79	J	
30	1980 & On	J	
40	1978 & On	K	
48	1977-79	L	
55	1986 & On	L	
60	1977-83	L	

Complete detailed service procedures for each carburetor are presented in separate sections.

SPECIAL GOOD WORDS

Carburetor repair kits, available at the local service dealer, contain the necessary parts to perform a proper carburetor overhaul. In most cases, an illustration showing the parts contained in the package is included in the Cleaning and Inspecting portion for each carburetor.

4-10 FUEL

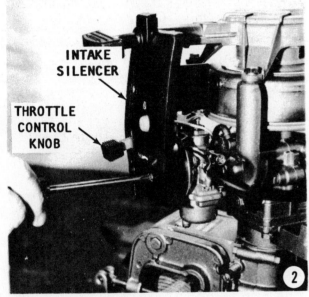

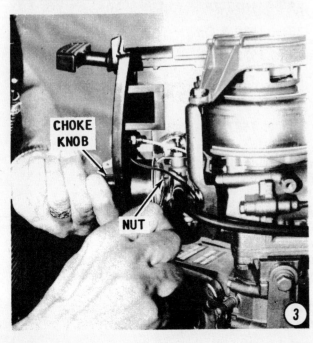

4-6 CARBURETOR "A"
USED ON MODEL 2HP POWERHEADS

This section provides complete detailed procedures for removal, disassembly, cleaning and inspecting, assembling including bench adjustments, installation, and operating adjustments for Carburetor "A", as used originally on the Model 2hp unit. This carburetor is a single-barrel, float feed type with a manual choke. Fuel to the carburetor is gravity fed from a fuel tank mounted at the rear of the powerhead.

REMOVAL AND DISASSEMBLING

FIRST, THESE WORDS

Good shop practice dictates a carburetor repair kit be purchased and new parts be installed any time the carburetor is disassembled.

Make an attempt to keep the work area organized and to cover parts after they have been cleaned. This practice will prevent foreign matter from entering passageways or adhering to critical parts.

1- Remove the screw on one half of the spark plug cover. Remove four more screws securing one-half of the cowling. Separate the cowling half from which the screws were removed. Remove the four screws securing the other half of the cowling and remove it from the engine. The spark plug cover will remain attached to one of the cowling halves.

SPECIAL WORDS

Observe the different length screws used to secure the cowling halves to the powerhead. Remember their location as an aid during assembling.

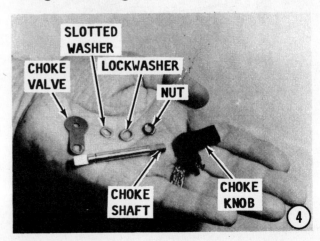

CARBURETOR "A"   4-11

2- Remove the two Philips screws securing the intake silencer and the screw securing the throttle control knob to the throttle lever. Remove the knob.

3- Hold the choke knob with one hand and remove the nut on the back side of the intake silencer with a wrench and the other hand. Slide the lockwasher free of the choke shaft.

4- Pull the choke knob with the choke shaft attached out through the hole in the intake silencer. The choke valve and slotted washer will come out with the shaft. Place these small items in order on the workbench as an aid during installation.

5- CAREFULLY separate the WHITE lead (engine stop button lead), from the quick disconnect fitting anchored to the powerhead. Disconnect the black lead, ground, at the horseshoe bracket. Thread the small screw back into the bracket as a precaution against the screw being lost. Lift the intake silencer free of the engine with the wire leads attached to the stop button.

6- Loosen the bolt on the clamp securing the carburetor to the powerhead.

7- Close the fuel valve. Be prepared to catch a small amount of fuel with a cloth when the fuel line is disconnected. Disconnect the fuel line at the carburetor. Remove the carburetor from the reed valve housing. The clamp will come off with the carburetor.

8- Pry off the circlip, and then remove the pivot pin attaching the throttle control lever to the carburetor. Remove the screw, spring, washer, and nut from the carburetor top bracket.

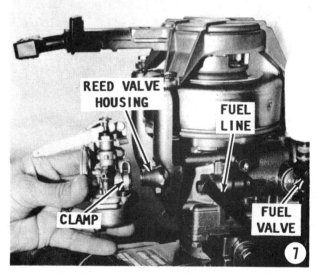

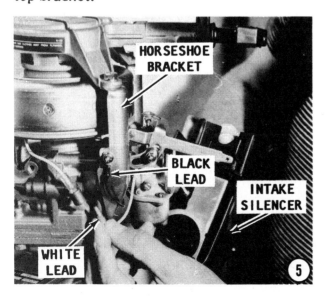

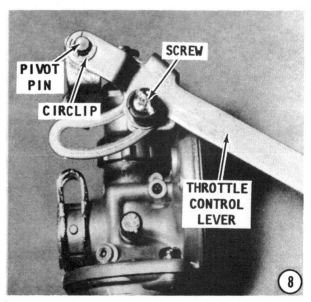

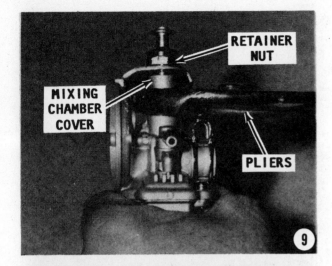

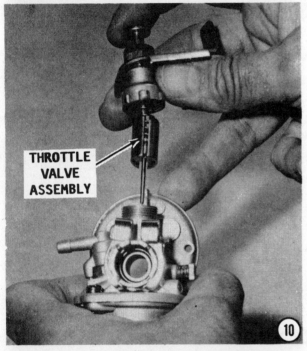

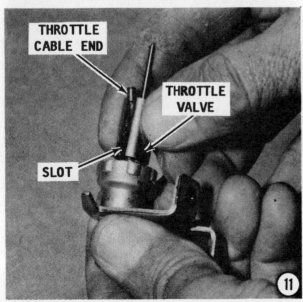

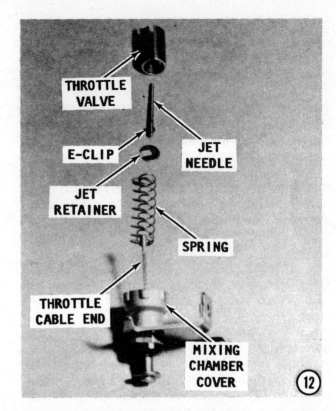

9- Loosen, but **DO NOT** remove the top retainer brass nut on the carburetor. Hold the carburetor securely with one hand and unscrew the mixing chamber cover with the other, using a pair of pliers.

10- Lift out the throttle valve assembly.

11- Compress the spring in the throttle valve assembly to allow the throttle cable end to clear the recess in the base of the throttle valve and to slide down the slot.

12- Disassemble the throttle valve consisting of the throttle valve, spring, jet needle (with an E-clip on the second groove), jet retainer, and throttle cable end.

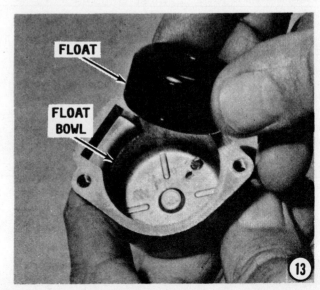

CARBURETOR "A" 4-13

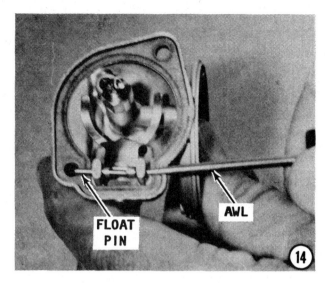

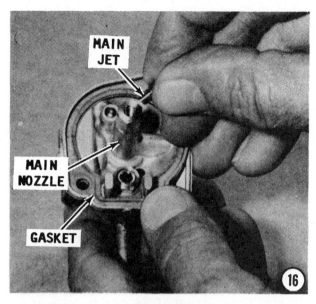

13- Remove and discard the two screws securing the float bowl to the carburetor body. Remove the float bowl. Lift out the float. A new pair of float bowl screws are provided in the carburetor rebuild kit.

14- Push the float pin free using a fine pointed awl.

15- Lift out the float arm and needle valve. Slide the needle valve free of the arm.

16- Unscrew the main jet from the main nozzle. Remove and **DISCARD** the float bowl gasket.

17- Count and record the number of turns required to **LIGHTLY** seat the idle speed screw. The number of turns will give a rough adjustment during installation. Back out the idle speed screw and **DISCARD** the screw, but **SAVE** the spring. A new screw is provided in the carburetor rebuild kit to ensure a damaged screw is not used again. This screw is most important for maximum performance.

18- Pry out and **DISCARD** the carburetor sealing ring.

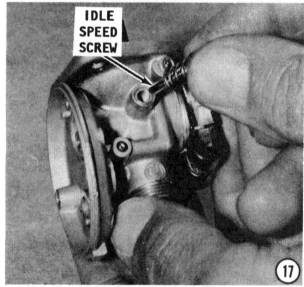

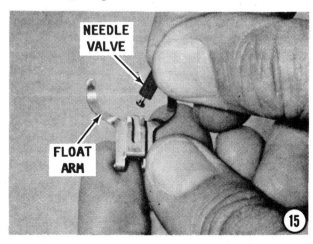

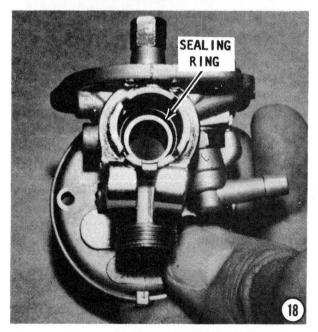

4-14 FUEL

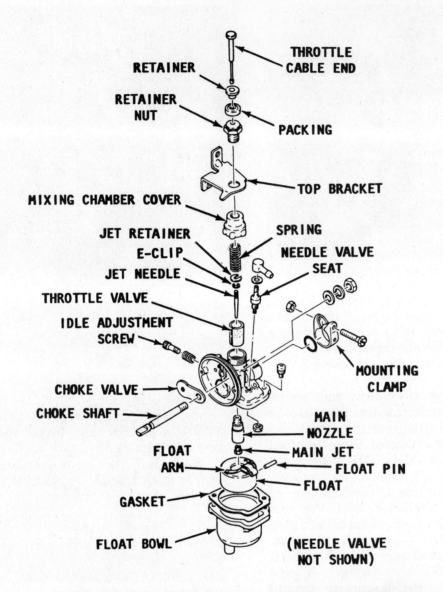

Exploded drawing of Carburetor "A", with major parts identified.

GOOD WORDS

It is not necessary to remove the E-clip from the jet needle, unless replacement is required or if the powerhead is to be operated at a significantly different elevation.

CLEANING AND INSPECTING

NEVER dip rubber parts, plastic parts, diaphragms, or pump plungers in carburetor cleaner. These parts should be cleaned **ONLY** in solvent, and then blown dry with compressed air.

Place all metal parts in a screen-type tray and dip them in carburetor cleaner until they appear completely clean, then blow them dry with compressed air.

Blow out all passages in the castings with compressed air. Check all parts and passages to be sure they are not clogged or contain any deposits. **NEVER** use a piece of wire or any type of pointed instrument to clean drilled passages or calibrated holes in a carburetor.

Move the throttle shaft back and forth to check for wear. If the shaft appears to be too loose, replace the complete throttle body because individual replacement parts are **NOT** available.

Inspect the main body, mixing chamber and gasket surfaces for cracks and burrs which might cause a leak. Check the float for deterioration. If any part of the float is damaged, the unit must be replaced. Check

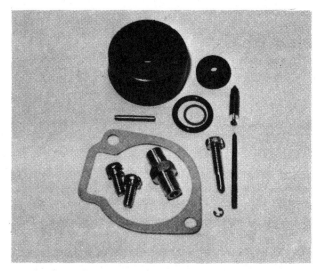

Parts included in a repair kit for Carburetor "A".

the float arm needle contacting surface and replace the float if this surface has a groove worn in it.

Inspect the tapered section of the idle adjusting needle and replace the needle if it has developed a groove.

Most of the parts which should be replaced during a carburetor overhaul are included in overhaul kits available from your local marine dealer. One of these kits will contain a matched fuel inlet needle and seat, if removable. This combination should be replaced each time the carburetor is disassembled as a precaution against leakage.

After all plastic and rubber parts have been removed, Carburetor "A" may be immersed in a carburetor cleaning solution for thorough cleaning.

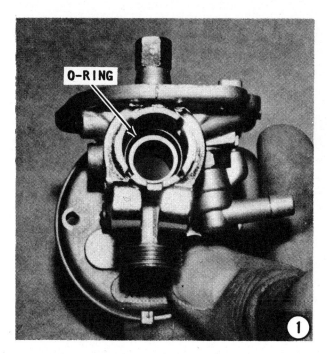

ASSEMBLING

1- Install a **NEW** carburetor O-ring into the caburetor body.

2- Apply an all-purpose lubricant to a **NEW** idle speed screw. Install the idle speed screw and spring.

3- Install the main jet into the main nozzle and tighten it just "snug" with a screwdriver.

4- Slide a **NEW** needle valve into the groove of the float arm.

5- Lower the float arm into position with the needle valve sliding into the needle valve seat. Now, push the float pin through the holes in the carburetor body and hinge using a small awl or similar tool.

4-16 FUEL

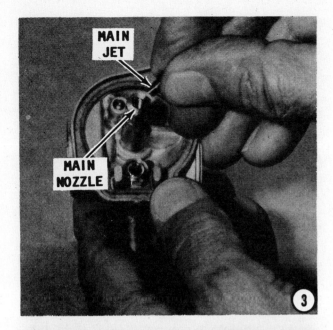

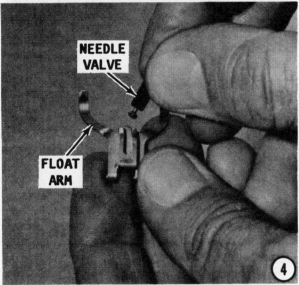

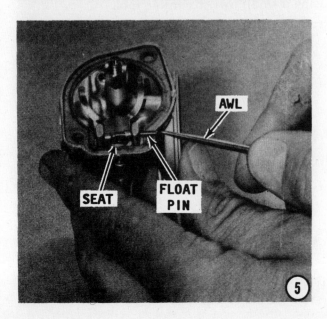

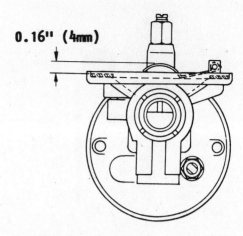

6- Hold the carburetor body in a perfect upright position. Check the float hinge adjustment. The vertical distance between the top of the hinge and the top of the gasket should be 0.16" (4mm). **CAREFULLY**, bend the hinge, if necessary, to achieve the required measurement.

7- Position a new float bowl gasket in place on the carburetor body. Install the float into the float bowl. Place the float bowl in position on the carburetor body, and then secure it with the two Phillips head screws.

8- If the E-clip on the jet needle is lowered, the carburetor will cause the powerhead to operate "rich". Raising the E-clip will cause the powerhead to operate "lean". Higher altitude -- raise E-clip to compensate for rarefied air. Begin to assemble the throttle valve components by inserting the E-clip end of the jet needle into the throttle valve (the end with the recess for the throttle cable end). Next, place the needle retainer into the throttle valve over the E-clip and align the retainer slot with the slot in the throttle valve.

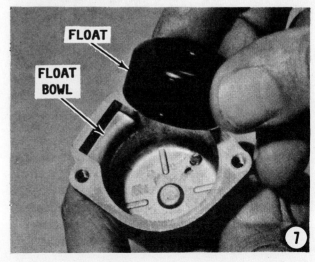

CARBURETOR "A" 4-17

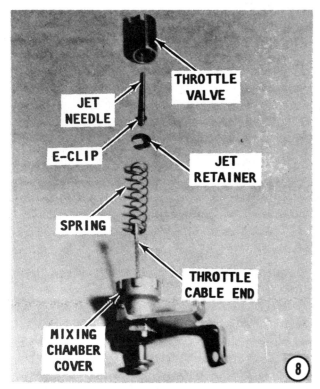

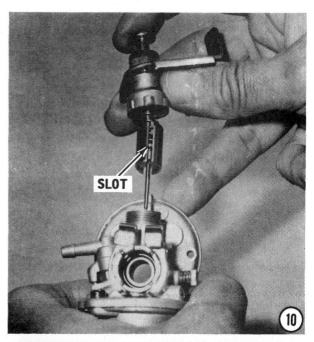

9- Thread the spring over the end of the throttle cable and insert the cable into the retainer end of the throttle valve. Compress the spring and at the same time, guide the cable end through the slot until the end locks into place in the recess.

10- Position the assembled throttle valve in such a manner to permit the slot to slide over the alignment pin while the throttle valve is lowered into the carburetor. This alignment pin permits the throttle valve to only be installed **ONE** way -- in the correct position.

11- **CAREFULLY** tighten the mixing chamber cover with a pair of pliers.

12- Slide the pivot pin through the top carburetor bracket and then through the upper hole in the throttle control lever. Secure the pin with the Circlip. Slide the

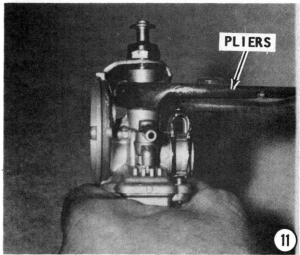

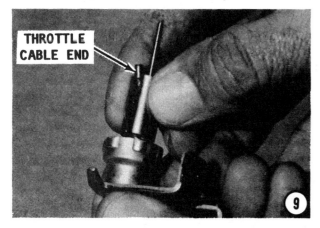

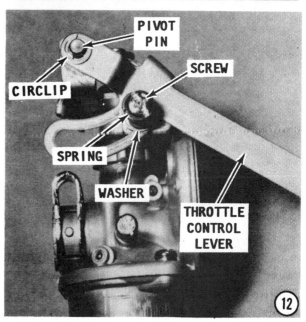

4-18 FUEL

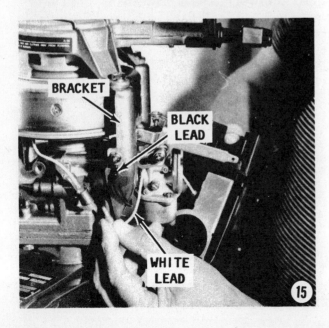

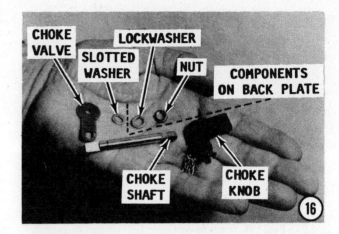

spring, then the washer over the screw and align the slot in the control lever with the threaded hole in the top carburetor bracket. Install the screw until the lever moves freely with just enough friction to provide the operator with a feel of the throttle opening.

13- Position the carburetor clamp over the carburetor mounting collar, with the locking bolt and nut in place in the clamp. Slide the carburetor into place on the reed valve housing.

14- Secure the carburetor in place by tightening the bolt and nut securely.

15- Hold the intake silencer up to the carburetor. Connect the white lead (engine stop button lead) to the quick disconnect fitting anchored at the top of the powerhead. Attach the black ground lead to the screw on the horseshoe powerhead bracket.

16- Push the choke knob onto the slotted end of the choke shaft. Place the threaded end through the intake silencer and slide the choke valve, followed by the slotted washer, onto the protruding threaded end past the threads onto the slotted part of the shaft.

Position the intake silencer up against the carburetor.

17- Hold the choke knob. Slide the lock washer onto the end of the choke shaft, and then thread on the nut.

WORDS FROM EXPERIENCE
Starting this nut is not an easy task. Holding the nut with an alligator clip may prove helpful.

Tighten the nut just "snug". Connect the fuel hose to the carburetor.

18- Install the throttle control knob onto the throttle lever and secure it with the Phillips head screw. Secure the intake silencer to the carburetor with the two Phillips head screws.

19- SLOWLY tighten the idle speed screw until it **BARELY** seats, then back it out the same number of turns recorded during disassembling. If the number of turns was not recorded, back the screw out 1-3/4 turns as a rough adjustment.

20- Install the two halves of the cowling around the powerhead.

Secure the cowling with the attaching screws. Eight screws hold the cowling halves in place plus one more for the spark plug cover.

SPECIAL WORDS
As noted during disassembling, the screws are different lengths. Ensure the proper size are used to the correct location.

Mount the outboard unit in a test tank, on the boat in a body of water, or connect a flush attachment and hose to the lower unit. Connect a tachometer to the powerhead.

NEVER, AGAIN, NEVER operate the engine at high speed with a flush device attached. The engine, operating at high speed

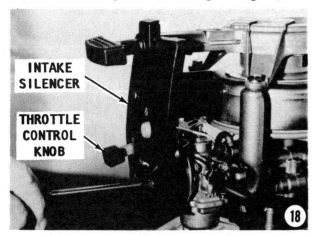

with such a device attached, would **RUNAWAY** from lack of a load on the propeller, causing extensive damage.

Start the engine and check the completed work.

CAUTION
Water must circulate through the lower unit to the powerhead anytime the powerhead is operating to prevent damage to the water pump in the lower unit. Just five seconds without water will damage the water pump impeller.

Idle Speed Adjustment
Allow the powerhead to warm to normal operating temperature. Adjust the idle speed screw until the powerhead idles between 900 and 1100 rpm for units with serial numbers 6A1-000100 and below, or between

1150 and 1250 rpm for units with serial numbers 6A1-000101 and above.

High Speed Adjustment

The only high speed change possible is by making a slight adjustment to the air/fuel mixture. A richer mixture will cause an increase in powerhead rpm -- to a certain point. Causing the mixture to become a bit more lean will result in a decrease in powerhead rpm.

A change in the air/fuel mixture and thus powerhead rpm, is made through movement of the E-clip on the jet needle. If the E-clip on the jet needle is lowered, the carburetor will cause the powerhead to operate "rich". Raising the E-clip will cause the powerhead to operate "lean". Higher altitude -- raise E-clip to compensate for rarefied air.

Detailed timing and synchronizing procedures are presented in Chapter 6.

4-7 CARBURETOR "B"
USED ON 3.5HP POWERHEADS

This section provides complete detailed procedures for removal, disassembly, cleaning and inspecting, assembling including bench adjustments, installation, and operating adjustments for Carburetor "B", as used originally on the Model 3.5hp unit. This carburetor is a single-barrel, float feed type with a manual choke. Fuel to the carburetor is gravity fed from a fuel tank mounted at the rear of the powerhead.

FIRST, THESE WORDS

Good shop practice dictates a carburetor repair kit be purchased and new parts be installed any time the carburetor is disassembled.

Make an attempt to keep the work area organized and to cover parts after they have been cleaned. This practice will prevent foreign matter from entering passageways or adhering to critical parts.

REMOVAL

1- Disconnect the fuel line and take adequate measures to prevent fuel from draining from the tank. Remove the fuel tank through the attaching hardware. Disconnect the carburetor linkage. Remove the intake cover and then tag and disconnect the two fuel lines.

Remove the two carburetor retaining bolts and lift the carburetor free of the powerhead.

DISASSEMBLING

2- Pry off the plunger cap cover and unscrew the plunger cap. Remove the starter lever plate spring and plunger from the carburetor body.

3- Remove the idle mixture screw and spring. Use the proper size screwdriver and back out the low speed jet. Invert the carburetor and the idle tube will fall free from the hole previously covered by the jet.

4- Pry the cap from the base of the float bowl. Remove the two screws and

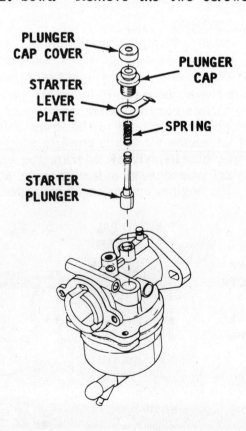

CARBURETOR "B"

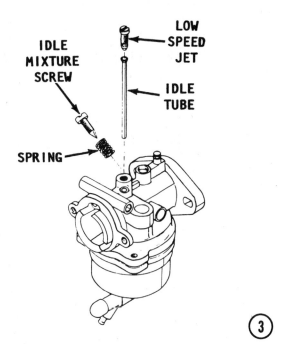

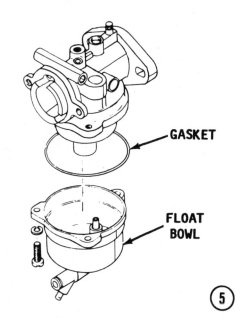

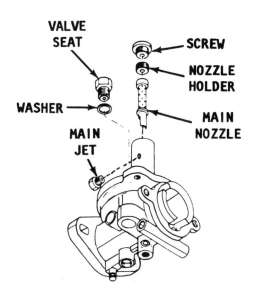

spring washers from the drain cup retaining plate. Remove the following in order: drain valve button, large washer, guide, spring, and finally the small washer.

5- Remove the three screws and washers securing the float bowl to the carburetor body. Remove and discard the gasket.

6- Push the float hinge pin free of the mounting posts. Once the end of the pin clears the float hinge, the float and needle valve may be lifted free.

7- Back off the screw at the top of the center turret of the mixing chamber and withdraw the nozzle holder and main nozzle. Use the correct size socket and remove the valve seat and washer. Remove the main jet from the side of the center turret.

A GOOD WORD

Further disassembly of the carburetor is not necessary in order to clean it properly.

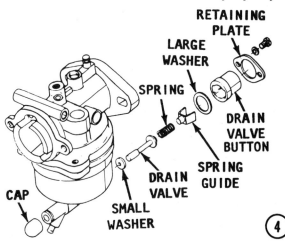

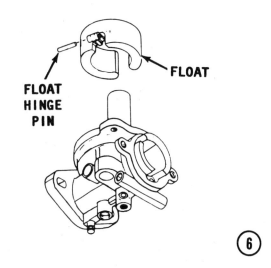

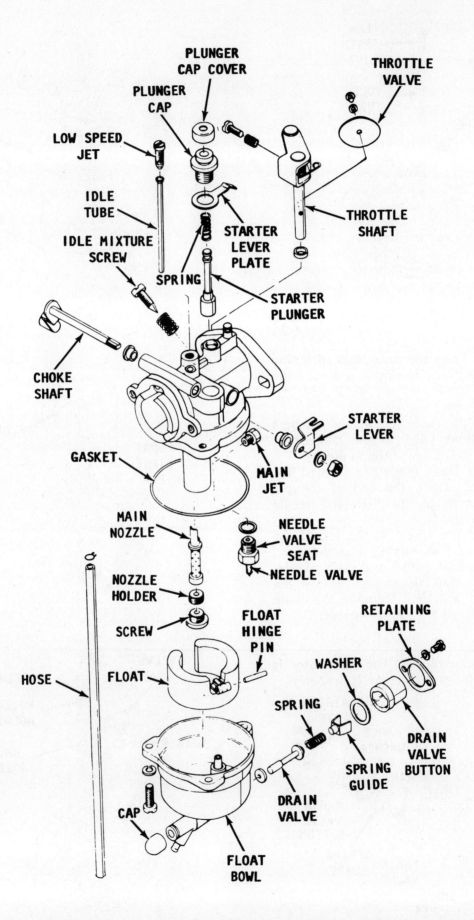

Exploded drawing of Carburetor "B", used on 3.5hp powerheads, with major parts identified.

CLEANING AND INSPECTING

NEVER dip rubber parts, plastic parts, diaphragms, or pump plungers in carburetor cleaner. These parts should be cleaned **ONLY** in solvent, and then blown dry with compressed air.

Place all metal parts in a screen-type tray and dip them in carburetor cleaner until they appear completely clean, then blow them dry with compressed air.

Blow out all passages in the castings with compressed air. Check all parts and passages to be sure they are not clogged or contain any deposits. **NEVER** use a piece of wire or any type of pointed instrument to clean drilled passages or calibrated holes in a carburetor.

Move the throttle shaft back and forth to check for wear. If the shaft appears to be too loose, replace the complete throttle body because individual replacement parts are **NOT** available.

Inspect the main body, mixing chamber, and gasket surfaces for cracks and burrs which might cause a leak. Check the float for deterioration. Check to be sure the float spring has not been stretched. If any part of the float is damaged, the unit must be replaced. Check the float arm needle contacting surface and replace the float if this surface has a groove worn in it.

Inspect the tapered section of the idle adjusting needle and replace the needle if it has developed a groove.

Most of the parts which should be replaced during a carburetor overhaul are included in overhaul kits available from your local marine dealer. One of these kits will contain a matched fuel inlet needle and seat, if removable. This combination should be replaced each time the carburetor is disassembled as a precaution against leakage.

ASSEMBLING

1- Insert the main nozzle into the mixing chamber center turret, followed by the nozzle holder. Install the screw over the holder and tighten it snugly. Slide a **NEW** gasket onto the shaft of the valve seat. Install and tighten the valve seat using the proper size thin walled socket. Install the main jet into the threaded hole on the side of the mixing chamber center turret.

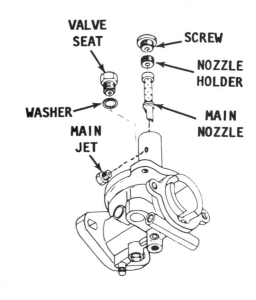

2- Insert the needle valve into the needle seat. Position the float between the two mounting posts. Push the hinge pin through the posts and hinge until the end of the pin is flush with the outer edge of the mounting posts.

CRITICAL WORDS

The float level adjustment on this carburetor is **NOT** made by measuring the distance between the top of the float and the carburetor body surface, as for most other carburetors. This carburetor has a unique method for determining the correct float level and this adjustment is performed **AFTER** the carburetor is fully assembled and the powerhead operating.

Unfortunately, if an adjustment is required, the float bowl must be drained and removed and then the float removed. Adjustments are then made to the float tab in the same manner as for other carburetors.

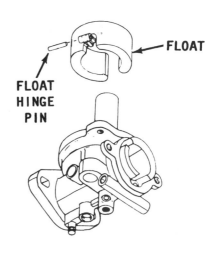

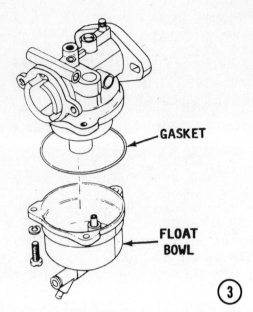

3- Position a **NEW** float bowl gasket in place over the mating surface of the carburetor. Install the float bowl and secure it with the three screws and washers. Tighten the screws evenly to prevent deforming the gasket sealing surface.

4- Refer to the accompanying illustration for the correct alignment and installation of the following items in the order given, into the base of the fuel bowl: small washer, drain valve, spring, guide, large washer, drain valve button, and finally the retaining plate. Secure the plate against spring pressure with the two retaining screws and washers.

Depress the drain valve button and check its return action. The button **MUST** return freely without binding. If the button binds and remains depressed, the drain system **MUST** be disassembled and repairs made as necessary. A binding button will cause the drain valve to leak continuously - causing a possible **FIRE HAZARD**.

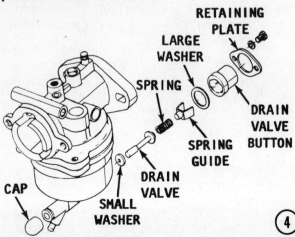

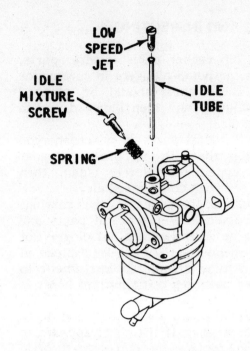

Install the cap at the other end of the drain system.

5- Lower the idle tube into the hole at the top of the carburetor. Install the low speed jet over the tube and tighten the jet just "snug".

Install the idle mixture screw and spring. Tighten the screw to a lightly seated position and then back the screw out 1 3/8 turns as a preliminary adjustment.

6- Slide the spring over the plunger and insert both pieces into the carburetor. Place the lever plate over the top of the plunger with the lever facing **AFT** and the lever end facing **DOWN**. Install the plunger cap over the plunger. Tighten the cap and snap the protective cover over the cap.

INSTALLATION

7- Place a **NEW** carburetor mounting gasket against the intake manifold. Install the carburetor and secure it in place with the two bolts and washers. Tighten the bolts alternately and evenly to a torque valve of 5.6 ft in (8Nm). Connect the two fuel lines at their original locations. Install the fuel tank with its attaching hardware.

Float Level Adjustment

Mount the outboard unit in a test tank, on the boat in a body of water, or connect a flush attachment and hose to the lower unit.

NEVER, AGAIN, NEVER operate the engine at high speed with a flush device attached. The engine, operating at high speed with such a device attached, would **RUNAWAY** from lack of a load on the propeller, causing extensive damage.

Start the engine and check the completed work.

CAUTION

Water must circulate through the lower unit to the powerhead anytime the powerhead is operating to prevent damage to the water pump in the lower unit. Just five seconds without water will damage the water pump impeller.

Operate the powerhead at approximately idle speed for a few minutes, then shut down the powerhead and perform the following task:

8- Obtain a length of approximately one foot (30cm) of transparent vinyl tubing of the same inner diameter as the overflow line attached to the lower fitting on the drain valve. Remove the existing overflow line and install one end of the vinyl tube over the fitting. Hold the other end of the tube up to prevent fuel from draining from the float bowl.

Rotate the fuel supply knob to the **"OPEN"** position. Depress the drain valve button and observe the level of fuel in the transparent vinyl tube.

The distance between the level of fuel in the line and the carburetor bore centerline must be 29/32" (23mm). If the distance is less than specified, the float is dropping down too far inside the float bowl. The float tab must be pushed toward the float "just a whisker" to correct the level. If the distance is greater than specified, the float level is too high inside the bowl and the float tab must be pulled away from the float "just a whisker" to correct the level.

Rotate the fuel supply knob to the **"STOP"** position, drain the remaining fuel from the bowl and perform Steps 1, 5 and 6 to free the float. Make the necessary adjustment to the float tab to correct the float level and then perform Steps 2, 3 and 7 to assemble and install the carburetor. Repeat Step 8 to check the float level once more.

GOOD WORDS ON TACHOMETER CONNECTIONS

The usual method of connecting a tachometer to a powerhead involves connecting the meter across the primary negative side of the ignition coil and a suitable powerhead ground. A connection to a suitable power-

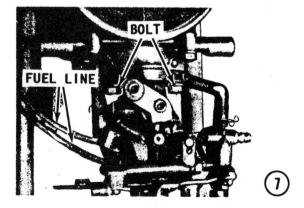

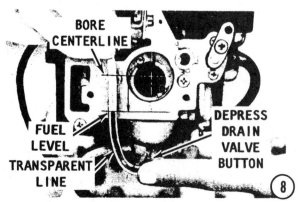

head ground is no problem on this powerhead, but a connection to the ignition coil **IS** a problem. The ignition coil is housed under the flywheel and the connection would be almost impossible to sustain under operating conditions.

Later powerheads may be equipped with a "kill switch". If so equipped, the "kill switch" may be temporarily disconnected and eliminated from the circuit and the tachometer input lead connected to the white lead from the powerhead to the switch.

After the necessary adjustments have been made, touch the white lead, with the tachometer lead still connected to ground, to shut down the powerhead by grounding the ignition. Then, remove the tachometer lead and reconnect the white "kill switch" leads together.

If the powerhead is not equipped with a "kill switch", obtain an inductive type tachometer, with a pickup lead which hooks around the spark plug lead in much the same manner as a timing light pickup.

If no such meter is available, the idle rpm must be estimated based on the operators past experience with the outboard unit.

Unit mounted in a test tank in preparation to making fine carburetor adjustments, as outlined in the text.

BEFORE POWERHEAD STARTUP

If the carburetor was **NOT** overhauled, rotate the idle mixture screw in to a lightly seated position and then back the screw out 1 3/8 turns as a preliminary adjustment.

Mount the outboard unit in a test tank, on the boat in a body of water, or connect a flush attachment and hose to the lower unit. Connect a tachometer to the powerhead.

NEVER, AGAIN, NEVER operate the engine at high speed with a flush device attached. The engine, operating at high speed with such a device attached, would **RUNAWAY** from lack of a load on the propeller, causing extensive damage.

CAUTION

Water must circulate through the lower unit to the powerhead anytime the powerhead is operating to prevent damage to the water pump in the lower unit. Just five seconds without water will damage the water pump impeller.

Idle Speed Adjustment

Allow the powerhead to warm to normal operating temperature. Shift the unit into **FORWARD** gear, and adjust the throttle stop screw, located at the top of the throttle shaft linkage, on the starboard side, until the powerhead idles between 1250 and 1350 rpm. Idle speed adjustment is accomplished by rotating the throttle stop screw inward to decrease engine speed or outward to increase engine speed.

High Speed Adjustment

There is no high speed adjustment possible on this type carburetor. A fixed main jet is located on the side of the mixing chamber center turret. The standard size of the main jet on this carburetor is gauge No. 75.

If the outboard is to be consistently operated in cold climates or at high speeds between 4000-5000 rpm, then the owner may consider replacing the main jet with a jet having a smaller gauge number.

If the outboard is to be consistently operated at low idle speeds or at elevations higher than 2500 ft above sea level, then the main jet should be replaced with one having a higher gauge number.

Detailed timing and synchronizing procedures are presented in Chapter 6.

CARBURETOR "C" 4-27

4-8 CARBURETOR "C"
WITH INTEGRAL FUEL PUMP
USED ON 4 HP AND 5 HP (WATER COOLED) POWERHEADS

Good shop practice dictates a carburetor rebuild kit be purchased and new parts, especially gaskets and O-rings be installed any time the carburetor is disassembled. A photograph of the parts included in the carburetor rebuild kit for carburetor "C" is shown in the Cleaning and Inspecting portion of this section.

REMOVAL

1- Turn off the fuel supply at the base of the fuel tank by turning the fuel knob to the **OFF** position.

2- Squeeze the wire type hose clamp on the fuel line, and then pull the hose free of the fuel inlet fitting. Plug the fuel line with a screw and slip the hose clamp in place. Tape the line to the tank to prevent loss of fuel.

3- Loosen the throttle wire retaining screw and pull the wire free of the brass barrel.

4- Remove the two Phillips screws securing the silencer to the carburetor body and move the silencer to one side out of the way. Pry the choke rod from the carburetor linkage using a small blade screwdriver.

5- Remove the starboard side carburetor mounting nut. Loosen but do **NOT** remove the port side mounting nut. Grasp the carburetor and pull it forward to clear the

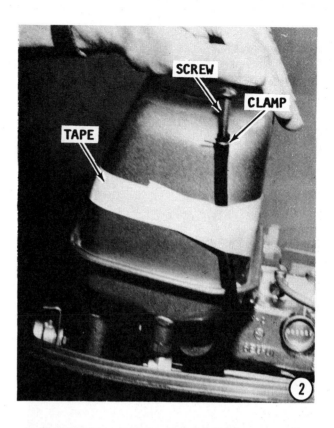

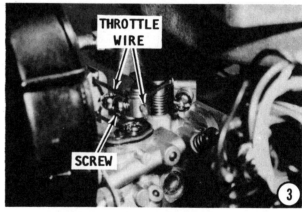

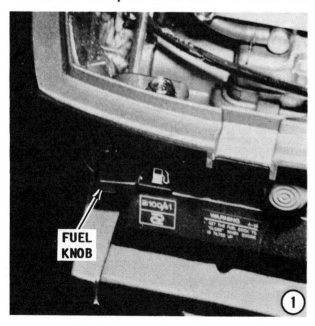

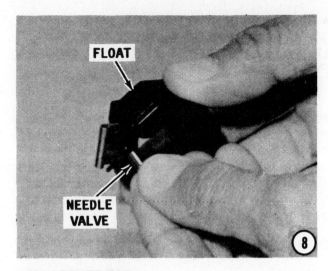

starboard stud, and then slide it to the left for the slot in the mounting flange to clear the port stud. Remove and discard the carburetor mounting gasket.

DISASSEMBLING

6- Remove the four Phillips screws securing the float bowl to the carburetor. Remove the float bowl. Lift off and discard the float bowl gasket.

7- Push out the hinge pin using an awl. Lift out the float. The needle valve will come out with the float.

8- Slide the needle valve free of the slot in the float.

9- Remove the main jet from the turret of the float bowl.

10- Unscrew and remove the main nozzle from deep inside the turret.

11- Remove the three Phillips screws securing the top cover. Remove the top cover and the gasket.

12- Back out the pilot screw and pilot jet. The number of turns need not be recorded because during installation a definite number of turns from the lightly seated position will be listed. Slide the pilot screw spring free.

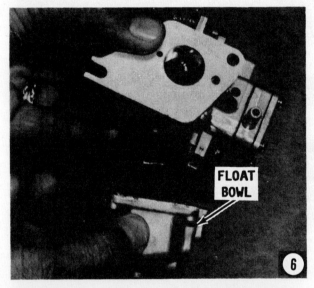

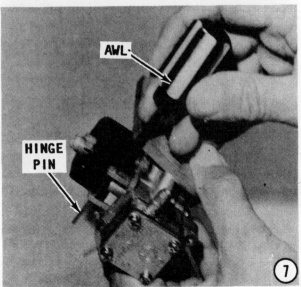

CARBURETOR "C" 4-29

13- Remove the four Phillips screws securing the fuel pump to the carburetor body.

If the fuel pump gaskets and diaphrams are to be used again, **CAREFULLY** disassemble the fuel pump. Use **CARE** during disassembly and note the order of the gaskets and diaphragms. The membranes are fragile and may be easily punctured or stretched. Disassembly involves removal in the following order: first, the cover, then the outer diaphragm, and finally the cover gasket.

14- Remove the fuel pump body. **TAKE CARE** not to lose the spring and plate as they are held under tension by the pump body. Remove the inner diaphragm, and then the pump body gasket.

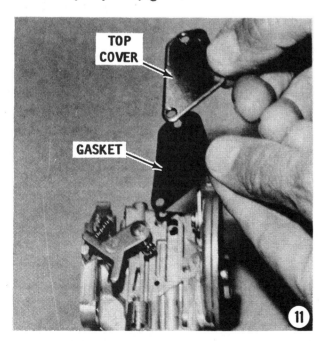

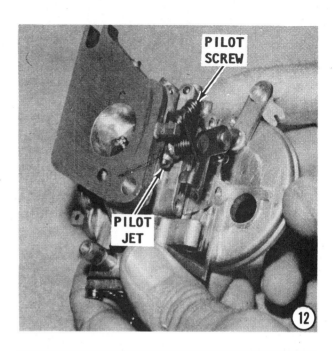

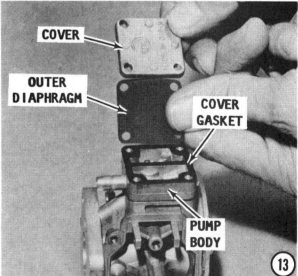

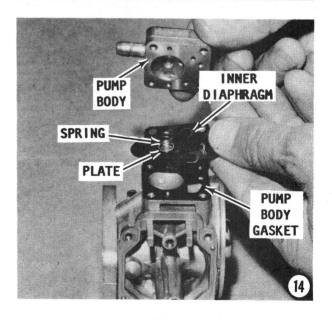

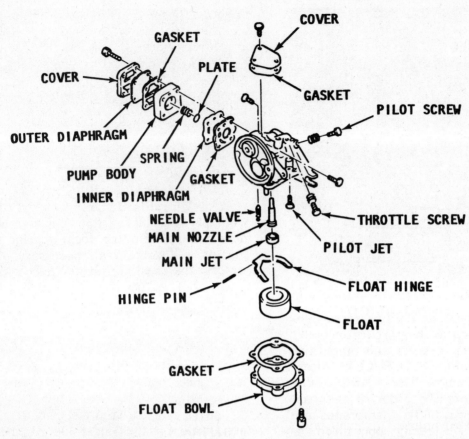

Exploded drawing of Carburetor "C", used on 4hp and 5hp water cooled powerheads, with major parts identified.

CLEANING AND INSPECTING

NEVER dip rubber parts, plastic parts, diaphragms, or pump plungers in carburetor cleaner. These parts should be cleaned **ONLY** in solvent, and then blown dry with compressed air.

Place all the metal parts in a screen-type tray and dip them in carburetor cleaner until they appear completely clean -- free of gum and varnish which accumulates from stale fuel. Blow the parts dry with compressed air.

Some parts of Carburetor "C" may be soaked in a carburetor cleaning solution while other parts may not, because they have rubber or plastic parts permanently attached. Such parts may be cleaned with a cloth dipped in a solution, and then immediately blown dry with compressed air.

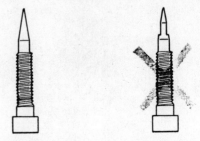

Comparison of two carburetor pilot screws. The left screw is new with a smooth taper. The worn screw on the right has developed a ridge, and is therefore unfit for further service.

CARBURETOR "C"   4-31

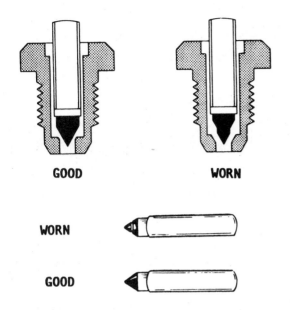

Needle and seat arrangement on the carburetor covered in this section, showing a worn and new needle for comparison.

Blow out all passageways in the castings with compressed air. Check all of the parts and passages to be sure they are not clogged or contain any deposits. **NEVER** use a piece of wire or any type of pointed instrument to clean drilled passages or calibrated holes in a carburetor.

Make a thorough inspection of the fuel pump diaphragms for the tiniest pin hole. If one is discovered, the hole will only get bigger. Therefore, the diaphragms must be replaced in order to obtain full performance from the powerhead.

Carefully inspect the casting for cracks, stripped threads, or plugs for any sign of leakage. Inspect the float hinge in the hinge pin area for wear and the float for any sign of leakage.

Examine the inlet needle for wear and if there is any evidence of wear, the inlet needle **MUST** be replaced.

ALWAYS replace any and all worn parts.

A carburetor service kit, as shown in the accompanying illustration, is available at modest cost from the local marine dealer. The kit will contain all necessary parts to perform the usual carburetor overhaul work.

ASSEMBLING AND INSTALLATION

1- Position the following parts onto the carburetor body in the order given: first, the pump body gasket, then the inner diaphragm, next the plate, then the spring. The plate and spring must index with the round hole on the underside of the pump body. When the pump body is placed over the spring, **ENSURE** all the parts are aligned and the spring seats in an upright position between the plate and the pump body. Hold it all together for the next step.

2- Place the cover gasket, the outer diaphragm and the cover over the pump body. Visually check to be sure the holes are aligned before threading the screws through the many layers of fuel pump parts. If the screws are forced through, one of the fragile parts may be damaged. Secure the

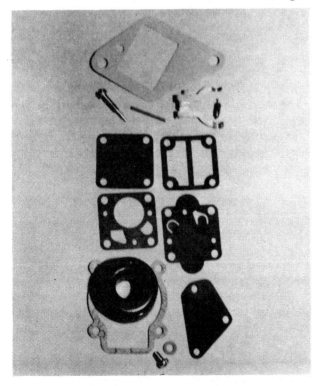

Parts included in a repair kit for Carburetor "C".

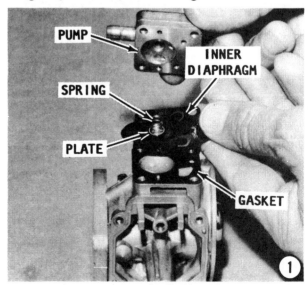

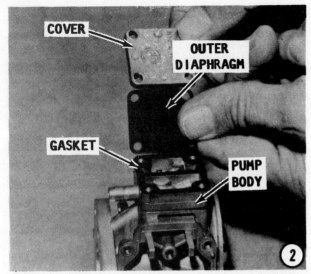

parts together with the four attaching screws.

3- Install the pilot jet snugly. Install the pilot screw and spring. The pilot screw adjusts the air/fuel mixture. SLOWLY ro-

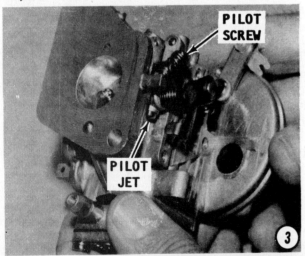

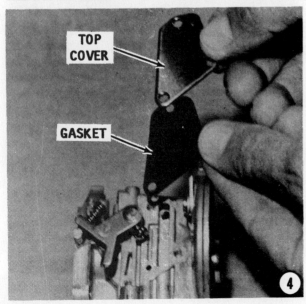

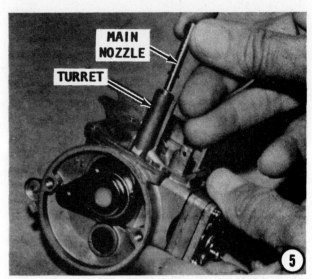

tate the pilot screw into the carburetor body until it BARELY seats. From this position, back it out 1-3/4 turns for the 4 hp model, and 1-1/4 turns for the 5 hp model. This screw setting should now not be disturbed. Idle speed adjustment is made later with the throttle stop screw in Step 15.

4- Install the gasket and top carburetor cover.

5- Thread the main nozzle into the float bowl turret and tighten it just snug.

6- Install the main jet over the main nozzle and tighten it snugly.

7- Slide the needle valve into the slot of the float. For models through 1985, slide the needle valve into the slot of the float hinge.

8- If servicing a model through 1985: lower the needle valve into the needle seat and position the float hinge between the mounting posts. Install the hinge pin through the posts securing the float hinge in place.

If servicing a model since 1985: lower the float and needle valve assembly over the

CARBURETOR "C" 4-33

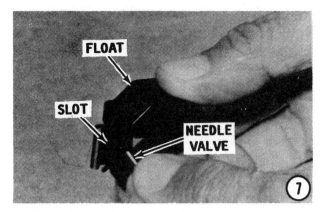

turret, engaging the needle valve into the needle seat. Position the float hinge between the mounting posts and install the hinge pin securing the float in place.

9- If servicing a model through 1985, refer to illustration No. 9A: Install the float bowl gasket. Invert the carburetor and measure the distance from the gasket surface to the top of the float arm. This distance should be 5/64" (2.0mm). **CAREFULLY** bend the float hinge arm, if necessary, to obtain the correct measurement.

If servicing a model since 1985, refer to illustration No. 9B: hold the carburetor inverted, as shown. Measure the distance between the carburetor body to the top of the float. This distance should be 7/8" (22.2mm). This dimension, with the carburetor inverted, places the "lower" surface of the float parallel to the carubretor body. If necessary, **CAREFULLY** bend the tab on the float to obtain the correct measurement.

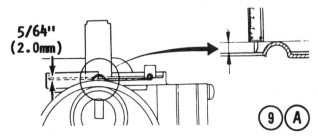

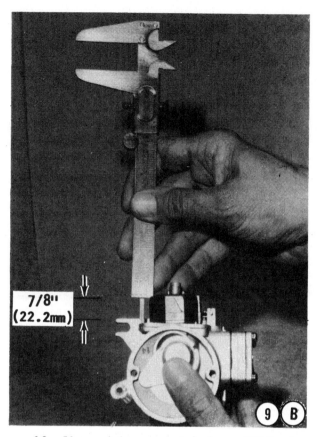

10- If servicing a model since 1985: install the float bowl gasket. On models through 1985, this gasket was installed in the previous step.

All models: install the float bowl to the carburetor body and secure it with the four Phillips screws.

11- Install the carburetor mounting gasket. Slide the port side slotted end of the carburetor mounting flange onto the stud first. The nut and washer should not have been removed during disassembling. Move

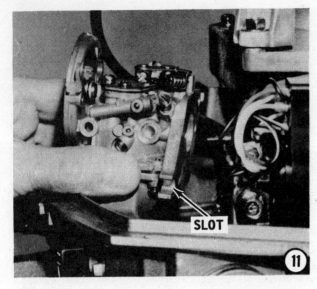

the starboard side of the carburetor mounting flange onto the left stud. Install the left nut and washer, and then tighten both nuts to 5.8 ft lbs. (8Nm).

12- Snap the choke link into place. Connect the fuel line to the fuel pump inlet fitting. Install the silencer to the carburetor body and secure it in place with the two Phillips screws.

13- Thread the throttle wire through the holder first, and then through the brass barrel. Tighten the screw to retain the wire. The wire should project out from the barrel approximately 0.12 - 0.16" (3 - 4mm).

14- Open the fuel valve at the base of the tank by turning the fuel knob to the **ON** position.

Idle Speed Adjustment

15- Mount the outboard unit in a test tank, on the boat in a body of water, or connect a flush attachment and hose to the lower unit.

Obtain a tachometer. Connect the Red tachometer lead to to the primary windings of the ignition coil (mounted on the starboard side of the block). Connect the Black tachometer lead to a suitable ground on the powerhead.

NEVER, AGAIN, NEVER operate the engine at high speed with a flush device attached. The engine, operating at high speed with such a device attached, would **RUNAWAY** from lack of a load on the propeller, causing extensive damage.

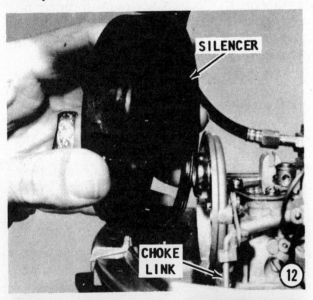

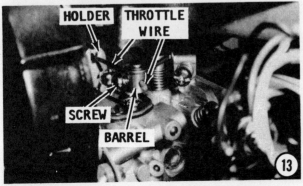

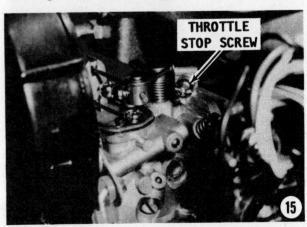

Start the engine and check the completed work.

REMEMBER, the powerhead will **NOT** start without the emergency tether in place behind the kill switch knob.

CAUTION
Water must circulate through the lower unit to the powerhead anytime the powerhead is operating to prevent damage to the water pump in the lower unit. Just five seconds without water will damage the water pump impeller.

Allow the powerhead to warm to normal operating temperature. Shift the unit into **FORWARD** gear, and adjust the throttle stop screw until the powerhead idles between 950 and 1050 rpm. Idle speed adjustment is accomplished by rotating the throttle stop screw inward to increase engine speed or outward to decrease powerhead speed.

High Speed Adjustment
There is no high speed adjustment possible on this type carburetor. A fixed main jet of standard size, is located on the mixing chamber center turret.

If the outboard is to be consistently operated in cold climates or at high speeds between 4500-5500 rpm, then the owner may consider replacing the main jet with a jet having a smaller gauge number.

If the outboard is to be consistently operated at low idle speeds or at elevations higher than 2500 ft above sea level, then the main jet should be replaced with one having a higher gauge number.

Detailed timing and synchronizing procedures are presented in Chapter 6.

4-9 CARBURETOR "D"
USED ON AIR COOLED 5HP POWERHEADS

This section provides complete detailed procedures for removal, disassembly, cleaning and inspecting, assembling including bench adjustments, installation, and operating adjustments for Carburetor "D", as used originally on the air cooled Model 5hp unit. This carburetor is a single-barrel, float feed type with a manual choke.

FIRST, THESE WORDS
Good shop practice dictates a carburetor repair kit be purchased and new parts be installed any time the carburetor is disassembled.

Make an attempt to keep the work area organized and to cover parts after they have been cleaned. This practice will prevent foreign matter from entering passageways or adhering to critical parts.

REMOVAL

1- Disconnect the fuel line and take adequate measures to prevent fuel from draining from the tank. Remove the choke knob and disconnect the throttle cable from the cable connector at the top of the throttle shaft. Loosen the single silencer mounting bolt and remove the silencer.

Remove the two carburetor retaining bolts and washers, and then lift the carburetor free of the powerhead. Remove and discard the mounting gasket.

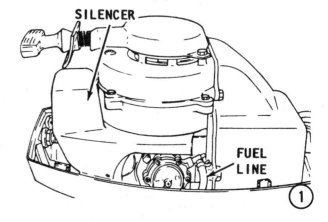

Outboard unit mounted in a test tank ready for fine carburetor adjustments to be made.

4-36 FUEL

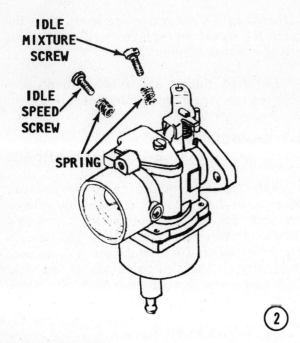

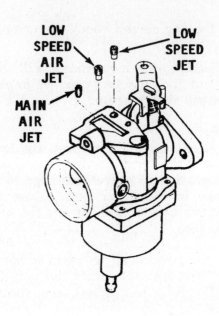

DISASSEMBLING

2- Two adjustment screws are located on the starboard side of the carburetor. The only screw which needs to be removed during carburetor overhaul is the idle mixture screw. This screw is threaded into the carburetor body. The other screw, the idle speed screw, is threaded into the carburetor linkage at the top of the throttle shaft and there is no need to disturb this screw.

Identify the idle mixture screw and remove the screw and spring.

3- Remove the two screws and lockwashers securing the mixing chamber cover. Remove the cover, and then remove and discard the gasket.

4- Identify the three jets in the mixing chamber: the main air jet, the low speed air jet and the low speed jet. These three jets perform different functions and are **NOT** interchangeable. Therefore, employ a suitable system to keep them separate and identified to ensure each will be installed in

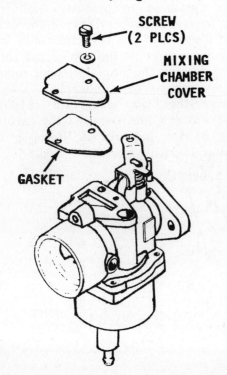

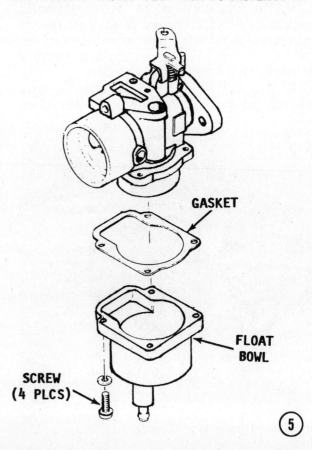

CARBURETOR "D"

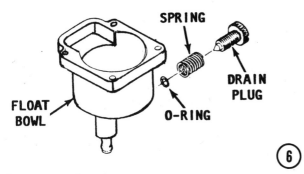

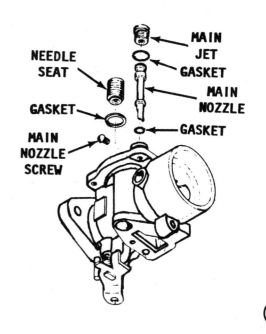

the proper location. Remove the jets one at a time and set each aside.

5- Remove the four screws and washers securing the float bowl to the carburetor. Remove the bowl. Lift off and discard the float bowl gasket.

6- Remove the drain valve spring and little O-ring from the base of the float bowl.

7- Lift off the float. Push the pin free of the mounting posts, using a long pointed awl. Remove the float hinge. Lift out the needle valve from the needle seat. **TAKE CARE** not to damage the pointed end of the valve if it is to be reused.

8- Using the proper size screwdriver, remove the needle seat. Remove and discard the gasket. Use a different screwdriver, if necessary, and remove the main jet and gasket beneath it. Remove the main nozzle securing screw from the side of the mixing bowl center turret. This screw **MUST** be removed **BEFORE** an attempt is made to remove the main nozzle.

Invert the carburetor and the main nozzle and gasket will fall free of the turret.

A GOOD WORD

Further disassembly of the carburetor is not necessary in order to clean it properly.

CLEANING AND INSPECTING

NEVER dip rubber parts, plastic parts, diaphragms, or pump plungers in carburetor cleaner. These parts should be cleaned **ONLY** in solvent, and then blown dry with compressed air.

Place all metal parts in a screen-type tray and dip them in carburetor cleaner until they appear completely clean, then blow them dry with compressed air.

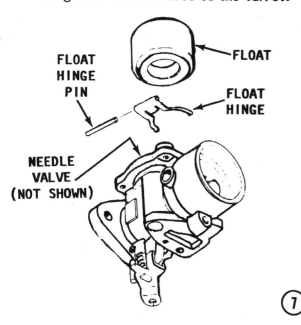

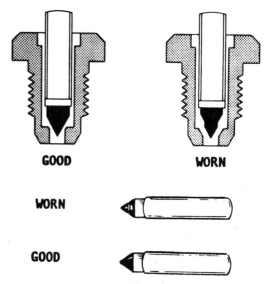

Needle and seat arrangement on the carburetor covered in this section, showing a worn and new needle for comparison.

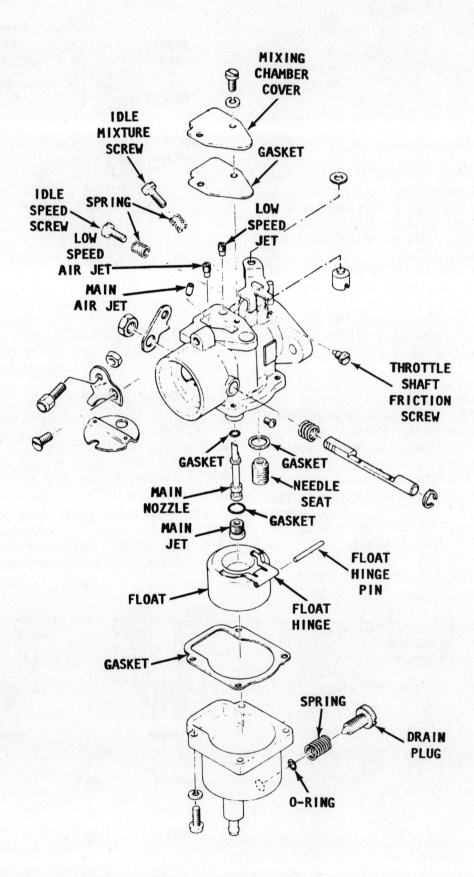

Exploded drawing of Carburetor "D", used on 5hp air cooled powerheads, with major parts identified.

Blow out all passages in the castings with compressed air. Check all parts and passages to be sure they are not clogged or contain any deposits. **NEVER** use a piece of wire or any type of pointed instrument to clean drilled passages or calibrated holes in a carburetor.

Move the throttle shaft back and forth to check for wear. If the shaft appears to be too loose, replace the complete throttle body because individual replacement parts are **NOT** available.

Inspect the main body, mixing chamber, and gasket surfaces for cracks and burrs which might cause a leak. Check the float for deterioration. Check to be sure the float spring has not been stretched. If any part of the float is damaged, the unit must be replaced. Check the float arm needle contacting surface and replace the float if this surface has a groove worn in it.

Inspect the tapered section of the idle adjusting needle and replace the needle if it has developed a groove.

Most of the parts which should be replaced during a carburetor overhaul are included in overhaul kits available from your local marine dealer. One of these kits will contain a matched fuel inlet needle and seat, if removable. This combination should be replaced each time the carburetor is disassembled as a precaution against leakage.

ASSEMBLING

1- Slide a **NEW** gasket onto the small end of the main nozzle. Insert the main nozzle into the center turret with the small end going in first. Install the main nozzle securing screw into the threaded hole on the mixing chamber center turret. Once this

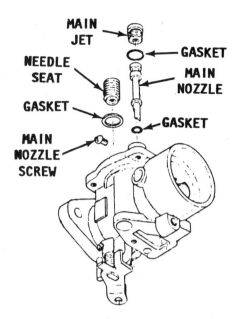

screw is installed, the main nozzle should be firmly held within the turret bore. Slide a **NEW** gasket over the main jet, and then install the jet over the main nozzle. Install the needle seat with a **NEW** gasket. Tighten the seat just "snug".

2- Insert the needle valve into the needle seat. Position the float hinge between the two mounting posts. Push the hinge pin through the posts and hinge until the end of the pin is flush with the outer edge of the mounting posts. Set the float on top of the hinge.

SPECIAL WORDS

The float level adjustment on this carburetor is **NOT** made by measuring the distance between the top of the float and the

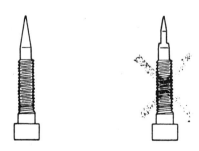

Comparison of two carburetor idle mixture screws. The left screw is new with a smooth taper. The worn screw on the right has developed a ridge, and is therefore unfit for further service.

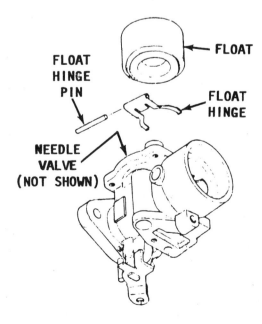

4-40 FUEL

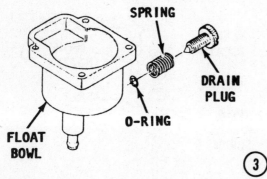

carburetor body surface, as for most other carburetors. This carburetor has a unique method for determining the correct float level and this adjustment is performed **AFTER** the carburetor is fully assembled, installed, and the powerhead operating.

Unfortunately, if an adjustment is required, the float bowl must be drained, removed, and then the float removed. Adjustments are then made to the float tab in the same manner as for other carburetors.

3- Slide the spring over the drain plug and install the little O-ring over the needle end of the plug. Thread the drain plug into the base of the float bowl until the needle end is lightly seated.

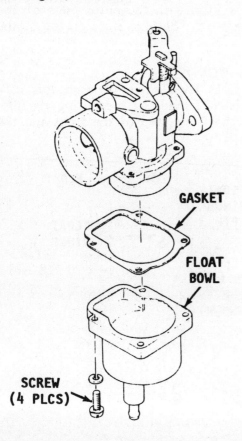

CRITICAL WORDS

The drain plug shuts off a passageway inside the fuel bowl casting, and performs the function of a valve. When the plug is properly seated, no fuel can pass. When the plug is open, fuel will flow from the bowl for carburetor adjustment or draining purposes. If fuel leaks from the fitting at the base of the fuel bowl but the plug is seated "snugly", either the O-ring or the needle end of the plug or both, are deformed. The damaged part **MUST** be replaced to prevent fuel from leaking during powerhead operation. Leaking fuel while the powerhead is running would create a definite **FIRE HAZARD**.

4- Position a **NEW** float bowl gasket in place over the mating surface of the carburetor. Install the float bowl and secure it with the four screws and lockwashers. Tighten the screws alternately and evenly to prevent deforming the float bowl sealing surface.

5- Install the main air jet, low speed air jet and low speed jet in their original locations in the mixing chamber. Tighten each jet just "snug".

6- Install a **NEW** gasket over the mixing chamber. Place the cover over the gasket and secure it with the two screws and lockwashers.

7- Slide the spring onto the idle mixture screw (the screw with the tapered end), and then install the screw into the carburetor body. Tighten the screw until it barely seats and then back it out 1 3/8 turns. This

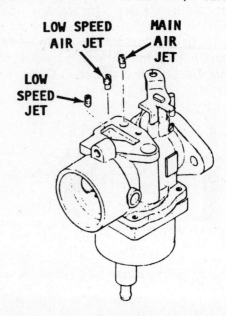

CARBURETOR "D" 4-41

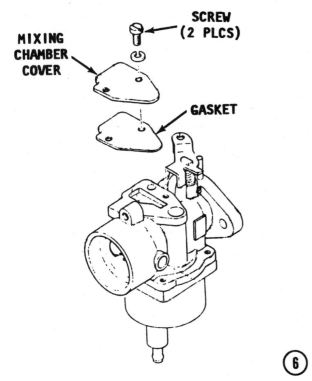

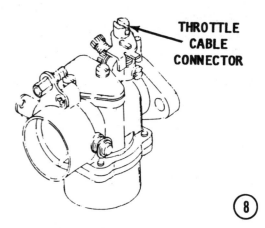

Tighten the bolts alternately and evenly to a torque value of 5.8 ft lb (8Nm).

Install the silencer over the carburetor and secure it with the single attaching bolt. Connect the throttle cable to the cable connector at the top of the throttle shaft. Connect the choke linkage and install the choke knob. Connect the fuel supply line from the pump to the carburetor. Connect the fuel line from the tank to the fuel joint.

screw remains as set. It is not disturbed for the idle speed adjustment. Therefore, it **MUST** be set correctly **NOW**.

If the idle speed screw was also removed during disassembling, slide the spring onto the screw and install the screw into the threaded hole in the throttle shaft linkage at the top of the carburetor. Thread the screw about half way into the linkage, as a preliminary adjustment. A fine adjustment will be made in a later step.

8- Position a **NEW** carburetor mounting gasket onto the intake manifold. Install the carburetor and secure it to the powerhead with the two attaching bolts and washers.

CLOSING TASKS

9- Mount the outboard unit in a test tank, on the boat in a body of water, or connect a flush attachment.

NEVER, AGAIN, NEVER operate the engine at high speed with a flush device attached. The engine, operating at high speed with such a device attached, would **RUNAWAY** from lack of a load on the propeller, causing extensive damage.

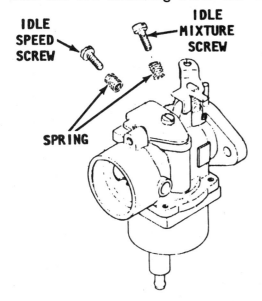

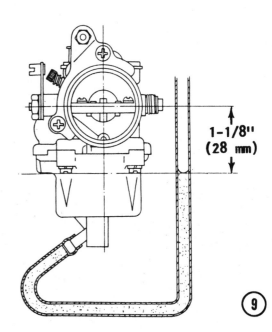

4-42 FUEL

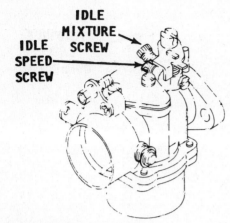

Line drawing of Carburetor "D", used on 5hp air cooled powerheads, with adjustment screws identified.

CAUTION
Water must circulate through the lower unit to the powerhead anytime the powerhead is operating to prevent damage to the water pump in the lower unit. Just five seconds without water will damage the water pump impeller.

Obtain a tachometer. Connect the Red tachometer lead to the primary windings of the ignition coil (mounted on the starboard side of the block). Connect the Black tachometer lead to a suitable ground on the powerhead.

Start the powerhead and allow it to idle for a few minutes, and then shut down the powerhead and remove the silencer.

Obtain about one foot (30cm) of 1/4" (6mm) inner diameter transparent vinyl tubing and connect one end to the fitting at the base of the float bowl. Hold the other end up to prevent fuel from draining from the float bowl. Back out the knurled drain screw next to the fitting on the bowl and allow fuel to drain into the tubing loop.

Unit in a test tank ready for fine carburetor adjustments to be made, after the cowling is removed.

The distance between the level of fuel in the line and the carburetor bore centerline must be 1 1/8" (28mm). If the distance is less than specified, the float is dropping down too far inside the float bowl. The float tab must be pushed toward the float "just a whisker" to correct the level. If the distance is greater than specified, the float level is too high inside the bowl and the float tab must be pulled away from the float "just a whisker" to correct the level.

If the float level is incorrect, the float bowl must be completely drained and removed. Make the necessary adjustments to the float tab and repeat this step until the float level matches specifications.

Idle Speed Adjustment

Connect a tachometer to the powerhead, as previously directed. If the carburetor was not overhauled, rotate the idle mixture screw (the screw threaded into the carburetor body, not to be confused with the idle speed screw, threaded into the throttle shaft linkage), to lightly seat the screw. Then, back the screw out 1 3/8 turns. This screw is left as set and must not be rotated to adjust powerhead rpm.

Allow the powerhead to warm to normal operating temperature. Shift the unit into **FORWARD** gear, and adjust the idle speed screw, located at the top of the throttle shaft linkage, on the starboard side, until the powerhead idles between 1550 and 1650 rpm.

High Speed Adjustment

There is no high speed adjustment possible on this type carburetor. A fixed main jet is located on the top of the mixing chamber center turret. The standard size of the main jet on this carburetor is gauge No. 82.

If the outboard is to be consistently operated in cold climates or at high speeds between 4500-5500 rpm, then the owner may consider replacing the main jet with a jet having a smaller gauge number.

If the outboard is to be consistently operated at low idle speeds or at elevations higher than 2500 ft above sea level, then the main jet should be replaced with one having a higher gauge number.

Detailed timing and synchronizing procedures are presented in Chapter 6.

4-10 CARBURETOR "E"
USED ON 8A AND 8HP POWERHEADS 1977-79

This section provides complete detailed procedures for removal, disassembly, cleaning and inspecting, assembling including bench adjustments, installation, and operating adjustments for Carburetor "E", as used originally on early 8A and 8hp units. This carburetor is a single-barrel, float feed type with a manual choke.

FIRST, THESE WORDS

Good shop practice dictates a carburetor repair kit be purchased and new parts be installed any time the carburetor is disassembled.

Make an attempt to keep the work area organized and to cover parts after they have been cleaned. This practice will prevent foreign matter from entering passageways or adhering to critical parts.

REMOVAL

1- Remove the two screws and the air box from the top of the carburetor. Squeeze the wire type fuel hose clamp and move it back along the fuel line. Work the fuel line free of the fitting. Disconnect the second fuel line to the carburetor in a similar manner. Plug the fuel line from the fuel pump to prevent loss of fuel.

Use two open end wrenches and back off the nut from the plunger cap. Remove the starter cable adjustment fitting from the plunger cap. Remove the nut and cap.

Reach in with a pair of needle nose pliers and remove the spring and starter plunger.

Remove the two securing nuts from the vertically mounted studs and lift off the carburetor. Remove and discard the mounting gasket.

DISASSEMBLING

FIRST, THESE SPECIAL WORDS

Seven screws are found on the exterior of this carburetor, not counting the four float bowl securing screws. Three screws are located above the float bowl parting line and four below the line. Take time to make arrangements to identify and tag each screw as it is removed, to ensure it is installed back in its original location.

Above the parting line on the forward face: identify a single pan head screw and lockwasher securing a plate to the carburetor. There is no reason to remove this screw during a carburetor overhaul.

Above the parting line on the starboard side: identify the idle jet plug, threaded into the carburetor body.

Above the parting line on the aft side: identify the main nozzle plug.

Below the parting line on the forward face: identify a screw with a knurled head and spring. This screw is the float bowl drain screw.

Below the parting line on the starboard side: identify the throttle stop screw and spring, threaded into a boss on the carburetor body.

Below the parting line on the starboard side: identify the idle mixture screw, threaded into the carburetor body.

Below the parting line on the port side: identify the throttle shaft friction screw, threaded into the carburetor body. This screw is not normally removed during a

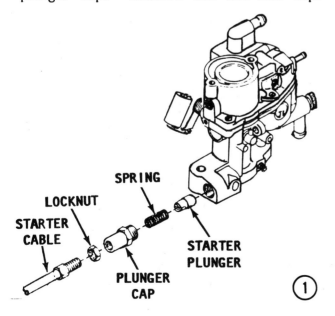

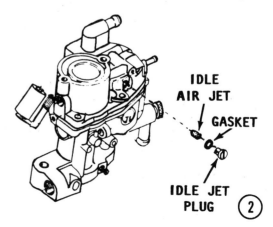

4-44 FUEL

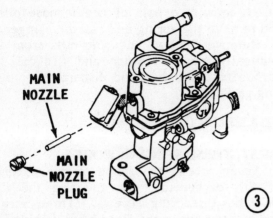

carburetor overhaul, unless the throttle shaft is to be removed.

2- Remove the idle jet plug. Remove and discard the gasket under the plug. Using the proper size screwdriver, remove the idle air jet.

3- Use the proper size screwdriver and remove the main nozzle plug. Tip the carburetor and the main nozzle, a long tube, should slide free of the hole. If the tube will not come out easily, do not use force because the tube will be deformed. Instead of using force, spray carburetor cleaner into the hole and leave the tube to soak for a few minutes, then try again.

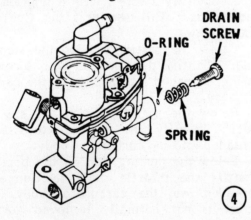

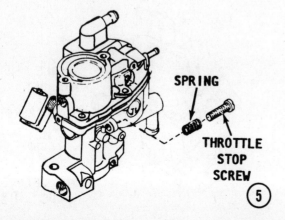

4- Remove the knurled drain screw, spring, and little O-ring. Inspect the tapered end of the screw. This screw MUST be replaced if any ridges appear on the tapered end.

CRITICAL WORDS

The drain plug performs the function of a valve by shutting off a passageway inside the fuel bowl casting. When the plug is properly seated, fuel cannot pass. When the plug is open, fuel will flow from the bowl for carburetor adjustment or draining purposes. If fuel leaks from the fitting at the base of the fuel bowl but the plug is seated "snugly", either the O-ring or the needle end of the plug or both, are deformed. The damaged part MUST be replaced to prevent fuel from leaking during powerhead operation. Leaking fuel while the powerhead is running would create a definite FIRE HAZARD.

5- Count the number of turns as the throttle stop screw is backed out free of the threaded boss. This will give a starting point for later idle speed adjustment.

6- Remove the idle mixture screw and spring. This screw also has a tapered end. If any ridges are found on the tapered end, proper adjustment of the idle mixture cannot be made. Therefore, a defective screw MUST be replaced.

7- Remove the four screws and washers securing the float bowl to the carburetor body. Separate the float bowl from the carburetor body. Remove and discard the gasket. Remove the starter jet from the carburetor body.

8- Lift off the float. The float hinge pin is "staked" in four places with special ma-

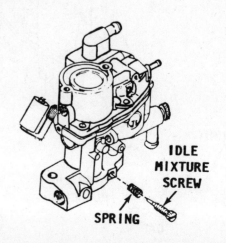

terial, as shown in the accompanying illustration. Remove the "staking" material and pry the pin and hinge free of the mixing chamber casting. Lift out the needle valve. Use a thin walled socket and remove the valve seat and gasket from the carburetor.

9- Remove the main jet and emulsion jet from the center turret of the mixing chamber.

CLEANING AND INSPECTING

NEVER dip rubber parts, plastic parts, diaphragms, or pump plungers in carburetor cleaner. These parts should be cleaned **ONLY** in solvent, and then blown dry with compressed air.

Place all metal parts in a screen-type tray and dip them in carburetor cleaner until they appear completely clean, then blow them dry with compressed air.

Blow out all passages in the castings with compressed air. Check all parts and passages to be sure they are not clogged or contain any deposits. **NEVER** use a piece of wire or any type of pointed instrument to clean drilled passages or calibrated holes in a carburetor.

Move the throttle shaft back and forth to check for wear. If the shaft appears to

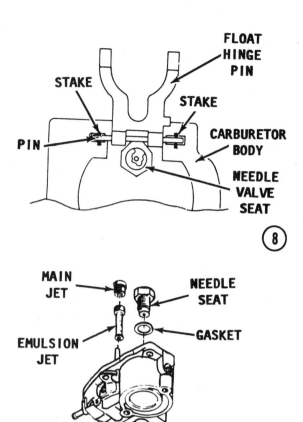

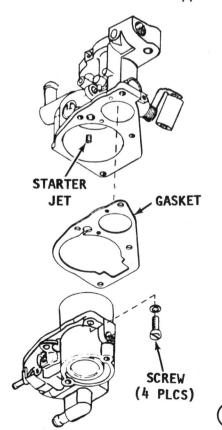

be too loose, replace the complete throttle body because individual replacement parts are **NOT** available.

Inspect the main body, mixing chamber and all gasket surfaces for cracks and burrs which might cause a leak. Check the float for deterioration. If any part of the float is damaged, the unit must be replaced. Check

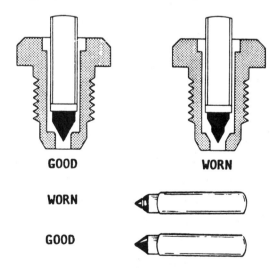

Needle and seat arrangement on the carburetor covered in this section, showing a worn and new needle for comparison.

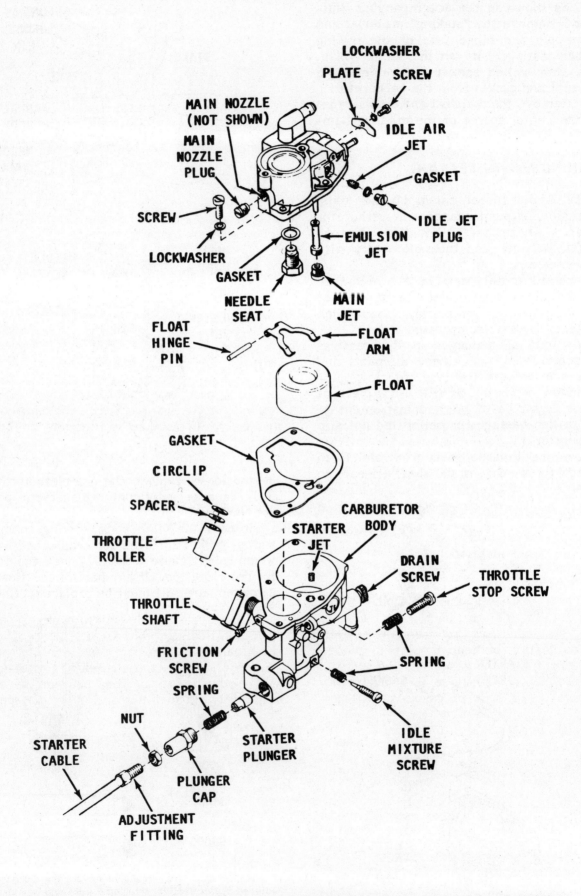

Exploded drawing of Carburetor "E", used on 8hp powerheads 1977-79, with major parts identified.

CARBURETOR "E"

the float arm needle contacting surface and replace the float if this surface has a groove worn in it.

Inspect the tapered section of the idle adjusting needle and replace the needle if it has developed a groove.

Most of the parts which should be replaced during a carburetor overhaul are included in overhaul kits available from your local marine dealer. One of these kits will contain a matched fuel inlet needle and seat. This combination should be replaced each time the carburetor is disassembled as a precaution against leakage.

ASSEMBLING

1- Insert the emulsion jet into the center of the mixing chamber and tighten the jet snugly using a slotted screwdriver. Install the main jet over the emulsion jet and again tighten the jet snugly.

2- Slide a **NEW** gasket onto the shaft of the needle seat. Install and tighten the needle seat just "snug", using the proper size socket. Insert the needle valve into the needle seat. Push the hinge pin through the hinge and install the pin into the mixing chamber casting with the bend in the hinge arms facing upwards after installation into the casting. (This will position the bend facing downward against the float upon final installation of the float bowl). "Stake" the hinge pin in four places as shown in the accompanying illustration.

GOOD WORDS

The "stake" material is the **ONLY** thing keeping the hinge pin in place. If not enough material can be transfered over from the casting to keep the pin in place, without distorting the gasket mating surface and consequently causing a fuel leak, consider applying a drop of gasoline resistant

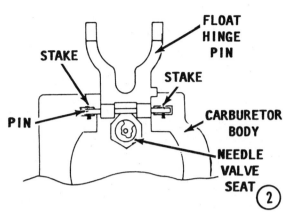

glue at each end of the pin. **TAKE CARE** not to glue the hinge to the pin!

3- Install a **NEW** gasket onto the mixing chamber surface, but **DO NOT** install the float, at this time. Hold the mixing chamber as shown in the accompanying illustration with the float hinge lightly resting on the needle valve. Measure the distance between the top on the bend in the hinge arm to the top of the gasket surface. This distance should be 1/32 to 3/64" (0.5-1.5mm). If necessary, bend each arm until the specified distance is obtained on both sides of the hinge.

4- Place the float onto the hinge arm. Install the starter jet into the carburetor body. Check the position of the gasket on the carburetor body to be sure all holes are aligned, including the hole in the gasket for the starter jet. Install the float bowl and secure it with the four screws and washers. Tighten the screws alternately and evenly to prevent distorting the gasket sealing surfaces.

5- Install the idle mixture screw and spring until the screw is lightly seated, and then back the screw out one complete turn. This setting will not change during later idle speed adjustments. This position of the screw sets the fuel mixture. The idle speed

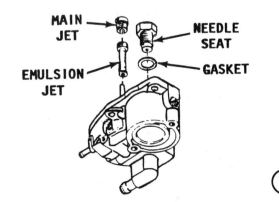

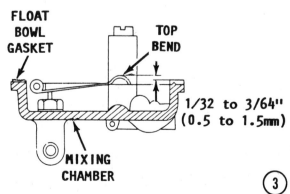

4-48 FUEL

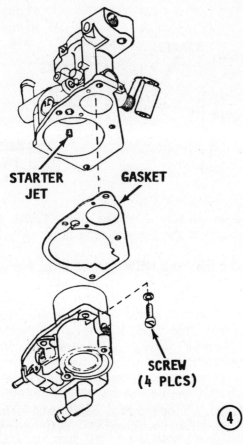

(4)

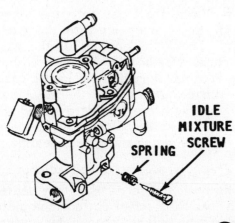

(5)

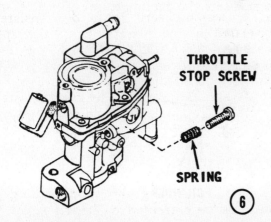

(6)

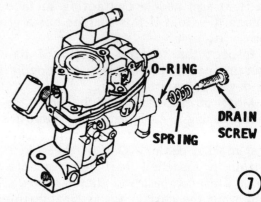

(7)

is set later, on the next page, using the idle stop screw on the throttle linkage.

6- Install the idle stop screw and spring. Thread the screw into the boss to place it in its original position, as recorded in Step 5 of disassembly. If no record was kept, install the screw midway into the boss.

7- Install the little O-ring over the tapered end of the drain screw. Slide the spring onto the screw and install the screw into the base of the float bowl. Tighten the knurled screw enough to prevent fuel from leaking from the drain fitting, but not tight enough to deform the tapered end of the screw.

8- Insert the main nozzle into the mixing chamber and install the main nozzle plug over the nozzle. No gasket is used at this location.

9- Use the correct size screwdriver and install the idle air jet. Slide a **NEW** gasket onto the jet plug, and then install the plug.

INSTALLATION

10- Insert the starter plunger followed by the spring, into the hole at the base of the float bowl. Install the plunger cap, nut and cable with the adjustment fitting. Pull out the starter knob at the other end of the

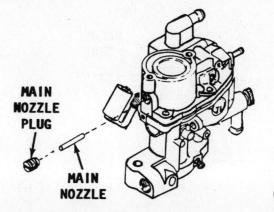

(8)

CARBURETOR "E" 4-49

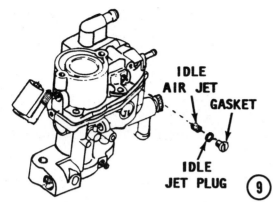

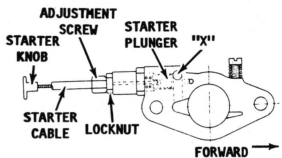

The starter plunger is adjusted to align with the aft edge of the hole marked "X", as described in the text.

starter cable as far as possible. Hold the knob in this position for the following adjustment. Set the adjusting screw and locknut to align the end of the starter plunger with the aft edge of the hole marked "X" in the accompanying illustration.

Place a **NEW** carburetor mounting gasket over the two studs. Install the carburetor and tighten the two securing nuts alternately and evenly to a torque value of 6 ft lb (8Nm).

Connect the two fuel lines to the carburetor and secure the lines with wire type fuel hose clamps. Install the air box with the attaching hardware.

CLOSING TASKS

Mount the outboard unit in a test tank, on the boat in a body of water, or connect a flush attachment and hose to the lower unit.

Connect a tachometer to the powerhead by making contact with the Red tachometer lead to the primary side of the ignition coil and with the Black tachometer lead to a suitable powerhead ground.

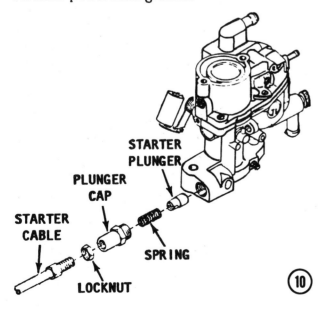

NEVER, AGAIN, NEVER operate the engine at high speed with a flush device attached. The engine, operating at high speed with such a device attached, would **RUNAWAY** from lack of a load on the propeller, causing extensive damage.

Start the engine and check the completed work.

CAUTION

Water must circulate through the lower unit to the powerhead anytime the powerhead is operating to prevent damage to the water pump in the lower unit. Just five seconds without water will damage the water pump impeller.

Idle Speed Adjustment

Allow the powerhead to warm to normal operating temperature. Shift into **FORWARD** gear and adjust the throttle stop screw until the powerhead idles between 1150 and 1250 rpm.

High Speed Adjustment

There is no high speed adjustment possible on this type carburetor. A fixed main jet is located on the side of the mixing chamber center turret. The standard size of the main jet on this carburetor is gauge No. 80.

If the outboard is to be consistently operated in cold climates or at high speeds between 4500-5500 rpm, then the owner may consider replacing the main jet with a jet having a smaller gauge number.

If the outboard is to be consistently operated at low idle speeds or at elevations higher than 2500 ft above sea level, then the main jet should be replaced with one having a higher gauge number.

Detailed timing and synchronizing procedures are presented in Chapter 6.

4-50 FUEL

Carburetor "F" removed from the powerhead, ready for cleaning, servicing, and new parts from an overhaul kit.

4-11 CARBURETOR "F" USED ON SOME 8B, 8W, MARATHON 8, 9.9C, AND 15HP POWERHEADS

This section provides complete detailed procedures for removal, disassembly, cleaning and inspecting, assembling including bench adjustments, installation, and operating adjustments for Carburetor "F", as used originally on some 8B, 8W, Marathon 8, 9.9C, and 15hp powerheads. This carburetor is a single-barrel, float feed type with a manual choke and integral fuel pump.

FIRST, THESE WORDS

Good shop practice dictates a carburetor repair kit be purchased and new parts be installed any time the carburetor is disassembled.

Make an attempt to keep the work area organized and to cover parts after they have been cleaned. This practice will prevent foreign matter from entering passageways or adhering to critical parts.

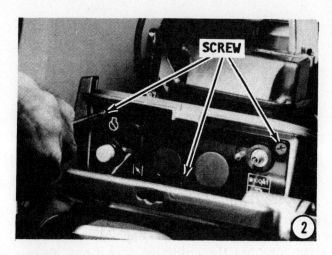

REMOVAL

1- Disconnect the fuel hose at the fuel joint. Protect the ends from contamination. Disconnect the fuel line at the fuel pump inlet fitting. Remove the two hoses from the fittings on the carburetor mounting flange.

On all 8hp models: Pry the cotter pin free, and then remove the washer from the choke link rod. Move the rod clear of the carburetor.

2- Remove the three Phillips head screws retaining the front panel to the lower cowling. The panel must be removed to gain access to the carburetor.

3- Pull the panel forward and set it to one side out of the way. It is not necessary to disconnect the "kill" switch harness or the fuel connection.

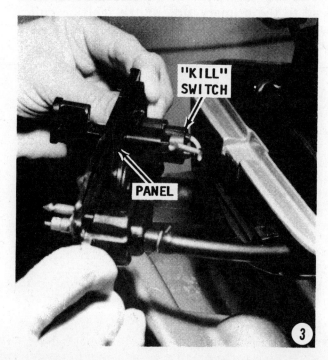

CARBURETOR "F"

4- Reach through the opening of the front panel and remove the two long bolts securing the silencer and the carburetor to the powerhead. Remove the silencer. On 9.9hp and 15hp models, the choke rod will slide out of the choke lever and the choke roller will slide out of the choke shaft cradle. Remove the carburetor. Remove and discard the carburetor mounting gasket.

DISASSEMBLING

SPECIAL WORDS

Three different design integral fuel pumps are used on this carburetor. There-fore, the pump being serviced may not resemble the pump shown in the illustration. The pump used on some 8hp powerheads is quite different in shape and has an additional pump diaphragm between the pump body and the carburetor. If servicing this different shape pump, refer to the exploded drawing on Page 4-55.

The other two integral fuel pumps are very similar, but have a different arrangement of check valves in the fuel pump body. Refer to the exploded diagrams on Page 4-54.

5- Remove the four screws securing the fuel pump to the carburetor body. Disassemble the fuel pump components in the following order: the pump cover, the outer gasket, the outer diaphragm, the pump body, the inner diaphragm, and finally the inner gasket.

6- Remove the screws from both sides of the fuel pump body. Remove the check valves.

7- Remove the idle mixture screw and spring. The number of turns out from a lightly seated position will be given during assembling.

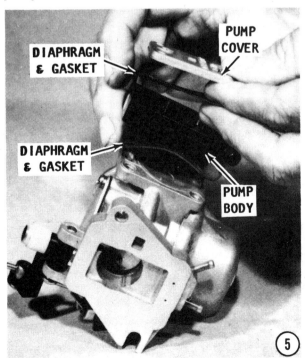

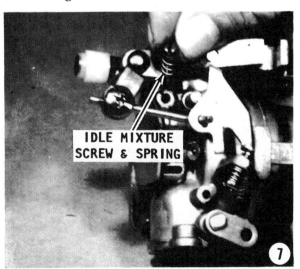

4-52 FUEL

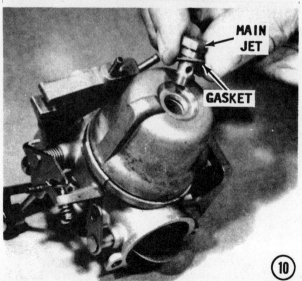

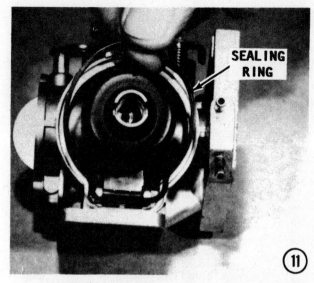

8- Remove the two screws securing the mixing chamber cover, and then lift off the cover. Remove and discard the gasket.

9- Use the correct size screwdriver and remove the pilot jet from the mixing chamber.

10- Turn the carburetor upside down and remove the bolt and gasket securing the fuel bowl. This bolt is actually the main fuel jet and is located in an easily accessible location for changing when operating the outboard at higher altitudes. Remove and discard the gasket under the main jet.

Remove the float bowl.

SPECIAL WORDS FOR ALL 8HP POWERHEADS

Some 8hp powerheads, specifically, those equipped with the different shape integral fuel pump, will also have an additional gas-

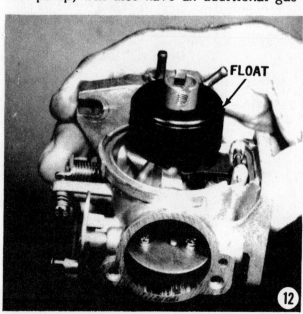

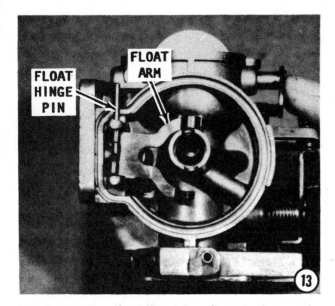

ket between the float bowl and the main nozzle. If the carburetor being serviced is equipped with this gasket, remove and discard the gasket.

11- Remove and discard the rubber sealing ring.
12- Lift off the float.
13- Push the hinge pin free of the float arm.
14- Lift the needle valve from the needle seat.
15- Use the proper size thin walled socket and remove the needle seat from the carburetor. Remove and discard the gasket under the seat.
16- Use the correct size screwdriver and remove the main nozzle from the center turret.

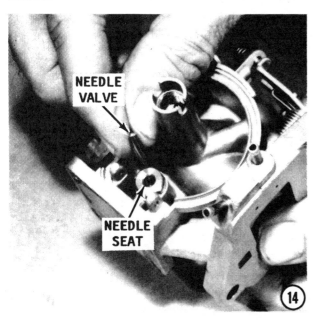

CLEANING AND INSPECTING

Inspect the check valves in the fuel pump for varnish build up as well as any deformity.

NEVER dip rubber parts, plastic parts, diaphragms, or pump plungers in carburetor cleaner. These parts should be cleaned **ONLY** in solvent, and then blown dry with compressed air.

Place all metal parts in a screen-type tray and dip them in carburetor cleaner until they appear completely clean, then blow them dry with compressed air.

4-54 FUEL

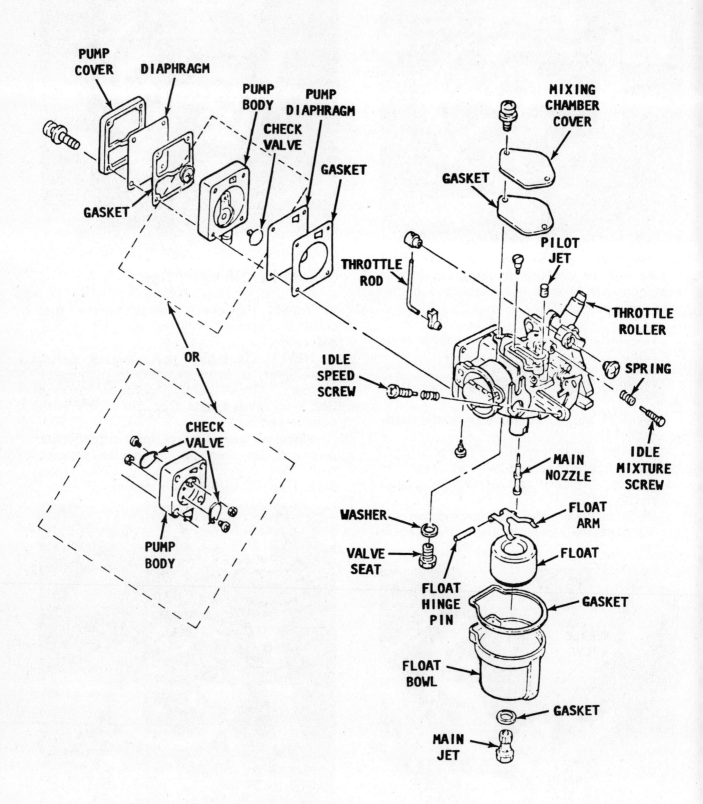

Exploded drawing of Carburetor "F", used on the powerheads listed in the heading of this section, showing two different designs of the integral fuel pump. Major parts are identified.

CARBURETOR "F" 4-55

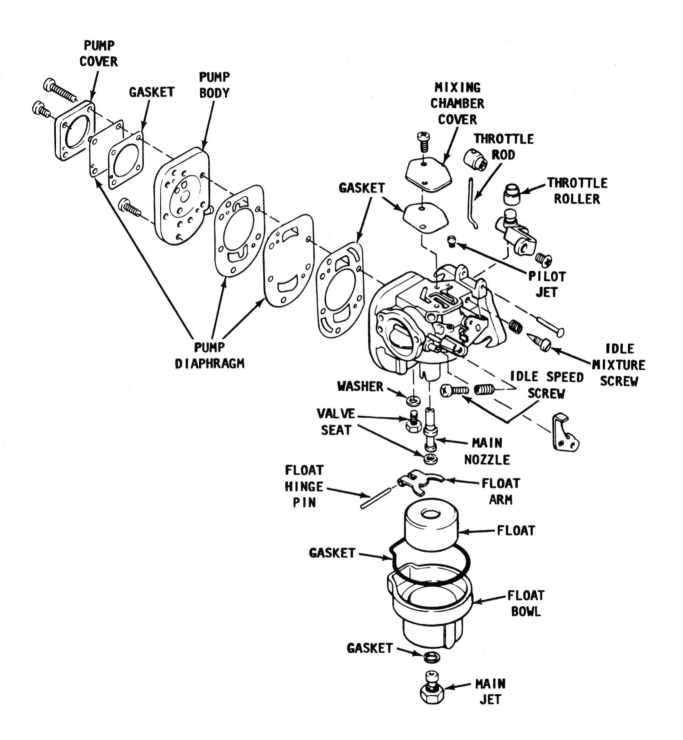

Exploded drawing of Carburetor "F", showing a third design of the integral fuel pump. This design is used only on 8hp powerheads. Major parts are identified.

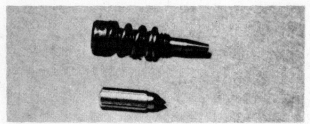

The idle mixture screw (top) and the inlet needle (bottom), must have smooth tapered ends. If the ends have developed grooves, the screw or needle must be replaced.

Blow out all passages in the castings with compressed air. Check all parts and passages to be sure they are not clogged or contain any deposits. **NEVER** use a piece of wire or any type of pointed instrument to clean drilled passages or calibrated holes in a carburetor.

Move the throttle shaft back and forth to check for wear. If the shaft appears to be too loose, replace the complete throttle body because individual replacement parts are **NOT** available.

Inspect the main body, mixing chamber, and gasket surfaces for cracks and burrs which might cause a leak. Check the float for deterioration. Check to be sure the float spring has not been stretched. If any part of the float is damaged, the unit must be replaced. Check the float arm needle contacting surface and replace the float if this surface has a groove worn in it.

Inspect the tapered section of the idle adjusting needles and replace any that have developed a groove.

Most of the parts which should be replaced during a carburetor overhaul are included in overhaul kits available from your local marine dealer. One of these kits will

contain a matched fuel inlet needle and seat, if removable. This combination should be replaced each time the carburetor is disassembled as a precaution against leakage.

ASSEMBLING

1- Install the main nozzle into the center turret and tighten the nozzle just "snug".

2- Slide a new gasket onto the shaft of the needle valve seat. Install, and then tighten the seat just "snug", using the correct size thin walled socket.

The float must be in good condition without any cracks or worn areas allowing fuel to enter. The slightest amount of fuel inside would defeat the purpose of the float.

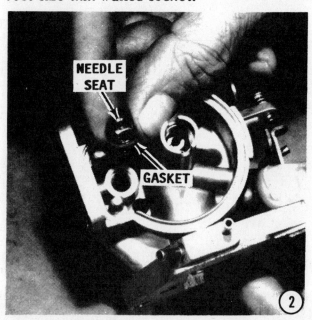

CARBURETOR "F" 4-57

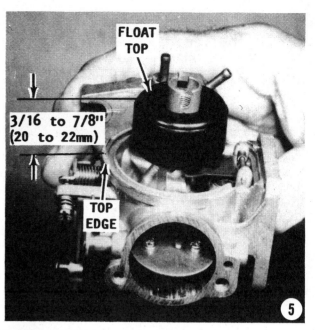

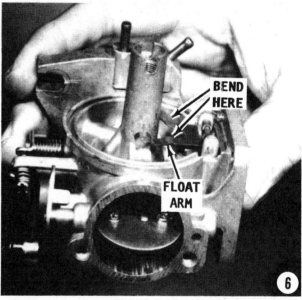

3- Insert the needle valve into the needle seat.

4- Hold the float arm between the two mounting posts in the position shown in the accompanying illustration. **TAKE CARE** - the float arm can be installed upside down by mistake.

Push the hinge pin through the posts and hinge until the end of the pin is flush with the outer edge of the mounting post. Install the float.

Float Adjustment

5- Invert the carburetor and allow the float hinge to rest on the needle valve. Measure the distance between the top of the float and the mixing chamber housing, as indicated in the accompanying illustration. The measurement should be 3/16-7/8" (20-

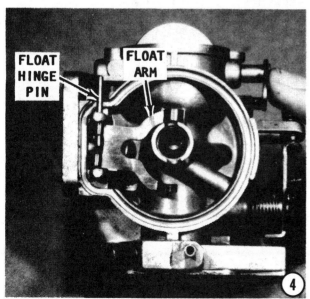

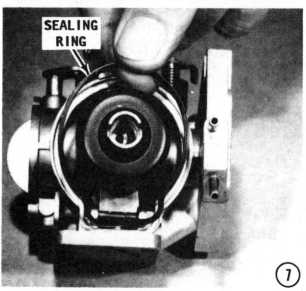

4-58 FUEL

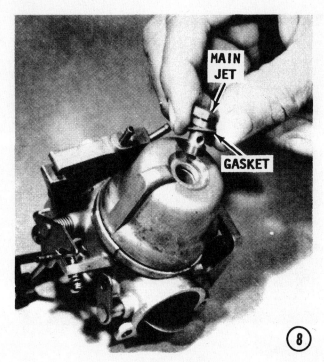

22mm). This dimension, with the carburetor inverted, places the "lower" surface of the float parallel to the carburetor body.

6- CAREFULLY bend the float arm, as required, to obtain a satisfactory measurement. Install the float.

7- Insert the rubber sealing ring into the groove in the float bowl.

8- If a gasket was used between the main nozzle and the float (only on some 8hp powerheads), position the gasket over the nozzle.

Install the float bowl and secure it in place with the main jet bolt and a new gasket.

9- Install the pilot jet into the mixing chamber. Tighten the jet just "snug".

10- Install a new gasket over the mixing chamber cover. Install the cover and secure it in place with the two Phillips head screws.

11- Slide the spring onto the idle mixture screw, and then install the screw. Tighten the screw until it **BARELY** seats, and then back the screw out 1-1/2 turns for all models.

This setting will not change during later idle speed adjustments. This position of the screw sets the fuel mixture. On this carbur-

CARBURETOR "F" 4-59

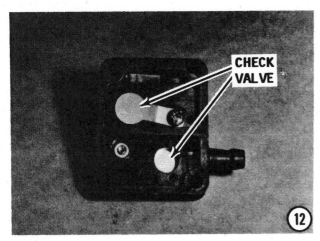

etor, the idle speed is set later, on Page 4-61, using the idle stop screw on the throttle linkage.

12- Place the check valves, one at a time, in position on both sides of the fuel pump body. Secure each valve with the attaching screw.

13- Assemble the fuel pump components onto the carburetor body in the following order: the inner gasket, the inner diaphragm, (one or two depending on the models being serviced), the pump body, the outer diaphragm, the outer gasket, and finally the pump cover. Check to be sure all the parts are properly aligned with the mounting holes. Secure it all in place with the four attaching screws.

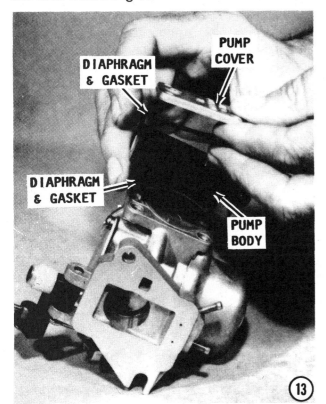

INSTALLATION

14- Place a new carburetor mounting gasket onto the two pins on the intake manifold.

For all 8hp units: Align the carburetor and the silencer with the two mounting holes. Start the bolts, and then tighten them, through the opening in the front panel, to a torque value of 5.8 ft lbs (8Nm).

15- For 9.9 and 15hp units: Move the carburetor into place. Guide the choke rod through the slot in the choke lever and at the same time guide the throttle roller into the throttle shaft cradle. Start the mounting nuts, and then tighten them, through the opening in the front panel, to a torque value of 5.8 ft lbs (8Nm).

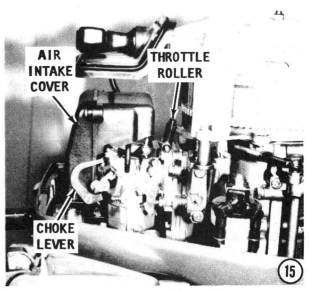

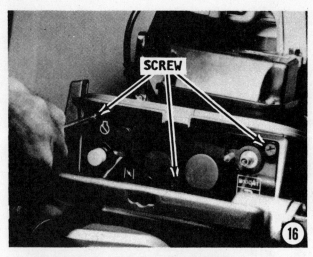

16- Install the front panel and secure it in place with the three Phillips screws.

For all 8hp powerheads only: Slide the washer over the choke shaft and connect the choke link rod to the choke shaft. Secure the link rod to the shaft using the little cotter pin.

Check the action of the choke knob to be sure there is no evidence of binding.

Throttle Rod Adjustment

17- Align the throttle roller centerline with the mark embossed on the throttle cam. Loosen the throttle rod retaining screw. Adjust the rod length to fully open the throttle shutter plate. Tighten the screw to hold this adjustment.

18- Connect the fuel line to the fuel pump. Connect the fuel line from the tank with the fuel joint.

CLOSING TASKS

Mount the outboard unit in a test tank, on the boat in a body of water, or connect a flush attachment and hose to the lower unit. Connect a tachometer to the powerhead.

Connect a tachometer to the powerhead by connecting the Red tachometer lead to the primary side of the ignition coil and the Black tachometer lead to a suitable powerhead ground.

NEVER, AGAIN, NEVER operate the engine at high speed with a flush device attached. The engine, operating at high speed with such a device attached, would **RUNAWAY** from lack of a load on the propeller, causing extensive damage.

Start the engine and check the completed work.

CAUTION

Water must circulate through the lower unit to the powerhead anytime the powerhead is operating to prevent damage to the water pump in the lower unit. Just five seconds without water will damage the water pump impeller.

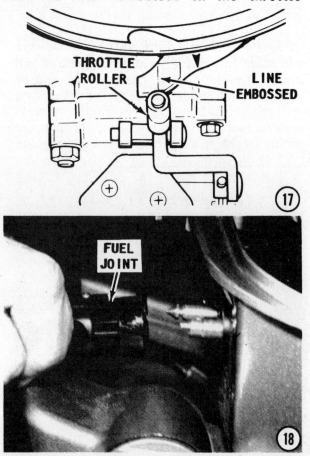

Unit in a test tank ready for fine carburetor adjustments to be made, after the cowling is removed.

CARBURETOR "G" 4-61

Idle Speed Adjustment

Allow the powerhead to warm to normal operating temperature. Shift into **FORWARD** gear and adjust the throttle stop screw by inserting a screwdriver through the opening in the silencer, until the powerhead idles between 600 and 700 rpm for all 8hp units or between 700 and 800 rpm for 9.9 and 15hp units.

High Speed Adjustment

There is no high speed adjustment possible on this type carburetor. A standard fixed main jet is located in an easily accessible location, for changing when operating the outboard at higher altitudes. The main jet is also the float bowl securing bolt.

If the outboard is to be consistently operated in cold climates or at high speeds between 4500-5500 rpm, then the owner may consider replacing the main jet with a jet having a smaller gauge number.

If the outboard is to be consistently operated at low idle speeds or at elevations higher than 2500 ft above sea level, then the main jet should be replaced with one having a higher gauge number.

Detailed timing and synchronizing procedures are presented in Chapter 6.

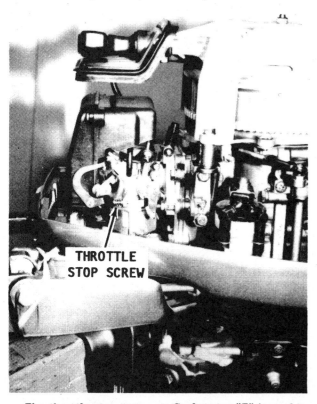

The throttle stop screw on Carburetor "F" is used to adjust the idle speed, as described in the text.

4-12 CARBURETOR "G"
USED ON SOME 8B, 8W, MARATHON 8, 9.9C, AND 15HP POWERHEADS AND ALL 8C POWERHEADS

Good shop practice dictates a carburetor rebuild kit be purchased and new parts, especially gaskets and O-rings be installed any time the carburetor is disassembled. A photograph of the parts included in the carburetor rebuild kit is shown in the Cleaning and Inspecting portion of this section.

FIRST, THESE WORDS

Carburetor "G" is used on 8, 9.9 and 15hp powerheads. The removal and installation procedures, plus the first step in disassembling vary slightly for 8hp, from the instructions for 9.9 and 15hp units. **THEREFORE**, these minor differences for 9.9 and 15hp units are emphasized with captioned illustrations.

REMOVAL

All Models

1- Disconnect the fuel hose at the fuel joint. Protect the ends from contamination. Disconnect the fuel line at the fuel pump inlet fitting.

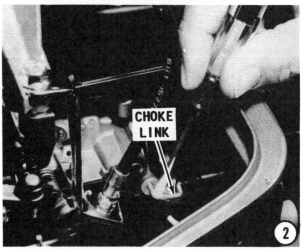

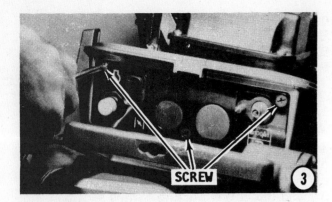

Model 8hp

2- Pry the small choke link from the plastic fitting on the carburetor using a small slotted screwdriver.

All Models

3- Remove the three Phillips screws retaining the front panel to the lower cowling.

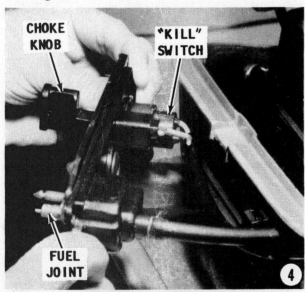

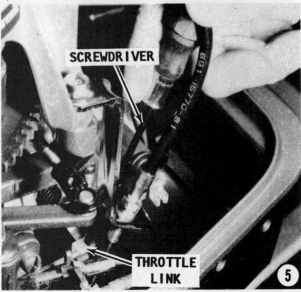

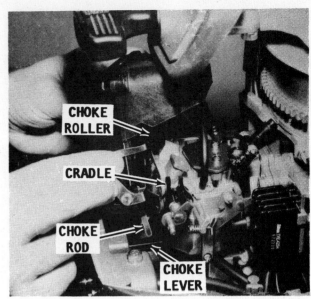

Carburetor "G" used on the 9.9hp and 15hp has a unique silencer and choke arrangement.

The panel must be removed to gain access to the carburetor.

4- Pull the panel forward and set it to one side out of the way. It is not necessary to disconnect the "kill" switch harness or the fuel connection.

Model 8hp

5- Pry the throttle link from the starboard side of the carburetor with a small slotted screwdriver.

SPECIAL WORDS
FOR 9.9HP AND 15HP UNITS

On 9.9 and 15hp units a throttle roller moving on the throttle cam is used instead of the throttle link. The roller arrangement cannot be disconnected from the carburetor.

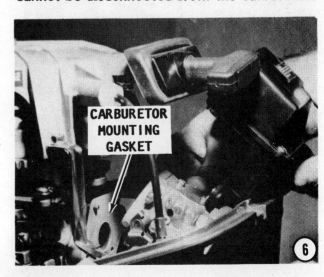

CARBURETOR "G" 4-63

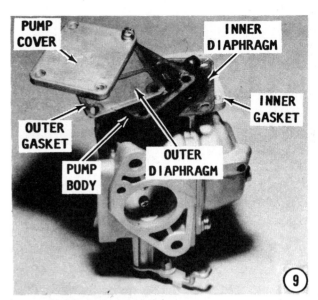

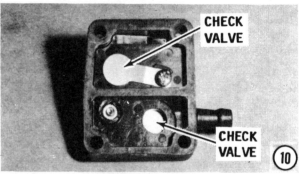

All Models

6- Reach through the opening of the front panel and remove the two long bolts which secure both the silencer and the carburetor to the powerhead. Remove the silencer. On 9.9 and 15hp models the choke rod will slide out of the choke lever and the choke roller will slide out of the choke shaft cradle. Remove the carburetor. Remove and discard the carburetor mounting gasket.

DISASSEMBLING

7- Remove the pilot screw and spring. The number of turns out from a lightly seated position will be given during assembling.

8- Remove the two screws securing the top cover and lift off the cover. Gently pry the two rubber plugs out with an awl. Remove the oval air jet cover and the round bypass cover.

9- Remove the four screws securing the fuel pump to the carburetor body. Disassemble the fuel pump components in the

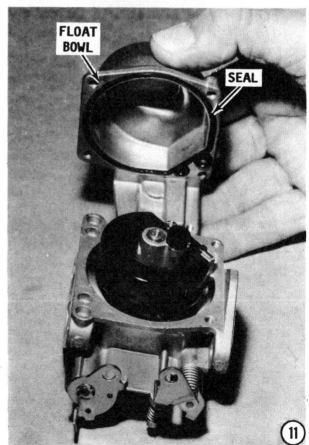

4-64 FUEL

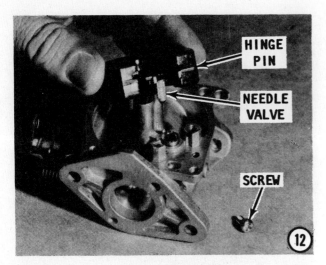

following order: the pump cover, the outer gasket, the outer diaphragm, the pump body, the inner diaphragm, and finally the inner gasket.

10- Remove the screws from both sides of the fuel pump body. Remove the check valves.

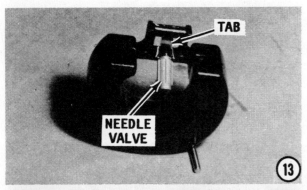

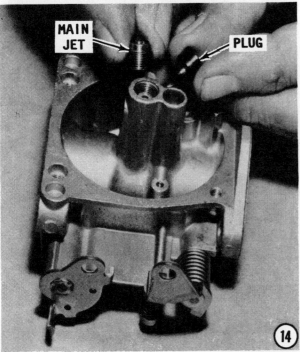

*All rubber and plastic parts **MUST** be removed before Carburetor "G" is immersed in a carburetor cleaning solution.*

11- Remove the four screws securing the float bowl cover in place. Lift off the float bowl. Remove and discard the rubber sealing ring.

12- Remove the small Phillips screw securing the float hinge to the mounting posts. Lift out the float, the hinge pin and the needle valve. Slide the hinge pin free of the float.

13- Slide the wire attaching the needle valve to the float free of the tab.

14- Use the proper size slotted screwdriver and remove the main jet, then unscrew the main nozzle from beneath the main jet. Remove the plug, and then unscrew the pilot jet located beneath the plug.

Parts included in a repair kit for Carburetor "G".

CARBURETOR "G" 4-65

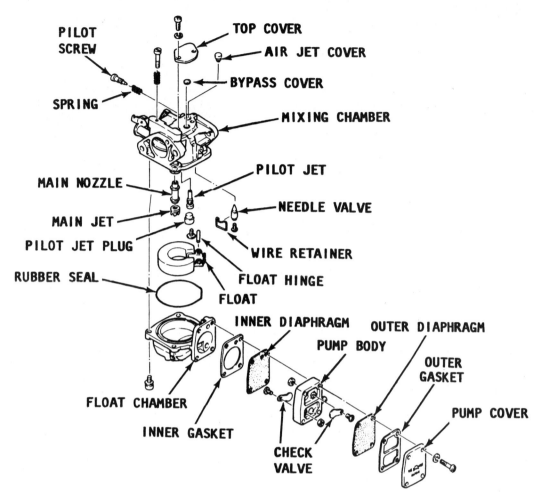

Exploded drawing of Carburetor "G", used on the powerheads listed in the heading of this section. Major parts are identified.

CLEANING AND INSPECTING

Inspect the check valves in the fuel pump for varnish build up as well as any deformity.

NEVER dip rubber parts, plastic parts, diaphragms, or pump plungers in carburetor cleaner. These parts should be cleaned **ONLY** in solvent, and then blown dry with compressed air.

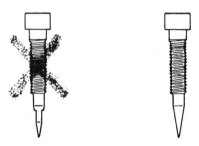

Comparison of two carburetor pilot screws. The left screw is new with a smooth taper. The worn screw on the right has developed a ridge, and is therefore unfit for further service.

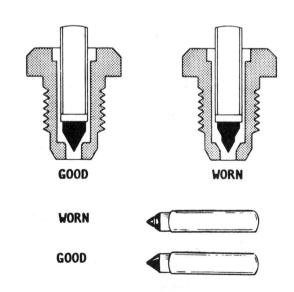

Needle and seat arrangement on the carburetor covered in this section, showing a worn and new needle for comparison.

4-66 FUEL

Place all metal parts in a screen-type tray and dip them in carburetor cleaner until they appear completely clean, then blow them dry with compressed air.

Blow out all passages in the castings with compressed air. Check all parts and passages to be sure they are not clogged or contain any deposits. **NEVER** use a piece of wire or any type of pointed instrument to clean drilled passages or calibrated holes in a carburetor.

Move the throttle shaft back and forth to check for wear. If the shaft appears to be too loose, replace the complete throttle body because individual replacement parts are **NOT** available.

Inspect the main body, mixing chamber, and gasket surfaces for cracks and burrs which might cause a leak. Check the float for deterioration. Check to be sure the float spring has not been stretched. If any part of the float is damaged, the unit must be replaced. Check the float arm needle contacting surface and replace the float if this surface has a groove worn in it.

Inspect the tapered section of the idle adjusting needles and replace any that have developed a groove.

Most of the parts which should be replaced during a carburetor overhaul are included in overhaul kits available from your local marine dealer. One of these kits will contain a matched fuel inlet needle and seat, if removable. This combination should

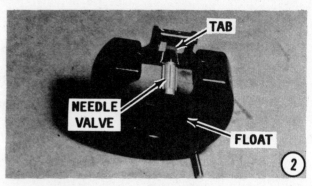

be replaced each time the carburetor is disassembled as a precaution against leakage.

ASSEMBLING

1- Install the main nozzle into the center hole and tighten it snugly. Install the main jet on top of the nozzle and tighten it snugly also. Install the pilot jet into the center hole and then the plug. Tighten the jet and the plug securely.

Model 8hp

2- Slide the wire attached to the needle valve onto the float tab. For 9.9 and 15hp units: Insert the needle valve into the needle seat.

3- Slide the hinge pin through the float hinge. Lower the float and needle assembly down into the float chamber and guide the needle valve into the needle seat. Check to be sure the hinge pin indexes into the mounting posts. Secure the pin in place with the small Phillips screw.

Model 9.9hp and 15hp

Place the float hinge arm between the mounting posts and secure the arm in place with the hinge pin.

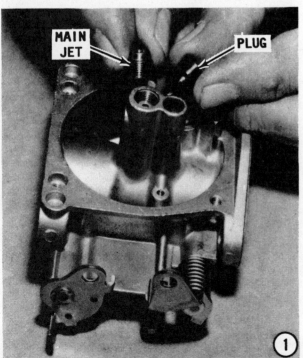

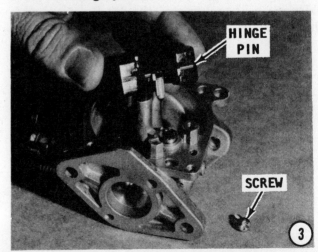

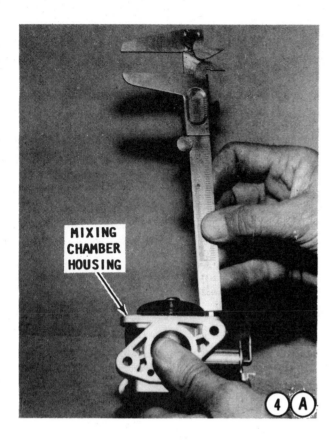

Float Adjustment

4- For 8hp units: Invert the carburetor and allow the float to rest on the needle valve. Measure the distance between the top of the float and the mixing chamber housing, illustration 4A. This distance should be 0.47-0.63" (12-16mm). This dimension, with the carburetor inverted, places the "lower" surface of the float parallel to the carburetor body. If the dimension is not within the limits listed, the needle valve must be replaced.

For 9.9 and 15hp units: Invert the carburetor and install the float bowl gasket. Measure the distance from top surface of the gasket to the top of the hinge, illustration 4B. The measurement should be 0.06-0.10" (1.5-2.5mm). **CAREFULLY** bend the

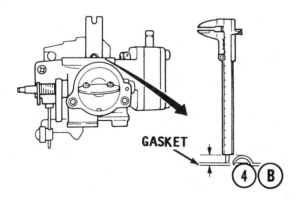

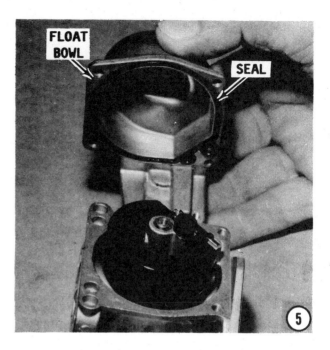

float arm, as required, to obtain a satisfactory measurement. Install the float.

All Models

5- Insert the rubber sealing ring into the groove in the float bowl. Install the float bowl and secure it in place with the four Phillips screws.

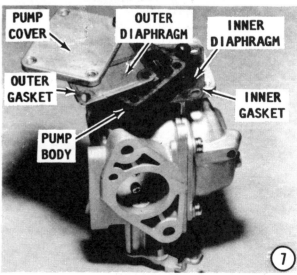

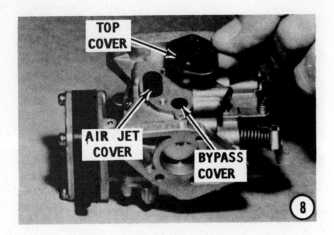

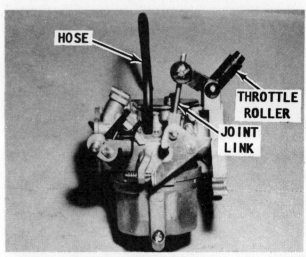

Before carburetor installation the hose and joint link are connected on the 9.9hp and 15hp powerheads. The joint link is adjusted in Chapter 6, Timing and Synchronizing.

6- Place the check valves, one at a time, in position on both sides of the fuel pump body. Secure each valve with the attaching screw.

7- Assemble the fuel pump components onto the carburetor body in the following order: the inner gasket, the inner diaphragm, the pump body, the outer diaphragm, the outer gasket, and finally the pump cover. Check to be sure all the parts are properly aligned with the mounting holes. Secure it all in place with the four attaching screws.

8- Install the oval air jet cover and the round bypass cover in their proper recesses. Place the top cover over them, no gasket is used. Install and tighten the two attaching screws.

9- Slide the spring over the pilot screw, and then install the screw. Tighten the screw until it **BARELY** seats and then back the screw out 1-1/4 turns for all models.

INSTALLATION

10- Place a new carburetor mounting gasket onto the two pins on the intake manifold, illustration 10A.

Model 8hp

Align the carburetor and the silencer with the two mounting holes. Start the bolts, and then tighten them, through the opening in the front panel, to a torque value of 5.8 ft lbs (8Nm).

Model 9.9hp and 15hp

Move the carburetor into place. Guide the choke rod through the slot in the choke lever and at the same time guide the throttle roller into the throttle shaft cradle,

CARBURETOR "G" 4-69

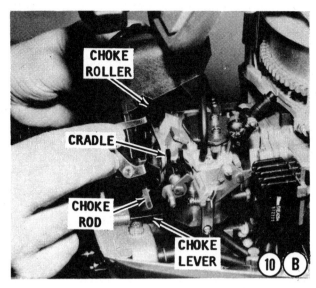

illustration 10B.. Start the mounting nuts, and then tighten them, through the opening in the front panel, to a torque value of 5.8 ft lbs (8Nm).

Model 8 hp

11- Snap the throttle link into the plastic fitting on the starboard side of the carburetor.

All Units

12- Install the front panel and secure it in place with the three Phillips screws.

13- Snap the small choke link into the plastic fitting on the starboard side of the carburetor. Check the action of the choke knob to be sure there is no evidence of binding.

14- Connect the fuel line to the fuel pump. Connect the fuel line from the tank with the fuel joint.

15- Mount the outboard unit in a test tank, on the boat in a body of water, or connect a flush attachment and hose to the lower unit. Connect a tachometer to the powerhead

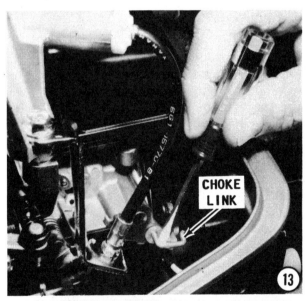

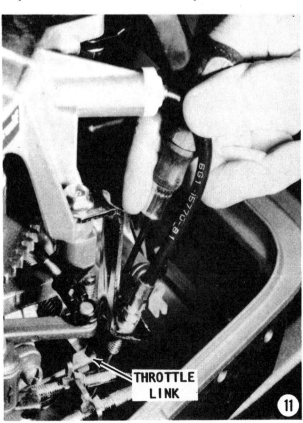

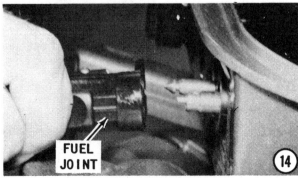

SPECIAL WORDS ON TACHOMETERS AND CONNECTIONS

The 8 to 30hp powerheads use a CDI system firing a twin lead ignition coil twice for each crankshaft revolution. If an induction tachometer is installed to measure powerhead speed, the tachometer will probably indicate **DOUBLE** the actual crankshaft rotation. Check the instructions with the tachometer to be used. Some tachometer manufacturers have allowed for the double reading and others have not.

For all models covered in this section using a Carburetor "G": connect the two tachometer leads to the two Green leads from the stator. These two Green leads are encased inside a sheath, but the connecting ends are exposed. On a manual start model, these two Green leads are not connected to anything. On the electric start models the leads are connected to a pair of Green female leads. Either tachometer lead may be connected to either green lead from the stator.

NEVER, AGAIN, NEVER operate the engine at high speed with a flush device attached. The engine, operating at high speed with such a device attached, would **RUNAWAY** from lack of a load on the propeller, causing extensive damage.

Start the engine and check the completed work.

REMEMBER, the powerhead will **NOT** start without the emergency tether in place behind the "kill" switch knob.

CAUTION

Water must circulate through the lower unit to the powerhead anytime the powerhead is operating to prevent damage to the water pump in the lower unit. Just five seconds without water will damage the water pump impeller.

Idle Speed Adjustment

16- Allow the powerhead to warm to normal operating temperature. Adjust the throttle stop screw until the powerhead idles between 750 and 850 rpm.

High Speed Adjustment

There is no high speed adjustment possible on this type carburetor. A fixed main jet is located on the side of the mixing chamber center turret. The standard size of the main jet on this carburetor is gauge No. 98.

If the outboard is to be consistently operated in cold climates or at high speeds

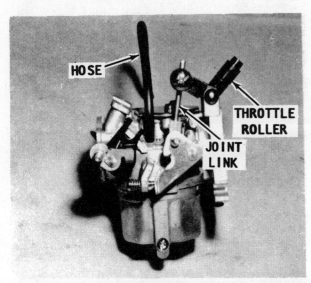

Carburetor "G" for the 9.9 and 15hp powerheads has a unique linkage arrangement, as described in the text.

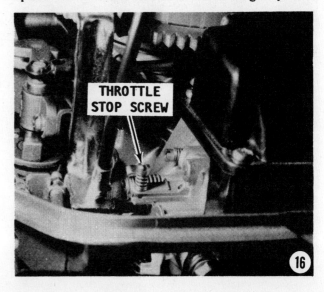

between 4500-5500 rpm, then the owner may consider replacing the main jet with a jet having a smaller gauge number.

If the outboard is to be consistently operated at low idle speeds or at elevations higher than 2500 ft above sea level, then the main jet should be replaced with one having a higher gauge number.

Detailed timing and synchronizing procedures are presented in Chapter 6.

4-13 CARBURETOR "H"
USED ON 15W AND 15HP POWERHEADS 1977-79

This section provides complete detailed procedures for removal, disassembly, cleaning and inspecting, assembling including bench adjustments, installation, and operating adjustments for Carburetor "H", as used originally on the Model 15hp unit. This carburetor is a single-barrel, float feed type with a manual choke.

FIRST, THESE WORDS

Good shop practice dictates a carburetor repair kit be purchased and new parts be installed any time the carburetor is disassembled.

Make an attempt to keep the work area organized and to cover parts after they have been cleaned. This practice will prevent foreign matter from entering passageways or adhering to critical parts.

REMOVAL

1- Disconnect the fuel hose at the fuel joint. Protect the ends from contamination. Disconnect the fuel line at the fuel pump inlet fitting.

Loosen, but do **NOT** remove, the screw securing the throttle link rod to the throttle arm. Loosen the screw enough to allow the rod to slide free of the barrel retainer. Using a pair of needle nose pliers, remove the little cotter pin securing the choke rod to the choke lever. Disconnect the rod from the lever.

Loosen, but do **NOT** remove, the two screws and washers securing the silencer to the carburetor. Lift off the silencer. Using an open end wrench, remove the two nuts from the mounting studs. Lift the carburetor free of the powerhead. Remove and discard the gasket.

DISASSEMBLING

2- Pry the plunger cap cover free, and then unscrew the plunger cap. Remove the starter lever plate spring and plunger from the carburetor body.

3- Remove the idle mixture screw and spring. It is not necessary to count the number of turns out from a lightly seated position at this time, because the number of turns will be given during assembling.

4- Take time to **COUNT** the number of turns as the throttle stop screw is backed out free of the threaded boss. The number of turns will give a starting point for later idle speed adjustment.

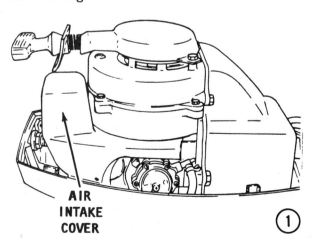

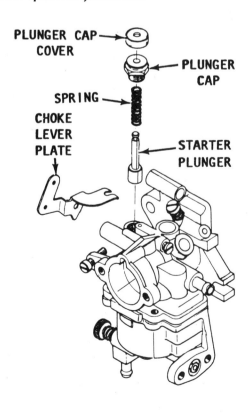

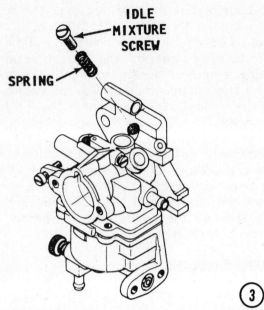

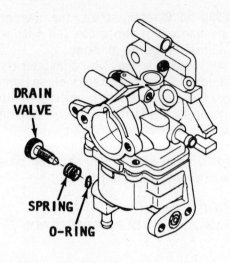

Remove the pilot jet from the other side of the carburetor using the proper size screwdriver. Reach into the carburetor throat with a small slotted screwdriver and remove the slow air jet.

5- Remove the knurled drain screw, spring and little O-ring. Inspect the tapered end of the screw. This screw **MUST** be replaced if any ridges appear on the tapered end.

CRITICAL WORDS

The drain plug performs the function of a valve by shutting off a passageway inside the fuel bowl casting. When the plug is properly seated, fuel cannot pass. When the plug is open, fuel will flow from the bowl for carburetor adjustment or draining purposes. If fuel leaks from the fitting at the base of the fuel bowl, but the plug is seated "snugly", either the O-ring or the needle end of the plug or both, are deformed. The damaged part **MUST** be replaced to prevent fuel from leaking during powerhead operation. Leaking fuel while the powerhead is running would create a definite **FIRE HAZARD**.

6- Remove the four screws securing the float bowl cover in place. Lift off the float bowl. Remove and discard the gasket.

Use the correct size screwdriver and remove the starter jet from the float bowl sealing surface.

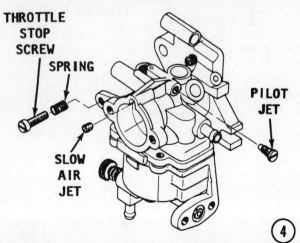

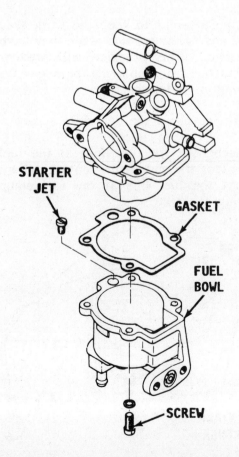

CARBURETOR "H" 4-73

7- Lift off the float. Push the pin free of the float arm and remove the float arm from between the mounting posts.

8- Lift the needle valve out of the needle seat. Use the proper size thin walled socket and remove the needle seat from the carburetor. Remove and discard the gasket under the seat.

9- Remove the main jet from the mixing chamber center turret. Unscrew and remove the main nozzle from the turret.

A GOOD WORD

Further disassembly of the carburetor is not necessary in order to clean it properly.

CLEANING AND INSPECTING

NEVER dip rubber parts, plastic parts, diaphragms, or pump plungers in carburetor cleaner. These parts should be cleaned **ONLY** in solvent, and then blown dry with compressed air.

Place all metal parts in a screen-type tray and dip them in carburetor cleaner until they appear completely clean, then blow them dry with compressed air.

Blow out all passages in the castings with compressed air. Check all parts and passages to be sure they are not clogged or contain any deposits. **NEVER** use a piece of wire or any type of pointed instrument to clean drilled passages or calibrated holes in a carburetor.

Move the throttle shaft back and forth to check for wear. If the shaft appears to be too loose, replace the complete throttle body because individual replacement parts are **NOT** available.

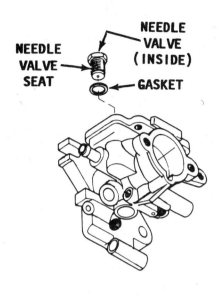

Inspect the main body, mixing chamber, and gasket surfaces for cracks and burrs which might cause a leak. Check the float for deterioration. Check to be sure the float spring has not been stretched. If any part of the float is damaged, the unit must be replaced. Check the float arm needle contacting surface and replace the float if this surface has a groove worn in it.

Inspect the tapered section of the idle adjusting needle and replace the needle if it has developed a groove.

Most of the parts which should be replaced during a carburetor overhaul are included in overhaul kits available from your local marine dealer. One of these kits will contain a matched fuel inlet needle and seat, if removable. This combination should be replaced each time the carburetor is disassembled as a precaution against leakage.

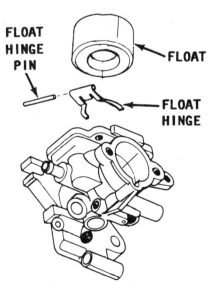

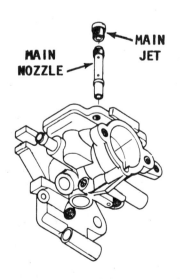

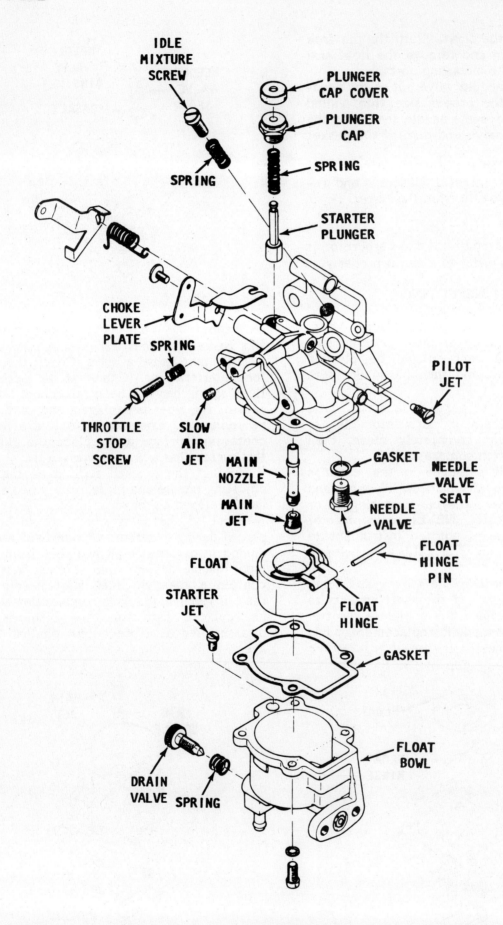

Exploded drawing of Carburetor "H", used on 15hp powerheads 1977-1979. Major parts are identified.

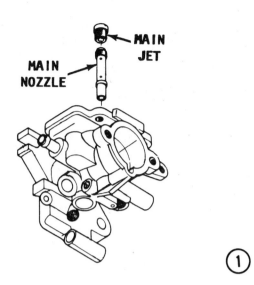

ASSEMBLING

1- Install the main nozzle into the center turret. Tighten the nozzle snugly. Install the main jet over the main nozzle. Tighten the jet snugly.

2- Slide a new gasket onto the shaft of the needle valve seat. Install and tighten the seat snugly, using a correct size thin walled socket. Insert the needle valve into the needle seat.

3- Position the float hinge between the two mounting posts. Push the hinge pin through the posts and hinge until the end of the pin is flush with the outer edge of the mounting post. Set the float in position on the hinge.

CRITICAL WORDS

The float level adjustment on this carburetor is **NOT** made by measuring the distance between the top of the float and the

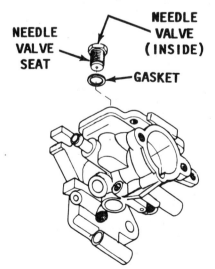

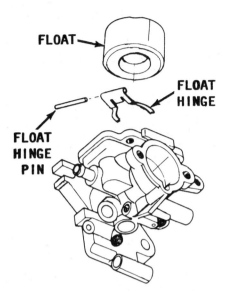

carburetor body surface, as for most other carburetors. This carburetor has a unique method for determining the correct float level and this adjustment is performed **AFTER** the carburetor is fully assembled, installed, and the powerhead is operating.

Unfortunately, if a correction is required, the float bowl must be drained and removed, and then the float removed. Adjustments are then made to the float tab in the same manner as for other carburetors.

4- Using the correct size screwdriver, install the starter jet into the recess on the

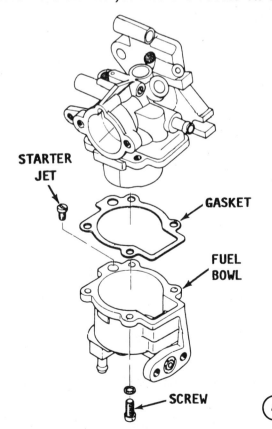

4-76 FUEL

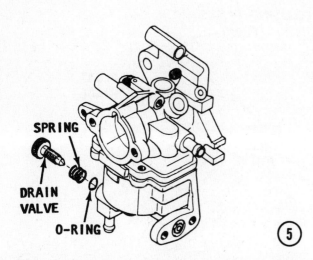

(5)

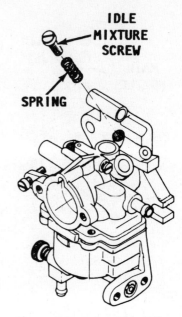

(7)

float bowl sealing surface. Position a **NEW** float bowl gasket in place over the mating surface of the carburetor. Install the float bowl and secure it with the four screws and washers. Tighten the screws alternately and evenly to prevent deforming the gasket sealing surface.

5- Install the little O-ring onto the tapered end of the drain screw. Slide the spring over the screw and install it into the base of the float bowl. Tighten the knurled screw enough to prevent fuel from leaking from the drain fitting, but not tight enough to deform the tapered end of the screw.

6- Install the slow air jet into the carburetor throat. Tighten the jet snugly. Install the pilot jet into the carburetor, using the proper size screwdriver. Again, tighten the jet snugly.

Install the throttle stop screw and spring. Thread the screw into the boss to place it in its original position, as recorded in Step 4 of disassembly. If no record was kept, install the screw midway into the boss. A correction will be made later.

7- Slide the spring onto the idle mixture screw, and then install the screw into the carburetor body. Tighten the screw until it barely seats and then back it out 1-1/4 turns. This screw position establishes the fuel mixture and remains as set. It is not disturbed for the idle speed adjustment. Therefore, it **MUST** be set correctly **NOW**. On this carburetor, the idle speed is set later using the throttle stop screw, as directed on Page 4-77.

8- Slide the spring onto the plunger shaft, and then insert both pieces into the

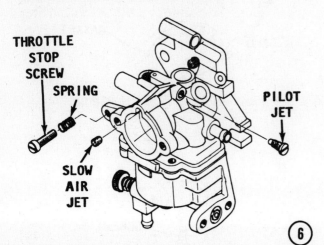

(6)

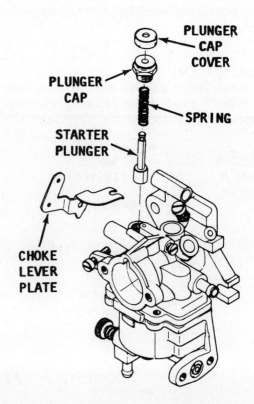

(8)

CARBURETOR "H"   4-77

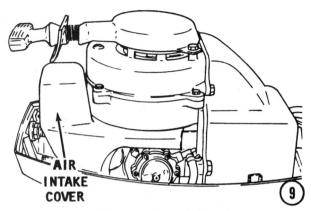

carburetor. Place the choke lever plate over the top of the plunger with the lever facing **AFT** and the lever end facing **DOWN**. Install the plunger cap over the plunger. Tighten the cap and snap the protective cover over the cap.

INSTALLATION

9- Place a **NEW** carburetor mounting gasket against the intake manifold. Install the carburetor and secure it in place with the two bolts and washers. Tighten the bolts alternately and evenly to a torque value of 5.6 ft in (8Nm). Connect the fuel line from the fuel pump to the carburetor.

Connect the choke link rod to the choke lever. Slide the washer over the end of the rod inserted into the lever and secure the pieces together with the little cotter pin through the hole in the rod. Slide the end of the throttle link into the barrel retainer on the accelerator rod. Hold the accelerator rod up against the stop on the pulley. With the throttle shutter valve fully **OPEN**, tighten the screw securing the rod to the barrel connector.

Install the silencer over the carburetor and secure it in place with the two attaching screws.

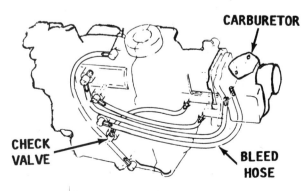

Bleed hose arrangement for Carburetor "H" when used as a single carburetor installation.

CLOSING TASKS

Mount the outboard unit in a test tank, on the boat in a body of water, or connect a flush attachment.

Obtain a tachometer. Connect the Red tachometer lead to the primary windings of the ignition coil (mounted on the starboard side of the block). Connect the Black tachometer lead to a suitable ground on the powerhead.

Start the powerhead and allow it to idle for a few minutes. Then shut down the powerhead and remove the silencer.

10- Obtain about one foot (30cm) of 1/4" (6mm) inner diameter transparent vinyl tubing and connect one end to the fitting at the base of the float bowl. Hold the other end up to prevent fuel from draining from the float bowl. Back out the knurled drain screw next to the fitting on the bowl and allow fuel to drain into the tubing loop.

The distance between the level of fuel in the line and the carburetor bore centerline must be 1-1/8" (28mm). If the distance is less than specified, the float is dropping down too far inside the float bowl. The float tab must be pushed toward the float "just a whisker" to correct the level. If the distance is greater than specified, the float level is too high inside the bowl and the float tab must be pulled away from the float "just a whisker" to correct the level.

If the float level is incorrect, the float bowl must be completely drained and removed. Make the necessary adjustments to the float tab and repeat this step until the float level matches specifications.

Idle Speed Adjustment

Connect a tachometer to the powerhead, as previously directed. If the carburetor

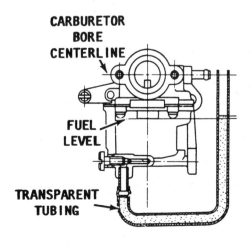

The idle mixture screw must be lightly seated; backed out 1-1/4 turns; then left undisturbed as set.

was not overhauled, rotate the idle mixture screw to lightly seat the screw. The idle mixture screw is threaded into the carburetor body, and is not to be confused with the throttle stop screw, threaded into a boss on the starboard side of the carburetor. Back the idle mixture screw out 1-1/4 turns. This screw is left as set and must not be rotated to adjust powerhead rpm.

Mount the outboard unit in a test tank, on the boat in a body of water, or connect a flush attachment and hose to the lower unit. Connect a tachometer to the powerhead.

NEVER, AGAIN, NEVER operate the engine at high speed with a flush device attached. The engine, operating at high speed with such a device attached, would **RUNAWAY** from lack of a load on the propeller, causing extensive damage.

CAUTION

Water must circulate through the lower unit to the powerhead anytime the powerhead is operating to prevent damage to the water pump in the lower unit. Just five seconds without water will damage the water pump impeller.

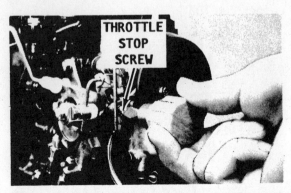

The throttle stop screw on Carburetor "H" is used to adjust idle rpm.

Allow the powerhead to warm to normal operating temperature. Shift the unit into **FORWARD** gear, and adjust the throttle stop screw, located on a threaded boss on the starbard side of the powerhead, until the powerhead idles between 1450 and 1550 rpm.

High Speed Adjustment

There is no high speed adjustment possible on this type carburetor. A fixed main jet is located on the top of the mixing chamber center turret. The standard size of the main jet on this carburetor is gauge No. 135.

If the outboard is to be consistently operated in cold climates or at high speeds between 4500-5500 rpm, then the owner may consider replacing the main jet with a jet having a smaller gauge number.

If the outboard is to be consistently operated at low idle speeds or at elevations higher than 2500 ft above sea level, then the main jet should be replaced with one having a higher gauge number.

Detailed timing and synchronizing procedures are presented in Chapter 6.

Unit in a test tank ready for fine carburetor adjustments to be made, after the cowling is removed.

4-14 CARBURETOR "J"
USED ON 20HP, 25 HP, 28HP, AND 30 HP POWERHEADS

Good shop practice dictates a carburetor rebuild kit be purchased and new parts, especially gaskets and O-rings, be installed any time the carburetor is disassembled. A photograph of parts included in the carburetor rebuild kit for Carburetor "J" is shown in the Cleaning and Inspecting portion of this section.

REMOVAL

1- Remove the powerhead cowling. Remove the Phillips head screws securing the silencer cover to the carburetor.

On models equipped with a manual choke: snap the choke link from the plastic fitting at the carburetor.

On models equipped with an electric choke solenoid: disconnect the solenoid lead at the quick disconnect fitting directly beneath the carburetor. Disconnect the ground lead at the screw on the crankcase. Remove the tiny metal ring from the choke solenoid link. This ring passes through a hole in the link and acts as a retaining ring. Therefore, **TAKE CARE** not to lose the ring. Remove the link. Remove the two screws securing the solenoid to the mounting bracket and remove the choke solenoid. Loosen

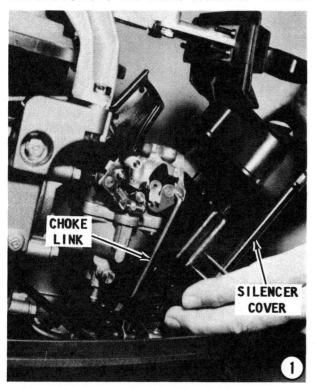

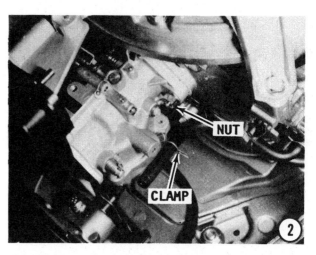

the screw at the barrel retaining the joint rod connecting the throttle arm to the accelerator arm. Pull the rod free of the barrel.

2- Squeeze the wire type fuel hose clamp and move it back on the fuel line. Work the fuel line free of the inlet fitting. Remove the two carburetor mounting nuts and lift the carburetor from the powerhead. Remove and discard the mounting gasket.

DISASSEMBLING

3- Remove the nut and washer from the long bolt. Slide the bolt and throttle cam free of the carburetor.

4- Remove the pilot screw and spring. Inspect the tapered end. Usually this part is supplied in the rebuild kit for this carburetor. The screw **MUST** be replaced if any ridges appear on the tapered end.

5- Remove the pilot jet on the opposite side of the carburetor.

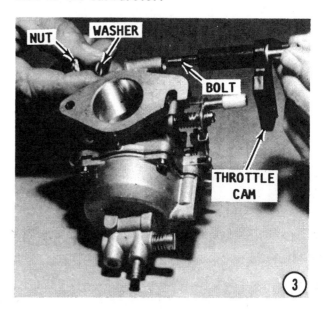

4-80 FUEL

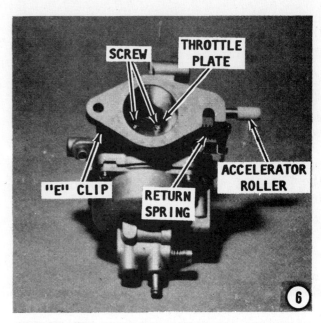

SPECIAL WORDS

Perform the next **TWO** steps **ONLY** if the throttle shaft or throttle plate is worn, bent, or damaged. Installing the two screws onto the throttle plate is **NOT** an easy task. If the accelerator roller is the only defective part, perform only the first part of the next step.

6- To remove the accelerator roller: first remove the screw next to the roller and the bracket will come free of the carburetor. Next, remove the circlip from the end of the arm and roller will slide free.

To remove the throttle plate: using a very small Phillips screwdriver, remove the two small screws securing the throttle plate to the throttle shaft. **TAKE CARE** not to strip the heads and do not lose the screws.

Jar the plate free of the carburetor throat. Pry the E-clip from the end of the throttle shaft.

7- Pull out the throttle shaft. Notice how one end of the return spring hooks around the bracket and the other end hooks around a post. Count the number of turns the spring unwinds when it is unhooked.

MORE SPECIAL WORDS

Perform the next step **ONLY** if the choke shaft or the choke plate is worn or bent or if the flapper valve or its return spring is defective. To assemble these parts is not an easy task.

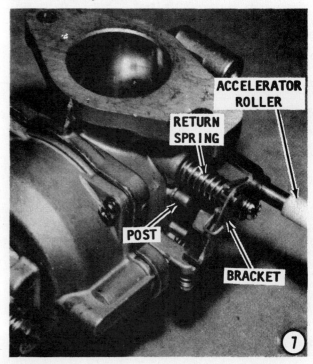

CARBURETOR "J" 4-81

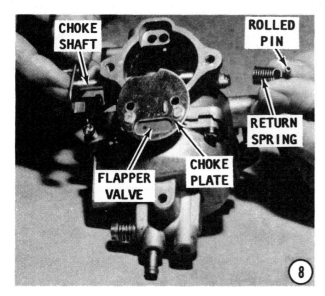

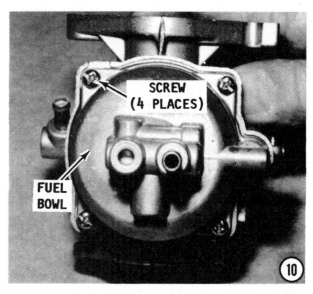

8- Using a very small Phillips screwdriver, remove the two screws securing the choke plate to the choke shaft. Count the number of turns the return spring is wound. Drive out the rolled pin from the choke shaft. Slide the spring free of the rolled pin. Pull out the choke shaft and unhook the other end of the return spring from the post on the carburetor.

9- Remove the drain valve and spring from the base of the float bowl.

10- Remove the four Phillips head screws securing the fuel bowl to the mixing chamber.

11- Lift the fuel bowl from the mixing chamber. Remove and discard the gasket.

12- Push the pin free of the mounting posts using a long pointed awl. The pin **MUST** be pushed out in the direction of the arrow embossed on one of the mounting posts. Once the end of the pin clears the

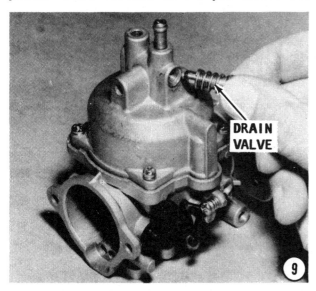

float hinge, the float may be lifted free, leaving the pin still in the other mounting post.

13- Lift out the needle valve from the needle seat. **TAKE CARE** not to damage the pointed end of the valve.

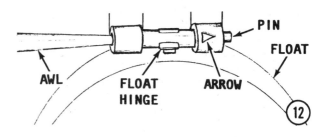

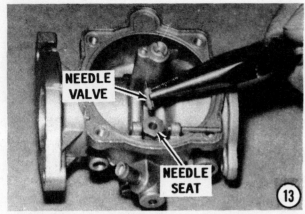

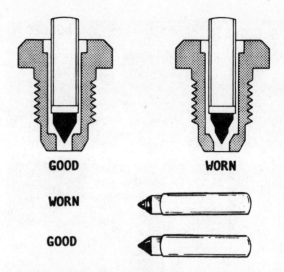

14- Using the proper size socket, remove the needle seat. Remove and discard the gasket.

15- Remove the main jet, and then the main nozzle from the center of the mixing chamber.

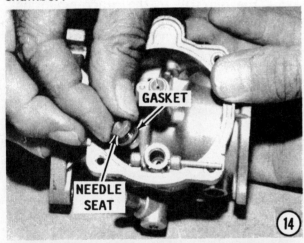

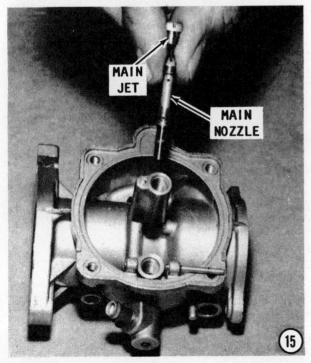

Needle and seat arrangement on the carburetor covered in this section, showing a worn and new needle for comparison.

CLEANING AND INSPECTING

NEVER dip rubber parts, plastic parts, diaphragms, or pump plungers in carburetor cleaner. These parts should be cleaned **ONLY** in solvent, and then blown dry with compressed air.

Place all metal parts in a screen-type tray and dip them in carburetor cleaner until they appear completely clean, then blow them dry with compressed air.

Blow out all passages in the castings with compressed air. Check all parts and passages to be sure they are not clogged or contain any deposits. **NEVER** use a piece of wire or any type of pointed instrument to clean drilled passages or calibrated holes in a carburetor.

Move the throttle shaft back and forth to check for wear. If the shaft appears to be too loose, replace the complete throttle body because individual replacement parts are **NOT** available.

Inspect the main body, mixing chamber, and gasket surfaces for cracks and burrs

Inspect the flapper valve to determine if the return spring functions properly to close the valve.

CARBURETOR "J"   4-83

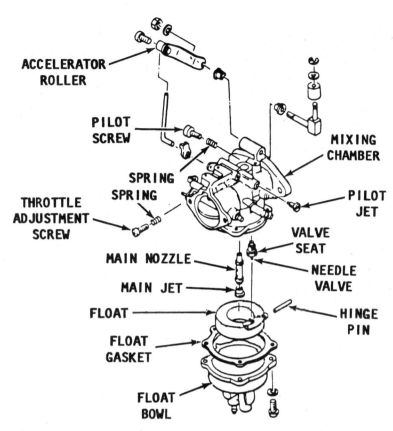

Exploded drawing of Carburetor "J" used on the powerheads listed in the heading of this section. Major parts are identified.

which might cause a leak. Check the float for deterioration. Check to be sure the float spring has not been stretched. If any part of the float is damaged, the unit must be replaced. Check the float arm needle contacting surface and replace the float if this surface has a groove worn in it.

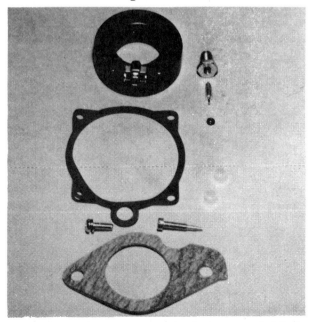

Parts included (at press time), in a repair kit for Carburetor "J".

*The mixing chamber of Carburetor "J" must **NOT** be immersed in any type carburetor cleaning solution. The chamber contains plastic bushings around the throttle shaft. These bushings are pressed into place and are not normally removed.*

4-84 FUEL

Inspect the tapered section of the idle adjusting needle and replace the needle if it has developed a groove.

Most of the parts which should be replaced during a carburetor overhaul are included in overhaul kits available from your local marine dealer. One of these kits will contain a matched fuel inlet needle and seat, if removable. This combination should be replaced each time the carburetor is disassembled as a precaution against leakage.

ASSEMBLING

1- Insert the main nozzle into the center of the mixing chamber and tighten the nozzle snugly using a slotted screwdriver. Install the main jet over the nozzle and tighten the jet snugly.

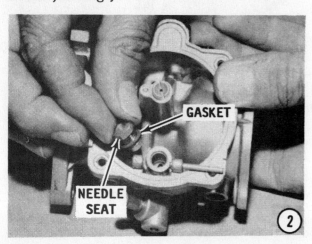

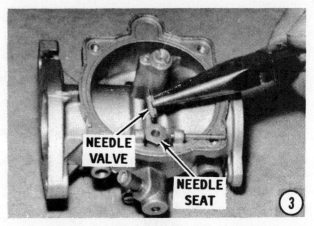

2- Slide a **NEW** gasket onto the shaft of the needle seat. Install and tighten the needle seat just "snug", using the proper size socket.

3- Insert the needle valve into the needle seat. Position the float between the two mounting posts. Push the hinge pin through the posts and hinge until the end of the pin is flush with the outer edge of the mounting post.

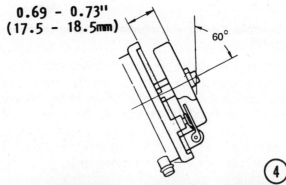

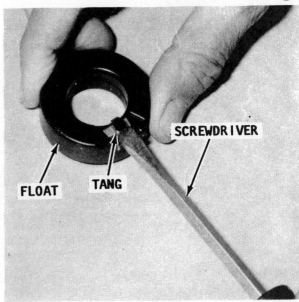

Using a screwdriver to **CAREFULLY** bend the float tang to obtain a satisfactory float drop measurement.

CARBURETOR "J" 4-85

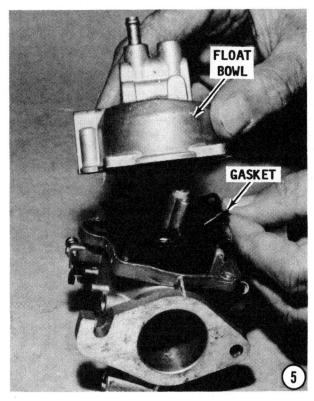

7- Slide the spring onto the drain valve shaft. Install the valve and spring into the base of the fuel bowl.

SPECIAL WORDS

If the choke shaft was removed during the disassembling procedures, perform Step 8. If the shaft was not removed, proceed directly to Step 9.

8- Insert the choke shaft through the carburetor. Rotate the shaft to permit the flattened surface of the shaft to face **UPWARD** in the carburetor throat. Lower the choke plate into the throat, aligning the notch with the air jet housing. Secure the plate to the shaft with the two small Phillips head screws. This is not an easy task. A magnetic screwdriver may prove very helpful in starting the screws in their holes.

Slide the return spring onto the choke shaft and hook the end onto the post on the

4- Invert (turn upside down) the carburetor. Now, rotate the carburetor until the axis of the float is about a 60° angle with vertical, as indicated in the accompanying illustration. With the carburetor in this position, measure the distance from the top edge of the mixing chamber to the top of the float. This distance should be 0.69-0.73" (17.5-18.5mm). If the measurement is not within these limits, **CAREFULLY** bend the tang to obtain a satisfactory measurement.

5- Position a **NEW** float bowl gasket in place on the mating surface of the carburetor. Install the float bowl and secure it with the four Phillips head screws.

6- Install the pilot jet. Tighten the jet until it seats snugly.

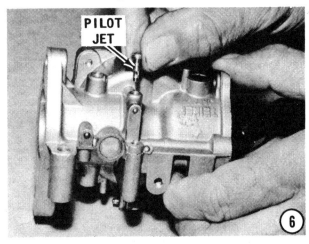

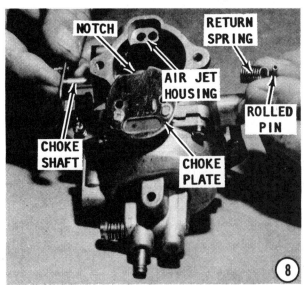

4-86 FUEL

(9)

carburetor. Wind the spring the same number of turns as recorded during Step 8 of disassembling. Hold tension on the spring and at the same time insert the roll pin. Test the return action of the plate when the choke shaft is rotated and then released. Adjust the tension on the spring if necessary to obtain a smooth and complete return of the plate when the shaft is released.

MORE SPECIAL WORDS

If the throttle shaft was removed as described in the disassembling procedures, perform Steps 9 and 10. If the shaft was not removed, proceed directly to Step 11.

9- Slide the return spring onto the throttle shaft and hook one end onto the bracket. Wind the spring the same number of turns recorded during Step 7 of disassembling. Insert the shaft through the carburetor and

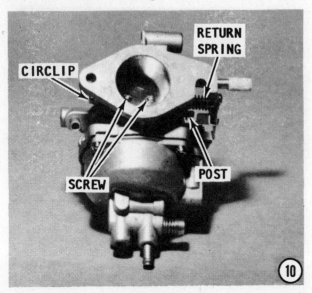

(10)

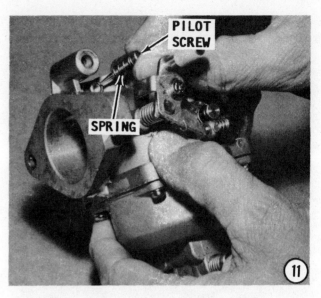

(11)

hook the tensioned end of the spring onto the post on the carburetor body. Test the return action of the plate when the throttle shaft is rotated and then released. Adjust the tension on the spring, by increasing the number of spring turns, if necessary, to obtain a smooth and complete return of the plate when the shaft is released.

10- Rotate the shaft slightly in either direction to allow the flattened surface of the shaft to face **UPWARD** in the carburetor throat. Lower the throttle plate into the carburetor throat and onto the shaft with the holes in the plate aligned with the holes in the shaft. Secure the plate with the two small Phillips head screws. A magnetic screwdriver may prove most helpful to guide the screws into their holes.

Install the circlip onto the end of the throttle shaft. If the accelerator roller was

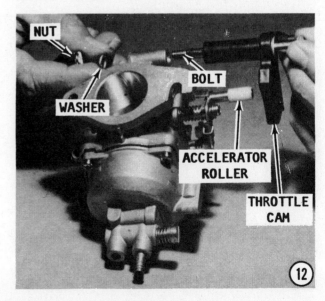

(12)

removed, slide a new roller onto the arm and secure it in place with a circlip.

11- Slide the spring onto the pilot screw shaft, and then install the screw. Tighten the pilot screw until it just **BARELY** seats. From this position, back the screw out 1-1/2 turns.

12- Slide the long bolt through the throttle cam and the carburetor. Install the washer, and then tighten the nut securely. Check the throttle cam action against the accelerator roller. The accelerator should roll smoothly on the throttle cam.

Throttle Rod Adjustment

Align the accelerator roller centerline with the mark embossed on the throttle cam. Loosen the accelerator rod retaining screw. Adjust the rod length to fully open the throttle shutter plate and tighten the screw to hold this adjustment.

INSTALLATION

13- Place a **NEW** carburetor mounting gasket on the powerhead studs. Install the carburetor and secure it in place with the washers and nuts. Tighten the nuts alternately and evenly to a torque value of 5.6 ft lbs (8Nm). Connect the fuel line onto the carburetor inlet fitting.

SPECIAL WORDS

The next step is divided into two parts. The first part for units with an electric solenoid for the electric choke and the second part for units with a manual choke.

Units with electric solenoid:

Slide the joint rod up into the barrel on the accelerator arm and tighten the screw to secure it in place. The length of this rod must be adjusted by first pushing down on the rod until the throttle arm makes contact with the wide open throttle stop. Hold this position and make sure the accelerator roller contacts the throttle cam. Tighten the screw at the barrel.

Install the solenoid to the mounting bracket on the carburetor. Secure it with the two screws. Connect the ground lead to the crankcase with the screw. Connect the solenoid lead at the quick disconnect fitting beneath the carburetor. Attach the choke solenoid link onto the choke arm. Secure the link with the tiny retaining ring.

Units with manual choke:

14- Snap the choke link into the plastic fitting on the carburetor.

All units: Install the silencer cover. Secure the cover in place with the Phillips head screws.

CLOSING TASKS

Mount the outboard unit in a test tank, on the boat in a body of water, or connect a flush attachment and hose to the lower unit. Connect a tachometer to the powerhead.

SPECIAL WORDS ON TACHOMETERS AND CONNECTIONS

The 8 to 30hp powerheads use a CDI system firing a twin lead ignition coil twice for each crankshaft revolution. If an induction tachometer is installed to measure

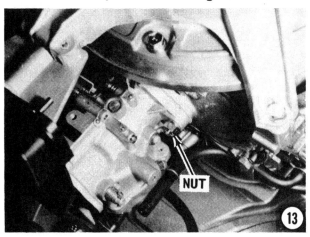

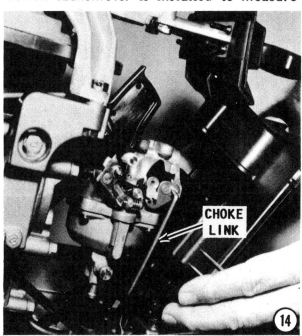

powerhead speed, the tachometer will probably indicate **DOUBLE** the actual crankshaft rotation. Check the instructions with the tachometer to be used. Some tachometer manufacturers have allowed for the double reading and others have not.

For all models covered in this section using a Carburetor "J", connect the two tachometer leads to the two Green leads from the stator. These two Green leads are encased inside a sheath, but the connecting ends are exposed. On a manual start model, these two Green leads are not connected to anything. On the electric start models the leads are connected to a pair of Green female leads. Either tachometer lead may be connected to either green lead from the stator.

NEVER, AGAIN, NEVER operate the engine at high speed with a flush device attached. The engine, operating at high speed with such a device attached, would **RUNAWAY** from lack of a load on the propeller, causing extensive damage.

Start the engine and check the completed work.

REMEMBER, the powerhead will **NOT** start without the emergency tether in place behind the kill switch knob, if so equipped.

CAUTION
Water must circulate through the lower unit to the powerhead anytime the powerhead is operating to prevent damage to the water pump in the lower unit. Just five seconds without water will damage the water pump impeller.

Idle Speed Adjustment
15- Allow the powerhead to warm to normal operating temperature. Shift into forward gear and adjust the throttle stop screw by inserting a screwdriver through the opening in the silencer, on some models, until the powerhead idles between:

- 900 and 1000 rpm for 20hp units equipped with Magneto type ignition
- 750 and 850 rpm for 20hp units equipped with CD type ignition
- 900 and 1000 rpm for 25hp units equipped with Magneto type ignition
- 650 and 750 rpm for 25hp units equipped with CD type ignition
- 1000 and 1100 rpm for all 28hp units
- 600 to 700 rpm for all 30hp units

High Speed Adjustment
There is no high speed adjustment possible on this type carburetor. A fixed main jet of standard size, is located on the mixing chamber center turret.

If the outboard is to be consistently operated in cold climates or at high speeds between 4500-5500 rpm, then the owner may consider replacing the main jet with a jet having a smaller gauge number.

If the outboard is to be consistently operated at low idle speeds or at elevations higher than 2500 ft above sea level, then the main jet should be replaced with one having a higher gauge number.

To time and synchronize the fuel system with the ignition system, see Chapter 6.

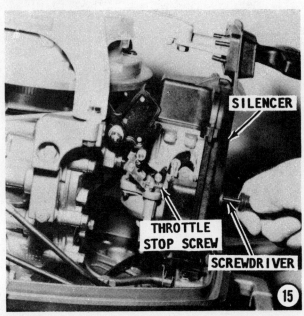

Outboard unit mounted in a test tank ready for fine carburetor adjustments to be made.

4-15 CARBURETOR "K"
USED ON 40HP POWERHEADS
SINGLE AND DUAL CARBURETOR
INSTALLATIONS

This section provides complete detailed procedures for removal, disassembly, cleaning and inspecting, assembling including bench adjustments, installation, and operating adjustments for Carburetor "K", as used originally on the Model 40hp unit, as a single or dual installation. This carburetor is a single-barrel, float feed type with a manual or electric choke.

REMOVAL

FIRST, THESE WORDS

Good shop practice dictates a carburetor repair kit be purchased and new parts be installed any time the carburetor is disassembled.

Make an attempt to keep the work area organized and to cover parts after they have been cleaned. This practice will prevent foreign matter from entering passageways or adhering to critical parts.

PRELIMINARY TASKS

Remove the screws securing the air silencer cover to the flame arrestor. Lift the silencer cover free of the flame arrestor. Use a small screwdriver and pry the choke link free of the plastic retainer at the bottom of the carburetor. On electric start models, disconnect the lead from the solenoid at the quick disconnect fitting and remove the choke solenoid from the air intake base. On manual start models, disconnect the choke lever from the choke knob.

All models: Remove the attaching hardware and then remove the air intake box.

Disconnect the fuel hose at the fuel joint. Protect the ends from contamination. Disconnect the fuel line at the fuel pump inlet fitting.

On dual carburetor installations: Make an identifying mark, "1" and "2" on the float bowl to ensure the carburetor will be installed in its original location. Remove the fuel lines and then remove the throttle and choke linkage between the carburetors.

Use an open end wrench and remove the nuts from the mounting studs, two per carburetor. Lift off the carburetor/s and remove and discard the gasket/s.

DISASSEMBLING

The following procedures pick up the work after the carburetor/s have been removed from the powerhead, as outlined in the previous paragraphs. The procedures for each of the carburetors is identical.

Therefore, perform the following procedures for each carburetor.

1- Remove the four Phillips head screws and lockwashers securing the fuel bowl to the mixing chamber.

2- Remove and discard the O-ring around the fuel bowl.

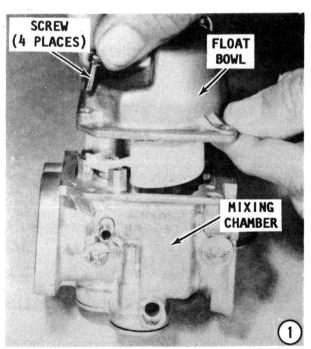

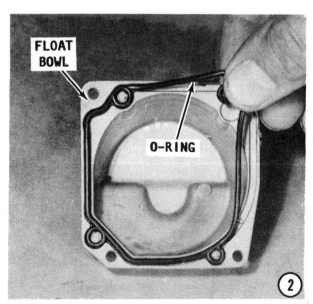

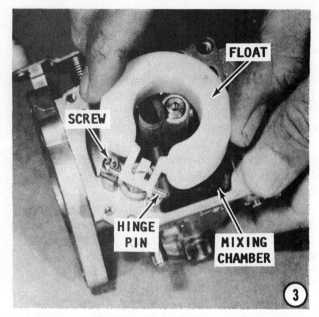

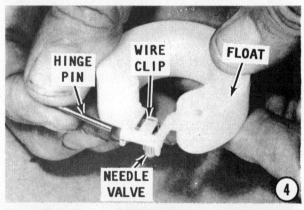

3- For single carburetor installations: Push the hinge pin through the mounting posts, to free the float hinge.

For dual carburetor installations: Loosen, but do not remove, the Phillips head screw retaining the hinge pin in its groove.

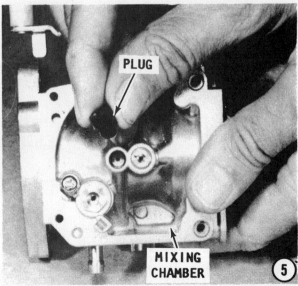

All models: Grasp the float and gently lift until the hinge pin can clear the retaining screw. The needle valve, attached to the tang on the float will also slide out of the needle seat.

4- Pull the hinge pin from the float. Unhook the wire clip and needle valve from the tang on the float.

5- Pry the small plastic plug from the center turret of the mixing chamber.

6- Use a suitable size screwdriver and remove the pilot jet located under the plug removed in the previous step.

7- Remove the main jet from the center turret of the mixing chamber.

8- Invert the mixing chamber and shake it, keeping a hand over the center turret. The main nozzle should fall free from the

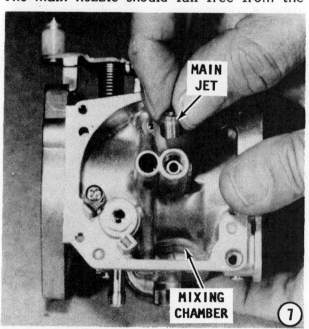

CARBURETOR "K"   4-91

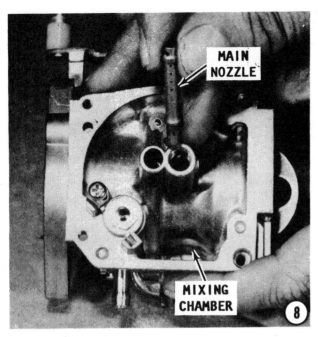

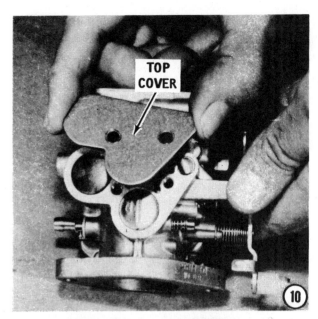

turret. If the nozzle refuses to fall out, gently reach in with a pick or similar instrument to raise the nozzle.

9- Obtain the correct size thin walled socket and remove the valve seat. Remove and discard the O-ring.

10- Remove the two Phillips head screws securing the top cover to the top of the mixing chamber. Lift off the cover.

11- Observe three flat rubber plugs, two large and one small. Lift off the large rubber plug closest to the throttle plate. Lift off the remaining large rubber plug and the metal plate beneath it. The first rubber plug does not and should not have a metal plate under it.

12- Lift off the small rubber plug and plate.

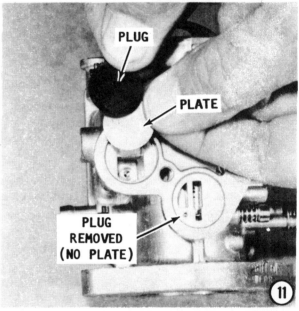

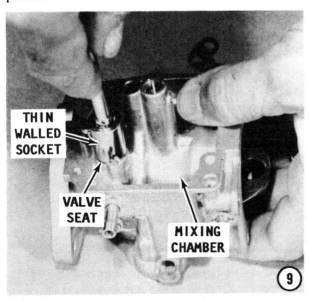

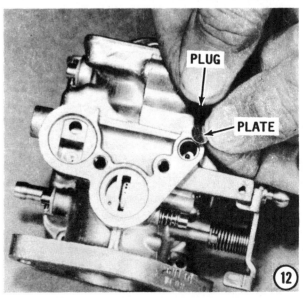

4-92 FUEL

13- Remove the pilot screw and spring from the carburetor. It is not necessary to count the number of "turns in to a lightly seated position" as a guide for installation, as the number of turns will be specified in the installation procedures.

GOOD WORDS

Normal carburetor overhaul work stops at this point. However, if noticeable wear has taken place on either the throttle shaft, the choke shaft, or the throttle or choke plates, the following two steps may be performed to remove any of the listed items.

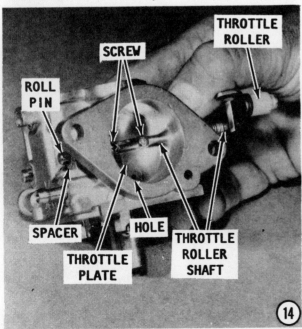

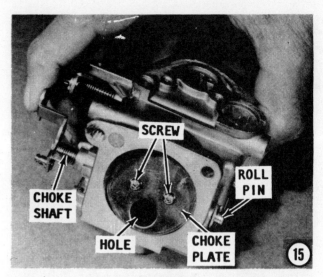

Bear in mind, it is unlikely any of these items can be purchased separately at the dealer. If they have worn beyond use and cannot be repaired, a new carburetor should be purchased.

14- Remove the two Phillips head screws securing the throttle shaft to the throttle plate. Rotate the shaft to provide sufficient clearance to permit the plate to slide from the shaft and out the carburetor bore. Use a punch and remove the roll pin from the end of the throttle shaft, and then remove the small white plastic spacer. Unhook the spring from the other end of the shaft and note the number of turns required to loosen or remove the spring. Pull the shaft from the carburetor body at the linkage end.

15- Remove the two Phillips head screws securing the choke plate to the choke shaft.

SPECIAL WORDS

The accompanying illustration shows a hole in the lower portion of the choke plate. Not all models have this hole, some have a perfectly solid choke plate.

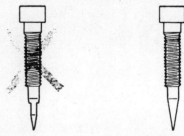

Comparison of two carburetor pilot screws. The left screw is new with a smooth taper. The worn screw on the right has developed a ridge, and is therefore unfit for further service.

CARBURETOR "K"   4-93

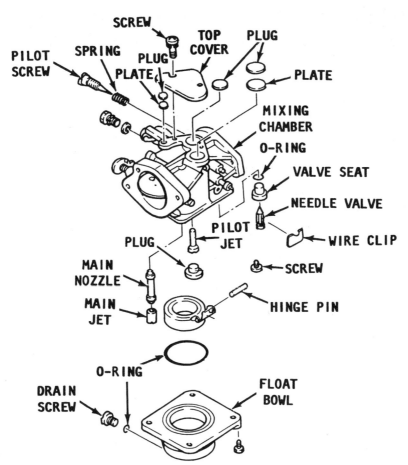

Exploded drawing of Carburetor "K" used on 40hp powerheads. This unit may be used as a single or dual installation. Major parts are identified.

Remove the choke plate. Use a punch and remove the roll pin from the end of the choke shaft and then remove the small white plastic spacer. Unhook the spring from the other end of the shaft and note the number of turns required to loosen or remove the spring. Pull the shaft from the carburetor body at the linkage end.

CLEANING AND INSPECTING

NEVER dip rubber or plastic parts, in carburetor cleaner. These parts should be cleaned **ONLY** in solvent, and then blown dry with compressed air.

Place all metal parts in a screen type tray and dip them in carburetor cleaner until they appear completely clean, then blow them dry with compressed air.

Blow out all passages in the castings with compressed air. Check all parts and passages to be sure they are not clogged or contain any deposits. **NEVER** use a piece of wire or any type of pointed instrument to clean drilled passages or calibrated holes in a carburetor.

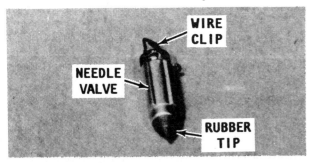

A wire clip secures the needle valve to the float. If this wire should break or slip free of the float, the fuel supply will be cut off.

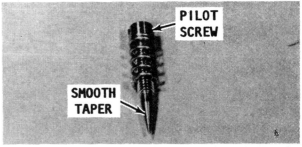

Inspect the taper on the end of the pilot screw for ridges or signs of roughness. Good shop practice dictates a new pilot screw be installed each time the carburetor is overhauled.

Move the throttle and choke shafts back and forth to check for wear. If the shaft appears to be too loose, replace the complete mixing chamber because individual replacement parts are **NOT** available.

Inspect the mixing chamber, and fuel bowl gasket surfaces for cracks and burrs which might cause a leak. Check the float for deterioration. Check to be sure the needle valve spring has not been stretched. If any part of the float is damaged, the float must be replaced. Check the needle valve rubber tip contacting surface and replace the needle valve if this surface has a groove worn in it.

Inspect the tapered section of the pilot screw and replace the screw if it has developed a groove.

As previously mentioned, most of the parts which should be replaced during a carburetor overhaul are included in an overhaul kit available from your local marine dealer. One of these kits will contain a matched fuel inlet needle and seat. This combination should be replaced each time the carburetor is disassembled as a precaution against leakage.

ASSEMBLING

Important Words

Perform the first two steps **ONLY** if steps 14 and 15 of Disassembling were performed. If the throttle and choke shafts were left intact on the carburetor, then proceed directly to Step 3.

1- Insert the choke shaft from the port side of the carburetor, and at the same time wind the spring the same number of turns it unwound in Step 15 of Disassembly. Hook the spring in place over the tab on the casting. If the spring was broken, missing or whatever, and a new spring is being installed, then estimate the tension and check the return action. Adjust the spring tension as necessary to obtain a smooth and complete return of the plate when the shaft is released.

Hold spring tension on the shaft while the roll pin is installed. Slide the white plastic retainer over the small end of the shaft. Push the roll pin into the hole. Rotate the shaft to permit the flattened surface of the shaft to face **UPWARDS** in the carburetor bore. Position the choke plate over the two screw holes. If there is a hole on the plate, the hole is positioned at the bottom of the carburetor bore. Thread the two small Phillips head screws firmly through the plate and into the shaft.

CRITICAL WORDS

No torque specification is available from the manufacturer for these two small screws, but imagine what damage would occur if one of these screws vibrated loose and chewed its way through the reeds and into the combustion chamber!

2- Insert the throttle shaft from the port side of the carburetor, and at the same

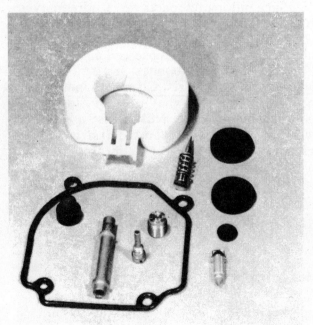

Parts included (at press time), in the repair kit for Carburetor "K".

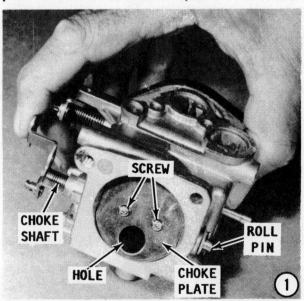

time, wind the spring the same number of turns it unwound in Step 14 of Disassembly. Hook the spring in place over the tab on the casting. If the spring was broken, missing or whatever, and a new spring is being installed, then estimate the tension and check the return action. Adjust the spring tension as necessary to obtain a smooth and complete return of the plate when the shaft is released.

Hold spring tension on the shaft while the roll pin is installed. Slide the white plastic retainer over the small end of the shaft. Push the roll pin into the hole. Rotate the shaft to provide sufficient clearance for the throttle plate to slide into the bore and behind the shaft. Align the plate with the screw holes. Thread the two small Phillips head screws through the shaft and into the throttle plate. Again, no torque specifications are provided by the manufacturer for tightening these two screws, but they could cause as much damage as loose choke shaft screws!

3- Slide a new spring over the pilot screw.

4- Install the pilot screw into the carburetor. Tighten the pilot screw until it **BARELY** seats. From this position, back out the screw 2 1/8 turns from a lightly seated position.

This setting will not change in later idle speed adjustments. This position of the screw sets the fuel mixture. On this carburetor, the idle speed is set using the idle speed screw, as directed later on Page 4-99.

5- Place the small metal plug in position over the smallest hole on top of the

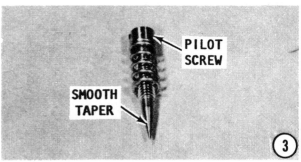

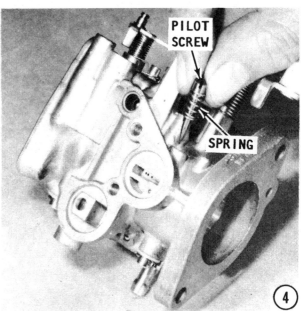

carburetor. Place the smallest rubber plug over the plate.

6- Place the large metal plug in position over the large hole on top of the carburetor, nearest the fuel fitting, and then place one of the large rubber plugs over the plate.

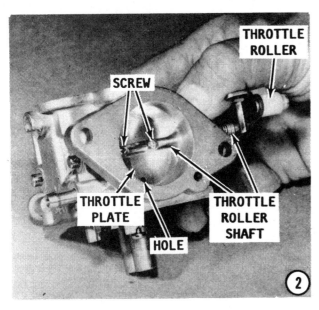

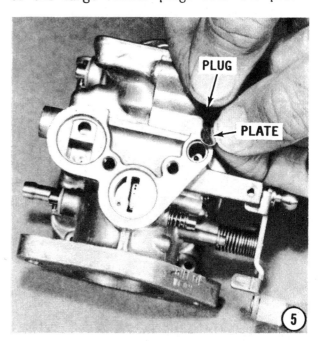

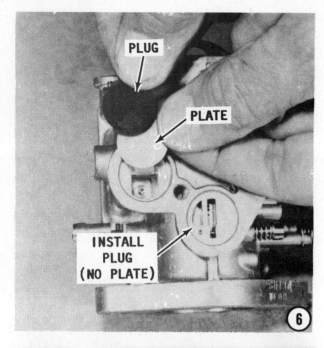

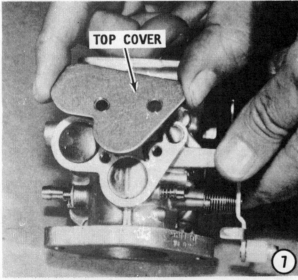

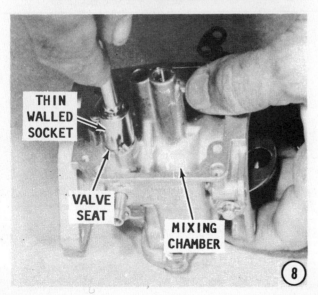

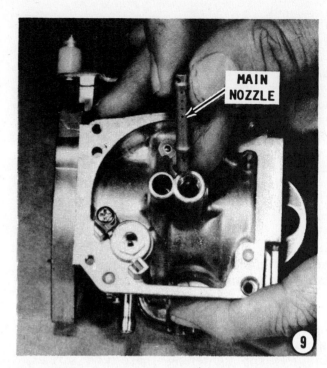

Place the remaining large rubber plug over the remaining hole. This plug is installed without a metal plate beneath it.

7- Position the top cover over all three plugs. Install and tighten the cover with the two Phillips head screws.

8- Slide a new O-ring over the shaft of the valve seat. Install and tighten the seat snugly, using a thin walled socket.

9- Insert the main nozzle into the aft hole on the center turret. Position the series of holes in the nozzle to face port and starboard when installed.

10- Install the main jet over the main nozzle. Tighten the jet until it seats snugly.

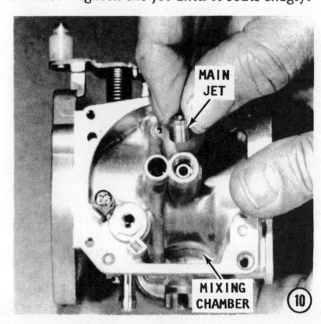

CARBURETOR "K" 4-97

11- Install the pilot jet into the forward hole on the center turret. Tighten the jet until it seats snugly.

12- Install the plug over the pilot jet. Push the plug in securely. A loose plug could wedge itself between the float and the float bowl.

13- Check to be sure the wire clip is securely in position around the needle valve. Slide the clip over the tang on the float, and check to see if the needle valve can be moved freely.

14- Slide the hinge pin through the hole in the float.

15- Lower the float assembly over the center turret, guiding the needle valve into the needle seat. Position the float hinge between the two mounting posts.

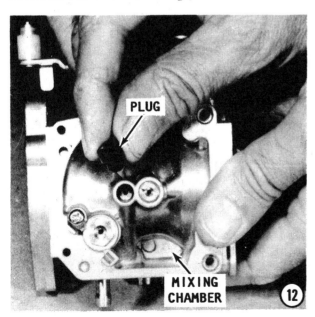

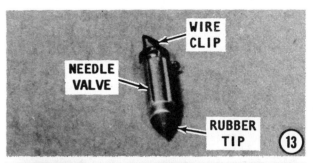

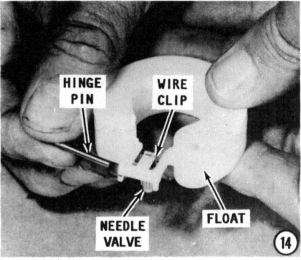

For single carburetor installations: Slide the hinge pin through the float hinge and mounting posts until the end of the pin is flush with the outer egdes of the posts.

For dual carburetor installations: Position the end of the hinge pin under its retaining screw. Tighten the screw securely.

16- Hold the mixing chamber in the inverted position, (as it has been held during

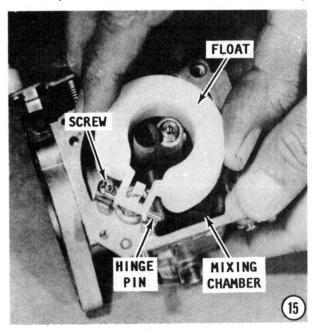

4-98 FUEL

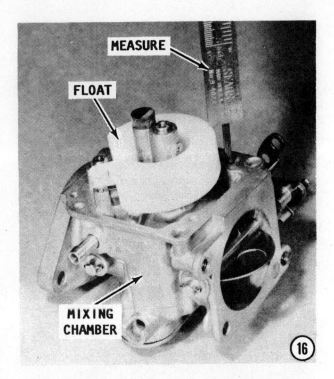

the past few steps). Observe the number stamped on the carburetor mounting flange. If the number stamped is "67900", then the upper and lower surfaces of the float **MUST** be parallel to the gasket sealing surface. For this carburetor, no actual float level measurement is given.

If the number stamped on the carburetor mounting flange is "67602" or "6E900", then proceed as follows:

Measure the distance between the top of the float and the gasket surface of the mixing chamber. This distance should be 25/32" (19mm).

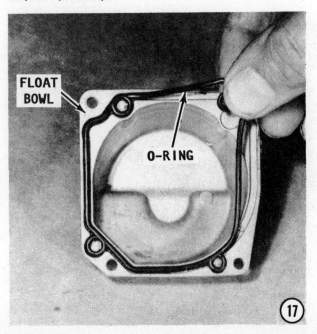

If the measurement is not as specified, inspect the valve seat, needle valve, and wire clip. Do not attempt to correct the reading by bending the tab on the float. The float is made of plastic and it will break before it bends.

17- Apply a coat of All Purpose Water Resistant Marine Grease to both sides of the fuel bowl O-ring. Install the O-ring into the groove of the fuel bowl.

18- Install the fuel bowl onto the mixing chamber, matching the one cutaway corner with the other. Install and tighten the four attaching screws.

INSTALLATION

Install new carburetor mounting gaskets over the studs on the intake manifold. On a dual carburetor installation, be sure each carburetor is installed in its original location. Slide the carburetor/s onto the mounting studs and tighten the nuts to a torque value of 6 ft lb (8Nm).

All models: Connect the fuel line/s.

On a dual carburetor installation: Install the choke and throttle link rods.

All models with manual choke: Snap the choke link back onto the plastic fitting on the lower carburetor. Check the action of the choke linkage on the starboard side of the carburetors by pulling the choke lever out and pushing it in.

All models: Install the air intake box over the carburetors and secure it in place with the attaching hardware.

On electric start models: Install the choke solenoid to the side of the air intake cover and connect the solenoid leads at

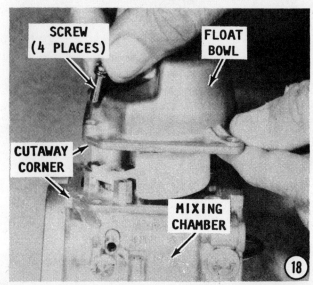

CARBURETOR "K" 4-99

their quick disconnect fittings. Install the air intake cover.

CLOSING TASKS

Mount the outboard unit in a test tank, on the boat in a body of water, or connect a flush attachment and hose to the lower unit. Connect a tachometer to the powerhead.

SPECIAL WORDS ON TACHOMETERS AND CONNECTIONS

Tachometer Connections
Manual Start Model
Connect the two tachometer leads to the two Green leads from the stator. These two Green leads are encased inside a sheath, but the connecting ends are exposed. The leads are connected to a pair of Green female leads. Either tachometer lead may be connected to either green lead from the stator.

Tachometer Connections
Electric Start Model
The following instructions apply to a model without a tachometer installed, or for a model with a tachometer installed, but a second meter is needed to assist in making adjustments.

Inside the control box, the Green lead with the female end connector is "input" or "signal" lead. The Black lead with either a male or female end connector is the tachometer "return" or "ground" lead.

Connect the tachometer to these two leads per the instructions with the meter, "input" and "ground".

NEVER, AGAIN, NEVER operate the engine at high speed with a flush device attached. The engine, operating at high speed

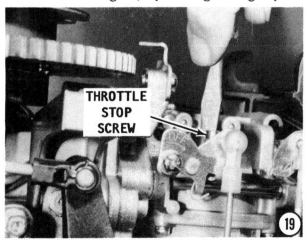

with such a device attached, would **RUNAWAY** from lack of a load on the propeller, causing extensive damage.

Start the engine and check the completed work.

CAUTION
Water must circulate through the lower unit to the powerhead anytime the powerhead is operating to prevent damage to the water pump in the lower unit. Just five seconds without water will damage the water pump impeller.

Idle Speed Adjustment
19- Allow the powerhead to warm to normal operating temperature and shift into forward gear. For single carburetor installations: Rotate the throttle twist grip to the "SLOW" position. Back off the throttle stop screw on the carburetor throttle arm, then rotate the screw **CLOCKWISE** until it barely makes contact with the arm. From this point, rotate the screw an additional two turns in the same direction to slightly open the throttle plate. Adjust the screw until the powerhead idles between 850 and 950 rpm for models 40A, 40B, and W40, or between 900 and 1000 rpm for model 40C.

For dual carburetor installations: Adjust the throttle stop screw until the powerhead idles between 850 and 950 rpm for models 40A, 40B, and W40, or between 900 and 1000 rpm for model 40C.

Rotating the throttle stop screw **CLOCKWISE** decreases powerhead speed, and rotating the screw **COUNTERCLOCKWISE** increases powerhead speed.

High Speed Adjustment
There is no high speed adjustment possible on this type carburetor. A standard size fixed main jet is located on the top of the mixing chamber center turret.

If the outboard is to be consistently operated in cold climates or at high speeds between 4500-5500 rpm, then the owner may consider replacing the main jet with a jet having a smaller gauge number.

If the outboard is to be consistently operated at low idle speeds or at elevations higher than 2500 ft above sea level, then the main jet should be replaced with one having a higher gauge number.

Detailed timing and synchronizing procedures are presented in Chapter 6.

4-16 CARBURETOR "L" USED ON 48HP, 55HP, AND 60HP POWERHEADS

This section provides complete detailed procedures for removal, disassembly, cleaning and inspecting, assembling including bench adjustments, installation, and operating adjustments for Carburetor "L", as used originally on Models 48, 55, and 60hp. This carburetor is a single-barrel, float feed type with a manual choke, or electric choke.

REMOVAL

PRELIMINARY TASKS

Electric Start Models

The cranking motor and solenoid assembly must first be removed to gain acess to the portside carburetor mounting nuts. Disconnect both positive and negative battery cables. Disconnect the large Black negative cable from the solenoid ground terminal. Disconnect the large Red positive cable, the small Red wire, and the small Orange wire (all on one eyelet connector) from the positive terminal on the solenoid. Disconnect the other small Red wire from the ignition terminal on the solenoid.

Remove the four cranking motor securing bolts and lift the motor free of the powerhead.

Disconnect the Blue wire from the choke solenoid, mounted on the starboard side of the air intake cover, at the quick disconnect fitting. Remove the two attaching screws and remove the choke solenoid, sliding the choke plunger from the solenoid.

Manual Start Models

Pry the choke rod from the plastic fitting on the choke lever on the starboard side of the lower carburetor.

All Models

Disconnect the fuel hose at the fuel joint. Protect the ends from contamination. Disconnect the fuel line at the fuel pump inlet fitting.

Disconnect the fuel supply line from the fuel pump at the "T" fitting between the two carburetors. Remove the eight nuts from the mounting studs and lift the carburetors and the intake cover free of the powerhead as an assembly. Remove and discard the two mounting gaskets.

Disconnect the long throttle link rod and the two shorter choke link rods. Make an identifying mark "1" and "2" on each carburetor to ensure it is installed in its original location after the overhaul work is complete. Remove the four bolts securing the carburetors to the intake cover, and then separate the cover from the carburetors.

NOW, THESE WORDS

Good shop practice dictates a carburetor repair kit be purchased and new parts be installed any time the carburetor is disassembled.

Make an attempt to keep the work area organized and to cover parts after they have been cleaned. This practice will prevent foreign matter from entering passageways or adhering to critical parts.

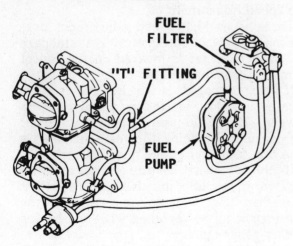

To remove Carburetor "L" from the powerhead, the electric cranking motor must be removed to gain access to the fuel lines and the carburetor mounting nuts.

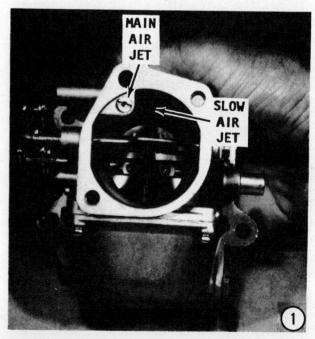

CARBURETOR "L" 4-101

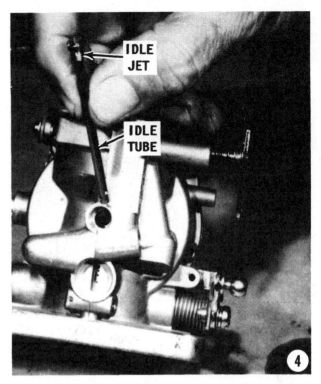

SPECIAL WORDS

Two jets are to be removed in the following step. These two jets have different shapes and perform different functions. Tag each one as it is removed to ensure the jets will be installed back in their original locations. The main air jet is "barrel" shaped, and the slow air jet is "screw" shaped.

1- Use a small slotted screwdriver and remove the main air jet from the forward face of the carburetor. Using the same screwdriver, reach into the carburetor throat and remove the slow air jet.

2- Remove the idle mixture screw and spring. It is not necessary to count the number of turns, at this time. The number of turns out from a lightly seated position will be given during assembling.

3- Remove the large brass plug over the mixing chamber. Notice the small drilled holes under the plug. After disassembly, when all the carburetor parts are being cleaned and inspected, **TAKE CARE** to clean these holes with carburetor cleaner until they appear completely clean, then blow them dry with compressed air.

Blow out all passages in the castings with compressed air. Check all parts and passages to be sure they are not clogged or contain any deposits. **NEVER** use a piece of wire or any type of pointed instrument to

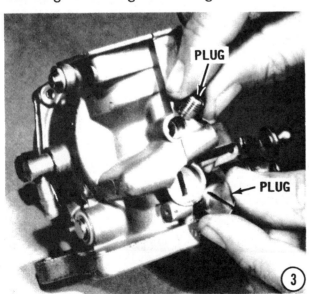

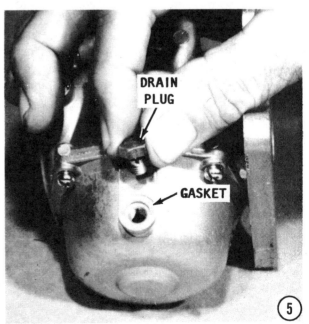

clean drilled passages or calibrated holes in a carburetor.

Use the proper size screwdriver and back out the plug over the idle jet and idle tube.

4- Back out the idle jet with the idle tube attached.

5- Remove the drain plug and gasket from the side of the float bowl. Remove and discard the gasket.

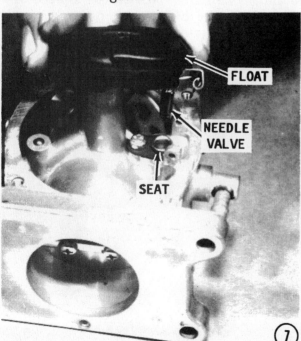

6- Remove the four attaching screws securing the float bowl to the carburetor body. Remove the bowl. Remove and discard the O-ring. Push the hinge pin free of the mounting posts, using a long pointed awl.

7- Lift off the float assembly, with the needle valve attached. The needle valve will slide out of the needle seat. **TAKE CARE** not to damage the pointed end of the valve if it is to be reused.

8- Remove the main jet from the mixing chamber center turret. Unscrew and remove the main nozzle from the turret.

Use a Phillips head screwdriver and remove the screw securing the needle seat retainer to the mixing chamber. Lift out the needle seat.

A GOOD WORD

Further disassembly of the carburetor is not necessary in order to clean it properly.

CLEANING AND INSPECTING

NEVER dip rubber parts, plastic parts, diaphragms, or pump plungers in carburetor cleaner. These parts should be cleaned **ONLY** in solvent, and then blown dry with compressed air.

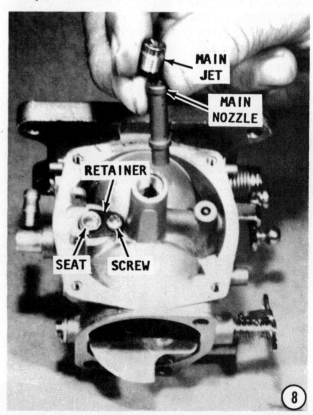

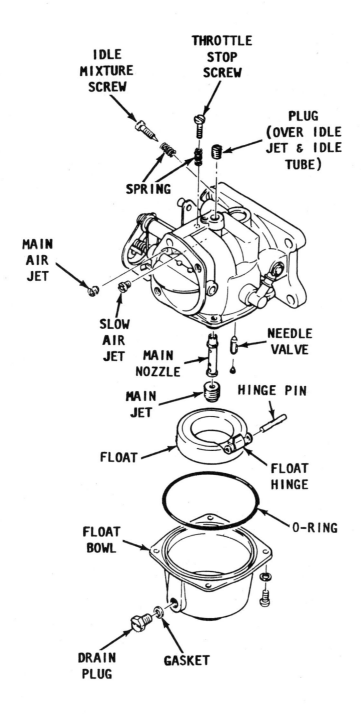

Exploded drawing of Carburetor "L" used on the powerheads listed in the heading of this section. Major parts are identified.

Example of an idle mixture screw (top), and a needle valve (bottom), in good condition. The taper in each case is smooth without any sign of a groove indicating wear.

Place all metal parts in a screen-type tray and dip them in carburetor cleaner until they appear completely clean, then blow them dry with compressed air.

Blow out all passages in the castings with compressed air. Check all parts and passages to be sure they are not clogged or contain any deposits. **NEVER** use a piece of wire or any type of pointed instrument to clean drilled passages or calibrated holes in a carburetor.

Move the throttle shaft back and forth to check for wear. If the shaft appears to be too loose, replace the complete throttle body because individual replacement parts are **NOT** available.

Inspect the main body, mixing chamber, and gasket surfaces for cracks and burrs which might cause a leak. Check the float for deterioration. Check to be sure the float spring has not been stretched. If any part of the float is damaged, the unit must be replaced. Check the float arm needle contacting surface and replace the float if this surface has a groove worn in it.

Inspect the tapered section of the idle adjusting needle and replace the needle if it has developed a groove.

Most of the parts which should be replaced during a carburetor overhaul are included in overhaul kits available from your local marine dealer. One of these kits will contain a matched fuel inlet needle and seat, if removable. This combination should be replaced each time the carburetor is disassembled as a precaution against leakage.

ASSEMBLING

1- Install the main nozzle into the center turret and tighten it snugly. Install the main jet over the main nozzle. Tighten the jet snugly.

Install a new needle seat into the mixing chamber. Place the seat retainer over the seat and secure it to the mixing chamber with the Phillips head screw. Tighten the screw snugly.

2- Slide the wire attached to the needle valve onto the float tab. Lower the float and needle assembly down into the mixing chamber and guide the needle into the needle seat. Position the float hinge between the two mounting posts. Push the hinge pin through the posts and hinge until the end of

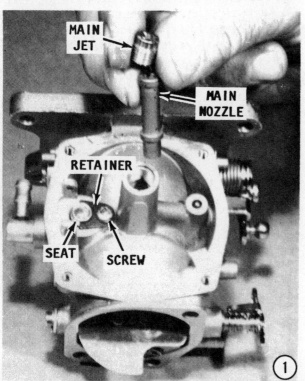

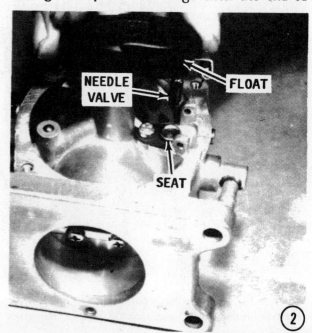

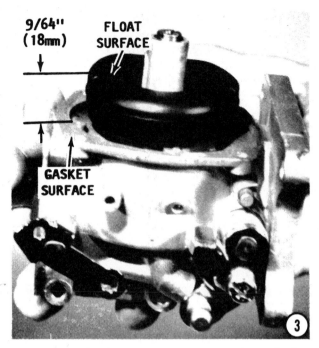

the pin is flush with the outer edge of the mounting post.

3- Hold the carburetor as shown in the accompanying illustration and allow the float hinge to rest on the needle valve. Measure the distance between the top of the float and the mixing chamber housing. This distance should be 3/4" (18mm). This dimension, with the carburetor inverted, places the lower surface of the float parallel to the carburetor body.

4- **CAREFULLY** bend the float arm, as required, to obtain a satisfactory measurement. Install the float.

5- Insert the O-ring into the groove in the float bowl. Install the float bowl and secure it in place with the four screws. Tighten the screws alternately and evenly to prevent deforming the sealing surface between the mixing chamber and the float bowl.

6- Place a **NEW** gasket over the drain plug opening. Install the plug into the side of the float bowl.

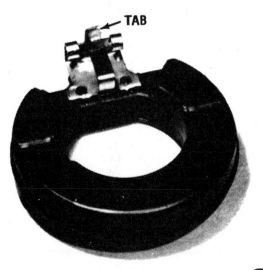

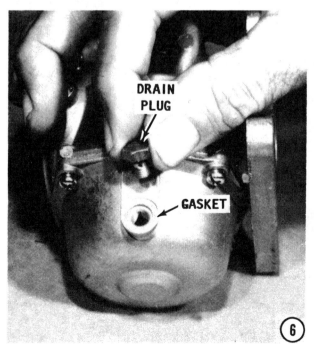

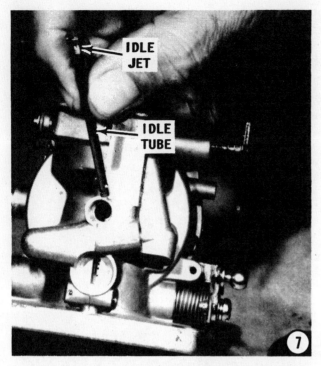

7- Lower the idle tube, with the idle jet attached, into the top of the carburetor. Tighten the jet snugly.

8- Install the plug over the idle jet and the large brass plug over the mixing chamber. Tighten both plugs snugly.

9- Slide the spring onto the idle mixture screw, and then install the screw. Tighten the screw until it **BARELY** seats, and then back the screw out 1-1/8 turns for all models.

This screw position establishes the fuel mixture and remains as set. It is not disturbed for the idle speed adjustment. Therefore, it **MUST** be set correctly **NOW**. On this carburetor, the idle speed is set using the throttle stop screw, as directed on the next page.

10- Identify the slow air jet, this jet is "screw" shaped. Install this jet deep into the carburetor throat using a small slotted screwdriver. The remaining "barrel" shaped jet is the main air jet. Install this jet into the front face of the carburetor. Tighten both jets snugly.

INSTALLATION

All Models

Place **NEW** carburetor mounting gaskets against the intake cover. Install the carburetors in their original locations and secure each in place with the four bolts and washers. Tighten the bolts alternately and evenly to a torque value of 5.6 ft lbs (8Nm). Place **NEW** carburetor mounting gaskets against the intake manifold. Install the carburetor assembly and secure it in place with the eight bolts and washers. Tighten the bolts alternately and evenly to a torque value of 5.6 ft lbs (8Nm). Connect the fuel line from the fuel pump to the "T" fitting between the carburetors.

Install the two short vertical choke link rods.

Back off the throttle stop screw on each carburetor, until the throttle shutter valves are fully closed. Adjust the length of the long vertical throttle link rod until the rod can be snapped onto both ball joints on the upper and lower throttle levers without disturbing the position of the throttle levers.

Manual Start Models

Snap the choke rod into the plastic fitting on the choke lever on the starboard side of the lower carburetor.

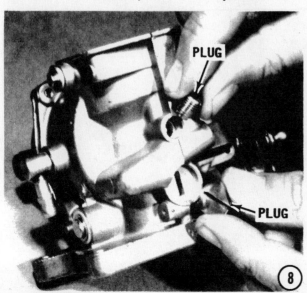

CARBURETOR "L" 4-107

Electric Start Models

Slide the choke plunger into the solenoid and position the choke solenoid over the two mounting holes on the starboard side of the intake cover. Secure the solenoid to the intake cover with the two attaching bolts. Connect the two Blue wires at their quick disconnect fittings.

Install the cranking motor and solenoid assembly to the powerhead with the four securing bolts. Tighten the bolts to a torque value of 13 ft lb (18Nm).

Connect the small Red wire from the starter switch to the ignition terminal on the solenoid. Connect the large Red cable, the small Orange wire and the other small Red wire (these are all grouped together on a single eyelet connector), to the positive terminal on the solenoid. Connect the large Black cable to the negative terminal on the solenoid. Make a final check of all connections before connecting the positive and negative cables to the battery.

CLOSING TASKS

Mount the outboard unit in a test tank, on the boat in a body of water, or connect a flush attachment and hose to the lower unit. Connect a tachometer to the powerhead.

Tachometer Connections
Manual Start Model

Connect the two tachometer leads to the two Green leads from the stator. These two Green leads are encased inside a sheath, but the connecting ends are exposed. The leads are connected to a pair of Green female leads. Either tachometer lead may be connected to either Green lead from the stator.

Tachometer Connections
Electric Start Model

The following instructions apply to a model without a tachometer installed, or for a model with a tachometer installed, but a second meter is needed to assist in making adjustments.

Inside the control box: the Green lead with the female end connector is "input" or "signal" lead. The Black lead with either a male or female end connector is the tachometer "return" or "ground" lead.

Connect the tachometer to these two leads per the instructions with the meter, "input" and "ground".

Idle Speed Adjustment

Connect a tachometer to the powerhead, as previously directed. If the carburetor was not overhauled, rotate the idle mixture screw to lightly seat the screw. Back the idle mixture screw out 1-1/8 turns. This screw **MUST** remain as set and must not be rotated to adjust powerhead rpm.

NEVER, AGAIN, NEVER operate the engine at high speed with a flush device attached. The engine, operating at high speed with such a device attached, would **RUNAWAY** from lack of a load on the propeller, causing extensive damage.

CAUTION

Water must circulate through the lower unit to the powerhead anytime the powerhead is operating to prevent damage to the water pump in the lower unit. Just five seconds without water will damage the water pump impeller.

Allow the powerhead to warm to normal operating temperature. Shift the unit into **FORWARD** gear. The idle speed is adjusted using the throttle stop screw on the lower carburetor linkage, on the starbard side of the powerhead. Rotate the throttle stop

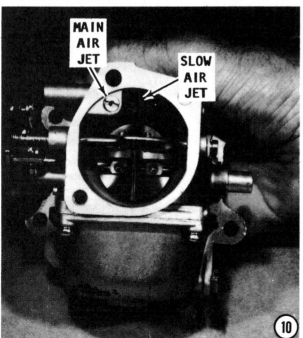

4-108 FUEL

Unit in a test tank ready for fine carburetor adjustments to be made, after the cowling is removed.

screw until the powerhead idles between 950 and 1050 rpm.

High Speed Adjustment

There is no high speed adjustment possible on this type carburetor. A fixed main jet is located on the top of the mixing chamber center turret. The standard size of the main jet on this carburetor is gauge No. 185 for 48hp powerheads, No. 190 for 55hp powerheads or No. 180 for 60hp powerheads.

If the outboard is to be consistently operated in cold climates or at high speeds between 4500-5500 rpm, then the owner may consider replacing the main jet with a jet having a smaller gauge number.

If the outboard is to be consistently operated at low idle speeds or at elevations higher than 2500 ft above sea level, then the main jet should be replaced with one having a higher gauge number.

Detailed timing and synchronizing procedures are presented in Chapter 6.

4-17 FUEL PUMP

DESCRIPTION AND OPERATION

The next few paragraphs briefly describe operation of the separate fuel pump used on powerheads covered in this manual. This description is followed by detailed procedures for testing the pressure, testing volume, removing, servicing, and installing the fuel pump.

The pump is a diaphragm displacement type. The pump is attached to the crankcase and is operated by crankcase impulses. A hand-operated squeeze bulb is installed in the fuel line to fill the fuel pump and carburetor with fuel prior to powerhead start. After the powerhead is operating, the pump is able to supply an adequate fuel supply to the carburetor to meet engine demands under all speeds and conditions.

The pump consists of a spring loaded inner diaphragm, a spring loaded outer diaphragm, two valves, one for inlet (suction) and the other for outlet (discharge), and a small opening leading directly into the crankcase. The suction and compression created as the piston travels up and down in the cylinder, causes the diaphragms to flex.

As the piston moves upward, the inner diaphragm will flex inward displacing volume on its opposite side to create suction. This suction will draw fuel in through the inlet valve.

When the piston moves downward, compression is created in the crankcase. This compression causes the inner diaphragm to flex in the opposite direction. This action causes the discharge valve to lift off its seat. Fuel is then forced through the discharge valve into the carburetor.

The function of the outer diaphragm is to absorb the pulsations of the fuel and allow smooth uninterrupted fuel flow.

This design fuel pump has the capacity to lift fuel two feet and deliver approximately five gallons per hour at four pounds pressure psi.

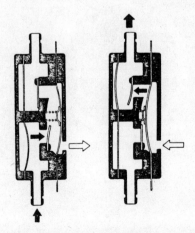

Cross-section drawing of a diaphragm displacement type fuel pump.

Problems with the fuel pump are limited to possible leaks in the flexible neoprene suction lines; a punctured diaphragm; air leaks between sections of the pump assembly; or possibly from the valves becoming distorted or not seating properly.

4-18 FUEL PUMP PRESSURE CHECK

FIRST, THESE WORDS

Lack of an adequate fuel supply will cause the powerhead to run lean, lose rpm, or cause piston scoring. If an integral fuel pump carburetor is installed, the fuel pressure **CANNOT** be checked.

Fuel pressure should be checked if a fuel tank, other than the one supplied by the outboard unit's manufacturer, is being used. When the tank is checked, be sure the fuel cap has an adequate air vent. Verify the size of the fuel line from the tank to be sure it is of adequate size to accommodate powerhead demands.

An adequate size line would be one measuring from 5/16" to 3/8" (7.94 to 9.52mm) ID (inside diameter). Check the fuel strainer on the end of the pickup in the fuel tank to be sure it is not too small and is not clogged. Check the fuel pickup tube. The tube must be large enough to accommodate the powerhead fuel demands under all conditions. Be sure to check the filter at the carburetor. Sufficient quantities of fuel cannot pass through into the carburetor to meet powerhead demands if this screen becomes clogged.

TO TEST

Mount the outboard unit in a test tank, or on the boat in a body of water.

NEVER, AGAIN, NEVER operate the engine at high speed with a flush device attached. The engine, operating at high speed with such a device attached, would **RUNAWAY** from lack of a load on the propeller, causing extensive damage.

Install the fuel pressure gauge in the fuel line between the fuel pump and the carburetor.

Start the engine and check the fuel pressure.

REMEMBER, the powerhead will **NOT** start without the emergency tether in place behind the kill switch knob, if the powerhead is so equipped.

CAUTION

Water must circulate through the lower unit to the powerhead anytime the powerhead is operating to prevent damage to the water pump in the lower unit. Just five seconds without water will damage the water pump impeller.

Operate the powerhead at full throttle and check the pressure reading. The gauge should indicate at least 2 psi.

Location of the separate fuel pump installed on Model 25hp and 30hp powerheads.

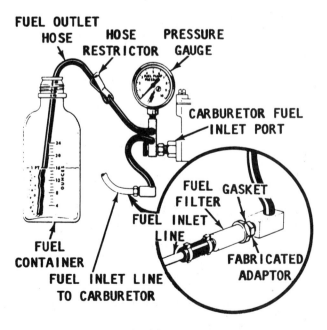

Test setup to check fuel pump pressure.

If the diaphragm in the fuel pump should rupture, an excessive amount of fuel would enter the cylinder and foul the spark plug, as shown.

4-19 SERVICING THE FUEL PUMP

Disassembly and assembling should be performed on a clean work surface. Make every effort to prevent foreign material from entering the fuel pump or adhering to the diaphragms.

REMOVAL AND DISASSEMBLING

Disconnect the fuel line fuel joint. Disconnect the inlet and outlet hose from the fuel pump. Remove the two bolts securing the pump to the crankcase.

Remove the pump and move it to a suitable clean work surface. Remove the screws securing the pump together. **TAKE CARE** not to let the spring fly out or to lose the cup.

Now, **CAREFULLY** separate the parts and keep them in **ORDER** as an assist in assembling. As a check valve is removed, **TAKE TIME** to **OBSERVE** and **REMEMBER**

Lack of adequate fuel, possibly a defective fuel pump, caused the burn condition damage to this piston.

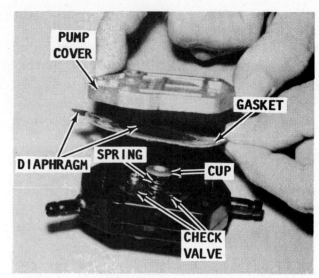

Disassembling a fuel pump, as described in the text.

how each valve faces, because it **MUST** be installed in exactly the same manner, or the pump will not function.

CLEANING AND INSPECTING

Wash all metal parts thoroughly in solvent, and then blow them dry with compressed air. **USE CARE** when using compressed air on the check valves. **DO NOT** hold the nozzle too close because the check valve can be damaged from an excessive blast of air.

Inspect each part for wear and damage. Verify that the valve seats provide a flat contact area for the valve. Tighten all check valve connections firmly as they are replaced.

Test each check valve by blowing through it with your mouth. In one direction the valve should allow air to pass through. In the other direction, air should not pass through.

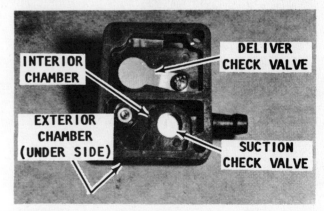

Integral fuel pump -- part of the carburetor -- with major parts and areas identified.

SERVICE FUEL PUMP 4-111

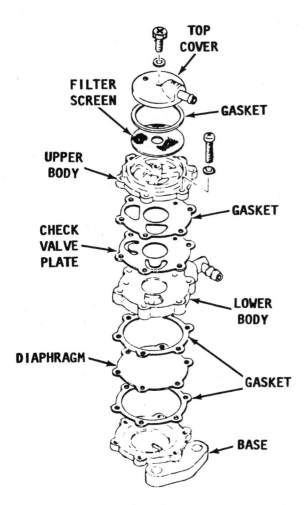

Exploded drawing of the fuel pump used on early 5hp and 8hp powerheads.

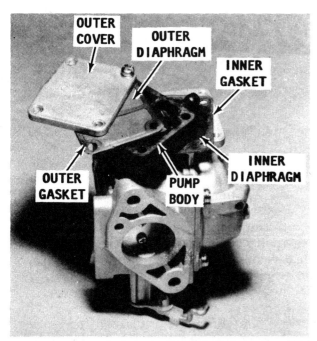

This carburetor has an integral fuel pump. The pump operates under crankcase pressure to deliver fuel to the float bowl.

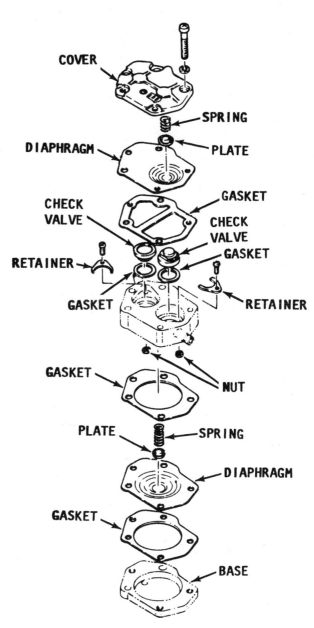

Exploded drawing of one of three possible fuel pump designs installed on late model powerheads covered in this manual. The other two are shown on the following page.

Check the diaphragms for pin holes by holding it up to the light. If pin holes are detected or if the diaphragm is not pliable, it **MUST** be replaced.

ASSEMBLING AND INSTALLATION

Proper operation of the fuel pump is essential for maximum powerhead performance. Therefore, always use **NEW** gaskets.

NEVER use any type of sealer on fuel pump gaskets.

Place the appropriate check valves on the appropriate sides of the pump body with

4-112 FUEL

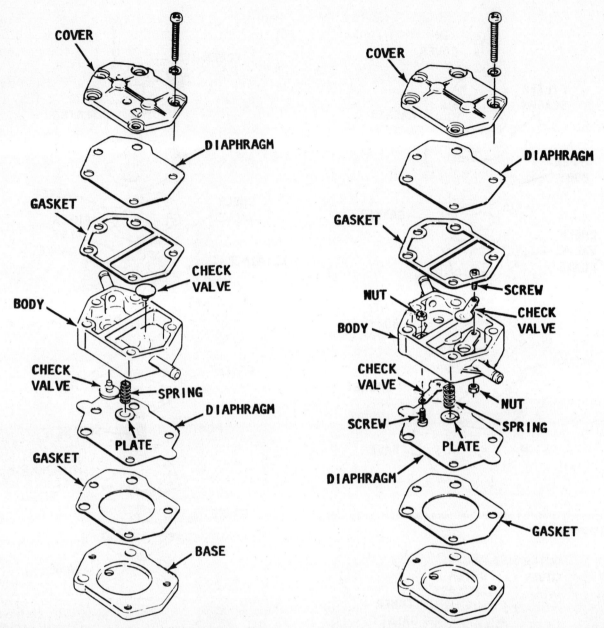

Exploded drawings of two fuel pump designs used on late model powerheads covered in this manual. An exploded drawing of a third design is shown on the previous page.

The words "IN" and "OUT" are embossed on the fuel pump cover to assist in making the proper connections.

the "fold" in the valve facing UP. TAKE CARE not to damage the very fragile and flat surface of the valve. Secure each check valve in place with a Phillips head screw. Tighten the screw securely.

Place the following parts in order on the pump body: the spring, the cup (on top of the spring), the diaphragm, the gasket, and finally the inner cover. Hold these parts together and turn the pump over. Install the diaphragm gasket and outer pump cover.

Check to be sure the holes for the screws are all aligned through the cover, diaphragms, and gaskets. If the diaphragms are not properly aligned, a tear would surely develop when the screws are installed.

Install the securing screws through the various parts and tighten the screws securely.

Position the mounting gasket onto the crankcase and the fuel pump against the gasket.

Secure the pump to the powerhead with the two bolts.

Connect the fuel line from the filter to the fitting embossed with the word **IN**. Connect the fuel line to the carburetor onto the fitting embossed with the word **OUT**.

Secure the fuel lines with the wire type clamps.

Connect the fuel supply at the fuel joint.

CAUTION

Water must circulate through the lower unit to the powerhead anytime the powerhead is operating to prevent damage to the water pump in the lower unit. Just five seconds without water will damage the water pump impeller.

Mount the outboard unit in a test tank, connect a flush attachment to the lower unit, or move the boat into a body of water.

Start the powerhead and check the completed work.

4-20 INTRODUCTION TO OIL INJECTION

Oil injection systems replace the age old method of manually mixing oil with the fuel for lubrication of internal moving parts in the powerhead.

Since outboard units have grown in number of cylinders with accompanying increases in horsepower, and because the size of the fuel tanks also grew to handle the increased demand of these larger powerheads, the requirement for a more sophisticated method of mixing oil with the fuel for internal lubrication became a primary design objective.

Almost all outboard manufacturers have now developed their own method to provide adequate oil delivery to the cylinders under all demands of the powerhead. Each system has its own trade name.

"Auto Blend"

Mercury/Mariner engineers designed and developed their oil injection system to "blend" the correct amount of oil with the fuel prior to delivery to the carburetor/s.

Therefore, the trade name used is "Auto Blend", because the blending is accomplished automatically for all powerhead demands and conditions.

Fuel from the tank moves to the "Auto Blend" unit where it mixes with the oil and is then pumped to the powerhead.

4-21 "AUTO BLEND" OIL INJECTION

DESCRIPTION

"Auto Blend" was the first oil injection system installed on the outboard units covered in this manual. Originally it was used on only the larger horsepower outboards. Over the years, the system has been upgraded including the addition of an electronic control module, movement of the warning horn, and other improvements.

"Auto Blend" is not considered an optional accessory by the manufacturer and is not recommended to be installed on units not equipped with the system from the factory.

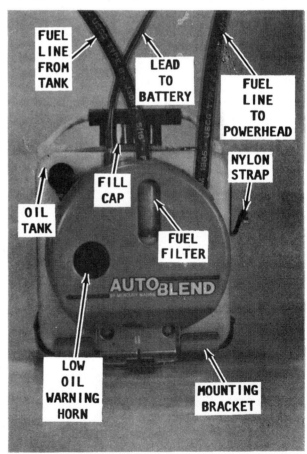

The Auto Blend unit ready to be secured in the bracket for service. Usually, short pieces of clear plastic support tubing (not shown), are placed over the end of the fuel lines to prevent kinking at the fittings.

The system consists of an oil reservoir (tank), oil screen, diaphragm vacuum-operated fuel "pump", low oil warning horn, fuel filter, and the necessary fittings and hoses for efficient operation. The reservoir and associated parts are supported in a bracket mounted on a bulkhead or the boat transom. The unit is secured in the bracket with a nylon strap and Velcro fastener. This arrangement provides quick and easy removal from the boat for refilling, testing, or for security reasons.

Oil Reservoir (Tank)

The tank is constructed of slightly transparent material and the quantity of oil can be determined by a quick glance at the tank. The tank has sight level lines in half quart (0.47L) increments. Total capacity is 3.5 quarts (3.3L). **ONLY** 2-cycle outboard oil with a BIA rating of TC-W should be used.

A screen installed in the tank filters the oil mixing with the fuel. Normally, this screen does not require service.

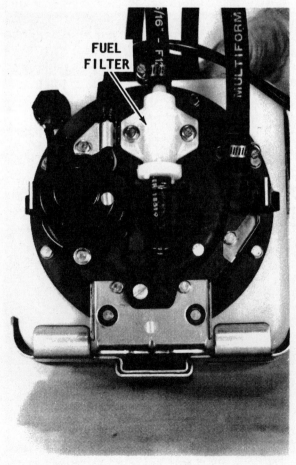

The fuel filter is transparent allowing visual inspection for foreign material. The filter may be removed and installed without the use of special tools.

Fuel "Pump"

A positive displacement diaphragm vacuum-operated fuel "pump" is mounted on the front of the oil tank. (In the strict sense of the word, it is not a pump because it is dependent on operation from another source.) The "pump" mixes oil with the fuel and is operated under vacuum supplied by the fuel pump mounted on the powerhead. The "pump" is provided with a drain plug.

Low Oil Warning Horn

A warning horn will sound to indicate one of two conditions:

a- The level of oil in the tank is dangerously low.

b- The oil screen in the tank has become clogged.

The horn circuit is connected to a 12-volt battery through leads and a harness plug. The **RED** lead is connected to the positive terminal of the battery. This lead has a 0.5 amp fuse installed as protection against damage to more expensive parts in the circuit. The black lead is connected to the negative battery terminal.

The harness plug should **NOT** be connected until the tank is filled with oil to prevent the horn from automatically sounding.

Fuel Lines

A 5/16" (7.87mm) I.D. hose is used to connect the fuel tank to the oil injection unit and also from the unit to the powerhead. The hose between the oil injection tank and the powerhead pump should never exceed 5 feet (1.50 meters). Clear plastic support tubing (not shown in the accompanying illustration) is usually used over the fuel lines and secured with standard hose clamps. These support tubes will prevent kinking at the fittings and subsequent restriction of fuel flow through the fuel lines.

The inlet fitting of the oil tank also serves as a fuel filter. With this arrangement, the fuel must pass through the filter before mixing with the oil in the tank. The filter is transparent, therefore, sediment can quickly be identified when it is present. The filter can be easily removed, cleaned, and installed, without any special tools.

A primer bulb must be installed between the oil injection unit and the fuel pump on the powerhead. **NEVER** connect the primer bulb between the fuel tank and the oil tank.

OIL INJECTION 4-115

This error could cause serious damage to the oil injection unit by providing excessive pressure.

OPERATION

While the powerhead is operating, the oil injection unit provides a variable fuel/oil mixture to the fuel pump in a ratio of 50:1 at full throttle. This is standard mixture for normal operation.

During the break-in period for a new or overhauled powerhead, oil should be added to the fuel tank in a ratio of 50:1. This ratio of oil in the fuel tank, added with the ratio of 50:1 from the oil injection unit will provide a mixture of 25:1 to the powerhead. This ratio is recommended by the manufacturer during the break-in period.

During normal operation, the level of oil in the tank will drop at a steady rate.

The unit must be positioned where the helmsperson may occasionally notice the decreasing amount in the tank. The decreasing oil level indicates the system is functioning properly and is supplying the correct proportions of fuel and oil to the powerhead.

TROUBLESHOOTING

Lack of Fuel

If the powerhead fails to operate properly and troubleshooting indicates a lack of fuel to the fuel pump, the problem may be blockage of fuel in the fuel lines, in the fuel passageway, in the oil injection unit, or a clogged fuel filter.

First, check the fuel lines to be sure they are free of stress, kinks, and nothing is laying on them, i.e.: tackle box, bait tank, etc.

Next, check the filter at the oil injection unit. Because the filter is transparent, any foreign material may be quickly discovered. The filter may be removed, cleaned, and installed quickly, without the use of any special tools.

Finally, the fuel passageway in the "Auto Blend" unit may need to be back-flushed. This is accomplished by simply kinking the line and at the same time slowly squeezing the primer bulb, as shown in the accompanying illustration. The primer bulb should only be squeezed a few times to prevent building up excessive pressure in the unit.

Warning Horn Sounds

If the horn should sound during operation of the outboard unit, shut down the unit immediately and make a couple of quick checks.

First, check the oil level in the tank and replenish as required.

Next, check the oil screen in the tank. If it is clogged with sediment, remove the filter, clean it, and install it back in the tank.

If the warning horn sounds when it should, but is intermittent, or weak, the indication is an excessive resistance or a loose connection in the circuit.

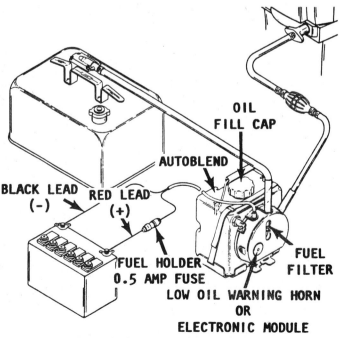

Functional diagram depicting complete hookup of the Auto Blend oil injection system.

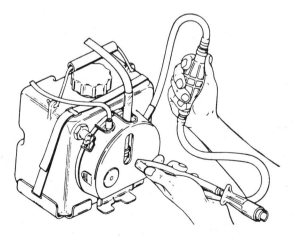

The outlet line and fitting may be back-flushed by kinking the hose and using the squeeze bulb, as explained in the text.

Warning Horn Fails To Sound
(When it should — low or no oil)

Models W/O Electronic Module

If the warning horn does not sound when the "Auto Blend" unit is inverted: First, check the 0.5 amp fuse in the in-line holder of the positive lead to the battery. Then, disconnect the two prong plug between the battery and the warning horn on the unit. Obtain and set a voltmeter to the 12V DC scale. Insert the two meter leads into the battery end of the disconnected plug. The meter should register at least 9V. If less than 9V is registered, the fault lies in the battery harness, the connector plug or the battery terminals. Correct or repair as necessary.

If the meter registers at least 9V, check for loose or dirty connections at the warning horn and repeat the test once more using two jumper leads in place of the battery harness.

Obtain and set an ohmmeter to the RX1000 scale. Insert one meter lead into the two prong connector leading from the warning horn. Make contact with the other meter lead to a suitable ground on the powerhead. The meter should register continuity. If continuity is not indicated, the horn is defective and must be replaced.

If continuity is indicated, check the battery connections and then the charge condition of the battery.

If the problem still persists, inspect the internal oil filter and finally replace the warning horn.

Models W/Electronic Module

If the warning horn does not sound a self test "beep" when the ignition switch is rotated to the "ON" position, proceed as follows: Identify the Purple lead between the main outboard harness connector and the three prong connector to the electronic module. Disconnect this lead at the quick disconnect fitting. Obtain and set a voltmeter to the 12V DC scale. With the ignition switch in the "ON" position, make contact with the Black meter lead to the negative battery terminal and the Red meter lead to the female part of the Purple lead connection.

If the meter fails to register at least 9V, the electronic module is not receiving battery voltage and therefore cannot signal the warning horn to sound. The problem lies elsewhere, probably in the battery connections. Perform corrections as required.

If the meter registers at least 9V, but still the warning horn does not sound a self test "beep", turn the ignition switch off. Connect a jumper cable between the Black and Tan or Tan/Blue leads at the female end of the three prong connector. This action bypasses the electronic module. If the horn now emits a "beep", the electric module is defective and must be replaced.

If the horn still fails to sound, obtain and set an ohmmeter to the RX1000 scale. Check for electrical continuity between the Tan or Tan/Blue lead at the female end of the three prong connector and the Tan or Tan/Blue lead from the temperature switch at the terminal block on the powerhead.

If continuity is not indicated, check for an open in the circuit or loose, dirty or corroded connections, terminals, or connector pins along the length of the Tan or Tan/Blue lead. If continuity is indicated, but still the horn fails to sound a self test "beep", then the fault lies in the ground connection. A Black lead is used to ground the terminal block to the powerhead. This Black lead extends between the grounding screw on the powerhead and the three prong connector. Somewhere along the length of this Black lead there is an open in the circuit or a loose, dirty or corroded connection, terminal, or connector pin.

Warning Horn Sounds Continuously
Powerhead Is Cold Or Not Operating
Models W/Electronic Module

With the ignition key in the "ON" position, disconnect the three prong connector. If the horn is silent, the electonic module is defective and must be replaced. If the horn continues sounding, disconnect all three Tan or Tan/Blue leads at the terminal block on the powerhead.

If the horn continues to sound, the problem is not in the oil injection system but elsewhere, possibly in the remote control box.

The horn should now be silent. Connect the Tan or Tan/Blue leads one by one to determine which lead was grounded and activated the horn. Trace and replace the defective lead.

If the defective lead is the one to the temperature switch, the switch and lead are replaced as a unit.

OIL INJECTION 4-117

Warning Horn Sounds Continuously During Powerhead Operation Oil Level Is Satisfactory Models W/Electronic Module

GOOD WORDS

Perform this test with the powerhead **NOT** operating, even though the symptoms appear when the powerhead is operating.

Rotate the ignition key to the **"ON"** position. Disconnect the two prong connector at the "Auto Blend" unit. If the warning horn continues to sound, the electronic module is defective and **MUST** be replaced.

If the horn is now silent, the fault lies in the internal oil filter. Replace the filter.

STORAGE
"AUTO BLEND" SYSTEM

Proper storage procedures are **CRITICAL** to ensure efficient operation when the unit is again placed in service.

First, disconnect the battery leads from the battery.

Next, disconnect and plug the fuel lines at the fuel tank and powerhead.

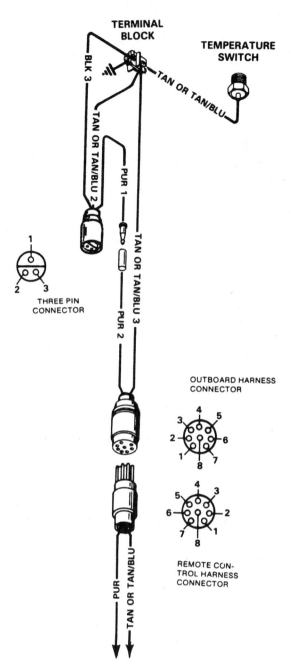

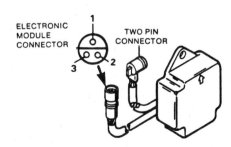

Auto Blend harness identification for resistance tests outlined in the text.

The electronic module and harnesses for the later version of Auto Blend.

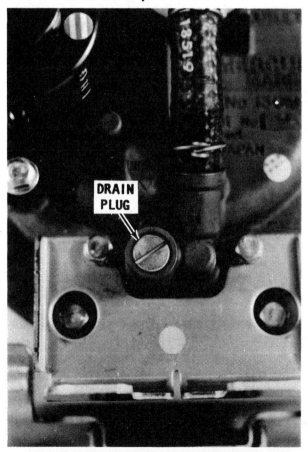

All fuel MUST be drained prior to placing the unit in storage, to prevent damage to the "pump" diaphragm.

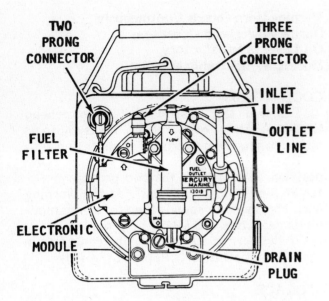

Line drawing to clearly identify major parts of an Auto Blend unit.

Now, drain all fuel from the unit. Remove the front cover of the unit by simultaneously pushing in on the cutaway tabs located on both sides of the cover, and at the same time pulling the cover away from the unit. Remove the drain plug and allow at least 5-minutes for all fuel to drain from the pump. Install the drain plug and tighten it securely.

CRITICAL WORDS

All fuel **MUST** be drained from the oil injection fuel "pump". The percentage of alcohol in modern fuels seems to increase each year. This alcohol in the fuel is a definite enemy of the diaphragm in the "pump". Therefore, if any fuel is left in the "pump" during storage the diaphragm will most likely be damaged.

Install the front cover by aligning the cover openings on both sides of the unit, and then pushing in on the cover until it snaps into place.

Oil may remain in the oil injection tank during storage without any harmful effects.

PREPARATION FOR USE "AUTO BLEND" SYSTEM

First, remove the front cover of the unit by simultaneously pushing in on the cutaway tabs located on both sides of the cover, and at the same time pulling the cover away from the unit. Check to be sure the fuel drain plug is tight. Replace the front cover by aligning the cover openings on both sides of the unit, and then pushing in on the cover until it snaps into place.

Next, fill the oil tank with 2-cycle outboard oil with a BIA rating of TC-W. Tighten the fill cap securely.

Remove any plugs in the fuel lines, and then connect the hoses to the fuel tank and the powerhead. Remember, the squeeze bulb **MUST** be in the hose between the oil injection unit and the fuel pump on the powerhead.

Connect the low oil warning wire harness to the battery. Connect the **RED** lead to the positive battery terminal and the **BLACK** lead to the negative battery terminal.

Check to be sure the low oil warning system is functioning correctly. First, verify the tank is full of oil, and then the fill cap is tightened securely. Now, turn the oil injection unit upside down. This position will allow the float to activate the horn.

If the horn sounds, immediately turn the unit rightside up and position it in the mounting bracket. Secure it in place with the strap and Velcro material.

If the horn does not sound, check the 0.5 amp fuse in the fuse holder of the positive battery lead. Check both the battery connections and the charge condition of the battery.

GOOD WORDS

The manufacturer recommends the fuel filter be replaced at the start of each season or at least once a year. The manufacturer also recommends oil be added to the fuel tank at the ratio of 50:1 for the first 6-gallons of fuel used after the unit is brought out of storage. The oil in the fuel tank plus the 50:1 oil mixture in the oil injection unit will deliver a mixture of 25:1 to the powerhead. This ratio will **ENSURE** adequate lubrication of moving parts which have been drained of oil during the storage period.

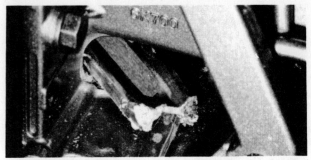

Inadequate lubrication -- rod through the block.

5
IGNITION

5-1 INTRODUCTION AND CHAPTER COVERAGE

The less an outboard engine is operated, the more care it needs. Allowing an outboard engine to remain idle will do more harm than if it is used regularly. To maintain the engine in top shape and always ready for efficient operation at any time, the engine should be operating every 3 to 4 weeks throughout the year.

The carburetion and ignition principles of two-cycle engine operation **MUST** be understood in order to perform a proper tuneup on an outboard motor.

If you have any doubts concerning your understanding of two-cycle engine operation, it would be best to study the operation theory section in the first portion of Chapter 7, before tackling any work on the ignition system.

Two different type ignition systems are used on the outboard units covered in this manual. The first sections of this chapter will be devoted to an explanation of each ignition system and its theory of operation. The latter sections will provide troubleshooting and repair instructions for the systems. For timing and synchronizing procedures, see Chapter 6.

For convenience, each ignition system has been identified with a code numeral either I or II. These code numerals will be used throughout this chapter.

IGNITION SYSTEMS

The Type I system is a Flywheel Magneto with Breaker Points under the flywheel and is used on all Model 2hp water-cooled, and the 3.5hp, and 5hp air-cooled to 1979. This

Outboard unit mounted on a stand ready for service work on the ignition system. A thick piece of wood clamped in a vise serves the same purpose as the stand.

*The carburetion and ignition system **MUST** be properly adjusted and synchronized for optimum powerhead performance.*

system was also used on the 8hp, 15hp, 20hp, and 28hp Models, prior to 1979.

The Type I system is described, serviced, and replaced in Section 5-4.

The Type II is a Capacitor Discharge Ignition system (CDI), and is used on all other models covered in this manual. This system is covered in Section 5-5.

5-2 SPARK PLUG EVALUATION

Removal

Remove the spark plug wires by pulling and twisting on only the molded cap. NEVER pull on the wire or the connection inside the cap may become separated or the boot damaged. Remove the spark plugs and keep them in order. TAKE CARE not to tilt the socket as you remove the plug or the insulator may be cracked.

Examine

Line the plugs in order of removal and carefully examine them to determine the firing conditions in each cylinder. If the side electrode is bent down onto the center electrode, the piston is traveling too far upward in the cylinder and striking the spark plug. Such damage indicates the piston pin or the rod bearing is worn excessively. In most cases, an engine overhaul is required to correct the condition. To verify the cause of the problem, turn the engine over by hand. As the piston moves to the full up position, push on the piston crown with a screwdriver inserted through the spark plug hole, and at the same time rock the flywheel back-and-forth. If any play in the piston is detected, the engine must be rebuilt.

Correct Color

A proper firing plug should be dry and powdery. Hard deposits inside the shell indicate too much oil is being mixed with the fuel. The most important evidence is the light gray to tan color of the porcelain, which is an indication this plug has been running at the correct temperature. This means the plug is one with the correct heat range and also that the air-fuel mixture is correct.

Rich Mixture

A black, sooty condition on both the spark plug shell and the porcelain is caused by an excessively rich air-fuel mixture, both at low and high speeds. The rich mixture lowers the combustion temperature so the spark plug does not run hot enough to burn off the deposits.

Deposits formed only on the shell is an indication the low-speed air-fuel mixture is too rich. At high speeds with the correct mixture, the temperature in the combustion chamber is high enough to burn off the deposits on the insulator.

Too Cool

A dark insulator, with very few deposits, indicates the plug is running too cool. This condition can be caused by low compression or by using a spark plug of an incorrect heat range. If this condition shows on only one plug it is most usually caused by low compression in that cylinder. If all of the plugs have this appearance, then it is probably due to the plugs having a too-low heat range.

This spark plug is foul from operating with an over-rich condition, possibly an improper carburetor adjustment.

This spark plug has been operating too-cool, because it is rated with a too-low heat range for the engine.

SPARK PLUG EVALUATION

Fouled

A fouled spark plug may be caused by the wet oily deposits on the insulator shorting the high-tension current to ground inside the shell. The condition may also be caused by ignition problems which prevent a high-tension pulse from being delivered to the spark plug.

Carbon Deposits

Heavy carbon-like deposits are an indication of excessive oil in the fuel. This condition may be the result of worn piston rings or excessive ring end gap.

Overheating

A dead white or gray insulator, which is generally blistered, is an indication of overheating and pre-ignition. The electrode gap wear rate will be more than normal and in the case of pre-ignition, will actually cause the electrodes to melt. Overheating and pre-ignition are usually caused by overadvanced timing, detonation from using too-low an octane rating fuel, an excessively lean air-fuel mixture, or problems in the cooling system.

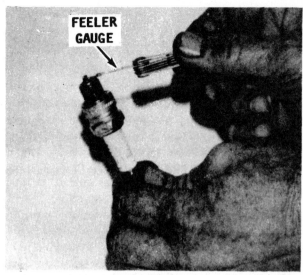

The spark plug gap should always be checked with a wire-type feeler gauge before installing new or used plugs.

Electrode Wear

Electrode wear results in a wide gap and if the electrode becomes carbonized it will form a high-resistance path for the spark to jump across. Such a condition will cause the engine to misfire during acceleration. If all of the plugs are in this condition, it can cause an increase in fuel consumption and very poor performance at high-speed operation. The solution is to replace the spark plugs with a rating in the proper heat range and gapped to specification.

Red rust-colored deposits on the entire firing end of a spark plug can be caused by water in the cylinder combustion chamber. This can be the first evidence of water entering the cylinders through the exhaust manifold because of an accumulation of scale. This condition **MUST** be corrected at

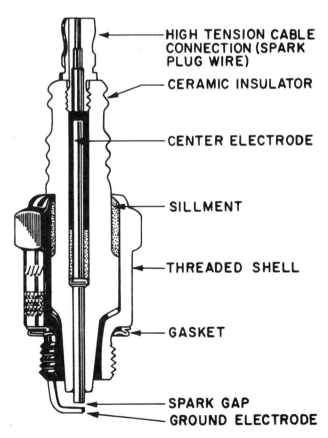

Cutaway drawing of a typical spark plug with principle parts identified.

*Damaged spark plugs. Notice the broken electrode on the left plug. The missing part **MUST** be found and removed before returning the powerhead to service. The missing part could cause serious and expensive damage to internal parts.*

the first opportunity. Refer to Chapter 7, Powerhead.

5-3 POLARITY CHECK

Coil polarity is extremely important for proper battery ignition system operation. If a coil is connected with reverse polarity, the spark plugs may demand from 30 to 40 percent more voltage to fire, or on most CDI systems, there will be **NO** spark. Under such demand conditions, in a very short time the coil would be unable to supply enough voltage to fire the plugs. Any one of the following three methods may be used to quickly determine coil polarity.

1- The polarity of the coil can be checked using an ordinary D.C. voltmeter set on the maximum scale. Connect the positive lead to a good ground. With the engine running, momentarily touch the negative lead to a spark plug terminal. The needle should swing upscale. If the needle swings downscale, the polarity is reversed.

2- If a voltmeter is not available, a pencil may be used in the following manner: Disconnect a spark plug wire and hold the metal connector at the end of the cable about 1/4" (6.35mm) from the spark plug terminal. Now, insert an ordinary pencil tip between the terminal and the connector. Crank the engine with the ignition switch ON. If the spark feathers on the plug side and has a slight orange tinge, the polarity is correct. If the spark feathers on the cable connector side, the polarity is reversed.

3- The firing end of a used spark plug can give a clue to coil polarity. If the ground electrode is "dished", it may mean polarity is reversed.

5-4 TYPE I IGNITION SYSTEM MAGNETO WITH CONTACT BREAKER POINTS

Description

The Type I system is a Flywheel Magneto with Breaker Points under the flywheel and is used on all Model 2hp, and the 3.5hp, and 5hp (air-cooled) powerheads. This system was also used on the 8hp, 15hp, 20hp, and 28hp Models, prior to 1979.

The Flywheel Magneto unit consists of a stator plate, and permanent magnets built into the flywheel. The condenser and breaker points are mounted on the stator plate. Single cylinder models have one set of breaker points and one condenser. Two cylinder models have two sets of breaker points and either one or two condensers.

On smaller single cylinder models, the ignition coil is also mounted under the flywheel. On larger two cylinder models, an ignition source coil is mounted under the flywheel and two ignition coils are mounted on the powerhead.

The ignition source coil functions in much the same manner as an ignition coil. Under the influence of the rotating magnetic field, voltage is induced into the windings of the source. As the field collapses, voltage is transfered to the ignition coils. The ignition coil steps up the voltage to the high tension required at the spark plug.

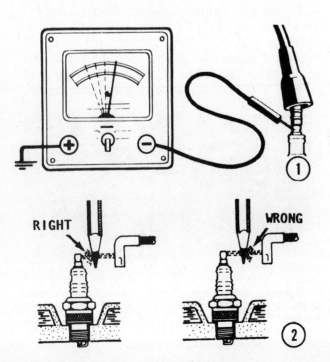

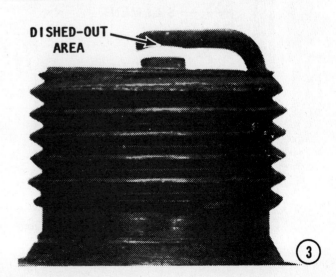

TYPE I SYSTEM   5-5

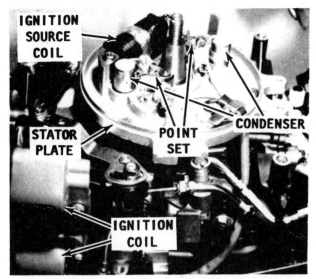

Flywheel removed from a 2-cylinder powerhead in preparation to servicing the ignition system.

As the pole pieces of the magnet pass over the heels of the coil mounted on the stator plate, a magnetic field is built up about the coil, causing a current to flow through the primary winding.

Now, at the proper time, the breaker points are separated by action of a cam designed into the collar of the flywheel and the primary circuit is broken. When the circuit is broken, the flow of primary current stops and causes the magnetic field

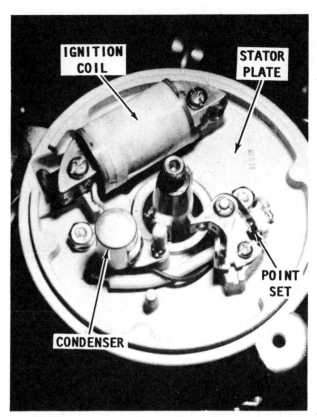

Flywheel removed from a single cylinder powerhead in preparation to servicing the ignition system.

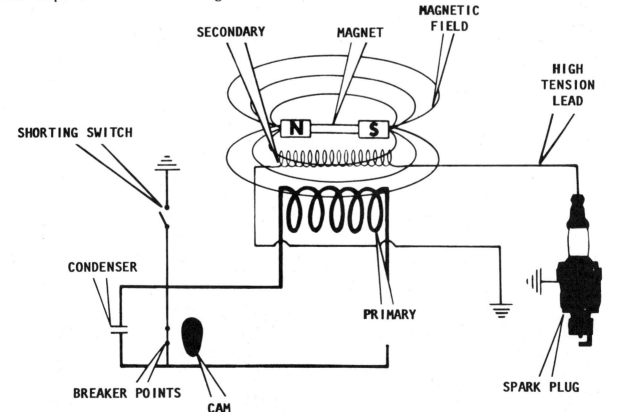

Schematic diagram of a magneto ignition system.

5-6 IGNITION

about the coil to break down instantly.

At this precise moment, an electrical current of extremely high voltage is induced in the fine secondary windings of the coil.

On models equipped with an ignition source coil and an ignition coil, the secondary voltage is passed from the source coil to the ignition coil. The ignition coil steps up the voltage and the high voltage is conducted to the spark plug where it jumps the gap between the points of the plug to ignite the compressed charge of air/fuel mixture in the cylinder.

An Interesting Fact

The time interval between the instant the points open to the "firing" of the spark plug is twenty five millionths of a second!

TROUBLESHOOTING
TYPE I IGNITION SYSTEM

Always attempt to proceed with the troubleshooting in an orderly manner. The "shot in the dark" approach will only result in wasted time, incorrect diagnosis, replacement of unnecessary parts, and frustration.

Begin the ignition system troubleshooting with the spark plug and continue through the system until the source of trouble is located.

Spark Plugs

1- Remove the cowling. On the 2hp Model: Remove the screw on one half of the spark plug cover. Remove four more screws securing one-half of the cowling. Separate the cowling half from which the screws were removed. Remove the four screws securing the other half of the cowling and remove it from the engine. The spark plug cover will remain attached to one of the cowling halves.

SPECIAL WORDS

Observe the different length screws used to secure the cowling halves to the powerhead. Remember their location as an aid during assembling.

2- Check the plug wire to be sure it is properly connected. Check the entire length of the wire from the plug to the magneto under the stator plate. If the wire is to be removed from the spark plug, **ALWAYS** use a pulling and twisting motion as a precaution against damaging the connection.

Attempt to remove the spark plug by hand. This is a rough test to determine if the plug is tightened properly. The attempt to loosen the plug by hand should fail. The plug should be tight and require the proper socket size tool. Remove the spark plug and evaluate its condition as described in Section 5-2.

3- Use a spark tester and check for spark. If a spark tester is not available, hold the plug wire about 1/4" (6.4mm) from the engine. Rotate the flywheel with the hand pull starter and check for spark. A strong spark over a wide gap must be observed when testing in this manner, because under compression a strong spark is necessary in order to ignite the air-fuel mixture

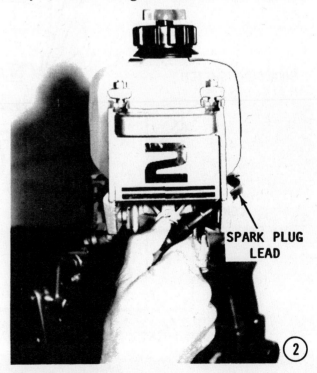

in the cylinder. This means it is possible to think a strong spark is present, when in reality the spark will be too weak when the plug is installed. If there is no spark, or if the spark is weak, the trouble is most likely under the flywheel in the magneto.

Compression

Before spending too much time and money attempting to trace a problem to the ignition system, a compression check of the cylinder should be made. If the cylinder does not have adequate compression, troubleshooting and attempted service of the ignition or fuel system will fail to give the desired results of satisfactory engine performance.

Remove the spark plug wire by pulling and twisting **ONLY** on the molded cap. **NEVER** pull on the wire because the connection inside the cap may be separated or the boot may be damaged. Remove the spark plug. Insert a compression gauge into the cylinder spark plug hole. Ground the spark plug lead to prevent damage to the ignition coil. If the lead is not grounded, the coil will attempt to match the demand created by the spark trying to jump from the electrode to the nearest ground. On a 2-cylinder model, ground both spark plug leads.

Crank the engine through several revolutions by pulling rapidly on the hand starter pull rope. Note the compression reading.

An acceptable pressure reading is 110 psi (760 kPa) or more.

Condenser

In simple terms, a condenser is composed of two sheets of tin or aluminum foil laid one on top of the other, but separated by a sheet of insulating material such as waxed paper, etc. The sheets are rolled into a cylinder to conserve space and then inserted into a metal case for protection and to permit easy assembling.

The purpose of the condenser is to prevent excessive arcing across the points and to extend their useful life. When the flow of primary current is brought to a sudden stop by the opening of the points, the magnetic field in the primary windings collapses instantly, and is not allowed to "fade away", which would happen if the points were allowed to arc.

The condenser stores the electricity that would have arced across the points and discharges that electricity when the points close again. This discharge is in the opposite direction to the original flow, and tends to "smooth out" the current. The more quickly the primary field collapses, the higher the voltage produced in the secondary windings and delivered to the spark plugs. In this way, the condenser (in the primary circuit), affects the voltage (in the secondary circuit) at the spark plugs.

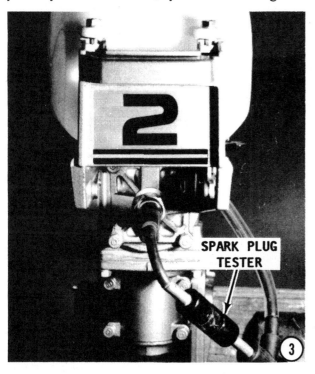

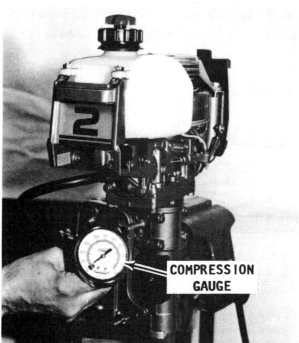

Taking a compression check should be done before major work or tune-up tasks are performed. Without adequate compression, other efforts may be wasted.

5-8 IGNITION

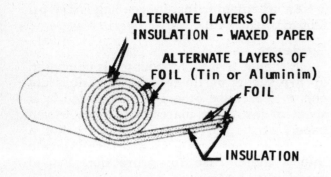

Rough sketch to illustrate how the waxed paper, aluminum foil, and insulation are rolled in the manufacture of a typical condenser.

Modern condensers seldom cause problems, therefore, it is not necessary to install a new one each time the points are replaced. However, if the points show evidence of arcing, the condenser may be at fault and should be replaced. A faulty condenser may not be detected without the use of special test equipment. Testing will reveal any defects in the condenser, but will **NOT** predict the useful life left in the unit.

The modest cost of a new condenser justifies its purchase and installation to eliminate this item as a source of trouble.

Breaker Points

The breaker points in an outboard motor are an extremely important part of the ignition system. A set of points may appear to be in good condition, but they may be the source of hard starting, misfiring, or poor

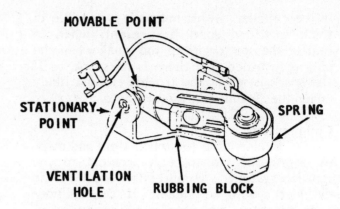

Line drawing of a typical point set with principle parts identified.

engine performance. The rules and knowledge gained from association with 4-cycle engines does not necessarily apply to a 2-cycle engine. The points should be replaced every 100 hours of operation or at least once a year. **REMEMBER**, the less an outboard engine is operated, the more care it needs. Allowing an outboard engine to remain idle will do more harm than if it is used regularly.

A breaker point set consists of two points. One is attached to a stationary bracket and does not move. The other point is attached to a movable mount. A spring is used to keep the points in contact with each other, except when they are separated by the action of a cam built into the flywheel

A normal set of breaker points used in a magneto will show evidence of a shallow crater and build-up after a few hours of operation. The left set of points is considered normal and need not be replaced. The set on the right has been in service for more than 450 hours and should be replaced.

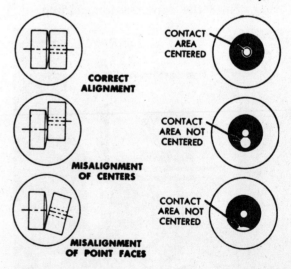

Before setting the breaker point gap, the points must be properly aligned (top). ALWAYS bend the stationary point, NEVER the breaker lever. Attempting to adjust an old worn set of points is not practical, when compared with the modest cost of a new set, thus eliminating this area as a possible cause of trouble. If a worn set of points is to be retained for emergency use, both contact surfaces of the set should be refaced with a point file.

collar which fits over the crankshaft. Both points are constructed with a steel base and a tungsten cap fused to the base.

To properly diagnose magneto (spark) problems, the theory of electricity flow must be understood. The flow of electricity through a wire may be compared with the flow of water through a pipe. Consider the voltage in the wire as the water pressure in the pipe and the amperes as the volume of water. Now, if the water pipe is broken, the water does not reach the end of the pipe. In a similar manner if the wire is broken the flow of electricity is broken. If the pipe springs a leak, the amount of water reaching the end of the pipe is reduced. Same with the wire. If the installation is defective or the wire becomes grounded, the amount of electricity (amperes) reaching the end of the wire is reduced.

SERVICING
TYPE I IGNITION SYSTEM

General Information

Magnetos installed on outboard engines will usually operate over extremely long periods of time without requiring adjustment or repair. However, if ignition system problems are encountered, and the usual corrective actions such as replacement of spark plugs does not correct the problem, the magneto output should be checked to determine if the unit is functioning properly.

Unfortunately, the breaker point set/s of the Type I ignition system are located under the flywheel. This location requires the hand rewind starter to be removed, and the flywheel to be "pulled" in order to replace the point set.

HOWEVER, the manufacturer made provisions for the point gap to be checked with a feeler gauge through one of two slots cut into the top of the flywheel for this purpose. Therefore, only the hand rewind starter needs to be removed to perform the task of point adjustment.

DISASSEMBLING
TYPE I IGNITION SYSTEM

The following instructions pickup the work after the cowling has been removed

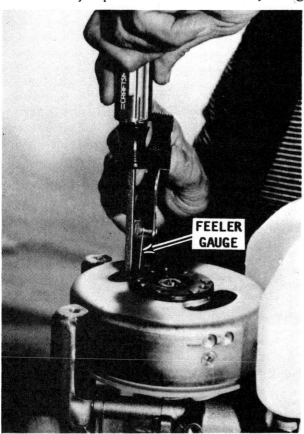

The flywheel on a single cylinder powerhead has a "window" to permit point gap adjustment using a screwdriver and feeler gauge, as shown.

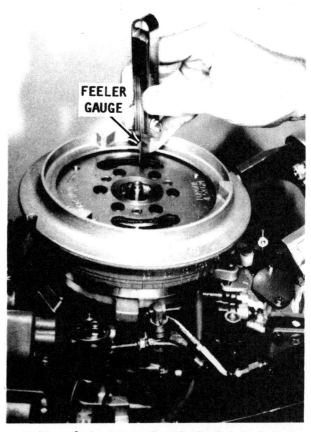

Using a feeler gauge through the "window" in the flywheel to check the point gap on a 2-cylinder powerhead.

5-10 IGNITION

and troubleshooting procedures indicate the breaker points require either adjustment or replacement.

1- Remove the three bolts securing the hand rewind starter to the powerhead. Lift the hand starter free.

2- Remove the three countersunk Phillips head screws. Use **CARE** not to strip the heads. One method is to apply a twisting **(COUNTERCLOCKWISE)** force with one hand on the screwdriver and at the same time lightly shock the screw by tapping the end of the screwdriver with a hammer, as shown. If necessary, have an assistant hold

the flywheel secure with the end of a large screwdriver, or similar tool, wedged in the starter pulley.

3- Lift off the starter puller, and then the flywheel cover.

SPECIAL WORDS ON BREAKER POINT SETS

When a point set is new, the fixed contact point surface is a hemispherical curve. The movable contact point surface is flat to aid in adjusting the point gap. The manufacturer states it is permissible to file the curved surface flat if the points are pitted. This can be accomplished with a small flat file through the opening in the flywheel.

4- Rotate the flywheel and observe the breaker point set through the openings in the flywheel. Stop the rotation when the cam follower is on the high point of the cam, allowing for maximum point gap opening. Loosen, but **DO NOT** remove the breaker point adjusting screw.

5- Insert a slotted head screwdriver into the point adjusting slot and pivot the point plate and at the same time measure the point gap with a feeler gauge. Adjust the gap as follows:

2hp	0.012-0.016"	(0.30-0.40mm)
3.5hp	0.012-0.016"	(0.30-0.40mm)
5hp	0.012-0.016"	(0.30-0.40mm)
8hp	0.010-0.018"	(0.25-0.45mm)
15hp	0.012-0.016"	(0.30-0.40mm)
20hp	0.010-0.018"	(0.25-0.45mm)
28hp	0.010-0.018"	(0.25-0.45mm)

Tighten the adjusting screw to maintain the adjustment before proceeding.

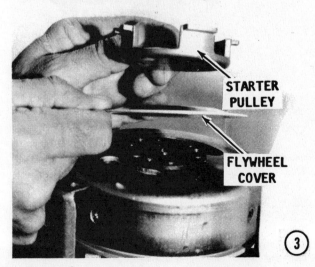

If servicing a 2-cylinder powerhead, continue to rotate the flywheel until the second set of breaker points have opened and repeat the measurement and necessary adjustments. The second breaker point set is adjusted to the same specifications as the first.

If a satisfactory point gap measurement can be obtained, no further work is required, proceed directly to Step 5 under Installation.

BAD NEWS

If the correct gap cannot be obtained, the flywheel will have to be "pulled" and a new point set installed, as outlined in the following steps.

6- Obtain special flywheel holder tool, P/N M-91-83163M. Insert the two indexing pins on the ends of the arms through the holes in the flywheel. Hold the flywheel steady with the special tool and remove the flywheel nut.

7- Obtain flywheel puller tool, P/N M-91-83164M. If this particular puller is not available a similar puller may be used **PROVIDED** it will pull from the bolt holes in the flywheel and **NOT** from around the perimeter of the flywheel.

NEVER attempt to use a puller which pulls on the outside edge of the flywheel.

Install the puller onto the flywheel, and then using puller tool take a strain on the

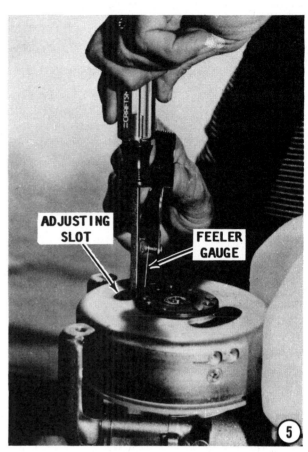

puller with the proper size wrench. Now, continue to tighten on the special tool and at the same time, **SHOCK** the crankshaft with a gentle to moderate tap with a hammer on the end of the special tool. This shock will assist in "breaking" the flywheel loose from the crankshaft.

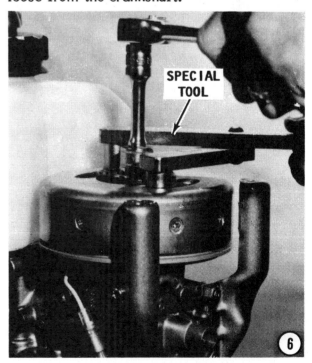

5-12 IGNITION

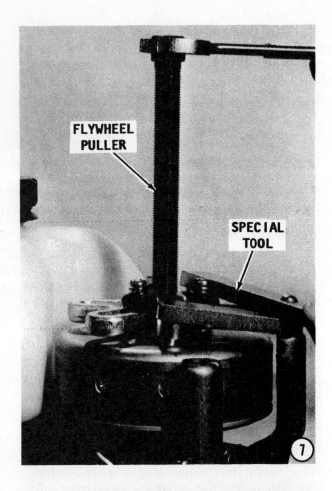

Lift the flywheel free of the crankshaft. Remove and **SAVE** the Woodruff key from the recess in the crankshaft.

IGNITION COIL TESTS

Primary Winding Check

8- Remove the bolts securing the stator plate to the powerhead and lift the plate free for bench testing. If servicing a 2-cylinder powerhead with the ignition coils mounted on the powerhead, disconnect the Black and the Orange or Grey leads and the high tension leads. Remove the coils from the powerhead for bench testing.

Obtain an ohmmeter and set it on the Rx1 scale. If servicing a single cylinder powerhead, make contact with the Red meter lead to the small coil lead between the coil and the breaker points. If servicing a 2-cylinder powerhead, make contact with the Red meter lead to the Gray or Orange coil lead.

Make contact with the Black meter lead to the ignition coil core. An acceptable resistance reading for all single cylinder models is 0.95 - 1.27 ohms. Resistance for all 2-cylinder models with an ignition source

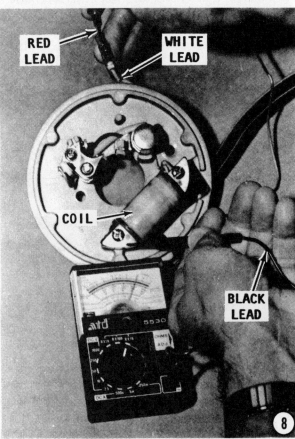

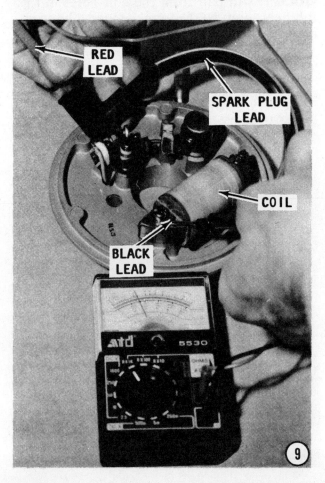

SERVICE TYPE I 5-13

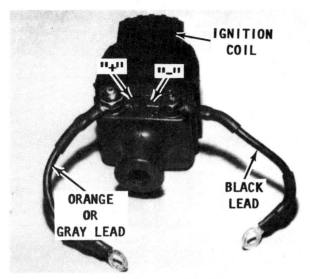

A typical powerhead mounted ignition coil removed and ready for bench testing, as described in the text.

coil mounted on the stator plate is 1.5-1.9 ohms. If the unit being serviced has powerhead mounted ignition coils, the resistance should be 1.35-1.65 ohms. If the reading is not within the acceptable range, the primary windings are defective. The coil cannot be serviced, it **MUST** be replaced.

GOOD WORDS

If servicing a two cylinder powerhead, the ignition source coil mounted on the stator plate does not have secondary windings. Therefore, no resistance test is included in the following step.

Secondary Winding Check

9- Set the meter to the Rx1000 scale. Insert the Red tester lead into the spark plug lead. Make contact with the black meter lead to the ignition coil core. The resistance reading should be 5.44-6.65K ohms for all single and 2-cylinder models. If the reading is not within the acceptable range, the ignition coil **MUST** be replaced.

Point Set Removal

10- Remove the nut securing the condensor lead and the coil lead to the point set. Pry off the circlip securing the point set to the pivot and remove the adjusting screws. Lift the point set free of the stator plate. Repeat these tasks for the other point set, if so equipped.

SPECIAL WORDS

It is considered good shop practice to replace the condenser whenever a new point set is installed.

CLEANING AND INSPECTING

Inspect the flywheel for cracks or other damage, especially around the inside of the center hub. Check to be sure metal parts have not become attached to the magnets. Verify each magnet has good magnetism by using a screwdriver or other tool.

Thoroughly clean the inside taper of the flywheel and the taper on the crankshaft to prevent the flywheel from "walking" on the crankshaft during operation.

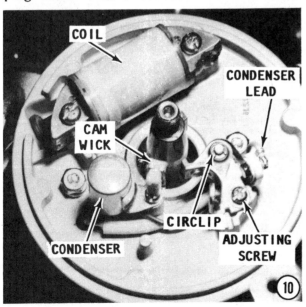

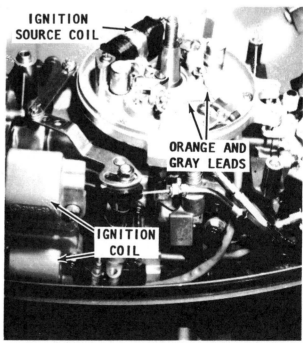

All 2-cylinder powerheads covered in this manual have an ignition source coil mounted on the stator plate in addition to the two powerhead mounted ignition coils.

5-14 IGNITION

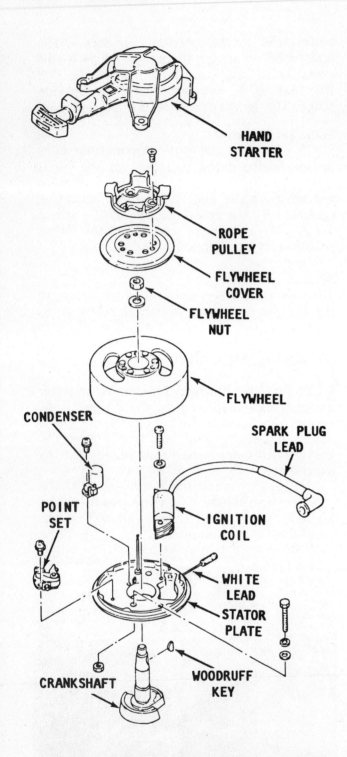

Exploded drawing of a Type I ignition system, with major components identified.

Check the top seal around the crankshaft to be sure no oil has been leaking onto the stator plate. If there is **ANY** evidence the seal has been leaking, it **MUST** be replaced. See Chapter 7.

Test the stator assembly to verify it is not loose. Attempt to lift each side of the plate. There should be little or no evidence of movement.

Lightly lubricate the cam wick with all purpose lubricant. Excess lubricant will shorten the breaker point life. Inadequate lubrication will quickly wear the rubbing block from the point set and thus alter the timing. Therefore, just use a dab of lubricant.

ASSEMBLING AND INSTALLATION TYPE I IGNITION SYSTEM

1- Install new condenser/s and secure them with to the stator plate with a screw through the mounting bracket.

Install a **NEW** point set. Install the adjusting screw but **DO NOT** tighten it at this time.

Align the slot in the set with the slot in the stator plate as a preliminary adjustment. The point gap will be set in Step 4, when the rubbing block contacts the flywheel cam.

Attach the coil and condenser leads to the point set. Install the circlip on the pivot post. If the model being serviced has two breaker point sets, install the second set in the same manner.

2- Place a tiny dab of thick lubricant on the curved surface of the Woodruff key to hold it in place while the flywheel is being installed. Press the Woodruff key into place in the crankshaft recess. Wipe away any excess lubricant to prevent the flywheel from "walking" during powerhead operation.

Check the flywheel magnets to ensure they are free of any metal particles. Double check the taper in the flywheel hub and

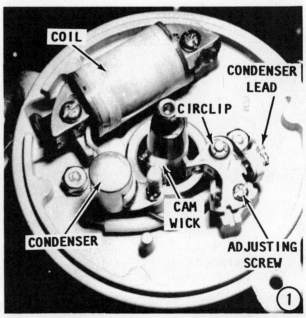

3- Slide the washer onto the crankshaft, and then thread the flywheel nut onto the crankshaft. Obtain special flywheel tool P/N M-91-83163M. With the pins on the ends of the holder arms indexed into the flywheel holes, tighten the flywheel nut to the following torque value.

2hp	23 ft lbs	(31Nm)
3.5hp	23 ft lbs	(31Nm)
5hp	23 ft lbs	(31Nm)
8hp	23 ft lbs	(31Nm)
15hp	54 ft lbs	(73Nm)
20hp	54 ft lbs	(73Nm)
28hp	54 ft lbs	(73Nm)

Breaker Point Adjustment

4- Rotate the flywheel and observe the breaker point set through the openings in the flywheel. Stop the rotation when the cam follower is on the high point of the cam, allowing maximum point gap opening. Insert a slotted head screwdriver into the point adjusting slot, pivot the point plate and at the same time measure the point gap with a feeler gauge. Adjust the gap as follows:

2hp	0.012-0.016"	(0.30-0.40mm)
3.5hp	0.012-0.016"	(0.30-0.40mm)
5hp	0.012-0.016"	(0.30-0.40mm)
8hp	0.010-0.018"	(0.25-0.45mm)
15hp	0.012-0.016"	(0.30-0.40mm)
20hp	0.010-0.018"	(0.25-0.45mm)
28hp	0.010-0.018"	(0.25-0.45mm)

the taper on the crankshaft to verify they are clean and contain no oil.

Now, slide the flywheel down over the crankshaft with the keyway in the flywheel aligned with the Woodruff key in place on the crankshaft. Rotate the flywheel **COUNTERCLOCKWISE** to be sure it does not contact any part of the stator plate or wiring.

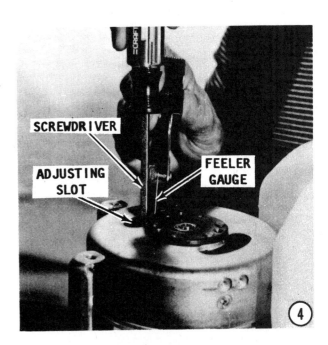

5-16 IGNITION

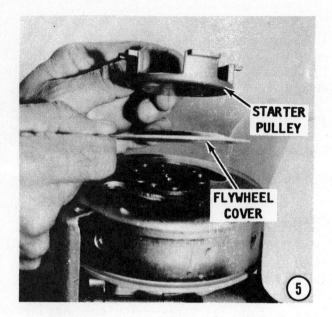

Tighten the adjusting screw to maintain the adjustment.

If servicing a 2-cylinder powerhead, continue to rotate the flywheel until the second set of breaker points has opened and repeat the measurement and necessary adjustments. The second breaker point set is adjusted to the same specifications as the first.

Make a final measurement after tightening the screw to ensure the adjustment was not disturbed.

5- Position the flywheel cover in place over the flywheel with the three holes in the cover aligned with the three holes in the flywheel. Position the hand rewind starter pulley in place on top of the cover. Install and tighten the three countersunk Phillips head screws. A torque value is not listed for these screws, but use some muscle and tighten them securely.

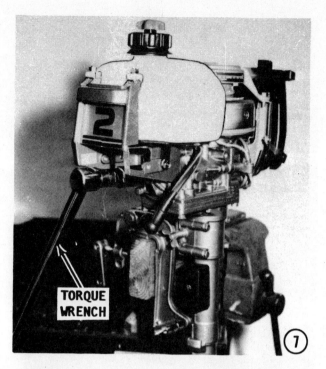

6- Install the hand rewind starter to the powerhead. Tighten the three attaching bolts to a torque value of 5.8 ft lbs (8Nm).

7- Install and tighten the spark plug to a torque value of 14 ft lbs (20Nm). Secure the spark plug lead to the spark plug.

8- Install the cowling. If servicing a 2hp model: Install the two halves of the cowling around the powerhead. Secure the cowling with the attaching screws. Eight screws hold the cowling halves in place plus one more for the spark plug cover.

As noted during disassembling, the screws are different lengths. Ensure the proper sizes are used in the correct location.

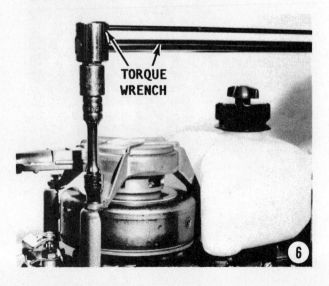

5-5 TYPE II IGNITION SYSTEM CDI (CAPACITOR DISCHARGE IGNITION)

The Type II CDI, capacitor discharge ignition system is used on all 2-cylinder and some single-cylinder powerheads covered in this manual since 1979. The system utilizes either an automatic advance, as on the 4hp, and 5hp models or a mechanical advance, as used on the 8hp, 9.9hp 15hp, 20hp, 25hp, and 30hp models.

DESCRIPTION AND OPERATION

Three major components are used in the Type II CDI system.

1- Charge coils, one for each cylinder, are housed under the flywheel. These coils provide a source of AC current and time the electrical pulses.

2- The CDI unit used to convert AC to DC current. The unit stores the DC current, then releases it and automatically controls the timing.

3- Ignition coil, boosts the DC voltage instantly to approximately 20,000 volts to the spark plugs.

Two basic circuits are used with the CDI system, a charge circuit and a pulsar circuit.

Charge Circuit

The charge circuit consists of the flywheel magnets, a charge coil, a diode, and a capacitor. As the flywheel magnet passes by the charge coil, a voltage is induced in the coil. As the flywheel continues to rotate and the coil is no longer influenced by the magnet, the magnetic field collapses. Therefore, an alternating current is produced at the charge coil. This AC current is changed to DC by a diode inside the CDI unit.

Special Words

A diode is a solid state unit which permits current to flow in one direction but prevents flow in the opposite direction. A diode may also be know as a rectifier.

The current then passes to a capacitor, also located inside the CDI unit, where it is stored.

Pulsar Circuit

The pulsar circuit has its own flywheel magnet, a pulsar coil, a diode, and a thyristor. A thyristor is a solid state electronic switching device which permits voltage to flow only after it is triggered by another voltage source. Pulsar coils are also known as trigger coils.

At the point in time when the ignition timing marks align, an alternating current is induced in the pulsar coil, in the same manner as previously described for the charge coil. This current is then passed to a second diode located in the CDI unit where it becomes DC current and flows on to the thyristor. This voltage triggers the thyristor to permit the voltage stored in the capacitor to be discharged. The capacitor voltage passes through the thyristor and on to the primary windings of the ignition coil.

In this manner, a spark at the plug may be accurately timed by the timing marks on the flywheel relative to the magnets in the flywheel and to provide as many as 100 sparks per second for a powerhead operating at 6000 rpm.

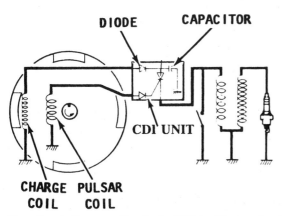

Functional diagram of a single CDI ignition system.

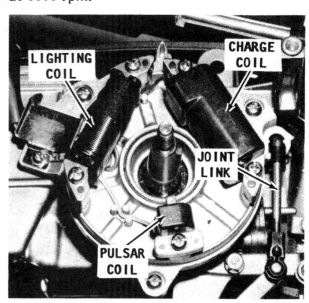

Stator plate of an 8hp powerhead with major parts identified. The stator plate is rotated through the joint link to advance the timing.

5-18 IGNITION

Timing Advance

Units equipped with an automatic advance type CDI system have no moving or sliding parts. Ignition advance is accomplished electronically -- by the electric waves emitted by the pulsar coils. These coils increase their wave output proportionately to engine speed increase, thus advancing the timing.

Units equipped with a mechanical advance type CDI system use a link rod between the carburetor and the ignition base plate assembly. At the time the throttle is opened, the ignition base plate assembly is rotated by means of the link rod, thus advancing the timing.

Charging System

The charging system consists of the flywheel magnet/s, a lighting coil (alternator), a rectifier, voltage regulator, and the battery.

The lighting coil, so named because it may be used to power the lights on the boat when used together with a voltage regulator, allows the powerhead to generate additional electrical current to charge the battery.

As the flywheel magnet/s rotate, voltage is induced in the lighting coil located next to the pulsar coils. This alternating current passes through a series of diodes and emerges as DC current. Therefore, it may be stored in the battery. A lighting coil may be identified by its clean laminated copper windings. All other coils are wrapped in tape and their windings are not visible.

The rectifier converts the alternating current into DC, which may then be stored in the battery. The rectifier is a sealed unit and contains four diodes. If one of the diodes is defective, the entire unit **MUST** be replaced.

The voltage regulator stabilizes the power output of the coils and extends the life of the light bulbs by preventing power surges.

TROUBLESHOOTING TYPE II IGNITION CDI SYSTEM AND CHARGING SYSTEM

Always attempt to proceed with the troubleshooting in an orderly manner. The "shot in the dark" approach will only result in wasted time, incorrect diagnosis, replacement of unnecessary parts, and frustration.

Begin the ignition system troubleshooting with the spark plug and continue through the system until the source of trouble is located.

Spark Plugs

1- Check the plug wires to be sure they are properly connected. Check the entire length of the wires from the plugs to the magneto under the stator plate. If the wires are to be removed from the spark plug, **ALWAYS** use a pulling and twisting motion as a precaution against damaging the connection.

Attempt to remove the spark plug by hand. This is a rough test to determine if the plug is tightened properly. The attempt to loosen the plug by hand should fail. The plug should be tight and require the proper socket size tool. Remove the spark plug and evaluate its condition as described in Section 5-2.

2- Use a spark tester and check for spark. If a spark tester is not available, hold the plug wire about 1/4" (6.4mm) from the engine. Rotate the flywheel with the hand pull starter and check for spark. A strong spark over a wide gap must be observed when testing in this manner, because under compression a strong spark is necessary in order to ignite the air-fuel mixture in the cylinder. This means it is possible to think a strong spark is present, when in reality the spark will be too weak when the plug is installed. If there is no spark, or if the spark is weak, the trouble is most likely under the flywheel in the magneto.

Compression

3- Before spending too much time and money attempting to trace a problem to the ignition system, a compression check of the

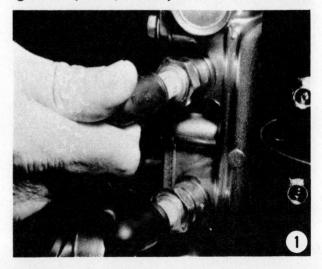

cylinder should be made. If the cylinder does not have adequate compression, troubleshooting and attempted service of the ignition or fuel system will fail to give the desired results of satisfactory engine performance.

Remove the spark plug wire by pulling and twisting **ONLY** on the molded cap. **NEVER** pull on the wire because the connection inside the cap may be separated or the boot may be damaged. Remove the spark plug. Insert a compression gauge into the cylinder spark plug hole. Ground the spark plug leads to prevent damage to the ignition coil. If a lead is not grounded, the coil will attempt to match the demand created by the spark trying to jump from the electrode to the nearest ground.

Crank the powerhead through several revolutions by pulling rapidly on the hand starter pull rope, or with the electric cranking motor, if the powerhead is so equipped. Note the compression reading.

An acceptable pressure reading for a powerhead covered in this manual is 110 psi (760 kPa) or more.

TESTING TYPE II COMPONENTS
GENERAL INFORMATION

Due to the complex nature of the circuitry and high energy output pulses of microsecond duration, conventional testing devices such as a Volt/Ohm/Ammeter will not measure electrical output with the degree of accuracy required. A Mariner Ignition Tester, as used by the local Mariner dealer is an electrical device capable of measuring the peak energy output of capacitor discharge ignition units, magneto charge and the pulsar coils. This instrument was designed to troubleshoot the Type II ignition system. Unfortunately this tester may not be easily accessible. Therefore, this chapter includes only those tests which may be performed with instruments normally and easily available.

If a component, must be replaced, in most cases, the hand rewind starter must first be removed, see Chapter 11, and the flywheel "pulled", see Section 5-8 at the end of this chapter.

WORDS FROM EXPERIENCE

During the tests, if a reading is slightly different from the specifications, but the powerhead still operates, then there is no real need to replace the affected component, until it actually fails. Bear in mind, in **MOST** cases electrical components are not returnable, once the item leaves the store. Therefore, make every attempt to avoid the **BUY** and **TRY** method of troubleshooting.

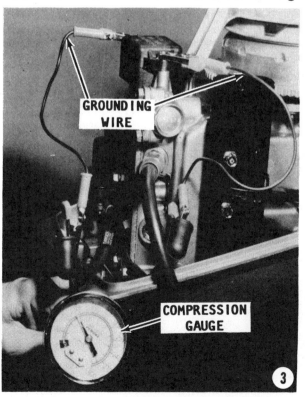

IGNITION

Such a practice will lead only to wasted time, unneeded cash outlay, and frustration.

Single-Cylinder 4hp and 5 hp Models

The Type II ignition system for these units has **TWO** pulsar/trigger coils, instead of one, as on the Type I system.

One pulsar coil, the low speed coil, is located on the outer perimeter of the stator plate. This coil is used during powerhead startup and during low speed operation, up to approximately 1700 rpm.

The other pulsar coil, the high speed coil, is mounted on the stator plate, close to the center magnets on the flywheel. This high speed pulsar coil is used during powerhead operation above approximately 1700 rpm. This second pulsar coil advances the ignition timing with increased powerhead speed.

SIMPLE REASONING

If the powerhead fails to start, the low speed pulsar may be defective.

If the powerhead will start and operate at low rpm, but cuts out and dies if the throttle is advanced for speeds above 1700 rpm, the high speed pulsar coil may be defective.

5-6 TESTING TYPE II COMPONENTS ONE-CYLINDER POWERHEADS

Low Speed Pulsar Coil Test

1- Obtain an ohmmeter and select the Rx100 scale. Disconnect the White/Green and the Black leads coming from the stator plate. Slide the Black probe from the ohmmeter up into the plastic lead connector to make contact with the black lead for the next two steps. This meter lead may now be left in place.

Make contact with the other meter lead to the White/Green lead from the stator plate. The meter should register 280-340 ohms at 68°F (20°C). If the resistance is not within the given limits, the low speed pulsar coil should be replaced. To replace the coil, the hand rewind starter must first be removed, see Chapter 11, and the flywheel "pulled", see Section 5-8 at the end of this chapter.

If the test is successful, disconnect the meter lead and connect the two White/Green leads.

High Speed Pulsar Coil Test

2- Select the Rx1 scale on the ohmmeter. Retain the Black meter probe in the Black stator lead. Disconnect the White/Red lead from the stator plate. Make contact with the other meter lead to the White/Red lead. The meter should register 30-36 ohms at 68°F (20°C). If the resistance is not within the limits listed, the pulsar coil must be replaced. To replace the coil, the hand rewind starter must first be removed, see Chapter 11, and the flywheel "pulled", see Section 5-8 at the end of this chapter. If the test is successful, connect the White/Red lead back to the lead from the stator plate.

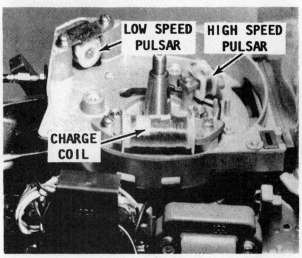

Location of the low speed and high speed pulsars on a 4hp or 5hp powerhead.

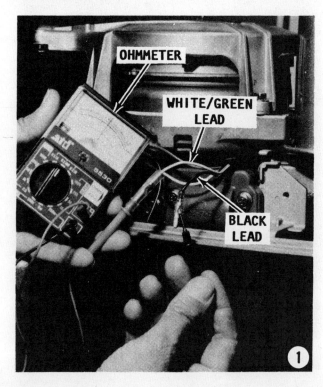

Charging Coil Test

3- Select the Rx100 scale on the meter. Retain the black meter probe in the Black stator lead. Disconnect the Brown lead from the stator plate and make contact with the other meter lead to the Brown lead. The meter should register 248-302 ohms at 68°F (20°C). If the reading is not within the limits given, the charging coil must be replaced. To replace the coil, the hand rewind starter must first be removed, see Chapter 11, and the flywheel "pulled", see Section 5-8 at the end of this chapter. If the test is successful, connect the Brown lead back to the lead from the stator plate and the Black lead to the Black lead from the stator plate.

SPECIAL WORDS

There may be two Green leads in the stator harness. These leads are not tested. The leads come from a lighting coil. The lighting coil is considered optional equipment for United States models. If the lighting coil is installed, the Green leads will be connected to a rectifier to supply power for boat accessories.

Primary Coil Test

4- Select the Rx1 scale on the meter. Disconnect the White/Black lead from the coil. Make contact with the Red meter probe to this White/Black lead. Make contact with the Black meter probe to a suit-

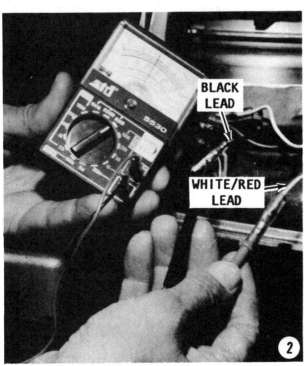

able ground on the powerhead, the coil mounting bolt will do. The meter should register 0.23-0.27 ohms at 68°F (20°C). If the reading is questionable, reconnect the White/Black lead and install a spark plug tester onto the powerhead. A weak spark will get even weaker. Therefore, replacement of the primary coil may be in order.

Secondary Winding Coil Test

5- Select the Rx1000 scale on the meter. Disconnect the spark plug lead from the spark plug. Insert the Red meter probe into the spark plug lead. Check to be sure the probe contacts the inside terminal. Make contact with the Black meter probe to a good ground on the powerhead, the coil mounting bolt will do. The meter should

5-22 IGNITION

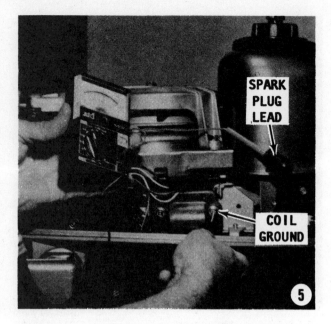

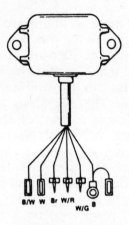

B/W - BLACK/WHITE
W - WHITE
Br - BROWN
W/R - WHITE/RED
W/G - WHITE/GREEN
B - BLACK

White -- no continuity
Brown -- no continuity
White/Green -- 16-36K ohms
Black -- 7.5-17.5K ohms

register 7.5-9.5K ohms at 68°F (20°C). If the meter reading is not within the limits given, the coil must be replaced.

CDI Unit Resistance Test

6- Disconnect all six leads from the CDI unit. Select the Rx1000 scale on the ohmmeter.

a- White stop lead test: Make contact with the Black meter lead to the White lead. Make contact with the Red meter lead -- one at a time -- to each of the following leads. The results should be as listed.

   Brown -- continuity
   White/Red -- 12-28K ohms
   White/Green -- 12-28K ohms
   Black -- 4.0-4.4K ohms

b- Brown charge lead test: Make contact with the Black ohmmeter lead to the Brown lead, them make contact with the Red meter lead -- one at a time -- to each of the following leads. The results should be as listed.

   White -- continuity
   White/Red -- 12-28K ohms
   White/Green -- 12-28K ohms
   Black -- 4.0-4.4K ohms

c- White/Red high speed pulsar lead test: Make contact with the Black meter lead to the White/Red lead and then make contact with the Red meter lead -- one at a time -- to each of the following leads. The results should be as listed.

d- White/Green low speed pulsar lead test: Make contact with the Black meter lead to the White/Green lead and then make contact with the Red meter lead -- one at a time -- to each of the following leads. The results should be as listed.

   White -- no continuity
   Brown -- no continuity
   White/Red -- no continuity
   Black -- no continuity

e- Black ground lead test: Make contact with the Black meter lead to the Black lead from the CDI unit and then make contact with the Red meter lead -- one at a time -- to each of the following leads and the results should be as listed.

   White -- no continuity
   Brown -- no continuity
   White/Red -- 7.5-17.5K ohms
   White/Green -- 4.5-14.5K ohms

f- Black/White ignition lead test: Make contact with the Black meter lead to the Black/White ignition lead and then make contact with the Red meter lead -- one at a time -- to each of the following leads. The results should be as listed.

CRITICAL WORDS

Read, believe, and obey: Between each test, touch the Black CDI lead to the Black/White lead to allow the internal capacitor circuit to discharge. Applying a small current, using the test ohmmeter, charges the

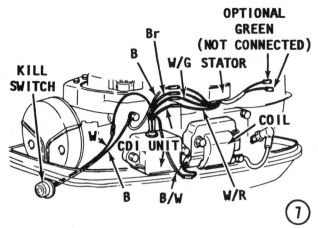

internal capacitor, which must then be discharged before proceeding to the next lead test.

When each test is made, the meter needle will deflect to 200-500K ohms and then return to "no continuity". With the ohmmeter leads **REVERSED**, the same results may be expected, **EXCEPT** for the Black meter lead contacting the White/Green lead and the Red meter lead on the White/Black lead. The ohmmeter should register **NO** continuity which is the only exception for the ignition lead test.

If the readings are not within the limits listed and the powerhead fails to operate, replace the CDI unit.

If the readings are questionable or marginal and the powerhead fails to operate it is possible and probable the CDI unit with **NO Load** could pass the resistance tests but fail to operate the powerhead. In this case, if the CDI unit is doubtful, take the unit to the nearest Mariner dealer for testing under a load with a Mariner Ignition Tester.

If the tests are successful, reconnect the color coded leads -- same color to same color.

7- Visually check all wiring connections before installing the cowling.

SPECIAL NOTE

Testing the kill switch is covered in Chapter 8, Electrical.

5-7 TESTING TYPE II COMPONENTS 2-CYLINDER POWERHEADS

Pulsar/Trigger Coil Test

1- Obtain an ohmmeter and select the appropriate scale. Disconnect the White/Red and the Black leads from the stator plate. If servicing Models 48hp, 55hp, or 60hp, disconnect the Yellow lead from the second Pulsar/Trigger coil. Make contact with the Red meter lead to the White/Red lead. Make contact with the Black meter lead to the Black stator lead.

Models 8hp, 9.9hp and 15hp: Meter should register 92-112 ohms at 68°F (20°C).

Models 20hp and 25hp: Meter should register 95-115 ohms at 68°F (20°C).

Models 30hp and 40hp: Meter should register 12.6-15.4 ohms at 68°F (20°C).

Models 48hp, 55hp, and 60hp: Meter should register 800-900 ohms at 68°F (20°C) Move the Red meter lead to the Yellow lead of the second Pulsar/Trigger coil. The meter should register 20-24 ohms at 68°F (20°C).

If the resistance is not within the limits given and the powerhead does not operate, the pulsar coil must be replaced.

To replace the pulsar coil, the hand rewind starter must first be removed, see Chapter 11, and the flywheel "pulled", see Section 5-8 at the end of this chapter.

If the test is successful, connect the White/Red and Yellow leads back to the leads from the stator plate.

Charge Coil Test

2- Select the appropriate scale on the ohmmeter.

Connection for Black meter lead:

All models **EXCEPT** the 48hp, 55hp, and 60hp: Retain the Black meter lead on the Black stator lead as directed in the previous step.

Models 48hp, 55hp, and 60hp: Disconnect the Blue stator lead and make contact

with the Black meter lead to the Blue stator lead.

Connection for Red meter lead:

All models: Disconnect the Brown charge coil lead and make contact with the Red meter lead to the Brown charge coil lead. The meter should register as follows:

8hp, 9.9hp, and 15hp -- 82-99 ohms
20 and 25hp -- 212-258 ohms
30 and 40hp -- 121-147 ohms
48hp, 55hp, and 60hp -- 225-245 ohms

Reconnect the Brown and Black or Blue leads.

If the resistance is not within the limits listed and the powerhead does not operate, the charge coil must be replaced.

To replace the charge coil, the hand rewind starter must first be removed, see Chapter 11, and the flywheel "pulled", see Section 5-8 at the end of this chapter.

Lighting Coil Test

3- Select the appropriate scale on the ohmmeter. Disconnect the two green leads between the stator and the rectifier at the rectifier. Make contact with the meter leads (either meter lead to either Green lead). The meter reading should be as follows:

8hp, 9.9hp, and 15hp -- 0.36-0.44 ohms
20hp and larger -- 0.25-0.31 ohms

Reconnect the two leads to the rectifier (either lead to either terminal).

If the resistance is not within the limits listed; the battery will not hold a charge; and the boat accessories depend on this coil for power, the charge coil must be replaced.

CRITICAL WORDS

Never attempt to verify the charging circuit by operating the powerhead with the battery disconnected. Such action would force current (normally directed to charge the battery), back through the rectifier and damage the diodes in the rectifier.

To replace the charge coil, the hand rewind starter must first be removed, see Chapter 11, and the flywheel "pulled", see Section 5-8 at the end of this chapter.

Primary Winding Coil Test

4- Select the Rx1 scale on the ohmmeter. Disconnect the Orange and Black leads from the primary coil. Make contact with the Red meter lead to the Orange lead and the Black meter lead to the small Black lead. The meter should register the following reading at 68°F (20°C).

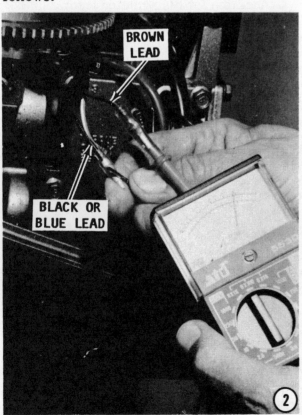

8hp, 9.9hp, and 15hp -- 0.12-0.18 ohms
20hp & 25hp -- 0.12-0.18 ohms
30hp and 40hp -- 0.08-0.10 ohms
48hp, 55hp, and 60hp -- 0.22-0.27 ohms

Reconnect the Orange and Black leads.

If the reading is questionable but the powerhead operates, install a spark plug tester and check the strength of the spark.

If the spark is acceptable, it is not necessary to replace the coil. However, keep this area in mind for possible future trouble.

If the spark is weak it will get weaker, therefore, replace the coil.

Secondary Winding Coil Test

5- Select the Rx1000 scale on the ohmmeter. Disconnect the high tension leads from both spark plugs. Insert the meter probes into the boots of the leads. Check to be sure contact is made between the inside boot terminal and the meter probe. The meter reading should be as follows at 68°F (20°C).

8hp, 9.9hp, and 15hp -- 2.8-4.2K ohms
20hp and 25hp -- 2.8-4.2K ohms
30hp and 40hp -- 3.0-4.0K ohms
48hp, 55hp, and 60hp -- 2.0-3.0K ohms

Reconnect the two high tension leads to the spark plugs.

If the reading is questionable but the powerhead operates, install a spark plug tester and check the strength of the spark.

If the spark is acceptable, it is not necessary to replace the coil. However, keep this area in mind for possible future trouble.

If the spark is weak it will get weaker, therefore, replace the coil.

CDI Unit Resistance Test
Models 8hp, 9.9hp, 15hp, 20hp,
25hp, 30hp, and 40hp
Also 48hp, 55hp, and 60hp with Specific Colored Leads

COLORED LEAD IDENTIFICATION

All the tests outlined in Step 6 also apply to Models 48hp, 55hp, and 60hp **PROVIDED** the leads emerging from the CDI unit are White, Brown, White/Red, Orange, and Black. Therefore, if servicing these three models, check the lead colors to be sure the proper tests are being performed.

6- Disconnect the White, Brown, White/Red, Orange, and any one of the three Black leads from the CDI unit. Select the Rx1000 scale on the ohmmeter.

a- White stop lead test: Make contact with the Black meter lead to the White lead and then make contact with the Red meter lead in turn to each of the following leads. The results should be as listed.

Black -- no continuity
Brown -- no continuity
White/Red -- no continuity
Orange -- no continuity

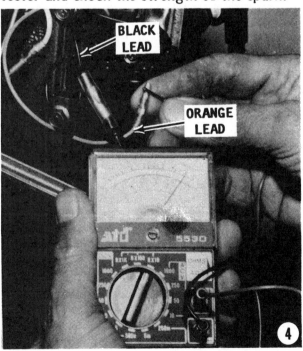

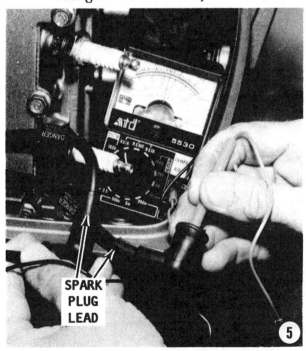

5-26 IGNITION

b- Brown charge lead test: Make contact with the Black meter lead to the Brown lead, and then make contact with the Red meter lead in turn to each of the following leads. The results should be as listed.

White -- no continuity
Black -- 64-96K ohms
White/Red -- no continuity

c- White/Red pulsar lead test: Make contact with the Black meter lead to the White/Red lead, and then make contact with the Red meter lead in turn to each of the following leads. The results should be as listed.

White -- 9-13K ohms
Black -- 14-22K ohms
Brown -- 30-46K ohms

d- Orange ignition lead test:

CRITICAL WORDS

Read, believe, and obey: Between each test, touch the Black CDI lead to the Orange lead to allow the internal capacitor circuit to discharge. Applying a small current, using the test ohmmeter, charges the internal capacitor, which must then be discharged before proceeding to the next lead test.

During each test the meter needle will deflect to between 200 and 300K ohms and then return to "no continuity".

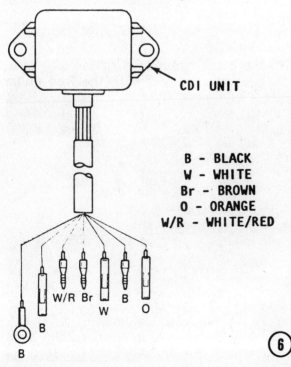

Make contact with the Black ohmmeter lead to the Orange lead, and then make contact with each of the remaining leads in turn with the Red meter lead. In each test the meter should indicate no continuity.

Next, make contact with the Red meter lead to the White lead. The meter should indicate no continuity. Keep the Red meter lead in contact with the Orange lead and make contact with the Black meter lead in turn to each of the following leads: Black, Brown, and the White/Red.

As explained in the Critical Words above, the meter needle will deflect to 200 - 300K ohms and then return to "no continuity".

e- Black ground wire test: Make contact with the Black meter to the Black CDI lead, and then in turn to each of the following leads. The results should be as indicated.

White -- no continuity
Brown -- 7.6-11.2K ohms
White/Red -- no continuity

If the readings are not within the limits listed and the powerhead fails to operate, replace the CDI unit.

If the readings are questionable or marginal and the powerhead fails to operate it is possible and probable the CDI unit with **NO Load** could pass the resistance tests but fail to operate the powerhead. In this case, if the CDI unit is doubtful, take the unit to the nearest Mariner dealer for testing under a load with a Mariner Ignition Tester.

If the tests are successful, reconnect the color coded leads -- same color to same color.

Visually check all wiring connections before installing the cowling.

Testing the "kill" switch and other electrical safety switches are covered in Chapter 8.

CDI Unit Resistance Test
Models 48hp, 55hp, and 60hp with Specific Colored Leads

COLORED LEAD IDENTIFICATION

All the tests outlined in Step 7 apply to Models 48hp, 55hp, and 60hp **PROVIDED** the leads emerging from the CDI unit are Black, Black/White, Brown, Blue, Yellow, White/Red, White/Black, Gray, & Orange. Therefore, if servicing these three models, check the lead colors to be sure the proper tests are being performed.

7- Disconnect the Black/White, Black, Brown, Blue, White/Red, White/Black, Yellow, Gray, and Orange leads from the CDI unit. Select the Rx1000 scale on the ohmmeter.

A Good Word

The Black/White lead is a Black lead with a White tracer and the White/Black lead is a White lead with a Black tracer.

Select the Rx1000 scale on the ohmmeter.

a- Black/White stop lead test: Make contact with the Black meter lead to the Black/White lead, and then make contact with the Red meter lead -- one at a time -- to each of the following leads. The results should be as indicated.

Black -- no continuity
Brown -- no continuity
Blue -- no continuity
White/Red -- no continuity
White/Black -- no continuity
Yellow -- no continuity
Gray -- no continuity
Orange -- no continuity

CRITICAL WORDS

Read, believe, and obey: Between each test, touch the Black CDI lead to the Orange lead and then to the Gray leads to allow the internal capacitor circuit to discharge. Applying a small current, using the test ohmmeter, charges the internal capacitor, which must then be discharged before proceeding to the next lead test.

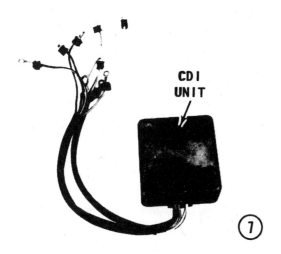

b- Black ground lead test: Make contact with the Black meter lead to the Black CDI lead, and then make contact with the Red meter lead in turn -- one at a time -- to each of the following leads. The results should be as indicated.

Black/White -- 10-14K ohms
Brown -- 3-9K ohms
Blue -- 3-9K ohms
White/Red -- no continuity
White/Black -- no continuity
Yellow -- no continuity
Gray -- See Note below
Orange -- See Note below

Note: During each test the meter needle will deflect to between 200 and 300K ohms and then return to "no continuity".

c- Brown charge lead test: Make contact with the Black meter lead to the Brown lead, and then make contact with the Red meter lead -- one at a time -- to each of the following leads. The results should be as indicated.

Black/White -- 3-9K ohms
Black -- 100-200K Ohms
Blue -- 160-200K ohms
White/Red -- no continuity
White/Black -- no continuity
Yellow -- no continuity
Gray -- See Note below
Orange -- See Note below

Note: During each test the meter needle will deflect to between 200 and 300K ohms and then return to "no continuity".

d- Blue charge lead test: Make contact with the Black meter lead to the Blue lead, and then make contact with the Red meter lead -- one at a time -- to each of the following leads. The results should be as indicated.

Black/White -- 3-9K ohms
Black -- 100-200K Ohms
Brown -- 160-200K ohms
White/Red -- no continuity
White/Black -- no continuity
Yellow -- no continuity
Gray -- See Note below
Orange -- See Note below

Note: During each test the meter needle will deflect to between 200 and 300K ohms and then return to "no continuity".

e- White/Red pulsar lead test: Make contact with the Black meter lead to the

White/Red lead, and then make contact with the Red meter lead -- one at a time -- to each of the following leads. The results should be as indicated.

> Black/White -- 35-45K ohms
> Black -- 15-20K Ohms
> Brown -- 25-30K ohms
> Blue -- 25-30K ohms
> White/Black -- no continuity
> Yellow -- no continuity
> Gray -- See Note below
> Orange -- See Note below

Note: During each test the meter needle will deflect to between 200 and 300K ohms and then return to "no continuity".

f- White/Black pulsar lead test: Make contact with the Black meter lead to the White/Black lead, and then make contact with the Red meter lead -- one at a time -- to each of the following leads. The results should be as indicated.

> Black/White -- 35-45K ohms
> Black -- 15-20K Ohms
> Brown -- 25-30K ohms
> Blue -- 25-30K ohms
> White/Red -- no continuity
> Yellow -- no continuity
> Gray -- See Note below
> Orange -- See Note below

Note: During each test the meter needle will deflect to between 200 and 300K ohms and then return to "no continuity".

g- Yellow pulsar lead test: Make contact with the Black meter lead to the Yellow lead, and then make contact with the Red meter lead -- one at a time -- to each of the following leads. The results should be as indicated.

> Black/White -- 35-45K ohms
> Black -- 15-20K Ohms
> Brown -- 25-30K ohms
> Blue -- 25-30K ohms
> White/Red -- no continuity
> White/Black -- no continuity
> Gray -- See Note below
> Orange -- See Note below

Note: During each test the meter needle will deflect to between 200 and 300K ohms and then return to "no continuity".

h- Gray ignition lead test: Make contact with the Black meter lead to the Gray lead, and then make contact with the Red meter lead -- one at a time -- to each of the following leads. The results should be as indicated.

> Black/White -- 20-28K ohms
> Black -- 3-9K Ohms
> Brown -- 12-16K ohms
> Blue -- 12-16K ohms
> White/Red -- no continuity
> White/Black -- no continuity
> Yellow -- no continuity
> Orange -- See Note below

Note: During each test the meter needle will deflect to between 200 and 300K ohms and then return to "no continuity".

j- Orange Ignition lead: Make contact with the Black meter lead to the Orange lead, and then make contact with the Red meter lead -- one at a time -- to each of the following leads. The results should be as indicated.

> Black/White -- 20-28K ohms
> Black -- 3-9K Ohms
> Brown -- 12-16K ohms
> Blue -- 12-16K ohms
> White/Red -- no continuity
> White/Black -- no continuity
> Yellow -- no continuity
> Gray -- See Note below

Note: During each test the meter needle will deflect to between 200 and 300K ohms and then return to "no continuity".

If the readings are not within the limits listed and the powerhead fails to operate, replace the CDI unit.

If the readings are questionable or marginal and the powerhead fails to operate it is possible and probable the CDI unit with **NO Load** could pass the resistance tests but fail to operate the powerhead. In this case, if the CDI unit is doubtful, take the unit to the nearest Mariner dealer for testing under a load with a Mariner Ignition Tester.

If the tests are successful, reconnect the color coded leads -- same color to same color.

Visually check all wiring connections before installing the cowling.

Testing the "kill" switch and other electrical safety switches are covered in Chapter 8.

TESTING TYPE II

CDI Unit Resistance Test
Models 48hp, 55hp, and 60hp with Specific Colored Leads

COLORED LEAD IDENTIFICATION

All the tests outlined in Step 8 apply to Models 48hp, 55hp, and 60hp **PROVIDED** the leads emerging from the CDI unit are Brown, White/Red, White/Black, White, Black, and two Black/White leads. Therefore, if servicing these three models, check the lead colors to be sure the proper tests are being performed.

A Good Word

The Black/White lead is a Black lead with a White tracer and the White/Black lead is a White lead with a Black tracer.

8- Disconnect the Brown, White/Red, White/Black, White, Black, and two Black/White leads from the CDI unit. Select the Rx1000 scale on the ohmmeter.

a- Brown charge lead test: Make contact with the Black meter lead to the Brown lead, and then make contact with the red meter lead -- one at a time -- to each of the following leads. The results should be as indicated.

    White/Red -- no continuity
    White/Black -- no continuity
    White -- no continuity
    Either Black/White -- 20-32K ohms
    Other Black/White -- 20-32K ohms
    Black -- 13-21K ohms

b- White/Red pulsar lead test: Make contact with the Black meter lead to the White/Red lead, and then make contact with the Red meter lead -- one at a time -- to each of the following leads. The results should be as indicated.

    Brown -- no continuity
    White/Black -- no continuity
    White -- 3-6K ohms
    Either Black/White -- 28-44K ohms
    Other Black/White -- no continuity
    Black -- no continuity

c- White stop lead test: Make contact with the Black meter lead to the White lead, and then make contact with the Red meter lead -- one at a time -- to each of the following leads. The results should be as indicated.

    Brown -- no continuity
    White/Red -- no continuity
    White/Black -- no continuity
    Either Black/White -- no continuity
    Other Black/White -- no continuity
    Black -- no continuity

d- Either Black/White ignition lead test: Make contact with the Black meter lead to the Black/White lead, and then make contact with the Red meter lead -- one at a time -- to each of the following leads. The results should be as indicated.

    Brown -- no continuity
    White/Red -- no continuity
    White/Black -- no continuity
    White -- no continuity
    Other Black/White -- no continuity
    Black -- no continuity

e- Other Black/White ignition lead: Make contact with the Black meter lead to the Black/White lead, and then make contact with the Red meter lead -- one at a time -- to each of the following leads. The results should be as indicated.

    Brown -- no continuity
    White/Red -- no continuity
    White/Black -- no continuity
    White -- no continuity
    Other Black/White -- no continuity
    Black -- no continuity

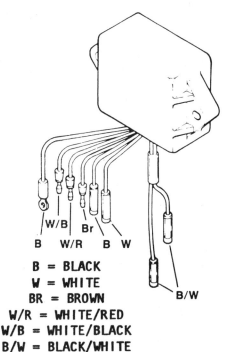

B = BLACK
W = WHITE
BR = BROWN
W/R = WHITE/RED
W/B = WHITE/BLACK
B/W = BLACK/WHITE

f- Black ground lead test: Make contact with the Black meter lead to the Black lead, and then make contact with the Red meter lead -- one at a time -- to each of the following leads. The results should be as indicated.

- Brown -- 3-6K ohms
- White/Red -- no continuity
- White/Black -- no continuity
- White -- no continuity
- Either Black/White -- 3-5K ohms
- Other Black/White -- 3-5K ohms

CDI Unit Resistance Test
Models 48hp, 55hp, and 60hp with Specific Colored Leads

COLORED LEAD IDENTIFICATION

All the tests outlined in Step 9 apply to Models 48hp, 55hp, and 60hp **PROVIDED** the leads emerging from the CDI unit are Brown, White/Red, Black, Yellow, and Pink. Therefore, if servicing these three models, check the lead colors to be sure the proper tests are being performed.

9- Disconnect the Brown, White/Red, Black, Yellow, and Pink leads emerging from the CDI unit. Select the Rx1000 scale on the ohmmeter.

a- Brown charge lead test: Make contact with the Black meter lead to the Brown lead, and then make contact with the Red meter lead -- one at a time -- to each of the following leads. The results should be as indicated.

- White/Red -- no continuity
- Black -- 14-19K ohms
- Yellow -- 21-29K ohms
- Pink -- 51-69K ohms

b- White/Red pulsar lead test: Make contact with the Black meter lead to the White/Red lead, and then make contact with the Red meter lead -- one at a time -- to each of the following leads. The results should be as indicated.

- Brown -- 36-49K ohms
- Black -- 16-21K ohms
- Yellow -- 36-48K ohms
- Pink -- 36-48K ohms

c- Black ground lead test: Make contact with the Black meter lead to the Black lead, and then make contact with the Red meter lead -- one at a time -- to each of the following leads. The results should be as indicated.

- Brown -- 14-19K ohms
- White/Red -- no continuity
- Yellow -- 5-7K ohms
- Pink -- 25-34K ohms

d- Yellow pulsar lead test: Make contact with the Black meter lead to the Yellow lead, and then make contact with the Red meter lead -- one at a time -- to each of the following leads. The results should be as indicated.

- Brown -- no continuity
- White/Red -- no continuity
- Black -- no continuity
- Pink -- no continuity

e- Pink ignition lead test: Make contact with the Black meter lead to the Pink lead, and then make contact with the Red meter lead -- one at a time -- to each of the following leads. The results should be as indicated.

- Brown -- no continuity
- White/Red -- no continuity
- Black -- no continuity
- Yellow -- no continuity

All Electric Start Models

10- At the quick disconnect fitting, disconnect the two Green, the two Red, and the Black lead coming from the rectifier. Select the Rx1000 scale on the ohmmeter. Make contact with the Black meter lead to the first Green lead. Now, make contact with the Red meter lead -- one at a time -- to each of the following leads. The results should be as indicated.

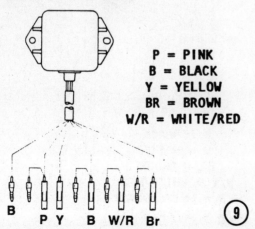

P = PINK
B = BLACK
Y = YELLOW
BR = BROWN
W/R = WHITE/RED

Second Green -- no continuity
Either Red -- continuity
Black -- no continuity

Shift the Black meter lead to the second Green lead. Make contact with the Red meter lead -- one at a time -- to each of the following leads. The results should be as indicated.

First Green -- no continuity
Either Red -- continuity
Black -- no continuity

Make contact with the black meter lead to either Red lead. Make contact with the Red meter lead -- one at a time -- to each of the following leads. The results should be as indicated.

First Green -- no continuity
Second Green -- no continuity
Black -- no continuity

Make contact with the Black meter lead to the Black rectifier lead. Make contact with the Red meter lead -- one at a time -- to each of the following leads. The results should be as indicated.

First Green -- continuity
Second Green -- continuity
Either Red -- continuity

If the meter readings are not satisfactory, the rectifier is defective and must be replaced. The rectifier is externally mounted and is easily removed and installed through the attaching hardware.

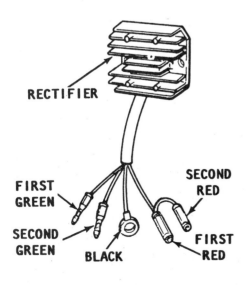

5-8 FLYWHEEL AND STATOR PLATE SERVICE

The following short section lists the procedures required to "pull" the flywheel and remove the stator plate in order to service the ignition system. Removal and installation of the stator plate is necessary in order to gain access to the wiring harness retainer underneath the stator plate. Cleaning and Inspecting procedures in addition to proper assembling and installation steps are also included.

Because so many different models are covered in this manual, it would not be feasible to provide an illustration of each and every unit covered.

Therefore, the accompanying illustrations are of a "typical" unit. The unit being serviced may differ slightly in appearance due to engineering or cosmetic changes but the procedures are valid. If a difference should occur, it will be clearly indicated for the model affected.

"PULLING" THE FLYWHEEL

1- Remove the cowling from the powerhead. Remove the spark plug/s. Remove the mounting hardware securing the hand rewind starter to the powerhead. Move the rewind starter to one side out of the way.

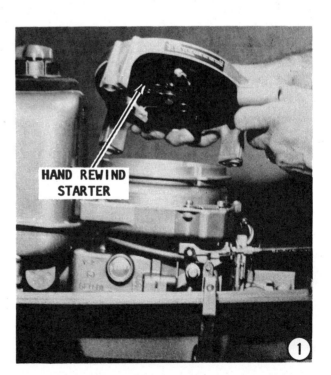

5-32 IGNITION

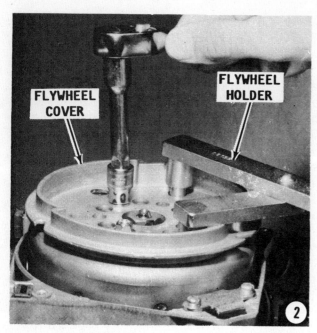

SPECIAL WORDS

Do not remove the handle from the starter rope because the rope would immediately rewind inside the starter. Such action would require considerable time and effort to correct.

2- Obtain special flywheel holder tool, P/N M-91-83163M. Insert the two indexing pins on the ends of the arms through the holes in the flywheel cover. Hold the cover steady with the special tool and remove the cover attaching bolts.

3- Before lifting the cover from the flywheel, scribe a mark on the cover and a matching mark on the flywheel to **ENSURE** the cover is installed in the same position from which it is removed.

4- Position the same special tool as used in Step 1 into the flywheel holes. Hold the flywheel from rotating with the special tool and at the same time remove the flywheel nut.

5- Obtain flywheel puller tool, P/N M-91-83164M. If this particular puller is not available a similar puller may be used **PROVIDED** it will pull from the bolt holes in the flywheel and **NOT** from around the perimeter of the flywheel.

NEVER attempt to use a puller which pulls on the outside edge of the flywheel.

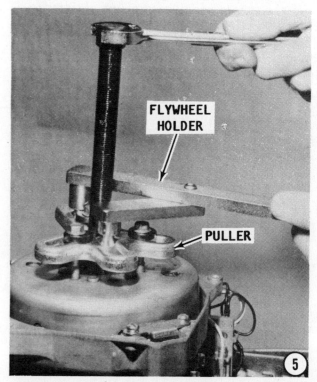

SERVICE FLYWHEEL & STATOR 5-33

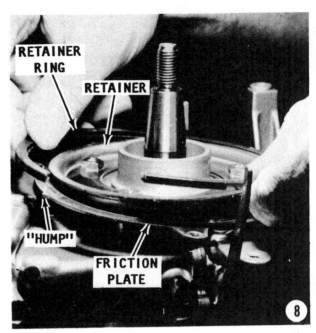

Install the puller onto the flywheel, and then using puller tool P/N M-91-83164M take a strain on the puller with the proper size wrench. Now, continue to tighten on the special tool and at the same time, **SHOCK** the crankshaft with a gentle to moderate tap with a hammer on the end of the special tool. This shock will assist in "breaking" the flywheel loose from the crankshaft.

6- Lift the flywheel free of the crankshaft. Remove and **SAVE** the Woodruff key from the recess in the crankshaft.

Stator Plate Removal

7- Unplug the stator harness at the quick disconnect fittings. Remove the bolts securing the stator plate to the powerhead and lift the stator plate free of the crankshaft and powerhead.

Some 2-cylinder Models

8- The stator is bolted to a friction plate beneath a retainer. Once the stator plate is removed, this friction plate is free to rotate. **DO NOT DISTURB** the position of this friction plate in relation to the stator plate. Movement of the friction plate will affect powerhead timing.

If it is necessary to remove and replace the friction plate, **TAKE TIME** to scribe a mark on the powerhead opposite the small timing "hump", as shown in the accompanying illustration.

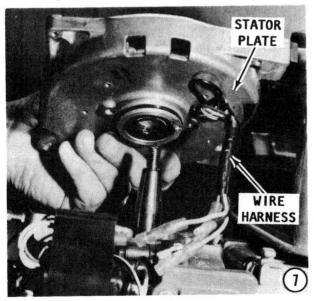

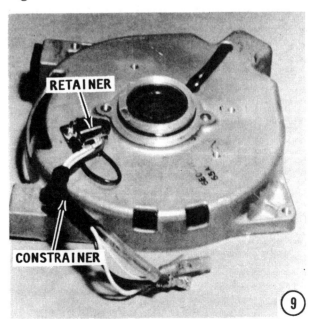

5-34 IGNITION

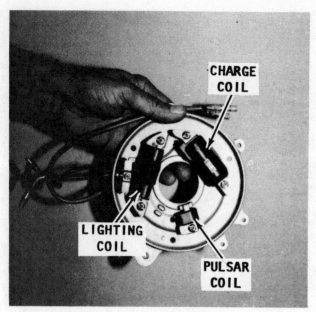

Stator plate from an 8hp powerhead ready for installation. Major parts are identified.

All Models

9- Place the stator plate on a suitable work surface. Remove the screw securing the harness retainer to the plate and unwind the black plastic harness "constrainer". Now, any component mounted on top of the stator plate and tested defective may be removed.

CLEANING AND INSPECTING

Inspect the flywheel for cracks or other damage, especially around the inside of the center hub. Check to be sure metal parts have not become attached to the magnets.

Always check the flywheel carefully to be sure particles of metal have not become stuck to the magnets.

Verify each magnet has good magnetism by using a screwdriver or other suitable tool.

Thoroughly clean the inside taper of the flywheel and the taper on the crankshaft to prevent the flywheel from "walking" on the crankshaft during operation.

Check the top seal around the crankshaft to be sure no oil has been leaking onto the stator plate. If there is **ANY** evidence the seal has been leaking, it **MUST** be replaced. See Chapter 7.

Test the stator assembly to verify it is not loose. Attempt to lift each side of the plate. There should be little or no evidence of movement.

Inspect the stator plate oil seal and the O-ring on the underside of the plate.

Some 2-cylinder models have a retainer, a retainer ring, and a friction plate located under the stator. The retainer ring is a guard around the retainer, subject to cracking and wear. Inspect the condition of this guard and replace if it is damaged.

SPECIAL WORDS ON FLYWHEEL MAGNETS

The outer ends of any magnet are called poles. One end is the north pole and the other end is the south pole. The magnetic field surrounding a magnet is concentrated around these two poles.

Some flywheel magnets are fairly long and curved around the outer perimeter of the underside of the flywheel. Others are short and are mounted around the center hub, depending on the location of the coils mounted on the stator. Magnets are usually installed in pairs with the north pole of one adjacent to the south pole of its neighbor and so on. In this manner continuous magnetic field surrounds the inside of the flywheel.

If a flywheel is accidently dropped, not only could the teeth be damaged, but the impact will weaken the magnetic strength of all the magnets housed in the flywheel.

If one or more of the magnets should break or fracture, two new magnetic poles will be created. A long magnet with two poles will become two short magnets with four poles. The new poles will possess only half the magnetic strength of an original pole. The overall magnetic field will be altered. The new field of the shorter magnets will not extend to cover the area of the flywheel.

SERVICE FLYWHEEL & STATOR 5-35

Serious consequences apply to a CD type ignition system in the event of a flywheel magnet fracture. Pulsar/Trigger coils evenly spaced around the perimeter of the stator plate are energized by the concentrated magnetic field at the magnet poles. If new poles are suddenly created, the pulsar/trigger will receive conflicting signals from the magnets and may even attempt to fire the cylinder twice in one revolution.

All these reasons require the flywheel to be handled with **CARE**.

ASSEMBLING AND INSTALLATION

Some 2-cylinder Models

1- Place the friction plate down over the crankshaft. Rotate the plate until the "hump" on the outer edge is aligned with the mark scribed on the powerhead prior to removal. Install the retainer on top of the friction plate and secure it in place with the attaching bolts. Stretch the retainer ring around the retainer with the outer edge of the retainer indexed into the ring groove.

SPECIAL WORDS

When the stator plate is installed, the mounting bolts thread into the friction plate instead of into the powerhead, as on the other models.

Make a final check to be sure the "hump" on the friction plate is still aligned with the scribed mark on the powerhead.

All Models

2- Position the stator plate in place over the crankshaft. Secure the plate with the attaching bolts. Tighten the bolts alternately and evenly to a torque value of 5.9 ft lbs (8Nm). Connect the stator wire harness wire by wire, color to color.

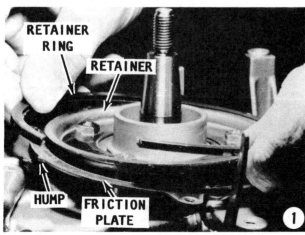

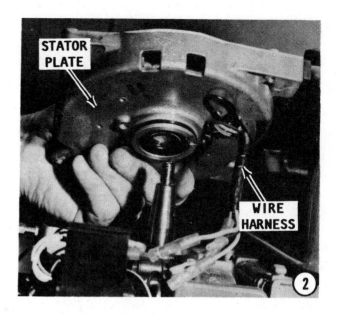

3- Place a tiny dab of thick lubricant on the curved surface of the Woodruff key to hold it in place while the flywheel is being installed. Press the Woodruff key into place in the crankshaft recess. Wipe away any excess lubricant to prevent the flywheel from "walking" during powerhead operation.

Check the flywheel magnets to ensure they are free of any metal particles. Double check the taper in the flywheel hub and the taper on the crankshaft to verify they are clean and contain no oil.

Now, slide the flywheel down over the crankshaft with the keyway in the flywheel aligned with the Woodruff key in place on the crankshaft. Rotate the flywheel **COUNTERCLOCKWISE** to be sure it does not contact any part of the stator plate or wiring.

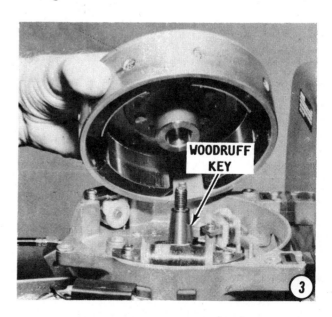

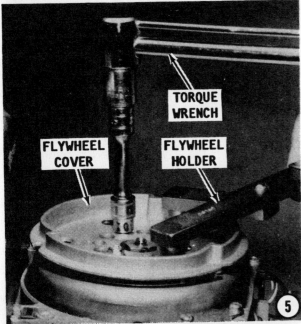

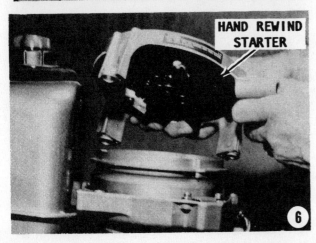

4- Slide the washer onto the crankshaft, and then thread the flywheel nut onto the crankshaft. Obtain special flywheel tool P/N M-91-83163M With the pins on the ends of the holder arms indexed into the flywheel holes, tighten the flywheel nut to the following torque value for the models listed.

4hp, 5hp, & 8hp -- 32 ft lbs (44Nm)
9.9hp, & 15hp -- 76 ft lbs (103Nm)
20hp, & 25hp -- 54 ft lbs (73Nm)
30hp, & 40hp -- 116 ft lbs (158Nm)
48hp, 55hp, & 60hp -- 108 ft lbs (147Nm)

5- Position the flywheel cover in place over the flywheel with the holes in the cover aligned with the holes in the flywheel. Install and tighten the bolts to a torque value of 5.8 ft lbs (8Nm).

6- Install the hand rewind starter to the powerhead. Tighten the three attaching bolts to a torque value of 5.8 ft lbs (8Nm).

7- Install and tighten the spark plug/s to a torque value of 14 ft lbs (20Nm). Secure the spark plug lead/s to the spark plug/s.

Install the cowling to the powerhead.

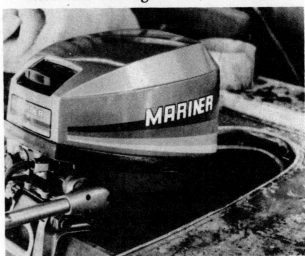

Unit mounted in a test tank ready for operation and final check of the completed service work.

6
TIMING AND SYNCHRONIZING

6-1 INTRODUCTION AND PREPARATION

Timing and the synchronization on an outboard engine is extremely important to obtain maximum efficiency. The powerhead cannot perform properly and produce its designed horsepower output if the fuel/carburetion and ignition systems have not been precisely adjusted.

Synchronization

In simple terms, synchronization is timing the carburetion to the ignition. This means, as the throttle is advanced to increase powerhead rpm, the carburetor and the ignition systems are both advanced equally and at the same rate.

Therefore, any time the fuel system or the ignition system on an engine is serviced to replace a faulty part, or any adjustments are made for any reason, powerhead timing and synchronization **MUST** be carefully checked and verified.

For this reason the timing and synchronizing procedures have been separated from all others and presented alone in this chapter.

Before making adjustments with the timing or synchronizing, the ignition system should be thoroughly checked according to the procedures outlined in Chapter 5, and the fuel system verified as in good working order per Chapter 4.

Timing

The timing on many models is set at the factory and there is no way in heaven, or on earth, to make an adjustment. However, on other powerheads, the timing may be adjusted. All outboard engines have some type of synchronization between the carburetion and ignition systems.

Many models have timing marks on the flywheel and CDI magneto base. A timing light is normally used to check the timing **DYNAMICALLY** -- with the powerhead operating.

An alternate method is to check the timing **STATICALLY** -- with the powerhead not operating. This second method requires the use of a dial indicator gauge.

Various models have unique methods of checking ignition timing. These differences are explained in detail and supported with illustrations whenever possible.

PREPARATION

Timing and synchronizing the ignition and fuel systems on an outboard motor are critical adjustments. Therefore, the following equipment is essential and is called out repeatedly in this chapter. This equipment must be used as described, unless otherwise instructed. Naturally, they are removed following completion of the adjustments.

Dial Indicator

Top dead center (TDC) of the No. 1 (top) piston must be precisely known before the timing adjustment can be made. TDC can only be determined through installation of a dial indicator into the No. 1 spark plug opening.

Timing Light

During many procedures in this chapter, the timing mark on the flywheel must be aligned with a stationary timing mark on the engine while the engine is being cranked, or is running. Only through use of a timing light connected to the No. 1 spark plug lead, can the timing mark on the flywheel be observed while the engine is operating.

Tachometer

A tachometer connected to the powerhead must be used to accurately determine engine speed during idle and high-speed adjustment.

The meter readings range from 0 to 6,000 rpm in increments of 100 rpm. Tachometers have solid state electronic circuits which eliminate the need for relays or batteries and contribute to their accuracy.

Refer to the Special Words on Tachometers in the next column, to make the proper connections.

Test Tank

The engine must be operated at various times during the procedures. Therefore, a test tank, or moving the boat into a body of water, is necessary.

CAUTION

Never operate the powerhead above a fast idle with a flush attachment connected to the lower unit. Operating the powerhead at a high rpm with no load on the propeller shaft could cause the powerhead to **RUNAWAY** causing extensive damage to the unit.

REMEMBER, the powerhead will **NOT** start without the emergency tether in place behind the kill switch knob.

CAUTION

Water must circulate through the lower unit to the powerhead anytime the powerhead is operating to prevent damage to the water pump in the lower unit. Just five seconds without water will damage the water pump impeller.

A dial indicator gauge is used to determine the TDC (Top Dead Center) position of the piston in the No. 1 (top), cylinder.

Flywheel Rotation

During the procedures listed in this chapter, the instructions may call for rotating the flywheel until certain marks are aligned with the timing pointer. When the flywheel must be rotated, **ALWAYS** move the flywheel in a **CLOCKWISE** direction.

If the flywheel should be rotated in the opposite direction, the water pump impeller tangs would be twisted backwards.

Should the powerhead be started with the pump tangs bent back in the wrong direction, the tangs may not have time to bend in the correct direction before they are damaged. The least amount of damage to the water pump will affect cooling of the powerhead.

6-2 CARBURETOR ADJUSTMENTS

SPECIAL WORDS ON TACHOMETERS AND CONNECTIONS

The 8hp to 30hp powerheads use a CDI system firing a twin lead ignition coil twice for each crankshaft revolution. If an induction tachometer is installed to measure powerhead speed, the tachometer will probably indicate **DOUBLE** the actual crankshaft rotation. Check the instructions with the tachometer to be used. Some tachometer manufacturers have allowed for the double reading and others have not.

Connections

For all models with the Type I (with points), ignition system, connect one tachometer lead (may be White, Red, or Yellow, depending on the manufacturer), to the primary negative terminal of the coil (usually a small black lead), and the other tachometer lead, Black, to a suitable ground on the powerhead.

For all models with the Type II (CDI) ignition system, connect the two tachometer leads to the two green leads from the stator. Either tachometer lead may be connected to either green lead.

Carburetor Adjustments

Due to local conditions, it may be necessary to adjust the carburetor while the outboard unit is running in a test tank or with the boat in a body of water. For maximum performance, the idle rpm should be adjusted under actual operating conditions.

CARBURETOR ADJUSTMENTS 6-3

Remove the cowling and attach a tachometer to the powerhead as directed under "Connections" above.

Start the engine and allow it to warm to operating temperature.

REMEMBER, if the powerhead is equipped with a "kill" switch knob, the powerhead will **NOT** start without the emergency tether in place behind the "kill" switch knob.

CAUTION
Water must circulate through the lower unit to the powerhead anytime the powerhead is operating to prevent damage to the water pump in the lower unit. Just five seconds without water will damage the water pump impeller.

NEVER, AGAIN, NEVER operate the engine at high speed with a flush device attached. The engine, operating at high speed with such a device attached, would **RUNAWAY** from lack of a load on the propeller, causing extensive damage.

All Units Except Model 2hp

The idle mixture screw, also known as the pilot screw, regulates the air/fuel mixture. The setting for this screw varies for each unit. The setting is given in Chapter 4, for each model covered in this manual. Check the Table of Contents, Chapter 4 for the model being serviced. Actually, this setting is **NOT** an adjustment, it is actually a specification.

The idle speed is regulated by the throttle stop screw which "sets" the position of the throttle plate inside the carburetor throat. Idle speed recommendations are given in Chapter 4. Rotating the throttle stop **CLOCKWISE** increases powerhead speed. Rotating the screw **COUNTERCLOCKWISE** decreases powerhead speed.

The idle rpm is adjusted under actual operation in Chapter 4 as follows:

HP MODEL	YEAR	CARB	IDLE ADJUST. PAGE NO.
2	1977 & On	A	4-19
3.5	1978-81	B	4-26
4	1982 & On	C	4-34
5	1977-79	D	4-42
5	1979 & On	C	4-34
8A & 8	1977-79	E	4-49
8B, 8W & Mrthn 8	1979 & On	F/G	4-61 or 4-70
9.9C & 15C	1977 & On	F/G	4-61 or 4-70
8C	1979 & On	G	4-70
15W & 15	1977-79	H	4-77
20	1977 & On	J	4-88
25	1980 & On	J	4-88
28	1977-79	J	4-88
30	1980 & On	J	4-88
40	1978 & On	K	4-99
48	1977-79	L	4-107
55	1986 & On	L	4-107
60	1977-83	L	4-107

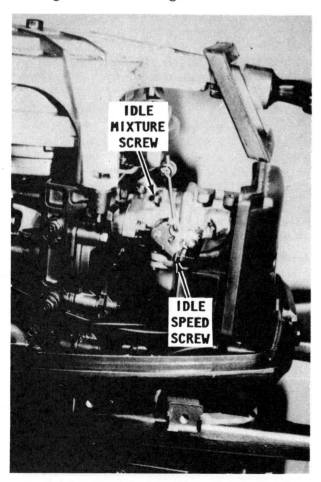

Typical location of carburetor adjustment screws for the powerheads covered in this manual.

6-3 IGNITION TIMING
TYPE 1 IGNITION SYSTEM
FLYWHEEL MAGNETO WITH BREAKER POINTS

The following timing adjustments are valid for the models listed:

2hp	1977 & on
3.5hp	1977-81 (air-cooled)
5hp	1977-79 (air-cooled)
8hp	1977-79
15hp	1977-79
20hp	1977-79
28hp	1977-79

Ignition Timing Adjustment
Using Ohmmeter and Dial Gauge

After the point gap is set, the ignition timing may be checked, using an ohmmeter and a dial gauge **WITHOUT** removing the hand rewind starter or the flywheel. The point gap is a dimension given by the manufacturer to provide optimum performance. Increasing the point gap will result in timing advance and decreasing the point gap will result in retarded timing.

First, remove the spark plug/s and install a dial gauge into the spark plug opening. If servicing a two cylinder powerhead, insert the dial gauge into the top spark plug opening. Rotate the flywheel **CLOCKWISE** until the piston reaches TDC (top dead center). Set the dial gauge at zero.

If servicing a single cylinder model: Disconnect the White lead at the quick disconnect fitting and remove the Black ground lead from the horseshoe bracket.

If servicing a two cylinder model: Disconnect the Grey and Black leads at their quick disconnect fittings. Perform the following tasks to check the timing of the upper cylinder and then reconnect the Grey lead at its quick disconnect fitting. To check the timing of the lower cylinder: disconnect the Orange lead at its quick disconnect fitting and repeat the tasks.

Set the ohmmeter on the Rx1000 scale. Make contact with the black ohmmeter lead to the Black ground lead and the red ohmmeter lead to the White or Grey lead from the stator plate.

When the points are **CLOSED**, the meter will indicate continuity by registering a "reading". The actual value of the reading is not important.

When the points are **OPEN**, creating an open in the circuit, the meter will register an infinite resistance -- an air gap. Therefore, the meter needle will swing either to the far right or to the far left, depending on the scale of the ohmmeter.

Now, **SLOWLY** rotate the flywheel **COUNTERCLOCKWISE** and observe the two positions of the meter needle. If the point gap is correct, the timing of the opening and closing of the points will register as the needle swings. The meter needle will swing from **OPEN** to **CLOSE** between the following limits BTDC (before top dead center):

2hp	0.047-0.079"	(1.19-2.00mm)
3.5hp	0.037-0.047"	(0.93-1.19mm)
5hp	0.041-0.073"	(1.05-1.85mm)
8hp	0.097-0.111"	(2.46-2.82mm)
15hp	0.083-0.114"	(2.11-2.89mm)
20hp	0.114-0.150"	(2.89-3.81mm)
28hp	0.114-0.150"	(2.89-3.81mm)

If the timing is incorrect, the point gap must be adjusted or the point set replaced.

Refer to the procedures outlined in the service section for the Type I ignition system in Chapter 5 to provide detailed instructions to adjust or replace the breaker point set/s.

Ignition Timing Adjustment
Adjustment by Matching Marks

The following procedures are valid only for models equipped with timing pointers and a single timing mark embossed on the flywheel:

8hp	1977-79
15hp	1977-79
20hp	1977-79
28hp	1977-79

The embossed timing mark may be either in the form of a punch mark, as on the 8hp model, or a short vertical line, as on models 15hp, 20hp, and 28hp.

After the point gap is set, the ignition timing may be checked, using the timing pointer and embossed marks, **WITHOUT** removing the hand rewind starter or the flywheel. The point gap is a dimension given by the manufacturer to provide optimum performance. Increasing the point gap will result in timing advance and decreasing the point gap will result in retarded timing.

If the powerhead being serviced is equipped with a link rod between the magneto base and the magneto lever, pry the link from the ball joint on the magneto base. **TAKE CARE** not to alter the length of the rod. Advance the throttle to the wide open position.

Disconnect the Gray and Black leads at their quick disconnect fittings. Remove both spark plugs. Insert a thin long shanked screwdriver into the spark plug hole and rotate the flywheel by hand to verify, the top cylinder is at the TDC position.

Set the ohmmeter on the Rx1000 scale. Make contact with the black ohmmeter lead to the black ground lead and the red ohmmeter lead to the Gray lead from the stator plate.

When the points are **CLOSED**, the meter will indicate continuity by registering a "reading". The actual value of the reading is not important.

When the points are **OPEN**, creating an open in the circuit, the meter will register an infinite resistance -- an air gap. Therefore, the meter needle will swing either to the far right or to the far left, depending on the scale of the ohmmeter.

Now, **SLOWLY** rotate the flywheel **COUNTERCLOCKWISE** and observe the meter needle. If the point gap is correct, the timing of the opening and closing of the points will register as the needle swings.

Push the angled magneto stop bracket aft to contact the timing pointer. Slowly rotate the flywheel **COUNTERCLOCKWISE** until the ohmmeter indicates the points have just closed. The meter needle will swing from infinity to "a reading". At this point, stop rotating the flywheel.

Connect the Gray lead at its quick disconnect fitting and disconnect the Orange lead at its quick disconnect fitting.

Insert a thin long shanked screwdriver into the lower spark plug hole and rotate the flywheel by hand to verify, the lower cylinder is at the TDC position.

Repeat the procedure using the timing mark at the opposite side of the flywheel. The alignment of the mark and the pointer **MUST NOT** be disturbed from the correction made earlier. If an adjustment is necessary, change the point gap of the secondary breaker point set (tracing the Orange lead, if necessary, to identify the secondary piont set). Increasing the point gap will result in timing advance and decreasing the point gap will result in retarded timing.

If the timing is correct, the forward side, for all models except the 15hp, will align with the embossed timing mark on the flywheel. On the 15hp model, the aft side of the timing pointer will be aligned with the timing mark.

If the mark is misaligned, loosen the two bolts on the underside of the magneto base stopper and adjust the position of the stopper until the contact face of the stopper and the correct side of the timing pointer are both aligned with the flywheel mark as required.

6-4 MODEL 4HP AND 5HP WATER-COOLED SINCE 1979

Ignition Timing

These units have a non-adjustable automatic spark advance system. The timing can only be checked. If powerhead performance is unacceptable to the point internal powerhead damage could be caused, then the CDI unit and the pulsar coil should be checked. Detailed procedures, supported with illustrations, to check these two items, are given in Chapter 5.

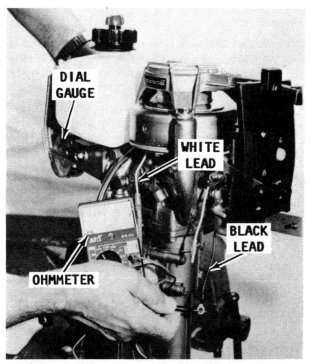

When servicing a powerhead with breaker point (Type I), ignition system, the timing may be checked using an ohmmeter and a dial indicator gauge, as explained in the text.

6-6 TIMING & SYNCHRONIZING

View through the window showing the timing mark on the flywheel.

If the timing mark appears in the specified window at the given rpm using a timing light, the ignition timing is correct and set at 5° BTDC at idle speed and 28° BTDC at WOT.

Mount the engine in a test tank, or on a boat in a body of water.

CAUTION
Never operate the powerhead above a fast idle with a flush attachment connected to the lower unit. Operating the powerhead at a high rpm with no load on the propeller shaft could cause the powerhead to **RUNAWAY** causing extensive damage to the unit.

REMEMBER, the powerhead will **NOT** start without the emergency tether in place behind the kill switch knob.

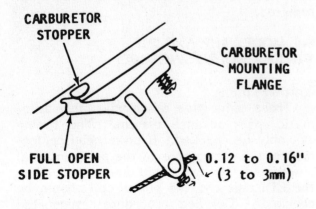

Top view line drawing depicting adjustment of the throttle cable end, as explained in the text.

Remove the cowling and connect a tachometer to the powerhead -- both tachometer leads connected to the two Green leads from the stator. Connect a timing light to the No. 1 spark plug lead.

CAUTION
Water must circulate through the lower unit to the powerhead anytime the powerhead is operating to prevent damage to the water pump in the lower unit. Just five seconds without water will damage the water pump impeller.

Observe the two windows on the stator plate. Identify the timing mark by rotating the flywheel **CLOCKWISE** by hand until a mark -- vertical line embossed on the lower edge of the flywheel -- is visible through one of the windows.

Start the engine. Aim the timing light at the left window. The timing mark should be centered in the window with the powerhead operating between 1150 and 1300 rpm.

Increase powerhead speed to approximately 4500. Aim the timing light at the right window. The timing mark should be centered in the window.

Throttle Cable Adjustment
With the powerhead shut down -- not operating -- rotate the throttle grip to the **FAST** position. Verify the full open side stopper for the throttle valve makes contact with the stopper on the carburetor. If the two stoppers fail to make contact before the throttle grip reaches the **FAST** position, proceed as follows: first, loosen the screw on the barrel retaining end of the throttle cable, and then adjust the length of the cable (wire) protruding from the barrel. The

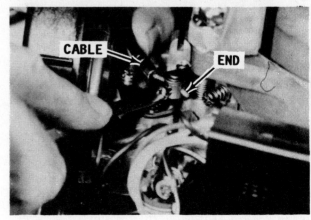

Loosening the screw on the barrel retaining end of the throttle cable.

amount beyond the barrel should be 0.12 - 0.16" (3 -4mm) with the throttle valve in the full open position.

6-5 MODEL 8C

Fully Advanced Ignition Timing

1- Pry off the link rod between the CDI magneto base and the magneto control lever at the ball joint. This link rod will remain disconnected until all adjustments have been completed.

Mount the engine in a test tank, or on a boat in a body of water.

CAUTION

Never operate the powerhead above a fast idle with a flush attachment connected to the lower unit. Operating the powerhead at a high rpm with no load on the propeller shaft could cause the powerhead to **RUNAWAY** causing extensive damage to the unit.

REMEMBER, the powerhead will **NOT** start without the emergency tether in place behind the "kill" switch knob.

CAUTION

Water must circulate through the lower unit to the powerhead anytime the powerhead is operating to prevent damage to the water pump in the lower unit. Just five seconds without water will damage the water pump impeller.

2- Remove the cowling and connect a timing light to the No. 1 spark plug lead.

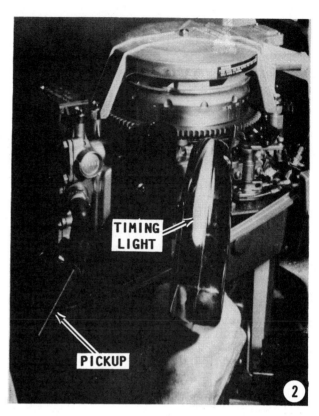

WARNING
BE CAREFUL NOT TO TOUCH THE SPINNING FLYWHEEL TEETH WHILE PERFORMING THE NEXT STEP. SUCH ACTION COULD CAUSE PERSONAL INJURY.

6-8 TIMING & SYNCHRONIZING

3- Start the engine. Rotate the CDI magneto base **COUNTERCLOCKWISE** until the base stopper on the left contacts the vertical timing pointer. This is the full advanced position.

GOOD WORDS

This adjustment automatically sets the fully retarded ignition timing.

Aim the timing light at the vertical timing pointer. The pointer should align between the 34° BTDC and 36° BTDC embossed marks on the flywheel edge.

If the pointer does not align, as described, loosen the bolt on the vertical pointer bracket and move the pointer and the stopper on the magneto base plate together, until the vertical pointer is properly aligned. Hold the magneto base plate and the pointer together with the pointer on the mark and tighten the bolt. Shut down the powerhead.

Throttle Linkage Adjustment
Wide Open Throttle

4- Twist the throttle grip to the **FAST (WOT)** position. Adjust the two locknuts on the outer throttle cable (wire) to bring the carburetor throttle lever into contact with the stopper. Tighten both locknuts against the throttle wire stay to hold this position.

Bring the left magneto base stopper in contact with the vertical timing pointer. Adjust the length of the link rod, which was removed in Step 1, to snap over the ball

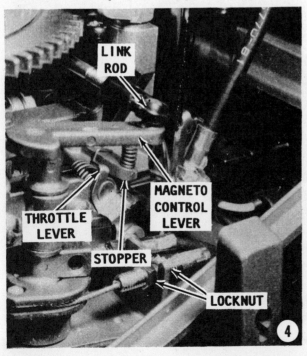

joint on the magneto control lever with two conditions existing: the throttle lever still contacts the stopper; the magneto base stopper still contacts the vertical timing pointer.

6-6 MODEL 8B, W8, MARATHON 8, 9.9HP AND 15Hp

GOOD WORDS

Many tasks may be performed without the outboard unit being mounted in a test tank or in a body of water. Actually, the powerhead will not be started until Step 9, when the work is almost completed.

Fully Advanced Ignition Timing

1- Shift the lower unit into **FORWARD** gear. Pry the link rod free from the ball joint on the magneto control lever.

2- Remove the cylinder head and secure a dial indicator gauge over the No. 1 piston dome. Rotate the flywheel **CLOCKWISE** until the piston is at top dead center (TDC). "Zero" the needle on the gauge. Rotate the flywheel **COUNTERCLOCKWISE** until the

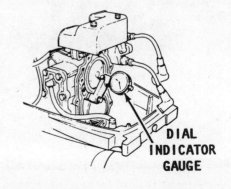

dial indicator gauge registers between 0.144 and 0.155" (3.66 and 3.94mm).

3- Rotate the CDI magneto base, under the flywheel, until the pointer on the base aligns between the triangle and 30° on the flywheel.

4- Loosen the two bolts on top of the throttle cam. Move the stopper, located underneath the throttle cam, until the stopper makes contact with the crankcase stopper. Tighten both bolts to hold this adjustment.

Fully Retarded Ignition Timing

5- Rotate the flywheel **CLOCKWISE** until the dial indicator gauge registers **Zero**

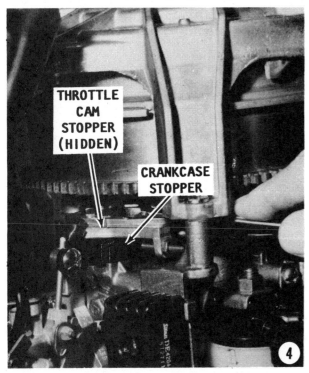

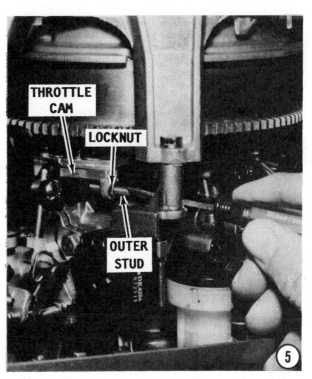

again. With the gauge at Zero, the piston will be at TDC. Now, rotate the flywheel **COUNTERCLOCKWISE** until the dial indicator registers between 0.003 and 0.007" (0.07 and 0.17mm). Loosen the locknut on the outer adjustment stud on the right side of the throttle cam. Rotate the CDI magneto base, under the flywheel, until the pointer on the base aligns between the 4° and 6° ATDC marks on the flywheel. Position the outer adjustment stud to just make contact with the crankcase stopper. Tighten the locknut to hold the adjustment.

6- Twist the throttle grip to the **FAST** position. Observe the location of the roller

6-10 TIMING & SYNCHRONIZING

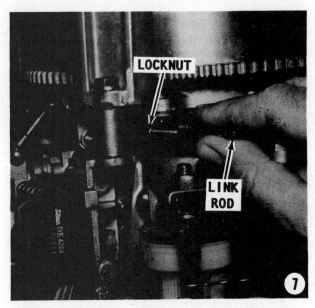

against the throttle cam. If necessary, move the throttle cam until the roller contacts the high point of the cam.

7- Loosen the locknut on the link rod. Adjust the length of the rod to enable the rod to snap onto the ball joint on the magneto control lever, **WITHOUT** any movement at the other end of the rod. Tighten the locknut to hold this adjustment. Rotate the throttle grip from **FAST** to the **IDLE** position. Observe and verify the crankcase stopper makes contact with the stopper beneath the throttle cam in the fully advanced position and then makes contact with the outer adjustment stud on the throttle cam in the fully retarded position.

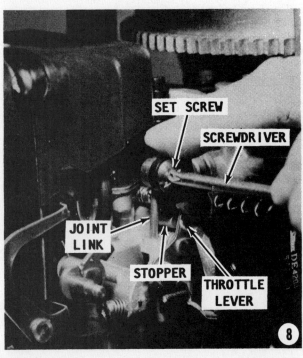

Throttle Linkage Adjustment at WOT In Forward Gear

8- Twist the throttle grip to the **FAST (WOT)** position. With the lower unit still in **FORWARD** gear, loosen the joint link set screw. Push down on the top of the joint link to bring the throttle lever into contact with the stopper and to position the throttle roller at the high point of the throttle cam. Tighten the joint link set screw to hold the adjustment.

9- Mount the engine in a test tank, or on a boat in a body of water. Remove the cowling. Connect a tachometer to the Green lighting coil leads. **DO NOT** start the powerhead at this time.

With the lower unit in **FORWARD** gear, twist the throttle grip to the **FAST (WOT)** position. Pry off the link rod from the ball joint at the top of the magneto control lever. Push in the lever until it makes contact with the stopper stay. Loosen the locknut on the link rod and adjust the length of the rod until it can be snapped back into place on the ball joint of the magneto control lever, **WITHOUT** any movement at the other end of the rod. When in place, tighten the locknut to hold the adjusted length. Return the throttle grip to the **START** position.

CAUTION
Never operate the powerhead above a fast idle with a flush attachment connected to the lower unit. Operating the powerhead at a high rpm with no load on the propeller shaft could cause the powerhead to **RUNAWAY** causing extensive damage to the unit.

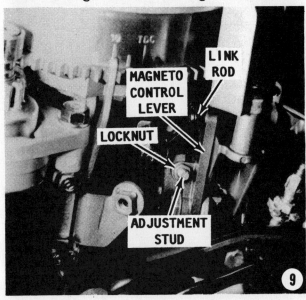

REMEMBER, the powerhead will **NOT** start without the emergency tether in place behind the "kill" switch knob.

CAUTION
Water must circulate through the lower unit to the powerhead anytime the powerhead is operating to prevent damage to the water pump in the lower unit. Just five seconds without water will damage the water pump impeller.

Start the powerhead and allow it to warm to operating temperature. After the powerhead has reached operating temperature, loosen the locknut and adjustment stud on the magneto control lever. Observe the tachometer and bring the powerhead rpm up to 1,000 to 1,600 rpm. Rotate the adjustment stud **CLOCKWISE** until it contacts the magneto control lever stopper. Hold this adjustment and at the same time tighten the locknut. Shut down the powerhead.

6-7 MODELS 20HP, 25HP, AND 30HP

FIRST, THESE WORDS
The magneto control lever is mounted on the starboard side of the powerhead. This lever is actually more like a disc than a lever, but the manufacturer's terminology is "lever". However, we call it a "disc".

Two link rods are snapped onto ball joints on small arms extending from the "disc". These link rods are of different lengths and are connected from the "disc" to other points on the powerhead.

One end of the horizontally mounted link rod is connected to the magneto control lever -- the other end to a ball joint on the underneath side of the CDI magneto base. This rod serves to advance the magneto base with increased throttle opening. This rod is called the magneto base to magneto control lever link rod. On Models 20hp and 25hp this rod is shaped like a "J". On the Model 30hp, this rod is perfectly straight.

One end of the vertically mounted link rod is connected to the magneto control lever -- the other end to the ball joint on the throttle control lever. The purpose of this longer rod is to limit throttle opening whenever the lower unit is in **NEUTRAL** gear. This rod is called the throttle control to carburetor control lever link rod.

The fully retarded ignition timing is automatically adjusted when the fully advanced ignition timing is set correctly.

Fully Advanced Ignition Timing Adjustment
1- Pry off the link rod between the magneto control lever and the CDI magneto base at the base ball joint. Remove the

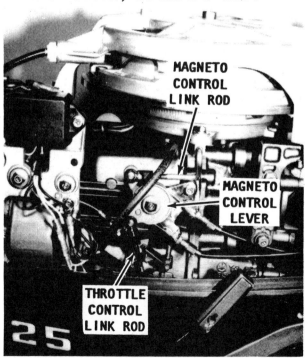

Location of the throttle and magneto control link rods on a 25hp powerhead. The 20hp and the 30hp setup is slightly different.

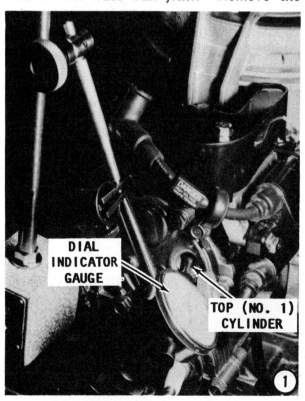

6-12 TIMING & SYNCHRONIZING

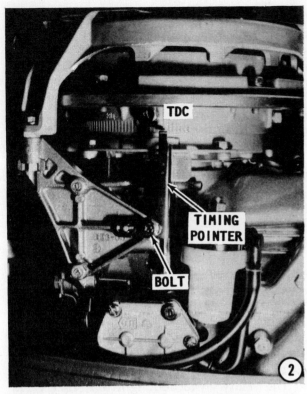

No. 1 (top) spark plug and install a dial indicator gauge into the cylinder on top of the piston crown. Rotate the flywheel **CLOCKWISE** until the dial indicator registers TDC (top dead center).

2- Observe the timing plate on the port side of the powerhead. Check to verify the timing pointer is aligned with the TDC mark embossed on the flywheel. If the pointer is not aligned, loosen the timing plate set bolt

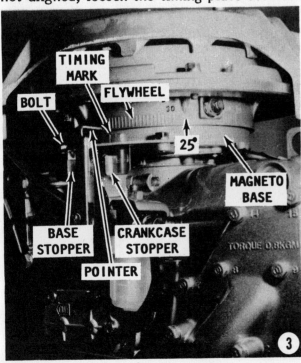

beneath the pointer and move the pointer to the correct location. Tighten the bolt to hold the pointer aligned with the TDC mark. Remove the dial indicator gauge and install the spark plug. Connect the high tension lead.

3- Rotate the flywheel **CLOCKWISE** until the timing pointer aligns with the appropriate mark embossed on the flywheel:

20hp -- 21° BTDC mark
25hp -- 24° BTDC mark
30hp -- 25° BTDC mark

Rotate the magneto base **COUNTERCLOCKWISE** to align the "single line" timing mark on the base with the TDC mark on the flywheel. Check to verify the magneto base stopper makes contact with the crankcase stopper. If they do not make contact or if they do make contact but the TDC mark is not aligned with the "single line" mark on the base, loosen the base stopper set bolt, align the marks, bring the stoppers into contact, and then tighten the bolt.

Throttle Cam and Throttle Link Adjustment at WOT

FIRST, THESE WORDS

The adjustments made in the following two steps, No. 4 and No. 5, are performed with the magneto base still set in the fully advanced position and with the magneto control lever link rod still disconnected as directed in the previous steps.

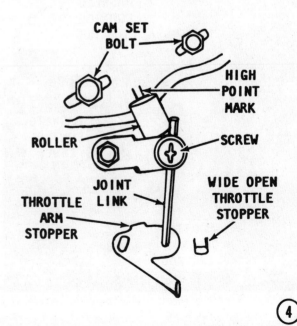

TIMING TYPE II IGNITION 6-13

4- Loosen the screw on the joint link to allow the throttle roller to slide up and down on the link. Loosen the two cam set bolts. Align the center of the roller with the two small embossed marks on the cam indicating the high point. Tighten the two cam set bolts to secure the throttle cam, but leave the joint link screw loose at this time. Push down on the joint link bringing the throttle arm stopper into contact with the wide open throttle stopper. At the same time, hold the throttle roller against the throttle cam and tighten the joint link screw to secure the roller in this position.

Throttle Linkage Adjustment
At WOT in Forward Gear

5- Shift the lower unit into **FORWARD** gear. Rotate the CDI magneto base until the stopper on the base contacts the crankcase stopper. This action sets the base to the fully advanced position. Loosen the locknut at the disconnected end of the short link rod.

On Models 20hp and 25hp: Adjust the length of the link rod until it can be snapped back onto the ball joint on the magneto control lever, **WITHOUT** any movement at the other end of the rod. Snap the link rod in place and tighten the locknut to hold the new adjusted length.

On the Model 30hp: Adjust the length of the rod until the measurement from the ball joint center to the ball joint center is 2 3/4" (69.6mm). After the measurement is satisfactory, snap the rod over the inside magneto control lever ball joint and tighten the locknut on the rod to hold this adjusted length.

Pry off the long link rod from the ball joint on the magneto control lever. Loosen the locknut at the disconnected end of the long link rod. The other end of this long link rod still remains connected to the throttle control lever. Push the throttle control lever downward until it contacts the stopper on the bottom cowling. Adjust the length of the long link rod until it can be snapped back onto the ball joint on the magneto control lever, **WITHOUT** any movement at the other end of the rod. Snap the link rod in place and tighten the locknut to hold the new adjusted length.

GOOD WORDS

Powerhead has been **STATICALLY** (powerhead not operating) timed for the fully advanced timing setting. There is no need to make any further adjustments for the fully retarded timing setting. The timing must be checked **DYNAMICALLY** (powerhead operating) and with the outboard unit mounted in a large test tank or on a boat in a body of water.

Throttle Linkage Adjustment
In Neutral Gear

6- Mount the engine in a test tank or on a boat in a body of water.

CAUTION WORDS

DO NOT use a flush attachment connected to the lower unit for this adjustment. The powerhead will be operated beyond an idle speed.

Remove the cowling. Connect a tachometer to the Green lighting coil leads.
REMEMBER, the powerhead will **NOT** start unless the emergency tether is in place behind the "kill" switch knob.

CAUTION

Water must circulate through the lower unit to the powerhead anytime the powerhead is operating to prevent damage to the

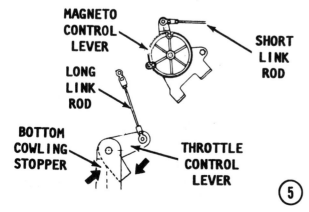

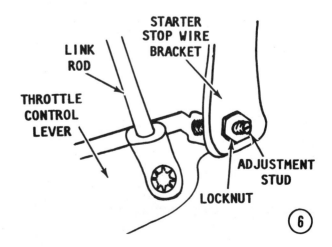

6-14 TIMING & SYNCHRONIZING

water pump in the lower unit. Just five seconds without water will damage the water pump impeller.

Start the engine. Loosen the locknut, then the adjustment stud on the starter stop wire bracket. Observe the tachometer.

Rotate the throttle grip until the tachometer indicates 3000 to 3500 rpm. Rotate the adjustment stud **CLOCKWISE** until the stud contacts the throttle control lever. Tighten the locknut to hold the stud adjustment.

WARNING
BE CAREFUL NOT TO TOUCH THE SPINNING FLYWHEEL TEETH WHILE PERFORMING THE NEXT STEP. SUCH ACTION COULD CAUSE PERSONAL INJURY.

7- Clip the pickup of a timing light around the upper (No. 1) cylinder high tension lead. Disconnect the link rod from the magneto control lever to the magneto base plate at the ball joint on the base plate. Rotate the base plate **COUNTERCLOCKWISE** until the stopper on the plate contacts the crankcase stopper. Loosen the screw on the carburetor link rod to allow the roller to move over the cam **WITHOUT** increasing powerhead speed.

Now, aim the timing light at the marks embossed on the flywheel and check the alignment of the vertical timing pointer against the appropriate mark on the flywheel:

20hp -- 21° BTDC mark
25hp -- 24° BTDC mark
30hp -- 25° BTDC mark

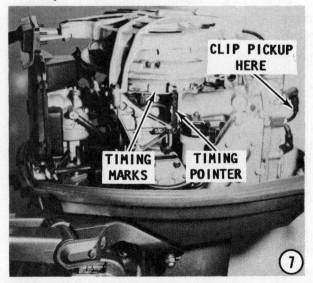

If the pointer aligns, the timing is set correctly. If the pointer does not align, loosen the magneto base stopper bolt and align the ignition timing mark with the vertical pointer.

Rotate the magneto base to allow the stopper on the base to contact the crankcase stopper, and then tighten the bolt. Shut down the powerhead. Repeat Step 4 in this section to correctly position the roller on the carburetor link rod.

Tiller Handle Models Only

8- Loosen the top locknut on the upper throttle wire and tighten the bottom locknut to take up the slack on the throttle wire. Loosen the top locknut on the lower throttle wire and tighten the bottom locknut to provide a deflection of 0.04 to 0.08" (1 to 2mm) at the mid point of the wire length.

6-8 MODEL 40HP SINGLE CARBURETOR INSTALLATIONS

SAFETY WORDS
Remove both spark plugs to prevent accidental startup of the powerhead. Shift the outboard into **FORWARD** gear.

Fully Advanced Ignition Timing Adjustment

1- Pry off the link rod between the arm at the top of the vertical shaft and the ball joint on the underside of the CDI magneto base, at the ball joint. Install a dial indicator gauge into the No. 1 (top) cylinder opening on top of the piston crown. Rotate the

TIMING TYPE II IGNITION 6-15

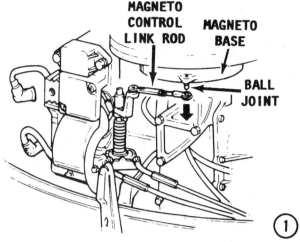

flywheel **CLOCKWISE** until the dial indicator registers TDC (top dead center).

2- Check to verify the timing pointer is aligned with the TDC mark embossed on the flywheel and also aligned with the TDC mark on the decal stuck to the flywheel. If the pointer is not aligned, relocate the decal to make alignment between the timing pointer and the TDC mark on the decal correct. The location of the timing pointer is not adjustable on the 40hp model. Remove the dial indicator gauge.

3- Rotate the flywheel **CLOCKWISE** until the timing pointer aligns with 22° BTDC on the timing decal. Rotate the magneto base **COUNTERCLOCKWISE** to align the "single line" timing mark on the base with the TDC mark on the decal. Check to verify the left stop on the magneto base stopper makes contact with the timing pointer. If they do not make contact or if they do make contact but the TDC mark is not aligned with the "single line" mark on the base, loosen the base stopper set bolts, align the marks, bring the stoppers into contact, and then tighten the bolts.

Idle Timing

4- Loosen the locknut on the idle timing adjustment screw. Continue to hold the

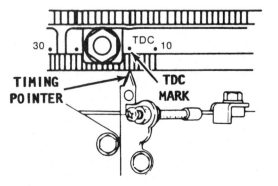

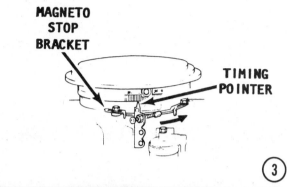

The single line embossed on the magneto base aligned with the TDC mark on the timing decal, as described in the text.

flywheel with the timing pointer aligned with the 22° BTDC mark on the timing decal.

Rotate the magneto base **CLOCKWISE** to bring the rear magneto stop bracket up against the idle timing adjustment screw.

Rotate the screw until the embossed line on the magneto base aligns with 22° BTDC timing mark on the timing decal. This action will set the idle timing at 4° BTDC. Tighten the locknut on the idle timing adjustment screw to hold this new adjustment.

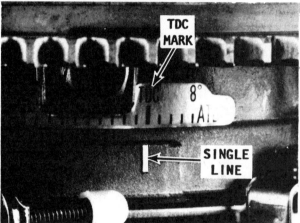

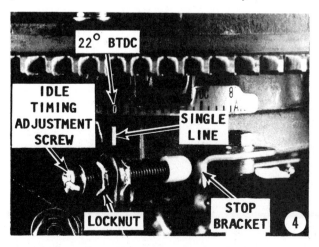

6-16 TIMING & SYNCHRONIZING

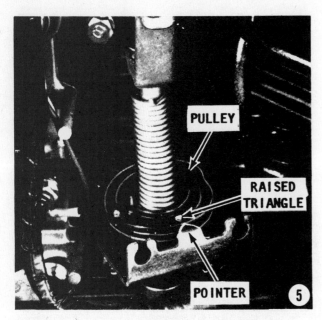

Throttle Control Linkage Adjustment

5- Continue to hold the flywheel as directed in Step 4 with the timing pointer aligned with the 22° BTDC mark on the timing decal and the rear magneto stop bracket up against the idle timing adjustment screw.

GOOD WORDS

Observe the two triangular marks embossed on the upper face of the vertical control linkage pulley. The mark on the left is the alignment mark for idle speed, and this mark is slightly raised, compared to the mark on the right side. This mark will be referred to as the "idle triangle". The triangular mark on the right side is the alignment mark at the maximum advanced position. This mark will be referred to as the "maximum advance triangle".

Rotate the twist grip to align the idle triangle with the pointer on the cable support bracket.

6- Loosen the locknut on the disconnected magneto control link rod. Adjust the length of the rod to enable the rod to snap onto the ball joint on the magneto base, **WITHOUT** any movement at the other end of the rod. Tighten the locknut to hold this adjustment. Shift the outboard into **NEUTRAL**.

SPECIAL WORDS

The tasks performed in the next two steps may throw off the adjustment made in Steps 5 and 6. Each step describes how to correct the adjustment. **DO NOT** attempt to made a correction by altering the length of the magneto control link rod, set in Steps 5 and 6.

Models Equipped with Twist Grip

7- Back off the locknuts on the turnbuckles on the twist grip cables.

Rotate the twist grip to the **SLOW** position. The idle triangle on the upper face of the pulley should align with the pointer on the cable support bracket. If not, rotate one or both turnbuckles on the twist grip cables to bring the pulley into alignment with the pointer.

Rotate the twist grip to the **FAST** position. The maximum advance triangle on the upper face of the pulley should align with the pointer on the cable support bracket. If not, rotate one or both turnbuckles on the twist grip cables to bring the pulley into alignment with the pointer.

These two adjustments should be accomplished with a maximum of 0.20" (5mm) play in the twist grip handle. Tighten the locknuts against the turnbuckles to hold this new adjustment.

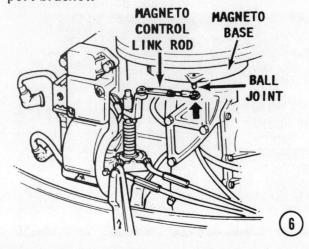

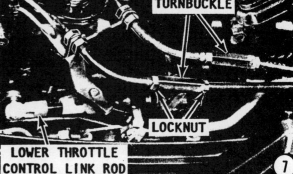

TIMING TYPE II IGNITION   6-17

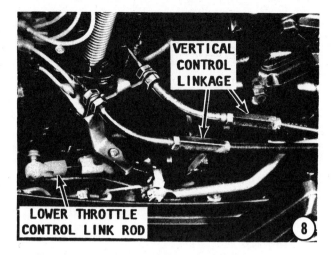

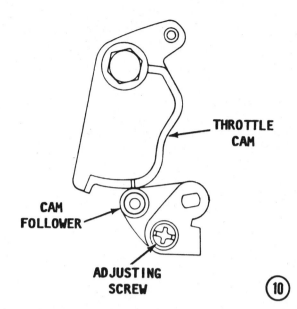

Models Equipped with Remote Control

8- Pry the short lower throttle control link rod from the ball joint on the throttle lever. This link rod extends from a lower arm of the vertical control linkage to a ball joint on the throttle lever inside the lower cowling. Rotate the vertical control linkage to align the maximum advance triangle on the upper face of the pulley with the pointer on the cable support bracket. Move the throttle lever on the starboard side inside the lower cowling until the lever contacts the stop block on the cowling.

Loosen the locknut on the disconnected throttle control link rod. Adjust the length of the rod to enable the rod to snap onto the ball joint on the throttle lever **WITHOUT** any movement at the other end of the rod. Tighten the locknut to hold this adjustment.

9- Disconnect the throttle cable from the throttle lever. Rotate the vertical control linkage to align the idle triangle on the upper face of the pulley with the pointer on the cable support bracket. With the remote control shift handle in the **NEU-**

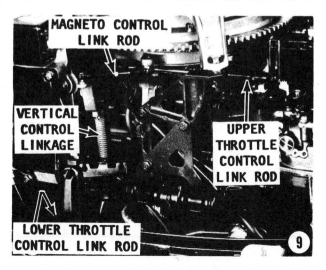

TRAL position, adjust the throttle cable length to allow the cable to be attached to the throttle lever without disturbing the position of the vertical control linkage pulley.

Pry the long upper throttle control link rod from the ball joint on the throttle cam. This rod extends from a ball joint on one of the upper arms of the vertical control linkage to a ball joint on the throttle cam.

Rotate the vertical control linkage to align the maximum advance triangle on the upper face of the pulley with the pointer on the cable support bracket.

Loosen the locknut on the disconnected throttle control link rod. Adjust the length of the rod to enable the rod to snap onto the ball joint on the throttle cam **WITHOUT** any movement at the other end of the rod. Tighten the locknut to hold this adjustment.

10- Adjust the throttle cam follower to eliminate any play in the throttle shaft when the follower is in the position shown in the accompanying illustration.

6-9 MODEL 40HP
DUAL CARBURETOR INSTALLATIONS

SAFETY WORDS
Remove both spark plugs to prevent accidental startup of the powerhead. Shift the outboard into **FORWARD** gear.

Fully Advanced Ignition
Timing Adjustment
1- Pry off the link rod between the arm at the top of the vertical shaft and the ball

6-18 TIMING & SYNCHRONIZING

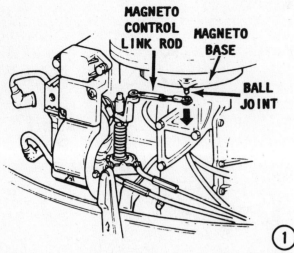

joint on the underside of the CDI magneto base, at the ball joint. Remove the No. 1 (top) spark plug and install a dial indicator gauge into the cylinder on top of the piston crown. Rotate the flywheel **CLOCKWISE** until the dial indicator registers TDC (top dead center).

2- Check to verify the timing pointer is aligned with the TDC mark embossed on the flywheel and also aligned with the TDC mark on the decal on the flywheel. If the pointer is not aligned, relocate the decal to make alignment between the timing pointer and the TDC mark on the decal correct. The location of the timing pointer is not adjustable on the 40hp model. Remove the dial indicator gauge.

3- Rotate the flywheel **CLOCKWISE** until timing pointer aligns with 22° BTDC mark on the timing decal. Rotate the magneto base **COUNTERCLOCKWISE** to align the "single line" timing mark on the base with the TDC mark on the decal. Check to verify the left stop on the magneto base stopper makes contact with the timing pointer. If they do not make contact or if they do make contact but the TDC mark is not aligned with the "single line"

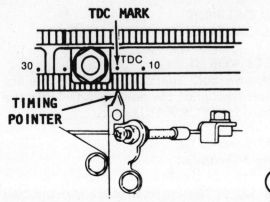

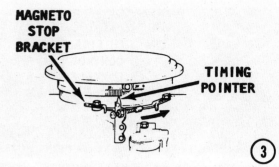

mark on the base, loosen the base stopper set bolts, align the marks, bring the stoppers into contact, and then tighten the bolts.

Idle Timing

4- Loosen the locknut on the idle timing adjustment screw. Rotate the flywheel **CLOCKWISE** until the timing pointer aligns with the 2° BTDC mark on the timing decal. Prevent movement of the flywheel and at the same time, rotate the magneto base **CLOCKWISE** to bring the rear magneto stop bracket up against the idle timing adjustment screw. Continue to hold the flywheel as directed with the timing pointer aligned with the 2° BTDC mark on the timing decal.

Rotate the screw until the embossed line on the magneto base aligns with the TDC timing mark on the timing decal. This action will set the idle timing at 2° TDC. Tighten the locknut on the idle timing adjustment screw to hold this new adjustment.

SPECIAL WORDS
LINK ROD IDENTIFICATION

Two throttle control link rods are used on this powerhead: A short link rod between the ball joint on the throttle lever and the ball joint on the lower arm of the vertical control linkage. This link rod will be referred to as the "short rod".

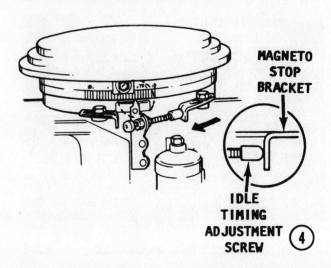

TIMING TYPE II IGNITION 6-19

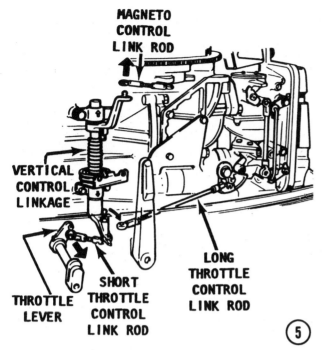

The other throttle control link rod is much longer and extends between the ball joint on the same lower arm of the vertical control linkage and the throttle cam at the carburetors. This link rod will be referred to as the "long rod".

Throttle Cam Adjustment

5- With the magneto control link rod still disconnected, as directed in Step 1, pry off the **LONG** throttle control rod from the ball joint on the vertical control linkage arm ball joint. Then, pry off the **SHORT** throttle control link rod from the ball joint on the throttle lever. Loosen the locknut on the **LONG** throttle control link rod at this time.

If the model being serviced is equipped with a steering handle, loosen both locknuts and turnbuckles to allow full unrestricted movement of the pulley.

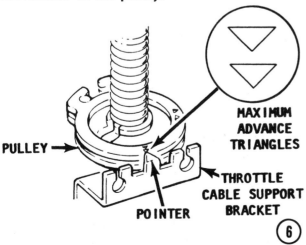

GOOD WORDS

Observe the two sets of triangular marks embossed on the upper face of the vertical control linkage pulley. The mark on the left is the alignment mark for idle speed, and this mark has two triangles side-by-side.

This mark is never used for alignment on 40hp models with dual carburetor installations.

The triangular mark on the right side is the alignment mark at the maximum advanced position. This mark has two triangles one over the other. This mark will be referred to as the "maximum advance triangle".

6- Rotate the vertical control linkage until the maximum advance triangles on the pulley align with the pointer on the throttle cable support bracket. Hold this alignment and at the same time position the WOT mark embossed on the throttle cam against the centerline of the throttle roller.

7- Adjust the length of the **LONG** throttle control link rod to enable the rod to snap onto the ball joint on the vertical control linkage arm **WITHOUT** any movement at the other end of the rod. Tighten the locknut to hold this adjustment.

Throttle Pickup Timing Adjustment

8- Rotate the flywheel **CLOCKWISE** to align the timing pointer with the 6°BTDC mark on the timing decal. Shift the out-

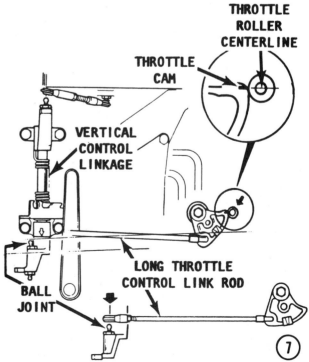

6-20 TIMING & SYNCHRONIZING

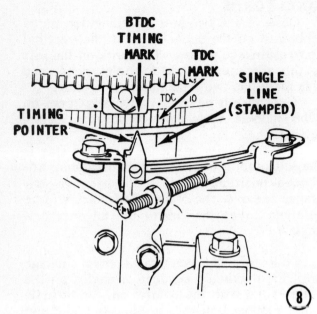

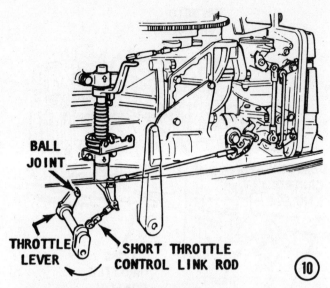

board into **FORWARD** gear. Rotate the magneto base **COUNTERCLOCKWISE** to align the "single line" timing mark on the base with the TDC mark on the decal.

9- Loosen the locknut on the magneto control link rod. Position the throttle cam to **BARELY** make contact with the throttle roller as shown in the accompanying illustration. Adjust the length of the magneto control link rod to enable the rod to snap onto the ball joint on the vertical control linkage arm **WITHOUT** any movement at the other end of the rod or disturbing the throttle cam, flywheel, or magneto base. Tighten the locknut to hold this adjustment.

Throttle Linkage Adjustment
10 With the outboard still in **FORWARD** gear and the **SHORT** throttle link rod still disconnected, rotate the vertical control linkage until the maximum advance triangles on the pulley align with the pointer on the throttle cable support bracket. Rotate the throttle lever, inside the lower cowling, **CLOCKWISE** to contact the stop.

Loosen the locknut on the disconnected **SHORT** throttle control link rod. Adjust the length of the rod to enable the rod to snap onto the ball joint on the throttle lever **WITHOUT** any movement at the other end of the rod. Tighten the locknut to hold this adjustment.

Throttle Cable Adjustment
Models Equipped with Steering Handle
11- Back off the locknuts on the turnbuckles on the throttle cables.

Rotate the twist grip to the **FAST** position. The maximum advance triangles on the upper face of the pulley should align with the pointer on the cable support bracket. If not, adjust the turnbuckle on the outermost throttle cable to bring the pulley into alignment with the pointer.

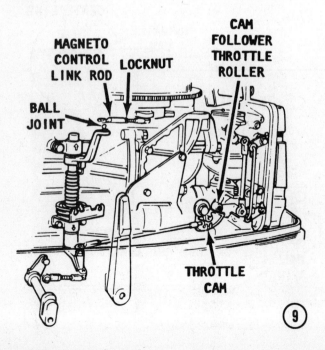

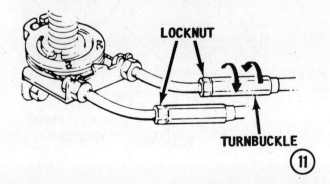

TIMING TYPE II IGNITION 6-21

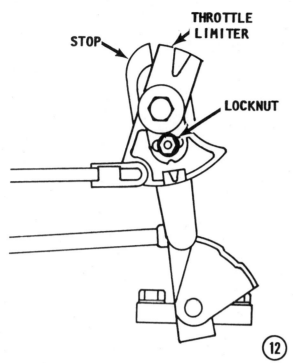

Rotate the twist grip to the **SLOW** position. Adjust the turnbuckle on the innermost throttle cable to hold the magneto base stop **LIGHTLY** against the idle adjustment timing screw. Tighten the locknuts against the turnbuckles on both cables to hold this new adjustment.

Neutral Throttle Limiter Adjustment for Models Equipped with Remote Control

12- Loosen the locknut on the throttle limiter stop. Install both spark plugs and high tension leads. Start the powerhead and allow it to reach operating temperature. Connect a tachommeter to the powerhead. See special words on tachommeter connections on Page 6-2.

REMEMBER, if the powerhead is equipped with a "kill" switch knob, the powerhead will **NOT** start without the emergency tether in place behind the "kill" switch knob.

CAUTION

Water must circulate through the lower unit to the powerhead anytime the powerhead is operating to prevent damage to the water pump in the lower unit. Just five seconds without water will damage the water pump impeller.

NEVER, AGAIN, NEVER operate the engine at high speed with a flush device attached. The engine, operating at high speed with such a device attached, would **RUNA-**

WAY from lack of a load on the propeller, causing extensive damage.

With the outboard in **NEUTRAL** gear, advance the throttle until powerhead speed is between 2900 and 3500 rpm.

WITHOUT disturbing the position of the throttle, shut down the powerhead by one of the following methods:

a- Shutting off the fuel supply by disconnecting the fuel joint.
b- Grounding both spark plugs using two large screwdrivers with **INSULATED** handles.
c- Twisting off the high tension leads.
d- Disconnecting the Black and Orange leads between the CDI unit and the ignition coil at their quick disconnect fittings.

Once the powerhead is not operating, position the throttle limiter against the throttle limiter stop and tighten the locknut to hold this position.

Start the powerhead once again.

REMEMBER, if the powerhead is equipped with a "kill" switch knob, the powerhead will **NOT** start without the emergency tether in place behind the "kill" switch knob.

Attempt to advance the throttle to its limit in **NEUTRAL** gear. The powerhead speed should not advance beyond 3500 rpm under this new adjustment.

6-10 MODELS 48HP, 55HP, AND 60HP

Magneto Control Link Rod Adjustment

1- Pry off the short magneto control link rod between the ball joint on the upper

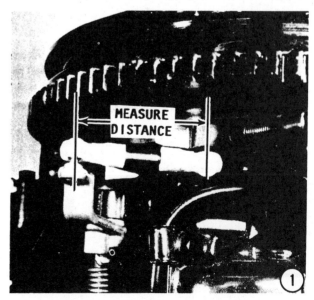

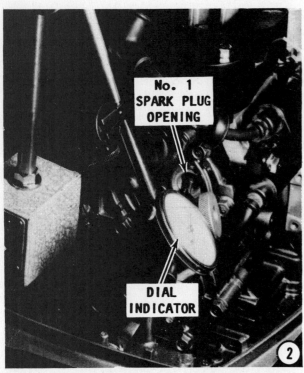

arm of the vertical control linkage and the ball joint on the underside of the magneto base.

Loosen the locknut and adjust the center-to-center distance to be 2-3/64" (52mm) for models prior to 1980 and 2" (50mm) for models 1980 and later. Tighten the locknut, and then snap the link back in place.

Ignition Timing Adjustment

2- Remove the No. 1 (top) spark plug and install a dial indicator gauge onto the top of the piston crown. Rotate the flywheel **CLOCKWISE** until the dial indicator registers TDC (top dead center).

3- Observe the timing plate on the aft side of the powerhead. Check to verify the timing pointer is aligned with the TDC mark on the plate.

If the pointer is not aligned, loosen the set bolt beneath the pointer and move the pointer to the correct location. Tighten the bolt to hold the pointer aligned with the TDC mark. Remove the dial indicator gauge and install the spark plug.

Idle Timing Adjustment

4- Rotate the flywheel to align the appropriate timing mark, with the timing pointer as follows:

48hp 4° ATDC
55hp 2° ATDC
60hp 2° ATDC

Observe the window on the upper face of the flywheel. A mark on the window edge must align with a raised line on the pulsar coil, mounted on the stator plate, when viewed through the window.

DO NOT MOVE the flywheel to make this adjustment, because the flywheel is already set to the idle timing specification. To make this adjustment, rotate the magneto base until the line on the flywheel and the line on the pulsar coincide.

5- Loosen the locknut on the magneto base stop screw and adjust the screw until its padded end contacts the magneto stop. This screw is located to the left of the vertical control linkage.

Tighten the locknut to hold this new adjustment.

Maximum Advance Timing Adjustment

6- Rotate the flywheel to align the appropriate timing mark with the timing pointer:

48hp 20° BTDC
55hp 26° BTDC
60hp 26° BTDC
(prior to 1980)
60hp 22° BTDC
(1980 and later)

TIMING TYPE II IGNITION

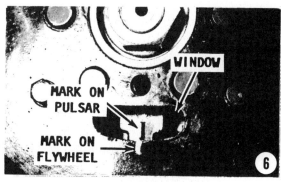

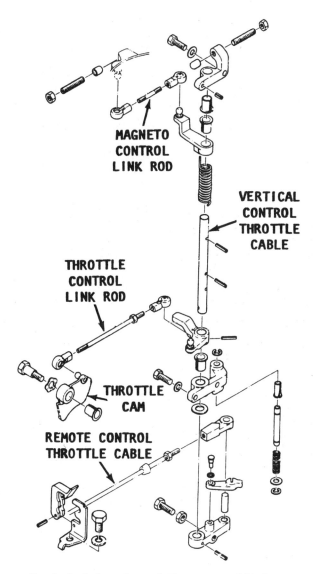

Exploded drawing of the control linkage.

Rotate the magneto base **COUNTERCLOCKWISE** and once again observe the window on the flywheel. Continue rotating the flywheel until the line on the window edge coincides with the line on the pulsar coil. Loosen the locknut on the magneto stop screw located to the right of the vertical control linkage. Adjust the screw until

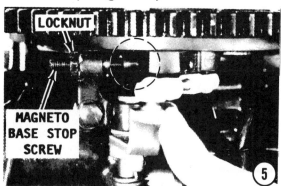

the padded end contacts the advance side of the magneto stop. Tighten the screw to hold this new adjustment.

Pickup Timing

7- Identify the **LONG** throttle control link rod, which extends from the magneto control lever to the throttle cam. Pry this **LONG** link rod free from the ball joint on the magneto control lever, and loosen the locknut on the rod.

Use this disconnected link rod to rotate the throttle cam until the cam begins to make contact with the throttle roller. At this point the throttle shutter valves are about to open.

Rotate the flywheel until the appropriate timing mark on the timing plate aligns with the timing pointer.:

$$
\begin{array}{ll}
48\text{hp} & 4° \text{ BTDC} \\
55\text{hp} & \text{TDC} \\
60\text{hp} & 5° \text{ ATDC}
\end{array}
$$

Adjust the length of the rod to enable it to be snapped back into place on the magneto control lever ball joint, **WITHOUT** disturbing the position of the magneto control lever or the throttle cam. Tighten the locknut to hold this new adjustment.

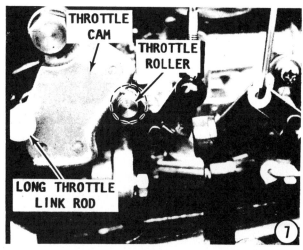

6-24 TIMING & SYNCHRONIZING

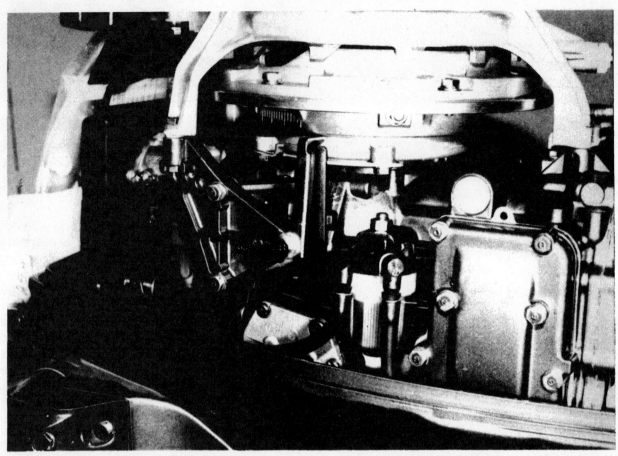

A clean, well serviced powerhead will contribute to the ease and accuracy of making proper adjustments during the timing and synchronizing operation described in this Chapter.

7
POWERHEAD

7-1 INTRODUCTION AND CHAPTER ORGANIZATION

This chapter is divided into eight main working sections as follows:
7-2 Two-Cycle Powerhead Description and operation.
7-3 Service Single-cylinder 2hp water-cooled Powerhead.
7-4 Service Single-Cylinder 3.5hp & 5hp air-cooled Powerheads.
7-5 Service Single-Cylinder 4hp & 5hp water-cooled Powerheads.
7-6 Service 2-Cylinder water-cooled, 8hp, 9.9hp, 15hp, 20hp, 25hp, 28hp, 30hp, 40hp, 48hp, 55hp, and 60hp Powerheads.
7-7 Cleaning & Inspecting all Powerheads

The carburetion and ignition principles of two-cycle operation **MUST** be understood in order to perform proper service work on the outboard powerheads covered in this manual.

Repair Procedures

Service and repair procedures will vary slightly between individual models, but the basic instructions are quite similar. Special tools may be called out in certain instances. These tools may be purchased from the local marine dealer.

Torque Values

All torque values must be met when they are specified. Torque values for various parts of each powerhead are given in the text.

A torque wrench is essential to correctly assemble the powerhead. **NEVER** attempt to assemble a powerhead without a torque wrench. Attaching bolts **MUST** be tightened to the required torque value in three progressive stages, following the specified tightening sequence. Tighten all bolts to 1/3 the torque value, then repeat the sequence tightening to 2/3 the torque value. Finally, on the third and last sequence, tighten to the full torque value.

Powerhead Components

Service procedures for the carburetors, fuel pumps, starter, and other powerhead components are given in their respective Chapters of this manual. See the Table of Contents.

Reed Valve Service

The reeds on two-cycle powerheads covered in this manual are contained in an externally mounted reed valve block. Therefore, the powerhead need not be disassembled in order to replace a broken reed.

Cleanliness

Make a determined effort to keep parts and the work area as clean as possible.

The exterior and interior of the powerhead must be kept clean, well-lubricated, and properly tuned and adjusted, if the owner is to receive the maximum enjoyment from the unit.

7-2 TWO-CYLCE POWERHEAD DESCRIPTION AND OPERATION

Intake/Exhaust

Two-cycle engines utilize an arrangement of port openings to admit fuel to the combustion chamber and to purge the exhaust gases after burning has been completed. The ports are located in a precise pattern in order for them to be opened and closed at an exact moment by the piston as it moves up and down in the cylinder. The exhaust port is located slightly higher than the fuel intake port. This arrangement opens the exhaust port first as the piston starts downward and therefore, the exhaust phase begins a fraction of a second before the intake phase.

Actually, the intake and exhaust ports are spaced so closely together that both open almost simultaneously. For this reason, the pistons of most two-cycle engines have a deflector-type top. This design of the piston top serves two purposes very effectively.

First, it creates turbulence when the incoming charge of fuel enters the combustion chamber. This turbulence results in more complete burning of the fuel than if the piston top were flat.

The second effect of the deflector-type piston crown is to force the exhaust gases from the cylinder more rapidly.

Lubrication

A two-cycle engine is lubricated by mixing oil with the fuel. Therefore, various parts are lubricated as the fuel mixture passes through the crankcase and the cylinder.

Physical Laws

The two-cycle engine is able to function because of two very simple physical laws.

One: Gases will flow from an area of high pressure to an area of lower pressure. A tire blowout is an example of this principle. The high-pressure air escapes rapidly if the tube is punctured.

Two: If a gas is compressed into a smaller area, the pressure increases, and if a gas expands into a larger area, the pressure is decreased.

If these two laws are kept in mind, the operation of the two-cycle engine will be easier understood.

TWO-CYCLE POWERHEAD OPERATION

Beginning with the piston approaching top dead center on the compression stroke: The intake and exhaust ports are closed by the piston; the reed valve is open; the spark plug fires; the compressed air/fuel mixture is ignited; and the power stroke begins. The reed valve was open because as the piston moved upward, the crankcase volume increased, which reduced the crankcase pressure to less than the outside atmosphere.

As the piston moves downward on the power stroke, the combustion chamber is filled with burning gases. As the exhaust port is uncovered, the gases, which are under great pressure, escape rapidly through the exhaust ports. The piston continues its

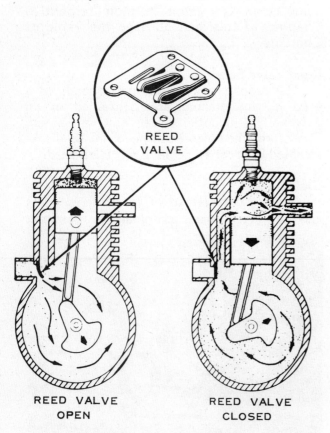

Reed valves are used to control the flow of air/fuel into the crankcase and eventually into the cylinder. As the piston moves upward in the cylinder, the resulting suction in the crankcase overcomes the spring tension of the reed. The reed is pulled free from its seat and the air/fuel mixture is drawn into the crankcase.

downward movement. Pressure within the crankcase increases, closing the reed valves against their seats. The crankcase then becomes a sealed chamber. The air/fuel mixture is compressed ready for delivery to the combustion chamber. As the piston continues to move downward, the intake port is uncovered. A fresh air/fuel mixture rushes through the intake port into the combustion chamber striking the top of the piston where it is deflected along the cylinder wall. The reed valve remains closed until the piston moves upward again.

When the piston begins to move upward on the compression stroke, the reed valve opens because the crankcase volume has been increased, reducing crankcase pressure to less than the outside atmosphere. The intake and exhaust ports are closed and the fresh fuel charge is compressed inside the combustion chamber.

Pressure in the crankcase decreases as the piston moves upward and a fresh charge of air flows through the carburetor picking up fuel. As the piston approaches top dead center, the spark plug ignites the air/fuel mixture, the power stroke begins and one full cycle has been completed.

TIMING

The exact time of spark plug firing depends on engine speed. At low speed the spark is retarded, fires later than when the piston is at or beyond top dead center. Engine timing is built into the unit at the factory.

At high speed, the spark is advanced, fires earlier than when the piston is at top dead center.

Summary

More than one phase of the cycle occurs simultaneously during operation of a two-cycle engine. On the downward stroke, power occurs above the piston while the ports are closed. When the ports open, exhaust begins and intake follows. Below the piston, fresh air/fuel mixture is compressed in the crankcase.

On the upward stroke, exhaust and intake continue as long as the ports are open. Compression begins when the ports are closed and continues until the spark plug ignites the air/fuel mixture. Below the piston, a fresh air/fuel mixture is drawn into the crankcase ready to be compressed during the next cycle.

7-3 SERVICING POWERHEAD "A" ONE-CYLINDER, TWO-CYCLE, 2HP

ADVICE

Before commencing any work on the powerhead, an understanding of two-cycle engine operation will be most helpful. Therefore, it would be well worth the time to study the principles of two-cycle engines, as outlined briefly in Section 7-1. A Polaroid, or equivalent instant-type camera is an extremely useful item, providing the means of accurately recording the arrangement of parts and wire connections **BEFORE** the disassembly work begins. Such a record is invaluable during assembling.

Preliminary Tasks

Remove the cowling, the hand starter, the flywheel, and the stator plate, see Chapter 5.

Remove the carburetor and fuel tank, see Chapter 4.

REMOVAL AND DISASSEMBLING

The following instructions pickup the work after the preliminary tasks listed above have been accomplished.

The link rod rotates the stator plate under the flywheel to achieve advanced or retarded timing. The link rod is activated by the throttle cable.

7-4 POWERHEAD

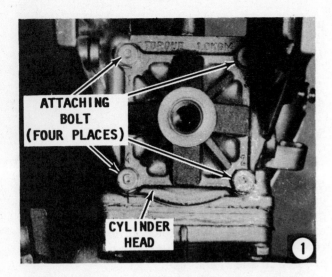

1- Remove the four bolts securing the cylinder head to the block. Lift the fuel tank tray free of the powerhead.

2- Remove the cylinder head and the gasket. The head may have to be tapped lightly with a soft mallet to "shock" the gasket seal free of the powerhead. Remove the gasket material from the mating surfaces of the head and the block.

3- Remove the six bolts securing the powerhead to the intermediate housing.

4- Tap the side of the powerhead with a soft head mallet to "shock" the gasket seal, and then lift the powerhead from the intermediate housing.

5- Remove the gasket, plate, and a second gasket from the intermediate housing. Remove and **TAKE CARE** not to drop or lose the dowel pin seated in the intermediate housing.

6- Remove the four bolts securing the horseshoe bracket and reed valve housing to the powerhead. Separate the horseshoe bracket from the block.

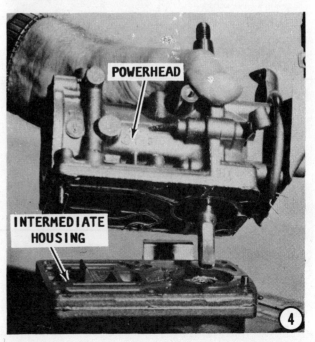

DISASSEMBLE 2HP   7-5

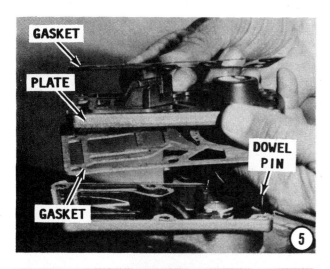

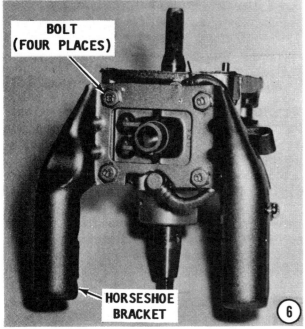

7- CAREFULLY lift off the valve reed housing and gasket. Note and remember the direction of the reeds as an aid during installation. It is possible to install this housing backwards by mistake.

8- Measure the reed stop clearance. The clearance should be 0.24" (6mm). If the measurement is not within 0.010" (0.25mm), the reed stop must be replaced. Check the reed valve for warpage. If the valve is warped more than 0.012" (0.3mm) the valve must be replaced.

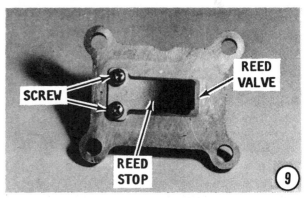

7-6 POWERHEAD

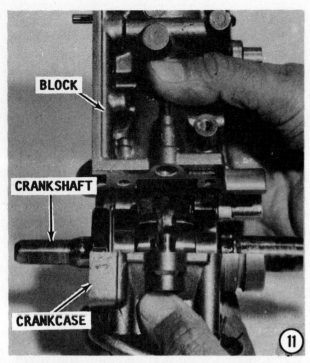

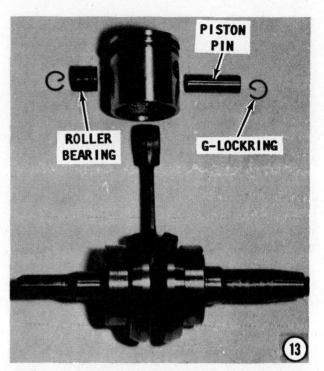

9- After the reed housing has been removed, inspect the reed valve. If the valve is distorted or cracked, a proper seal cannot be obtained. Therefore, the reed valve **MUST** be replaced.

10- Remove the two bolts holding the two halves of the crankcase together Note the position of the wire clip on one of these bolts as an aid during installation.

11- Separate the two crankcase halves. It may be necessary to "shock" one half with a softhead mallet in order to jar the gasket sealing qualities loose. Slowly pull the two halves apart. The crankshaft will remain with one half, and the piston assembly will come out of the cylinder in the other half.

12- Slide the upper and lower crankcase oil seals and the washer free of the crankshaft.

SAFETY WORDS

The piston pin lockrings are made of spring steel and may slip out of the pliers or pop out of the groove with considerable force. Therefore, **WEAR** eye protection glasses while removing the piston pin lockrings in the next step.

13- Remove the G-lockring from both ends of the piston pin using a pair of needle nose pliers. Slide out the piston pin, and then the piston may be separated from the connecting rod. Push out the caged roller bearings from the small end of the connecting rod.

CRITICAL WORDS

The rod and crankpin are pressed into the counterweights and can only be separated using a hydraulic press.

Slowly rotate both bearing races. If rough spots are felt, the bearings will have to be pressed free of the crankshaft.

14- Obtain a universal bearing separator tool and an arbor press. Press each bearing

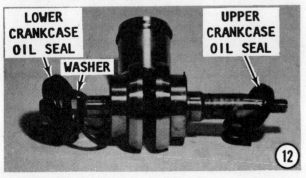

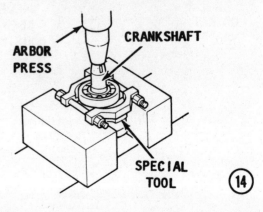

DISASSEMBLE 2HP   7-7

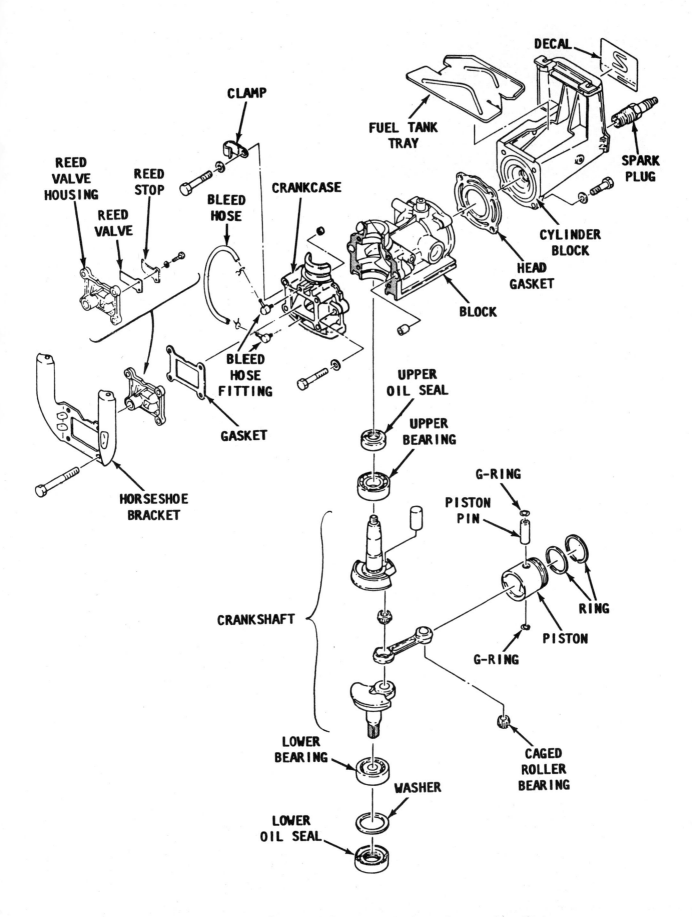

Exploded drawing of a Type "A" powerhead, with major parts identified.

7-8 POWERHEAD

from the crankshaft. Be sure the crankshaft is supported, because once the bearing "breaks" loose, the crankshaft is free to fall.

15- Gently spread the top piston ring enough to pry it out and up over the top of the piston. No special tool is required to remove the piston rings. Remove the lower ring in a similar manner. These rings are **EXTREMELY** brittle and have to be handled with care if they are intended for further service.

CLEANING AND INSPECTING

Detailed instructions for cleaning and inspecting all parts of the powerhead, including the block, will be found in the last section of this chapter, Section 7-7.

ASSEMBLING AND INSTALLATION
2 HP POWERHEAD

The following instructions pickup the work after all parts of the powerhead have been thoroughly cleaned and inspected according to the procedures outlined in the last section of this chapter.

1- Install a new set of piston rings onto the piston. No special tool is necessary for installation **HOWEVER**, take care to spread the ring only enough to clear the top of the

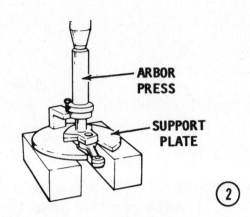

piston. The rings are **EXTREMELY** brittle and will snap if spread beyond their limit. Align the ring gap over the locating pin.

2- If the main bearings were removed, support the crankshaft and press the bearings onto the shaft one at a time. Take note of the bearing size embossed on one side of the bearing. The side with the marking **MUST** face **AWAY** from the crankshaft throw.

SAFETY WORDS

The piston pin lockrings are made of spring steel and may slip out of the pliers or pop out of the groove with considerable force. Therefore, **WEAR** eye protection glasses while installing the piston pin lockrings in the next step.

3- Apply a thin coating of multi-purpose water resistant lubricant to the inside surface of the upper end of the piston rod. Slide the caged roller bearings into the

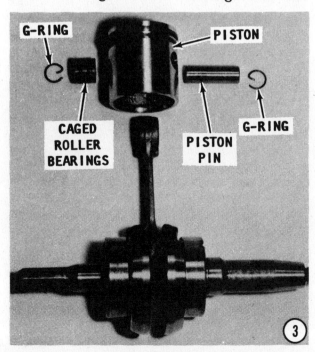

ASSEMBLE 2 HP 7-9

The word "UP" is embossed on the piston crown. This word MUST face the threaded end of the crankshaft when the piston/rod assembly is installed into the cylinder.

small end of the rod. Move the piston over the rod end with the word **UP** on the piston crown facing **TOWARD** the tapered end of the crankshaft. Shift the piston to align the holes in the piston with the rod end opening. Slide the piston pin through the piston and connecting rod. Center the pin in the piston. Install a G-ring at each end of the piston pin.

4- Install the large washer onto the squared end of the crankshaft. Apply a light coating of multi-purpose water resistant lubricant to both seals before installation. Slide the lower crankshaft oil seal onto the squared end of the crankshaft with the lip facing **TOWARD** the connecting rod. Slide the upper crankshaft oil seal onto the threaded end of the crankshaft with the lip of the seal facing **TOWARD** the connecting rod.

5- Coat the piston and rings with engine oil. Check to be sure the ring gaps are centered over the locating pins. Lower the piston into the cylinder bore and seat the crankshaft assembly onto the crankcase. The rings will slide into the cylinder, provided each ring end gap is centered properly over the locating pin. Rotate the two main bearings until the two indexing pins are recessed into the square notches at the mating surface of the crankcase, as shown.

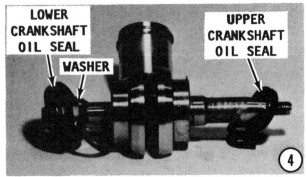

6- Lay down a bead of Loctite 514 P/N C-92-75505-1 to the crankcase mating surfaces. Check to be sure the dowel bushings are in place on either side of the crankcase. Press the two halves together. Wipe away any excess Loctite. Hold the two halves together and at the same time rotate the crankshaft. If any binding is felt, separate the two halves before the sealing agent has a chance to set. Verify the crankshaft is properly seated and the loca-

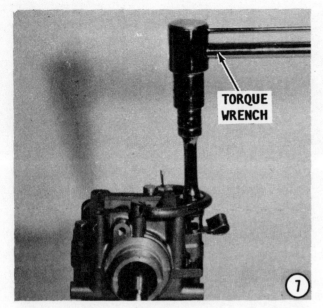

ting pins are in their recesses. Bring the two halves together as described in the first portion of this step.

7- Install the crankcase bolts. Install the bolt with the clip on the side of the crankcase which has the hose. Tighten the bolts alternately and evenly to a torque value of 7 ft lbs (10Nm).

A torque wrench is essential to correctly assemble the powerhead. **NEVER** attempt to assemble a powerhead without a torque wrench. Attaching bolts **MUST** be tightened to the required torque value in three progressive stages, following the specified tightening sequence. Tighten all bolts to 1/3 the torque value, then repeat the sequence tightening to 2/3 the torque value. Finally, on the third and last sequence, tighten to the full torque value.

8- Check to be sure each piston ring has spring tension. This is accomplished by **CAREFULLY** pressing on each ring with a screwdriver extended through the transfer port. **TAKE CARE** not to burr the piston rings while checking for spring tension. If spring tension cannot be felt (the ring fails to return to its orignial position), the ring was probably broken during the piston and crankshaft installation process. Should this occur, new rings must be installed.

9- If the reed valve stop or reed valve was replaced, a new valve and stop should be positioned with the "cut" in the lower corner when the notch in the housing is facing **UP**, as shown. Tighten the screws alternately to avoid warping the valve.

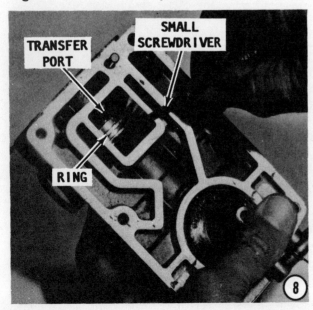

ASSEMBLE 2HP 7-11

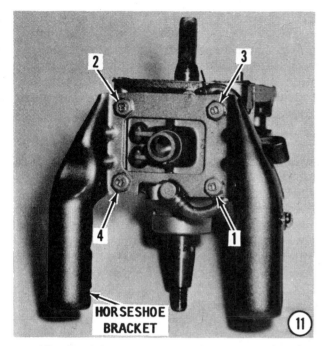

HORSESHOE BRACKET

10- Position a new gasket over the intake port and install the valve reed housing with the notch facing **UPWARD** toward the threaded end of the crankshaft.

11- Place the horseshoe bracket over the housing. Install the four bolts and tighten them alternately and evenly to a torque value of 7 ft lbs (10Nm), in the sequence shown.

12- Install the two gaskets and the plate on top of the upper casing, as shown. Apply a small amount of Loctite 514 P/N C-92-75505-1 to both surfaces of each gasket as the gasket is positioned in place. Check to be sure the dowel pin is in place on the upper casing. Each gasket can only be

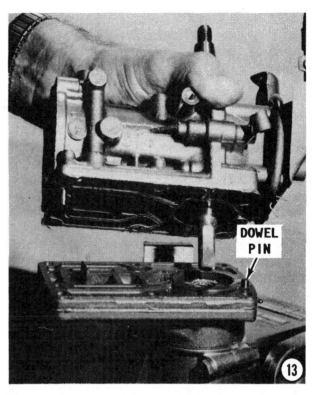

fitted one way to prevent blocking of water passages.

13- Apply a coating of multi-purpose water resistant lubricant to the lower end of the driveshaft. Install the powerhead onto the intermediate housing, using the dowel pin to align the two surfaces. The propeller may have to be rotated slightly to permit the crankshaft to index with the driveshaft and allow the powerhead to seat properly.

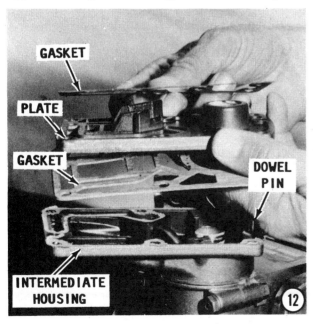

14- Apply Loctite to the threads of the six bolts securing the powerhead to the intermediate housing. Tighten the bolts in two stages starting with the center two bolts then the two on both ends. Tighten the bolts alternately and evenly to a torque value of 5.8 ft lbs (8Nm).

15- Install a **NEW** head gasket **WITHOUT** any sealing substance. Position the head in place over the gasket.

16- Coat the threads of the four attaching bolts with Loctite. Install and tighten the bolts to two stages to a torque value of 7.2 ft lbs (10Nm) in the sequence shown. The sequence is also embossed on the head.

CLOSING TASKS

After the powerhead is secured to the intermediate housing, complete the work by first following the procedures listed in Chapter 5 beginning with the stator plate installation.

Install the flywheel.

Install the carburetor and fuel tank, see Chapter 4.

Install the hand rewind starter.

Install the cowling.

Mount the engine in a test tank, on a boat in a body of water, or connect a flush attachment to the lower unit.

Start the engine and check the completed work.

CAUTION

Water must circulate through the lower unit to the powerhead anytime the powerhead is operating to prevent damage to the water pump in the lower unit. Just five seconds without water will damage the water pump impeller.

Attempt to start and run the engine without the cowling installed. This will provide the opportunity to check for fuel and oil leaks, without the cowling in place.

After the engine is operating properly, install both halves of the cowling and secure them in place with the attaching hardware. Follow the break-in procedures with the unit on a boat and with a load on the propeller.

Break-in Procedures

As soon as the engine starts, **CHECK** to be sure the water pump is operating. If the water pump is operating, a fine stream will be discharged from the exhaust relief hole at the rear of the drive shaft housing.

DO NOT operate the engine at full throttle except for **VERY** short periods, until after 10 hours of operation as follows:

a- Operate at 1/2 throttle, approximately 2500 to 3500 rpm, for 2 hours.

b- Operate at any speed after 2 hours **BUT NOT** at sustained full throttle until another 8 hours of operation.

c- Mix gasoline and oil during the break-in period, total of 10 hours, at a ratio of 25:1.

d- While the engine is operating during the initial period, check the fuel, exhaust, and water systems for leaks.

e- See Chapter 2 for tuning procedures.

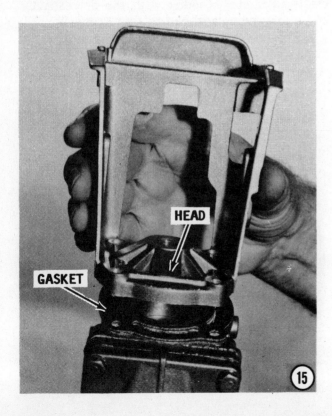

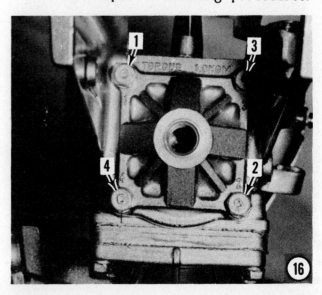

7-4 SERVICING POWERHEAD "B" SINGLE-CYLINDER AIR-COOLED 3.5HP AND 5HP MODELS

ADVICE

Before commencing any work on the powerhead, an understanding of two-cycle engine operation will be most helpful. Therefore, it would be well worth the time to study the principles of two-cycle engines, as outlined briefly in Section 7-1. A Polaroid, or equivalent instant-type camera is an extremely useful item, providing the means of accurately recording the arrangement of parts and wire connections **BEFORE** the disassembly work begins. Such a record is invaluable during assembling.

Preliminary Tasks

Remove the cowling and the air cooling shroud; the hand starter, see Chapter 11; the flywheel, and the stator plate, see Chapter 5; and the carburetor and fuel tank, see Chapter 4.

Disconnect all bleed hoses. On the 5hp model: disconnect the fuel line at the fuel pump mounted on the portside of the powerhead.

REMOVAL AND DISASSEMBLING

The following instructions pickup the work after the preliminary tasks listed above have been accomplished.

Model 3.5hp Only

1- Remove the steering handle securing bolt, and then remove the handle.

All Models

2- Remove the four bolts securing the reed block assembly to the powerhead. Remove and discard the gasket.

3- Remove the two bolts and washers securing the exhaust manifold to the powerhead and another two bolts securing the exhaust manifold to the intermediate housing.

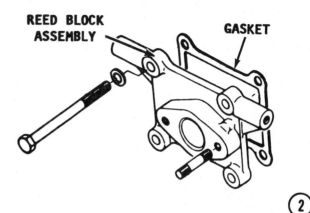
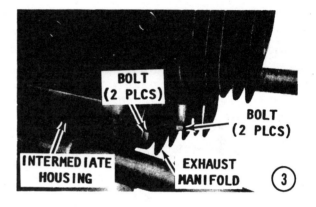
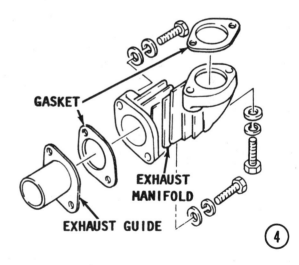
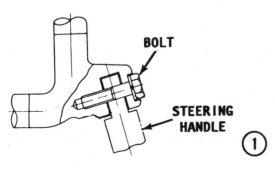
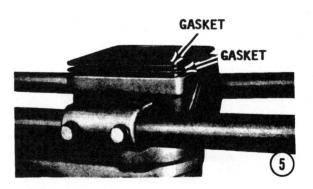

7-14 POWERHEAD

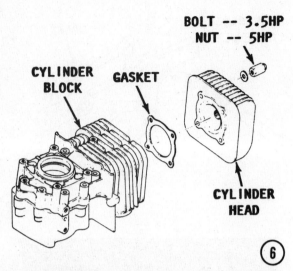

4- Remove the manifold from the powerhead and intermediate housing. Remove and discard the two gaskets. Lift out the exhaust guide from the exhaust port in the intermediate housing.

5- Remove the bolts securing the powerhead to the intermediate housing. Remove and discard the two powerhead-to-intermediate housing gaskets. Place the powerhead on a suitable work surface.

6- Remove the four long bolts (3.5hp), or the four nuts and washers (5hp), securing the cylinder head to the block. Remove the head and discard the head gasket.

Model 5hp
Remove the two bolts securing the fuel pump to the portside of the powerhead, and the remove the pump.

Model 3.5hp
7- Remove the two bolts securing the crankcase to the block, and then separate the two halves. **TAKE CARE** not to lose the two dowel pins used to align the two halves together.

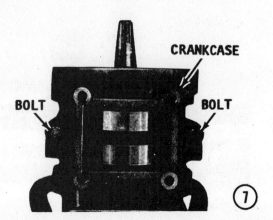

Model 5hp
This model has no dowel pins at this location. Slide the cylinder up and away from the piston and clear of the four long studs.

Model 5hp
Remove the eight bolts and washers securing the two halves of the crankcase together. Separate the two halves, taking care not to lose the two dowel pins at the mating surfaces. Lift off the two halves from the ends of the crankshaft.

The piston rod is an integral part of the crankshaft and cannot be separated.

Model 3.5hp
8- Tap the crankshaft lightly with a soft head mallet to unseat it from the crankcase.

9- Lift out the crankshaft assembly from the cylinder block. The piston rod is an integral part of the crankshaft and cannot be separated.

Model 3.5hp
and Late Model 5hp
10- Slide the upper and lower crankcase oil seals free of the crankshaft.

Early Model 5hp
Use a pair of snap ring pliers and remove the circlip from the lower end of the crankshaft. Slide the upper and lower crankcase oil seals free of the crankshaft.

SAFETY WORDS
The piston pin G-lockrings are made of spring steel and may slip out of the pliers or pop out of the groove with considerable force. Therefore, **WEAR** eye protective glasses while removing the piston pin lockrings in the next step.

DISASSEMBLE AIR-COOLED 7-15

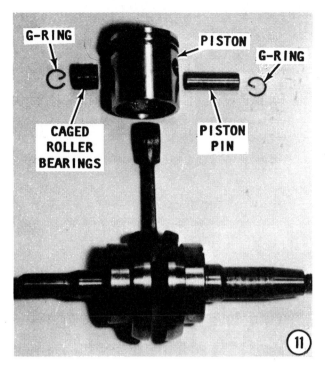

11- Remove the G-lockring from both ends of the piston pin using a pair of needle nose pliers. Slide out the piston pin, and then the piston may be separated from the connecting rod. Push the caged roller bearings free of the connecting rod end.

CRITICAL WORDS

The rod is an integral part of the crankshaft. The two are manufactured together and **CANNOT** be separated.

Slowly rotate both bearing races. If rough spots are felt, the bearings will have to be pressed free of the crankshaft.

Some early Model 5hp powerheads have two circlip retaining the upper and lower bearings onto the crankshaft. Remove this circlip using a pair of circlip pliers, prior to pressing the bearing free of the shaft.

Some late Model 5hp units have no circlip but a single oil slinger disc between the lower ball bearing and the oil seal.

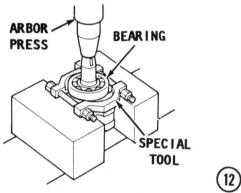

12- Obtain a universal bearing separater tool and an arbor press. Press each bearing from the crankshaft. Be sure the crankshaft is supported, because once the bearing "breaks" loose, the crankshaft is free to fall.

13- Gently spread the top piston ring enough to pry it out and up over the top of the piston. No special tool is required to remove the piston rings. Remove the lower

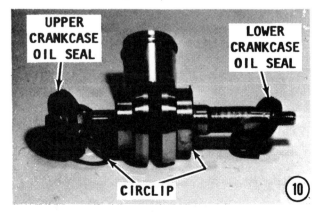

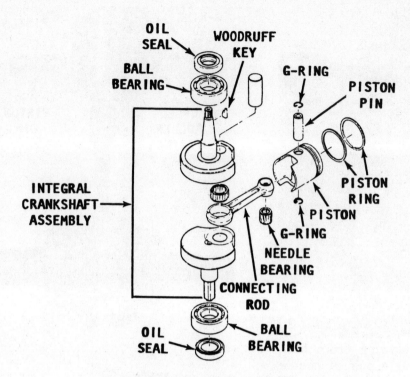

Exploded drawing of the crankshaft assembly of a 3.5hp air-cooled powerhead. Major parts are identified.

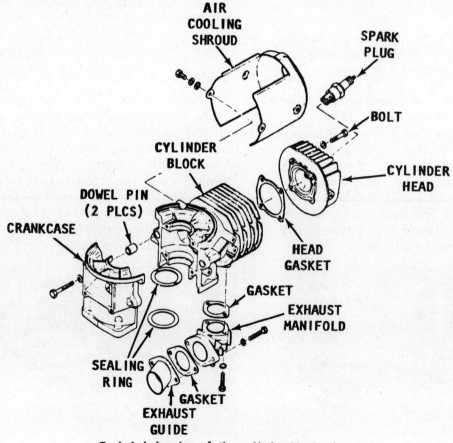

Exploded drawing of the cylinder block of a 3.5hp powerhead. Major parts are identified.

DISASSEMBLE AIR-COOLED 7-17

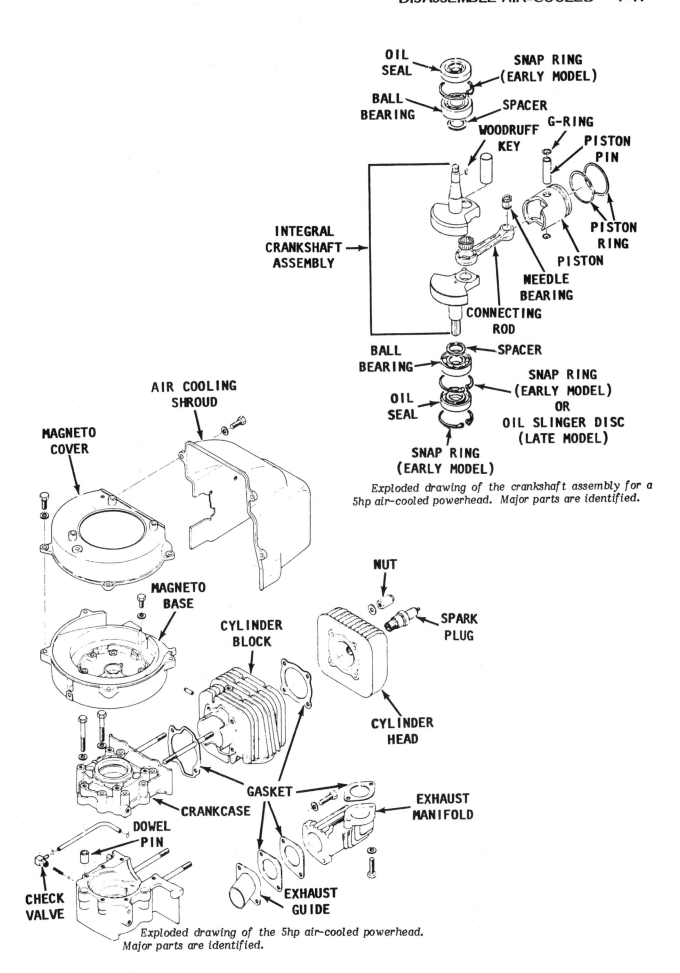

Exploded drawing of the crankshaft assembly for a 5hp air-cooled powerhead. Major parts are identified.

Exploded drawing of the 5hp air-cooled powerhead. Major parts are identified.

7-18 POWERHEAD

ring in a similar manner. These rings are **EXTREMELY** brittle and have to be handled with care if they are intended for further service.

CLEANING AND INSPECTING

Detailed instructions for cleaning and inspecting all parts of the powerhead, including the block, will be found in the last section of this chapter, Section 7-7, beginning on Page 7-67.

ASSEMBLING AND INSTALLATION
3.5HP & 5HP AIR-COOLED POWERHEAD

The following instructions pickup the work after all parts of the powerhead have been thoroughly cleaned and inspected according to the procedures outlined in the last section of this chapter.

1- Install a new set of piston rings onto the piston. No special tool is necessary for installation. **HOWEVER,** take care to spread the ring only enough to clear the top of the piston. The rings are **EXTREMELY** brittle and will snap if spread beyond their limit. Align the ring gap over the locating pin.

2- If the main bearings were removed, support the crankshaft and press the bearings onto the shaft one at a time. Take note of the bearing size embossed on one side of the bearing. The side with the marking **MUST** face **AWAY** from the crankshaft throw. Install the snap rings or oil slinger, if so equipped.

SAFETY WORDS

The piston pin lockrings are made of spring steel and may slip out of the pliers or pop out of the groove with considerable

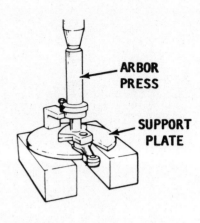

force. Therefore, **WEAR** eye protection glasses while installing the piston pin lockrings in the next step.

3- Apply a thin coating of multi-purpose water resistant lubricant to the inside surface of the upper end of the piston rod. Slide the caged roller bearings into the small end of the rod. Move the piston over the rod end with the word **UP** on the piston crown facing **TOWARD** the tapered end of the crankshaft. Shift the piston to align the holes in the piston with the rod end opening. Slide the piston pin through the piston and connecting rod. Center the pin in the piston. Install a G-ring at each end of the piston pin.

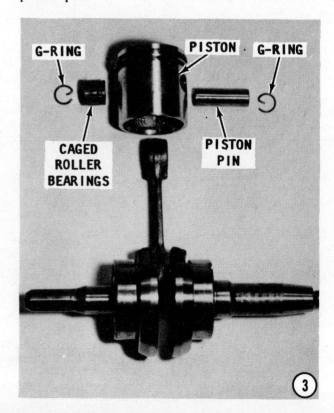

ASSEMBLE AIR-COOLED 7-19

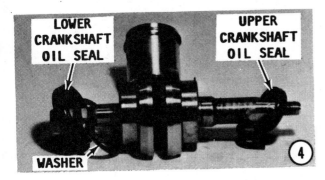

4- Apply a light coating of multi-purpose water resistant lubricant to both seals before installation. Slide the lower crankshaft oil seal onto the squared end of the crankshaft with the lip facing **TOWARD** the connecting rod. Slide the upper crankshaft oil seal onto the threaded end of the crankshaft with the lip of the seal facing **TOWARD** the connecting rod.

Model 5hp only

5- Lay down a bead of Loctite 514 P/N C-92-75505-1 onto the two crankcase mating surfaces. Slide the two crankcase halves over the ends of the assembled crankshaft, with the tapered crankshaft end through the upper crankcase half.

Rotate the two main bearings until the two indexing pins are recessed into the square notches machined into each crankcase half.

Check to be sure the two dowel pins are in place on both halves of the crankcase. Press the two halves together. Wipe away any excess Loctite.

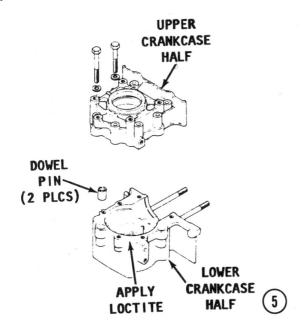

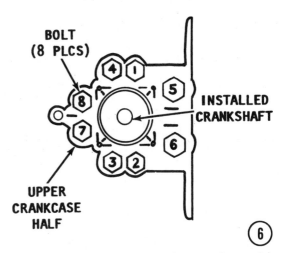

Hold the two halves together and, at the same time, rotate the crankshaft. If any binding is felt, separate the two halves before the sealing agent has a chance to set. Verify the crankshaft is properly seated and the locating pins are in their recesses. Bring the two halves together as described in the first portion of this step.

6- Install and tighten the eight securing bolts and washers in the sequence shown in the accompanying illustration. Tighten the bolts to a torque value of 7 ft lbs (10Nm).

All Models

7- Coat the piston and rings with engine oil. Check to be sure the ring gaps are centered over the locating pins. Lower the piston into the cylinder bore.

Model 3.5hp

Seat the crankshaft assembly onto the crankcase.

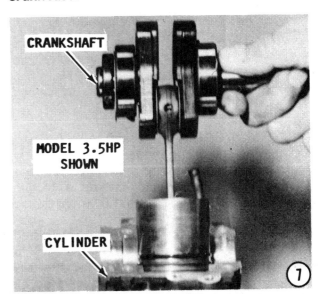

7-20 POWERHEAD

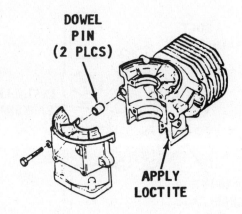

All Models
The rings will slide into the cylinder, provided each ring end gap is centered properly over the locating pin. Rotate the two main bearings until the two indexing pins, if so equipped, are recessed into the square notches at the mating surface of the crankcase.

8- Lay down a bead of Loctite on the crankcase-to-block mating surface.

Model 3.5hp
Check to be sure the two dowel pins on both halves of the crankcase are in place.

All Models
Press the two halves together. Wipe away any excess Loctite.

Model 3.5hp
Hold the two halves together and, at the same time, rotate the crankshaft. If any binding is felt, separate the two halves before the sealing agent has a chance to set. Verify the crankshaft is properly seated and the locating pins are in their recesses. Bring the two halves together as described in the first portion of this step.

CRITICAL WORDS
A torque wrench is essential to correctly assemble the powerhead. **NEVER** attempt to assemble a powerhead without a torque wrench. Attaching bolts **MUST** be tightened to the required torque value in two progressive stages, following the specified tightening sequence. Tighten all bolts to 1/2 the torque value, then repeat the sequence tightening to the full torque value.

Install the drain hose onto the crankcase, if so equipped.

Model 3.5hp
Install and tighten the two attaching bolts to a torque value of 4.3 ft lbs (6Nm) on the first sequence, and then to a torque value of 8.7 ft lbs (12Nm).

9- Position a new cylinder head gasket in place over the block.

Model 3.5hp
The gasket and the cylinder head are secured to the block with four lockwashers and four long bolts.

Model 5hp
The gasket and the cylinder head are secured to the block with four lockwashers and four nuts.

SEALING SURFACE WORDS
CRITICAL READING
Do **NOT** apply any form of sealant to the head gasket surfaces.

Because of the high temperatures and pressures developed, the sealing surfaces of the cylinder head and the block are the most prone to water leaks. No sealing agent is recommended **BECAUSE** it is almost impossible to apply an even coat of sealer. An even coat would be essential to ensure a air/water tight seal.

Some head gaskets are supplied with a "tacky" coating on both surfaces applied at the time of manufacture. This "tacky" substance will provide an even coating all around. Therefore, no further sealing agent is required.

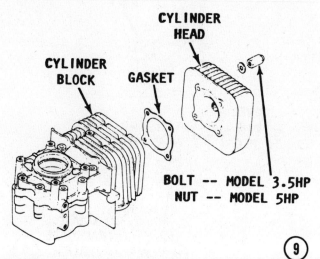

ASSEMBLE AIR-COOLED 7-21

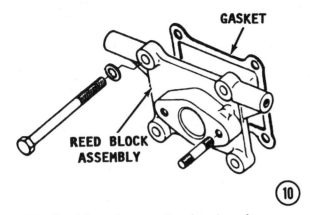

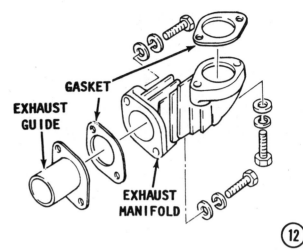

10- Position the reed valve housing onto the crankcase over the two studs. The reed valve opening **MUST** face the starboard side of the block. Install and tighten the two bolts and two nuts to 2.9 ft lbs (4Nm) on the first sequence, and then to 5.8 ft lbs (8Nm).

11- Check to be sure the two dowel pins on the intermediate housing are in place forward and aft of the driveshaft area. Coat both sides of the powerhead gasket with Permatex, or equivalent material. Position the two gaskets in place with the dowel pins passing up through the correct holes in the gasket.

Apply a coating of multi-purpose water resistant lubricant to the lower end of the driveshaft. Install the powerhead onto the intermediate housing using the dowel pins to align the two surfaces. The propeller may have to be rotated slightly to permit the crankshaft to index with the driveshaft and allow the powerhead to seat properly.

Apply Loctite 514 P/N C-92-75505-1 to the threads of the two bolts securing the powerhead to the intermediate housing. Tighten the bolts alternately and evenly to a torque value of 5.8 ft lbs (8Nm).

12- Install the exhaust guide into the exhaust port of the intermediate housing. Install the exhaust housing to the powerhead using new gaskets between the block and intermediate housing. Install and tighten the two pair of attaching bolts alternately and evenly to a torque value of 5.8 ft lbs (8Nm).

Model 3.5hp only

13- Install the steering handle. Apply a coating of multi-purpose lubricant to both sides of the washer. Install the washer and bolt. Tighten the bolt while checking the handle movement until an acceptable amount of resistance (friction) is felt.

CLOSING TASKS

After the powerhead is secured to the intermediate housing, complete the work by first following the procedures listed in Chapter 5 beginning with the stator plate installation.

Install the flywheel.

Install the carburetor and fuel tank, see Chapter 4.

Install the hand rewind starter. See Rewind Starter "B" in Chapter 11.

Install the air cooling shroud with the attaching hardware.

Mount the engine in a test tank, on a boat in a body of water, or connect a flush attachment to the lower unit.

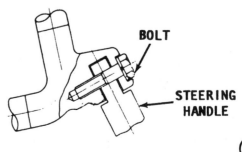

Start the engine and check the completed work.

CAUTION

Water must circulate through the lower unit to the powerhead anytime the powerhead is operating to prevent damage to the water pump in the lower unit. Just five seconds without water will damage the water pump impeller.

Attempt to start and run the engine without the cowling installed. This will provide the opportunity to check for fuel and oil leaks, without the cowling in place.

After the engine is operating properly, install the cowling. Follow the break-in procedures with the unit on a boat and with a load on the propeller.

Break-in Procedures

As soon as the engine starts, **CHECK** to be sure the water pump is operating. If the water pump is operating, a fine stream will be discharged from the exhaust relief hole at the rear of the drive shaft housing.

DO NOT operate the engine at full throttle except for **VERY** short periods, until after 10 hours of operation as follows:

a- Operate at 1/2 throttle, approximately 2500 to 3500 rpm, for 2 hours.

b- Operate at any speed after 2 hours **BUT NOT** at sustained full throttle until another 8 hours of operation.

c- Mix gasoline and oil during the break-in period, total of 10 hours, at a ratio of 25:1.

d- While the engine is operating during the initial period, check the fuel, exhaust, and water systems for leaks.

e- See Chapter 2 for tuning procedures.

7-5 SERVICING POWERHEAD "C" SINGLE-CYLINDER WATER-COOLED 4HP AND 5HP MODELS

ADVICE

Before commencing any work on the powerhead, an understanding of two-cycle engine operation will be most helpful. Therefore, it would be well worth the time to study the principles of two-cycle engines, as outlined briefly in Section 7-1. A Polaroid, or equivalent instant-type camera is an extremely useful item, providing the means of accurately recording the arrangement of parts and wire connections **BEFORE** the disassembly work begins. Such a record is extremely valuable during assembling.

Preliminary Tasks

Remove the cowling; the hand starter, see Chapter 11; the flywheel, and the stator plate, see Chapter 5; and the carburetor and fuel tank, see Chapter 4.

REMOVAL AND DISASSEMBLING

The following instructions pickup the work after the preliminary tasks listed above have been accomplished.

1- Loosen the outer nut on the cable to the starter and remove the cable end from the plastic barrel. Remove the three nuts shown in the accompanying illustration. Remove the metal and plastic brackets.

2- Remove the seven bolts, four on one side and three on the other side, securing the powerhead to the intermediate housing.

SPECIAL WORDS

Two dowel pins are used to mate the powerhead perfectly to the intermediate housing. These dowel pins may remain with the powerhead or they may stay with the intermediate housing. **TAKE CARE** when removing the powerhead, to prevent the dowel pins from falling down into the intermediate housing.

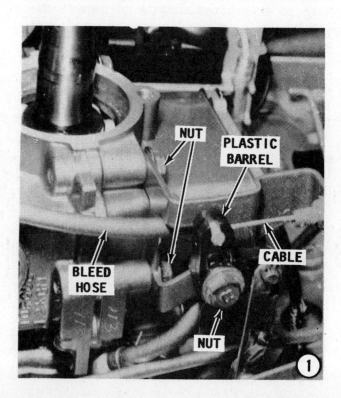

DISASSEMBLE 4HP & 5HP 7-23

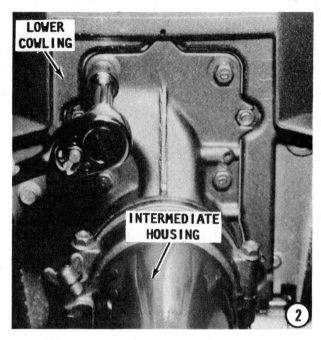

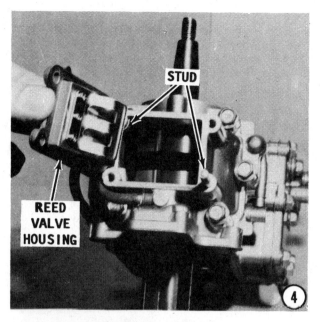

3- Grasp the powerhead with both hands, pull upward, and make an attempt to "rock" the powerhead to break the gasket seal. Lift the powerhead clear of the intermediate housing.

4- Remove the two bolts and the two nuts, and then lift the reed valve housing clear of the two studs. **NOTE** and remember the direction of the reeds as an aid to installation. Remove and discard the gasket.

5- Remove the bolt and washer securing the oil seal housing over the lower end of

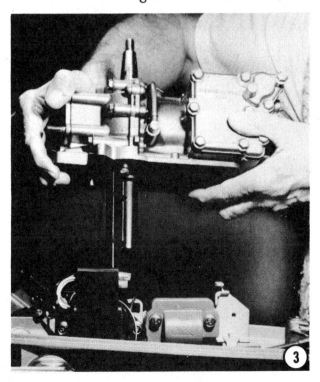

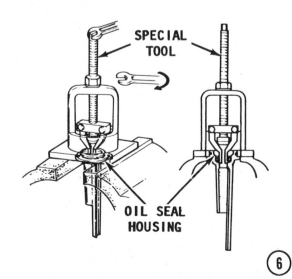

7-24 POWERHEAD

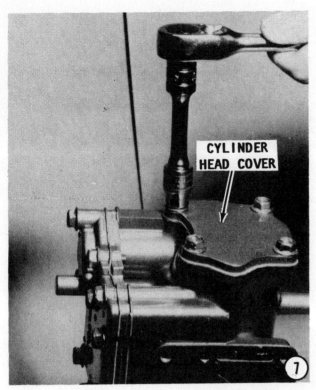

the crankshaft. Remove and discard the O-ring.

6- Obtain bearing/oil seal puller tool P/N M-91-83165M. Secure the oil seal housing in a vise equipped with soft jaws. Insert the expanding jaws of the tool under the edge of the oil seal. Tighten the top nut against the collar of the tool housing, and then rotate the center shaft to raise the jaws and pull the seal free.

7- Remove the four bolts securing the cylinder head cover to the block.

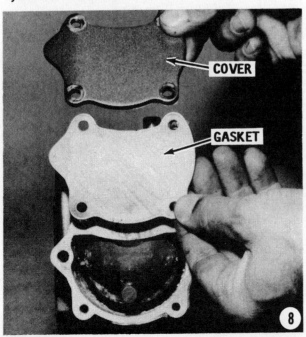

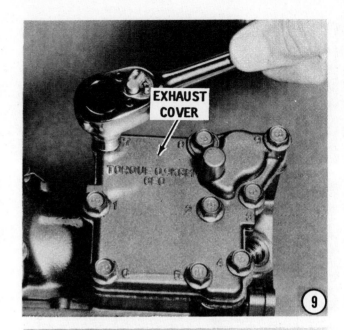

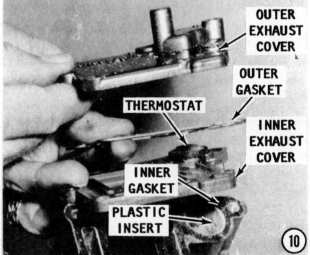

DISASSEMBLE 4HP & 5HP 7-25

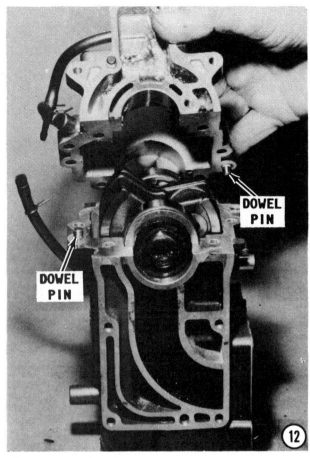

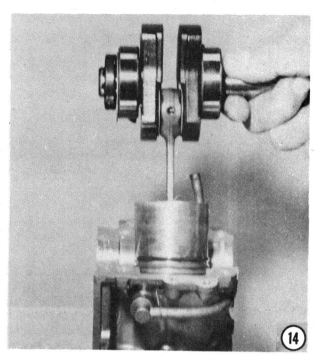

8- Lift off the cylinder head cover. Remove and discard the cover gasket.

9- Remove the six bolts securing the exhaust cover to the powerhead.

10- Lift off the outer exhaust cover, the outer gasket, the thermostat, the inner exhaust cover, the inner gasket and finally the plastic thermostat insert.

11- Remove the six bolts securing both halves of the crankcase together.

12- Insert two screwdrivers between the projections, provided for this purpose, on both sides of the crankcase. Pry on both sides at the same time to move the two crankcase halves apart. **TAKE CARE** not to lose the two dowel pins used to align the two halves perfectly.

13- Tap the crankshaft lightly with a soft head mallet to unseat it from the crankcase.

14- Lift out the crankshaft assembly from the cylinder block. The piston rod is an integral part of the crankshaft and cannot be separated.

15- Slide the upper and lower crankcase oil seals and the washer free of the crankshaft.

SAFETY WORDS

The piston pin G-lockrings are made of spring steel and may slip out of the pliers or pop out of the groove with considerable force. Therefore, **WEAR** eye protection glasses while removing the piston pin lockrings in the next step.

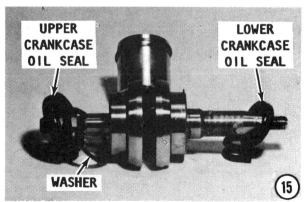

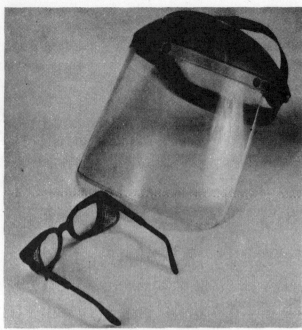

Protect your eyes with a face mask or safety glasses while working with the rewind spring, especially a used one. The spring is a "tiger in a cage", almost 13 feet (4 meters), of spring steel wound and confined into a space less than 4 inches (about 10 cm) in diameter.

16- Remove the G-lockring from both ends of the piston pin using a pair of needle nose pliers. Slide out the piston pin, and then the piston may be separated from the connecting rod. Push the caged roller bearings free of the connecting rod end.

CRITICAL WORDS

The rod is an integral part of the crankshaft. The two are manufactured together and **CANNOT** be separated.

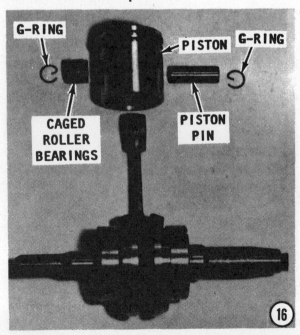

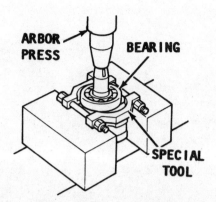

Slowly rotate both bearing races. If rough spots are felt, the bearings will have to be pressed free of the crankshaft.

17- Obtain special bearing separater tool, P/N C-91-37241 and an arbor press. Press each bearing from the crankshaft. Be sure the crankshaft is supported, because once the bearing "breaks" loose, the crankshaft is free to fall.

18- Gently spread the top piston ring enough to pry it out and up over the top of the piston. No special tool is required to remove the piston rings. Remove the lower ring in a similar manner. These rings are **EXTREMELY** brittle and have to be handled with care if they are intended for further service.

CLEANING AND INSPECTING

Detailed instructions for cleaning and inspecting all parts of the powerhead, including the block, will be found in the last section of this chapter, Section 7-7, beginning on Page 7-67.

ASSEMBLING AND INSTALLATION 4 HP AND 5 HP POWERHEAD

The following instructions pickup the work after all parts of the powerhead have

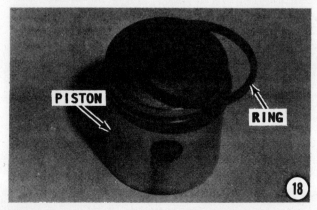

ASSEMBLE 4HP & 5HP   7-27

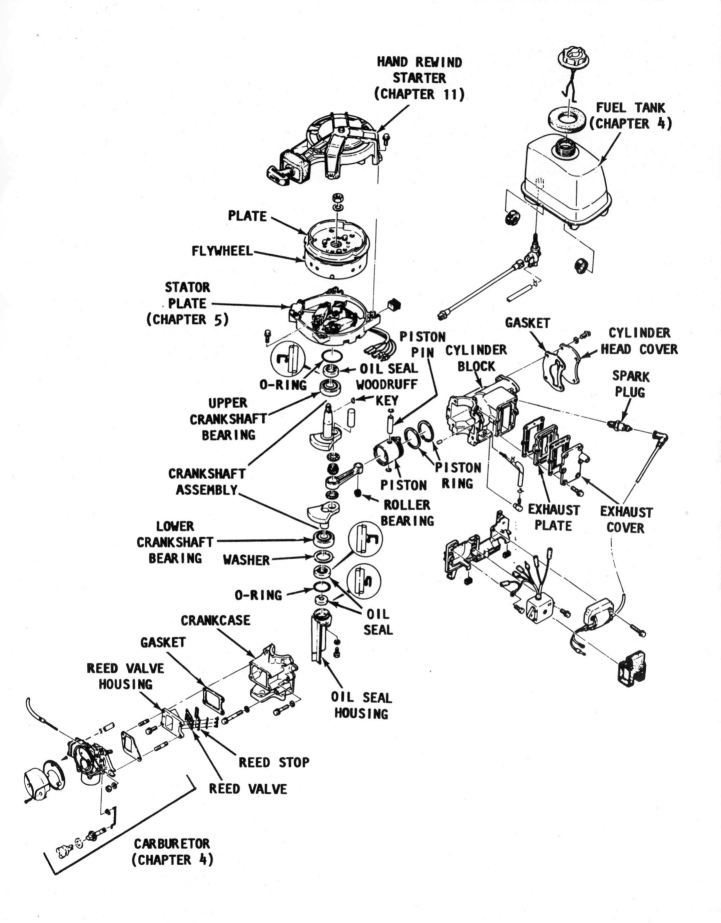

Exploded drawing of the Type "C" 4hp and 5hp water-cooled powerhead. Major parts are identified.

7-28 POWERHEAD

been thoroughly cleaned and inspected according to the procedures outlined in the last section of this chapter.

1- Install a new set of piston rings onto the piston. No special tool is necessary for installation. **HOWEVER**, take care to spread the ring only enough to clear the top of the piston. The rings are **EXTREMELY** brittle and will snap if spread beyond their limit. Align the ring gap over the locating pin.

2- If the main bearings were removed, support the crankshaft and press the bearings onto the shaft one at a time. Take note of the bearing size embossed on one side of the bearing. The side with the marking **MUST** face **AWAY** from the crankshaft throw.

SAFETY WORDS

The piston pin lockrings are made of spring steel and may slip out of the pliers or pop out of the groove with considerable force. Therefore, **WEAR** eye protection glasses while installing the piston pin lockrings in the next step.

3- Apply a thin coating of multi-purpose water resistant lubricant to the inside surface of the upper end of the piston rod.

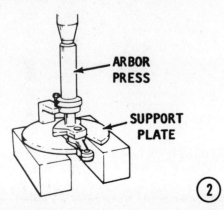

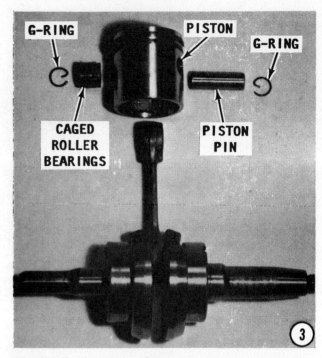

Slide the caged roller bearings into the small end of the rod. Move the piston over the rod end with the word **UP** on the piston crown facing **TOWARD** the tapered end of the crankshaft. Shift the piston to align the holes in the piston with the rod end opening. Slide the piston pin through the piston and connecting rod. Center the pin in the piston. Install a G-ring at each end of the piston pin.

4- Install the large washer onto the squared end of the crankshaft. Apply a light coating of multi-purpose water resistant lubricant to both seals before installation. Slide the lower crankshaft oil seal onto the squared end of the crankshaft with the lip facing **TOWARD** the connecting rod. Slide the upper crankshaft oil seal onto the threaded end of the crankshaft with the lip of the seal facing **TOWARD** the connecting rod.

5- Coat the piston and rings with engine oil. Check to be sure the ring gaps are

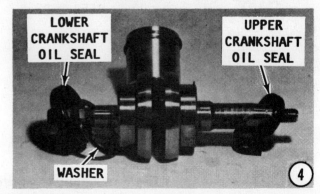

ASSEMBLE 4HP & 5HP 7-29

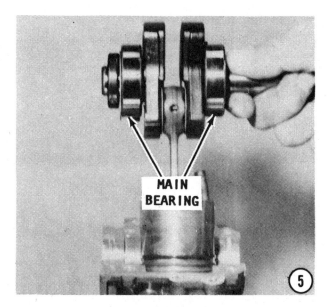

centered over the locating pins. Lower the piston into the cylinder bore and seat the crankshaft assembly onto the crankcase. The rings will slide into the cylinder, provided each ring end gap is centered properly over the locating pin. Rotate the two main bearings until the two indexing pins are recessed into the square notches at the mating surface of the crankcase.

6- Check to be sure each piston ring has spring tension. This is accomplished by **CAREFULLY** pressing on each ring with a screwdriver extended through the transfer port. **TAKE CARE** not to burr the piston rings while checking for spring tension. If spring tension cannot be felt (the ring fails to return to its orignial position), the ring was probably broken during the piston and crankshaft installation process. Should this occur, new rings must be installed.

7- Lay down a bead of Loctite 514 P/N C-92-75505-1 to the crankcase mating surfaces. Check to be sure the dowel bushings on both halves of the crankcase are in place on both sides of the crankcase.

8- Press the two halves together. Wipe away any excess Loctite. Hold the two halves together and at the same time rotate the crankshaft. If any binding is felt, separate the two halves before the sealing agent has a chance to set. Verify the crankshaft is properly seated and the locating pins are in their recesses. Bring the two halves together as described in the first portion of this step.

CRITICAL WORDS

A torque wrench is essential to correctly assemble the powerhead. **NEVER** attempt to assemble a powerhead without a torque wrench. Attaching bolts **MUST** be tightened to the required torque value in two progressive stages, following the specified tightening sequence. Tighten all bolts to

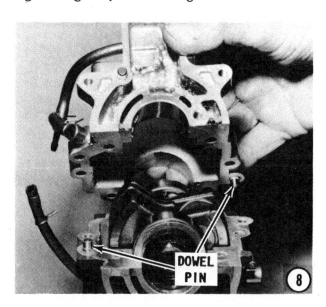

7-30 POWERHEAD

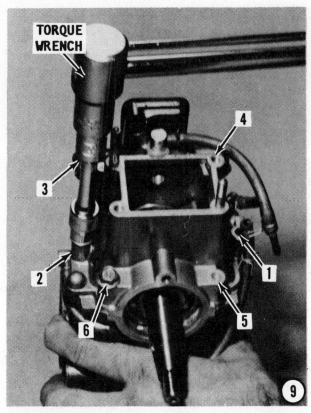

1/2 the torque value, then repeat the sequence tightening to the full torque value.

9- Install the drain hose onto the crankcase and position the metal bracket of the in-gear protection system against bolt holes No. 1 and No. 5. Install and tighten the attaching bolts to a torque value of 4.3 ft

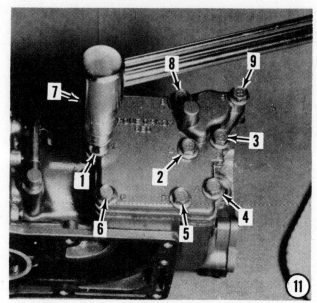

lbs (6Nm) on the first sequence, and then to a torque value of 8.7 ft lbs (12Nm).

10- Slide the plastic thermostat insert into the block. Install the following components in the order given: first, the inner gasket, then the inner plate, the thermostat, the outer gasket, and finally the outer cover.

11- Install and tighten the exhaust cover bolts to a torque value of 2.9 ft lbs (4Nm) on the first sequence, and then to torque value of 5.8 ft lbs (8Nm).

12- Position a new cylinder head cover gasket in place, and then install the cover. Secure the cover with the four attaching bolts. Tighten the bolts to a torque value of 2.9 ft lbs (4Nm) on the first sequence, and then to a torque value of 5.8 ft lbs (8Nm).

13- If the oil seal inside the oil seal housing was removed, as directed in Step 6

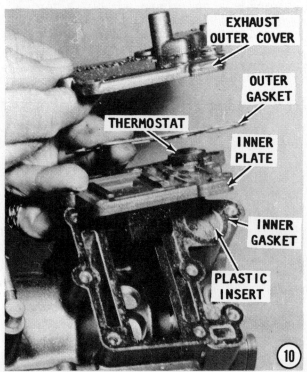

ASSEMBLE 4HP & 5HP 7-31

of Disassembling, coat the lip of a new seal with multi-purpose water resistant lubricant. Press the new seal into the housing using the appropriate mandrel, with the lip of the seal facing **DOWNWARD** as the seal is pressed into the housing. Install a new O-ring into the outer groove of the housing.

Coat the outer surface of the O-ring with the same grease as for the seal. Install the housing and tap it lightly with a soft head mallet to be sure it is fully seated. Secure the housing in place with the attaching bolt.

14- Position the reed valve housing onto the crankcase over the two studs. The reed valve opening **MUST** face the starboard side

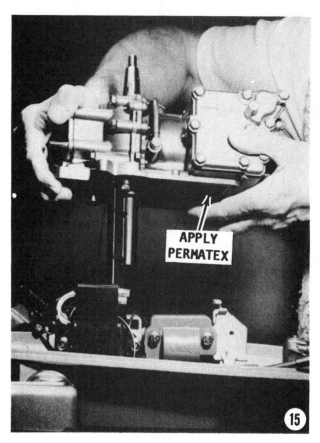

of the block. Install and tighten the two bolts and two nuts to 2.9 ft lbs (4Nm) on the first sequence, and then to 5.8 ft lbs (8Nm).

15- Check to be sure the two dowel pins on the intermediate housing are in place forward and aft of the driveshaft area. Coat both sides of the powerhead gasket with Permatex, or equivalent material. Position the gasket in place with the dowel

pins passing up through the correct holes in the gasket.

Apply a coating of multi-purpose water resistant lubricant to the lower end of the driveshaft. Install the powerhead onto the upper casing using the dowel pin to align the two surfaces. The propeller may have to be rotated slightly to permit the crankshaft to index with the driveshaft and allow the powerhead to seat properly.

16- Apply Loctite 514 P/N C-92-75505-1 to the threads of the seven bolts securing the powerhead to the intermediate housing. Tighten the bolts in two stages starting with the center two bolts then the two on both ends. Tighten the bolts alternately and evenly to a torque value of 5.8 ft lbs (8Nm).

17- Install the plastic bracket to the metal bracket on the starboard side of the powerhead. Place the end of the cable from the starter into the plastic barrel. Slide the cable onto the metal retainer with the one lock cable nut on each side of the retainer.

CLOSING TASKS

After the powerhead is secured to the intermediate housing, complete the work by first following the procedures listed in Chapter 5 beginning with the stator plate installation.

Install the flywheel.

Install the carburetor and fuel tank, see Chapter 4.

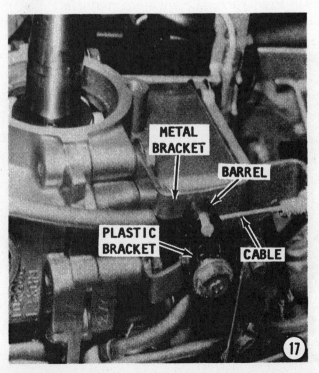

Install the hand rewind starter, see Chapter 11. After the starter is installed, check the action of the no start in-gear protection system. If adjustment is required, see the instructions at the end of the installation procedures for Rewind Starter "C" in Chapter 11.

Mount the engine in a test tank, on a boat in a body of water, or connect a flush attachment to the lower unit.

Start the engine and check the completed work.

CAUTION

Water must circulate through the lower unit to the powerhead anytime the powerhead is operating to prevent damage to the water pump in the lower unit. Just five seconds without water will damage the water pump impeller.

Attempt to start and run the engine without the cowling installed. This will provide the opportunity to check for fuel and oil leaks, without the cowling in place.

After the engine is operating properly, install the cowling. Follow the break-in procedures with the unit on a boat and with a load on the propeller.

Break-in Procedures

As soon as the engine starts, CHECK to be sure the water pump is operating. If the water pump is operating, a fine stream will be discharged from the exhaust relief hole at the rear of the drive shaft housing.

DO NOT operate the engine at full throttle except for **VERY** short periods, until after 10 hours of operation as follows:

a- Operate at 1/2 throttle, approximately 2500 to 3500 rpm, for 2 hours.

b- Operate at any speed after 2 hours **BUT NOT** at sustained full throttle until another 8 hours of operation.

c- Mix gasoline and oil during the break-in period, total of 10 hours, at a ratio of 25:1.

d- While the engine is operating during the initial period, check the fuel, exhaust, and water systems for leaks.

e- See Chapter 2 for tuning procedures.

DISASSEMBLE 2-CYLINDER 7-33

7-6 POWERHEAD TYPE "D"
TWO CYLINDER — WATER-COOLED
8HP, 9.9HP, 15HP, 20HP, 25HP, 28HP, 30HP, 40HP, 48HP, 55HP, AND 60HP

ADVICE

Before commencing any work on the powerhead, an understanding of two-cycle engine operation will be most helpful. Therefore, it would be well worth the time to study the principles of two-cycle engines, as outlined briefly in Section 7-1. A Polaroid, or equivalent instant-type camera is an extremely useful item, providing the means of accurately recording the arrangement of parts and wire connections **BEFORE** the disassembly work begins. Such a record is extremely valuable during assembling.

Preliminary Tasks

Remove the cowling; the hand starter, see Chapter 11; the flywheel, and the stator plate, see Chapter 5; disconnect the fuel line and remove the carburetor, see Chapter 4.

Models with electric cranking motor: remove the cranking motor and relay; disconnect the leads from the rectifier at the quick disconnect fittings, and remove the rectifier.

AUTHOR'S WORDS

Because so many different models are covered in this section, it would not be feasable or practical to provide an illustration of each and every unit.

Therefore, the accompanying illustrations are of a "typical" unit. The unit being serviced may differ slightly in appearance due to engineering or cosmetic changes but the procedures are valid. If a difference should occur, the models affected will be clearly identified.

FIRST, THESE WORDS

Procedural steps are given to remove and disassemble virtually all items of the powerhead. However, as the work moves along, if certain items, i.e. bearings, bushings, seals, etc. are found to be fit for further service, simply skip the disassembly steps involved. Proceed with the required tasks to disassemble the necessary components.

PERFORM only the steps for the model being serviced according to the headings in bold face type.

COMPLETE and **DETAILED** procedures to clean and service all parts of the powerhead are outlined in the last section of this chapter, Section 7-7.

POWERHEAD PREPARATION

The following instructions pickup the work after the preliminary tasks listed above have been completed.

Model 8C and All Other Model 8hp Prior to 1979

The following two steps apply only to the models listed in this heading.

1- Remove the collar around the wire harness from the stator plate and the CDI unit. Disconnect all wires at their quick disconnect fittings and any ground wires which may be secured to the block.

Disconnect the two bolts securing the CDI unit to the powerhead and remove the CDI unit. Remove the two bolts securing the ignition coil to the powerhead and remove the coil.

Identify the ends of the two throttle cables and a matching identification on the stays the ends slip into.

2- Loosen the two locknuts, one on each side of the two stays and lift each throttle cable free of the stays. Lift the ends of the cables up and out of the slots in the throttle

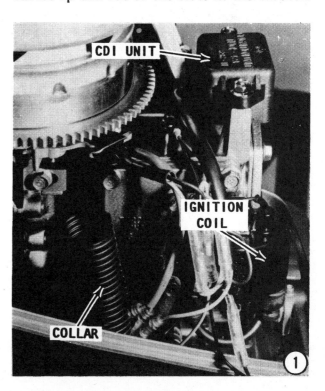

7-34 POWERHEAD

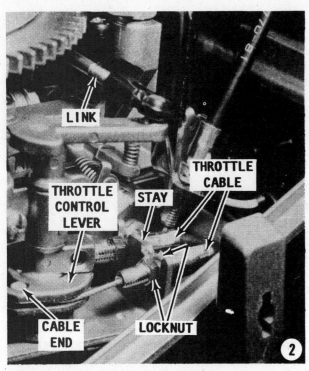

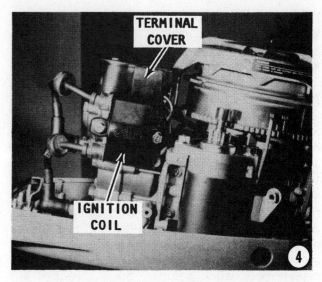

control lever. Use a small screwdriver and pry the link free of the ball joint.

Model 9.9hp, 15hp and
All Other Model 8hp 1979 and On
Except Model 8C

The following four steps apply only to the models listed in this heading.

3- Remove the hose from the exhaust cover. This hose may be moved out of the way while still attached to the pilot hole. Remove the bolts securing both the terminal cover and the CDI unit to the powerhead. Lift off the terminal cover and disconnect all wires leading to the CDI unit and the ignition coil. Remove the CDI unit.

4- Remove the two bolts securing the ignition coil to the powerhead and remove the coil. Remove the one remaining bolt securing the CDI unit bracket to the powerhead. Remove the bracket by pushing it **TOWARD** the flywheel.

5- Pry the link from the magneto control lever at the ball joint. Remove the bolt securing the magneto base contol lever. Loosen the locknuts on the starter cable, and then remove the cable end from the bracket.

6- Pry the link free of the ball joint at the top of the throttle control rod. Remove the inlet and outlet lines to the fuel filter. Plug both lines to prevent fuel spillage and contaminants from entering the lines. Fuel

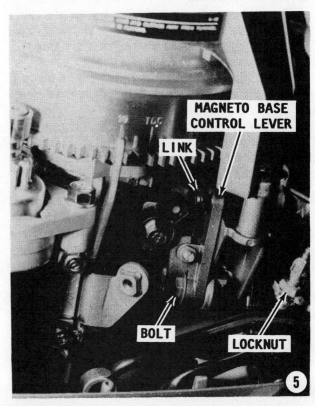

DISASSEMBLE 2-CYLINDER 7-35

in the bottom of the cowling may attack grommets and rubber mounts. Remove the fuel filter.

Model 20hp and Larger
The following two steps apply only to the models listed in this heading.

7- Disconnect all leads from the CDI unit and ignition coil at their quick disconnect fittings. Remove the attaching hardware, and then remove the CDI unit and the coil from the powerhead. Disconnect the ground lead from the powerhead to the CDI bracket. Remove the three bolts, and then remove the bracket.

8- Pry the magneto control, and the throttle control link rods free of their ball joints. Loosen the locknuts retaining the

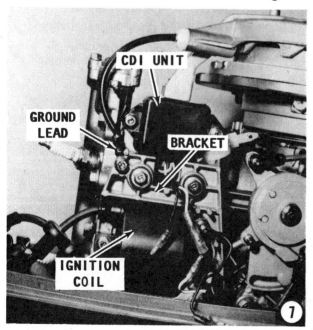

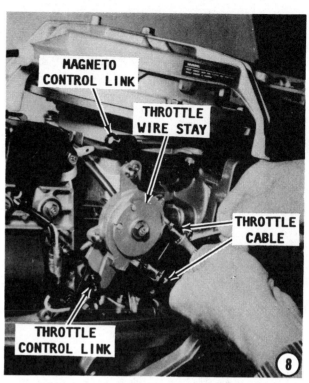

two throttle cables, and then lift their ends from the throttle wire stay. Remove the two bolts securing the throttle wire stay and magneto control lever, and then remove the stay and lever. Loosen the locknuts from the starter wire and remove the wire end from the bracket. Remove the attaching bolt, and then the bracket.

SPECIAL WORDS
The powerhead preparation work is now complete for the various models. The following steps, in general, apply to all models.

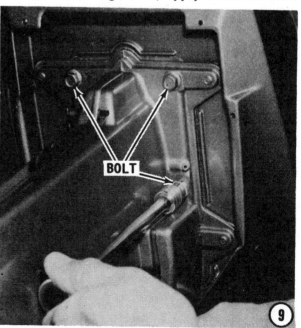

7-36 POWERHEAD

If differences should occur, the special task involved will be clearly indicated.

REMOVAL AND DISASSEMBLING

All Models

9- Tilt the lower unit to the full up position and lock it in place. Remove the

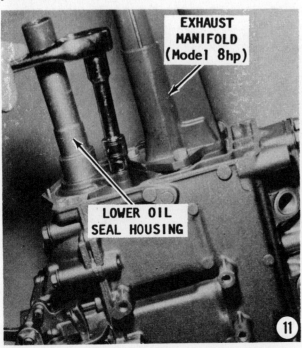

bolts securing the powerhead to the intermediate housing. The powerhead may have to be rotated to gain access to any front bolts. After the bolts have been removed, lower the unit to the full down position.

10- Lift the powerhead straight up and free of the intermediate housing.

BAD NEWS

If the unit is several years old, or if it has been operated in salt water, or has not had proper maintenance, or shelter, or any number of other factors, then separating the powerhead from the intermediate housing may not be a simple task. An air hammer may be required on the bolts to shake the corrosion loose; heat may have to be applied to the casting to expand it slightly, or other devices employed in order to remove the powerhead. One very serious condition would be the driveshaft "frozen" with the lower end of the crankshaft. In this case a circular plug type hole must be drilled and a torch used to cut the driveshaft.

A piece of wood may be inserted between the powerhead and the intermediate housing as a means of using leverage to force them apart.

Let's assume the powerhead will come free on the first attempt.

TAKE CARE not to lose the two dowel pins. The pins may come away with the powerhead or they may stay in the intermediate housing. Be **ESPECIALLY** careful not to drop them into the lower unit.

11- Remove the bolt/s securing the lower oil seal housing. Tap the housing **LIGHTLY** with a soft head mallet to jar it loose.

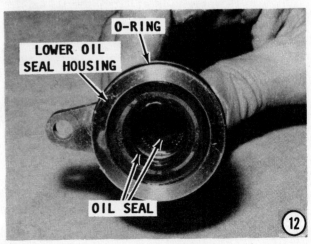

DISASSEMBLE 2-CYLINDER 7-37

Some units have a cylindrical exhaust manifold secured to the powerhead with three bolts. Remove the bolts and remove the manifold. It may be necessary to tap the manifold lightly with a soft head mallet to jar it loose.

12- Inspect the condition of the oil seals in the lower oil seal housing. Make a determination if they are fit for further service. If they are not, obtain slide hammer P/N C-91-27780 with an expanding jaw attachment. Use the slide hammer and remove the oil seals from the lower oil seal housing.

Some units have one large seal and a small seal behind the large seal.

Others have one large seal and two identical small seals behind the large seal. Remove and discard the outer O-ring, or gasket.

13- Remove the upper oil seal housing from the top of the crankshaft.

Some units are equipped with an O-ring in the bottom recess. Remove and discard the O-ring in the recess.

Some models do not have an upper oil seal housing. The upper crankshaft oil seal is contained in the stator plate.

If the unit being serviced is equipped with an upper oil seal housing, inspect the condition of the oil seal in the upper oil seal housing. Make a determination if the seal is fit for further service. The seal will be destroyed during removal, therefore, remove it only if it is damaged and has lost its sealing qualities.

To remove the seal, first obtain slide hammer puller P/N C-91-27780 with expanding jaw attachment, and then "pull" the seal.

Model 8C

14- Remove the bolt and then the limited throttle opening device next to the timing pointer. Pry the link from the free acceleration lever. Remove the two Phillips head screws from the bracket retaining the magneto control lever. Remove the lever, the pulley and the free acceleration lever from the throttle wire stay. Remove the attaching hardware, and then the throttle wire stay.

All Models

15- Remove the attaching bolts, and then the thermostat cover. If the cover is stuck fast, tap it lightly with a soft head mallet to jar it free. Lift out the thermostat and at the same time note the direction the thermostat faces as an aid during installation. Remove and discard the O-ring or washer. Some larger horsepower units are also equipped with a water pressure relief valve, located next to the thermostat. Remove this valve also.

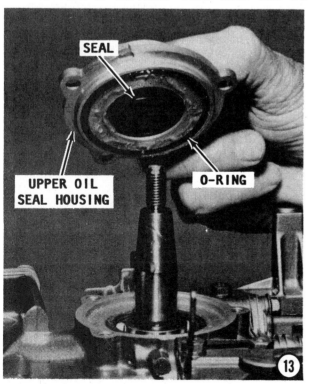

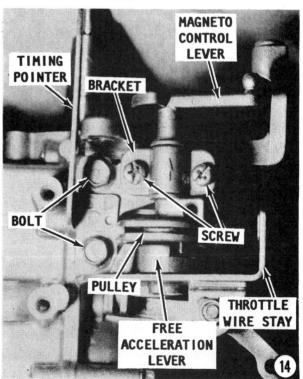

7-38 POWERHEAD

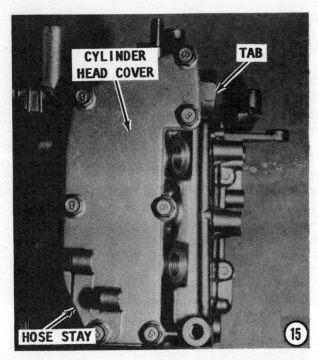

Some older and smaller horsepower units do not have a separate thermostat cover. On these units, the thermostat is located under the cylinder head cover.

Remove the cylinder head cover bolts, and then the cover. On some models a hose stay is secured by one of the cover bolts. Remember which bolt is used, as an aid during assembling. Remove and discard the gasket. If the cover is stuck to the cylinder head, insert a slotted screwdriver between the tabs provided for this purpose and pry the two surfaces apart. **NEVER** pry at gasket sealing surfaces. Such action would very likely damage the sealing surface of an aluminum powerhead.

SPECIAL WORDS ON CYLINDER HEADS

Some models have a cylinder head cover and a cylinder head with a gasket on both sides of the cylinder head. Others just have a cylinder head, with a single head gasket.

Some small horsepower models do not have a removable cylinder head. The block and head are cast as one piece. These units have only a cylinder head cover and gasket. Be sure all traces of gasket material are removed to provide the best sealing surface possible, to prevent the loss of compression or coolant.

16- If the model being serviced is equipped with a thermostat under the cylinder head cover, remove the collar and thermostat from beneath the cylinder head cover.

Model 40hp and Larger

These powerheads have an anode installed in the water passage. This anode can be removed by simply removing the securing Phillips head screw, and then the anode.

ADVICE

The exhaust cover should always be removed during a powerhead overhaul. Many times water in the powerhead is caused by a leaking exhaust cover gasket or plate.

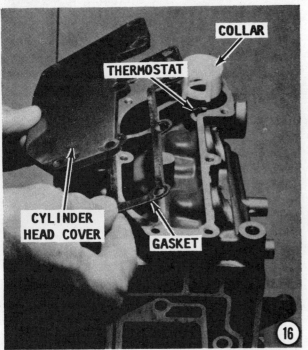

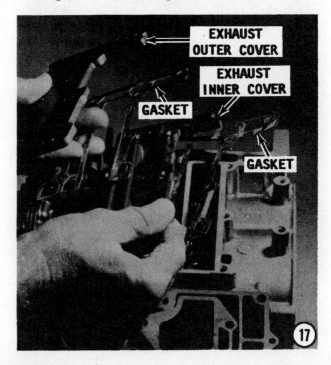

DISASSEMBLE 2-CYLINDER 7-39

17- Remove the bolts and then the exhaust outer cover, the gasket, the exhaust inner cover, and another gasket. Discard all gaskets. If the cover is stuck to the powerhead, insert a slotted screwdriver between the tabs provided for this purpose and pry the two surfaces apart. **NEVER** pry at gasket sealing surfaces. Such action would very likely damage the sealing surface of an aluminum powerhead.

SPECIAL WORDS

Take great care in the next step when handling reed valve assemblies. Once the assembly is removed, keep it away from sunlight, moisture, dust, and dirt. Sunlight can deteriorate valve seat rubber seals. Moisture can easily rust stoppers overnight. Dust and dirt -- especially sand or other gritty material -- can break reed petals if caught between stoppers and reed petals.

Make special arrangements to store the reed valves to keep them isolated from the elements while further work is being performed on the powerhead.

18- Remove the bolts and then the intake manifold. When removing the bolts, notice their lengths and locations as an aid during assembling. Remove and discard the gasket. Remove the reed valve assembly. Discard the gasket under the reed valve assembly. Some models have a wire stay attached by the middle left bolt and a fuel pipe clamp attached by the lower right bolt. Remember these locations as an aid during assembling.

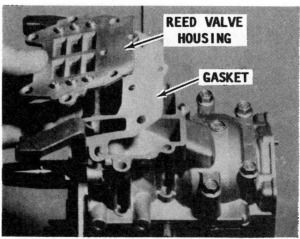

The reed valve housing MUST be handled with care to prevent accidental damage. The slightest injury could impare performance.

Some models do not have an intake manifold. The carburetor is mounted directly onto the reed valve assembly.

Remove the upper right, middle left, and two lower outer bolts. **DO NOT** remove the upper left stud, middle right stud, or center lower bolt. Remove the reed valve assembly. Discard the gasket under the assembly.

19- Remove the crankcase bolts. When removing the bolts, notice their lengths and locations as an aid during assembling. To remove the crankcase, insert a slotted screwdriver between the tabs provided for this purpose and pry the two surfaces apart. **NEVER** pry at gasket sealing surfaces. Such action would very likely damage the sealing surface of an aluminum powerhead.

20- Separate the two halves of the crankcase. **TAKE CARE** not to lose the two dowel pins. These pins may remain in either half when the crankcase is separated.

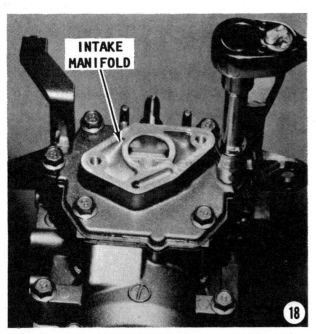

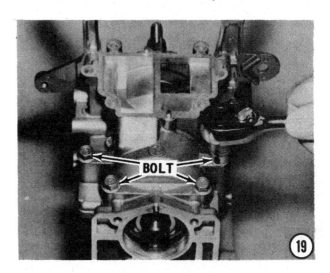

the cap in the original direction, during assemblying and also to ensure each rod will be installed in its original location. Numbers such as 1 and 1, 2 and 2, would be excellent.

Each rod cap **MUST** be kept with its connecting rod to ensure they remain as matched sets and the cap **MUST** be installed in its original direction -- **NOT** 180° out.

If new parts are being used, the connecting rod and rod cap must be installed in the same direction from which they were separated when removed from the package.

Remove the connecting rod bolts and rod caps. Lift out both sets of caged needle bearings from around the crankshaft journal. Keep the bearings with the connecting rod cap to ensure they will be installed in their original locations.

Tap the crankshaft **LIGHTLY** with a soft head mallet to jar it free from the block. Lift the crankshaft out of the block.

Pull each piston and connecting rod assembly from the **BOTTOM** -- not through the top of the block. A ridge might have formed on the top of the cylinder bore. This ridge may have to be removed with a ridge reamer.

Models Smaller than 40hp

21- Tap the tapered end of the crankshaft, with a soft head mallet, to jar it free from the block. **SLOWLY** lift the crankshaft assembly straight up and out of the block.

Model 40hp and Larger

The next step is rather lengthy, but it all applies only to Model 40hp and larger powerheads.

22- Obtain a marker and identify both halves of each connecting rod to ensure the mating halves will be brought together, with

All Models

23- Slide the labyrinth seal circlips from the crankshaft, one at each end. Scribe a mark on the inside of the piston skirt to identify the top and bottom piston **BEFORE** removal from the connecting rod as described in the next step.

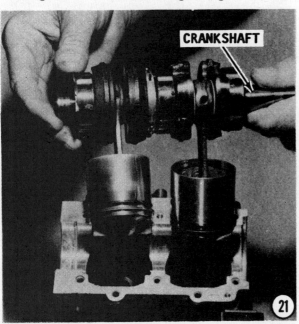

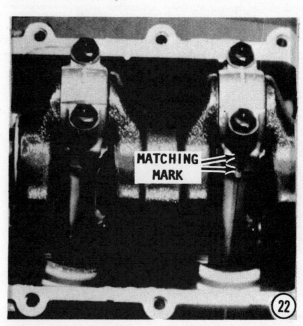

DISASSEMBLE 2-CYLINDER 7-41

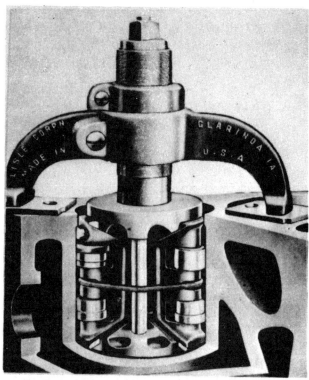

A ridge remover MUST be used to cut the ridge from the top of the cylinder walls. The stop under the blade prevents cutting into the walls too deeply. NEVER cut more than 1/32" below the bottom of the ridge.

SAFETY WORDS

The piston pin C-lockrings are made of spring steel and may slip out of the pliers or pop out of the groove with considerable force. Therefore, **WEAR** eye protection glasses while removing the piston pin lockrings in the next step.

24– Remove the C-lockring from both ends of the piston pin using a pair of needle nose pliers. Discard the C-lockrings. These rings stretch during removal and cannot be used a second time. The needle bearings and two retainers will fall away from the connecting rod small end.

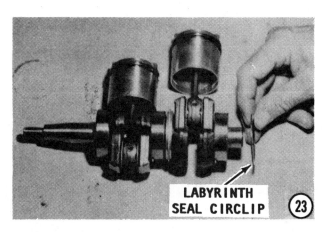

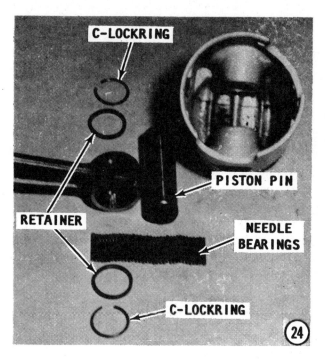

ADVICE

New needle bearings should be installed in the connecting rods, even though they may appear to be in serviceable condition. New bearings will ensure lasting service after the overhaul work is completed. If it is necessary to install the used bearings, keep them separate and identified to **EN-SURE** they will be installed into the same connecting rod from which they were removed.

CRITICAL WORDS
Models Less Than 40hp

The rod is an integral part of the crankshaft. The two are manufactured together and **CANNOT** be separated.

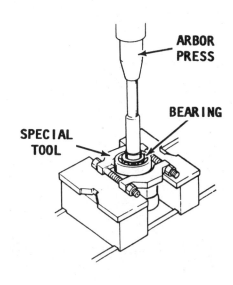

7-42 POWERHEAD

All Models

Slowly rotate both bearing races. If rough spots are felt, the bearings will have to be pressed free of the crankshaft.

25- Obtain a universal bearing separater tool and an arbor press. Press each bearing from the crankshaft. Be sure the crankshaft is supported, because once the bearing "breaks" loose, the crankshaft is free to fall.

GOOD WORDS

Good shop practice dictates to replace the rings during a powerhead overhaul. However, if the rings are to be used again, expand them **ONLY** enough to clear the piston and the grooves because used rings are brittle and break very easily.

26- Gently spread the top piston ring enough to pry it out and up over the top of the piston. No special tool is required to remove the piston rings. Remove the lower ring in a similar manner. These rings are **EXTREMELY** brittle and have to be handled with care if they are intended for further service.

CLEANING AND INSPECTING

COMPLETE and **DETAILED** instructions for cleaning and inspecting **ALL** parts of the powerhead, including the block, are presented in the last section of this chapter, Section 7-7, beginning on Page 7-67.

ASSEMBLING AND INSTALLATION

Detailed procedures are given to assemble and install virtually all parts of the

powerhead. Therefore, if certain parts were not removed or disassembled because the part was found to be fit for further service, simply skip the particular step involved and continue with the required tasks to return the powerhead to operating condition.

The following instructions pickup the work after all parts of the powerhead have been thoroughly cleaned and inspected according to the procedures outlined in Section 7-7, the last section in this chapter.

1- Install a new set of piston rings onto the piston. No special tool is necessary for installation. **HOWEVER**, take care to spread the ring only enough to clear the top of the piston. The rings are **EXTREMELY** brittle and will snap if spread beyond their limit. Align the ring gap over the locating pin.

2- If the main bearings were removed, support the crankshaft and press the bearings onto the shaft one at a time. Press only on the inner race. Pressing on the cage, the ball bearings, or the outer race, may destroy the bearing.

Take note of the bearing size embossed on one side of the bearing. The side with the marking **MUST** face **AWAY** from the crankshaft throw.

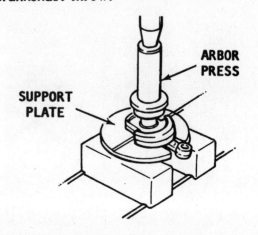

ASSEMBLE 2-CYLINDER   7-43

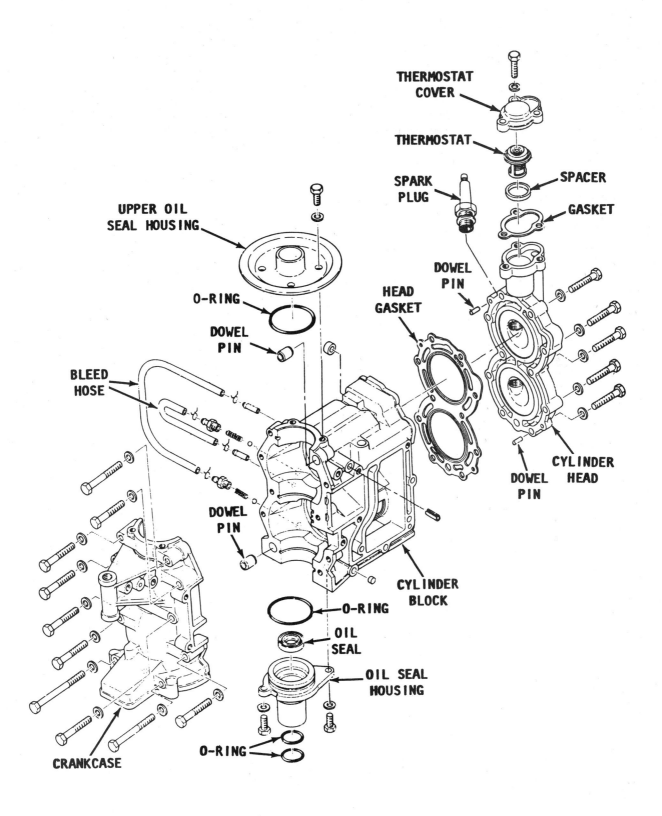

Exploded drawing of the Model 8hp powerhead prior to 1979. Major parts are identified.

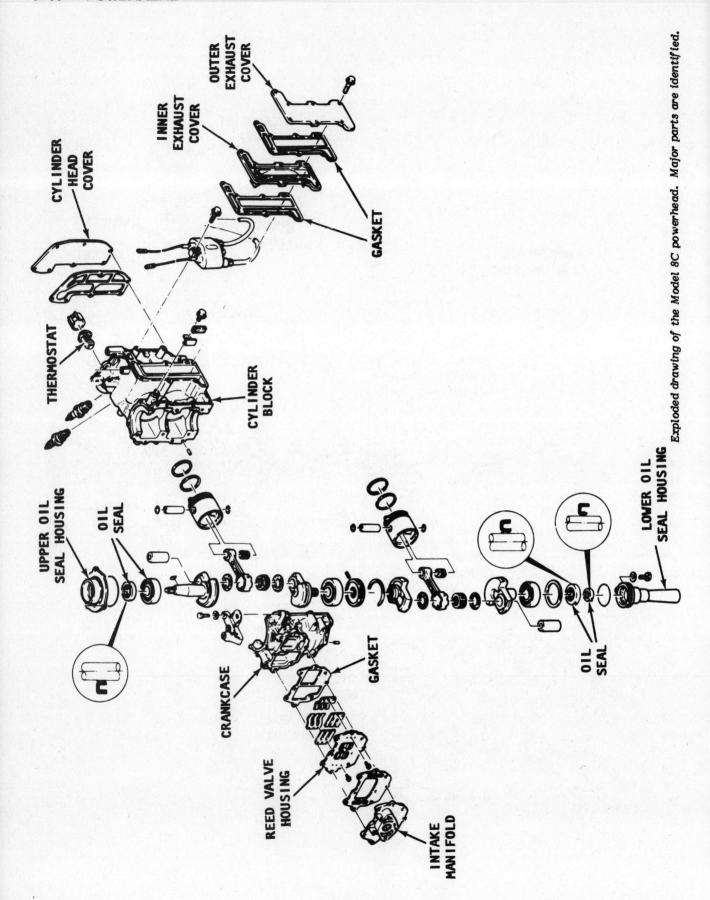

Exploded drawing of the Model 8C powerhead. Major parts are identified.

ASSEMBLE 2-CYLINDER   7-45

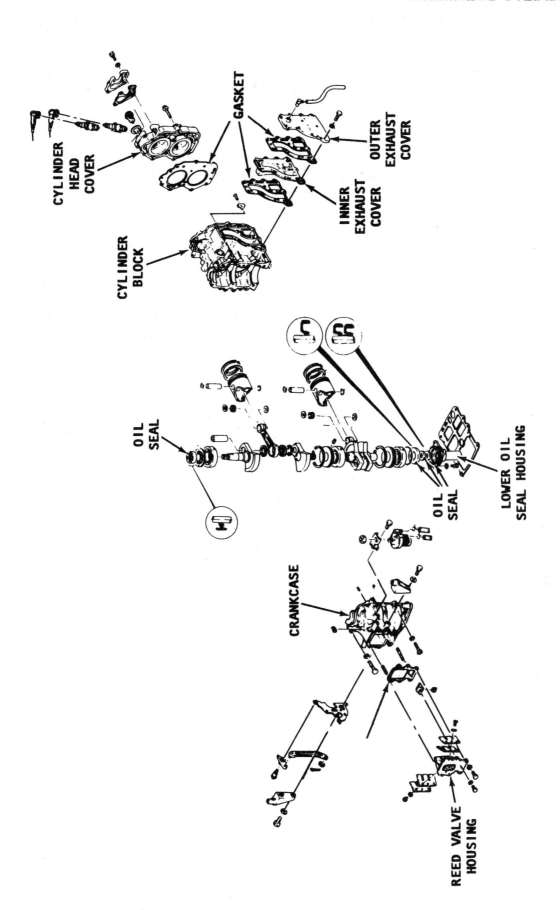

Exploded drawing of the powerhead for the following units: 8B, W8, Marathon 8, 9.9hp, and 15hp. Major parts are identified.

7-46 POWERHEAD

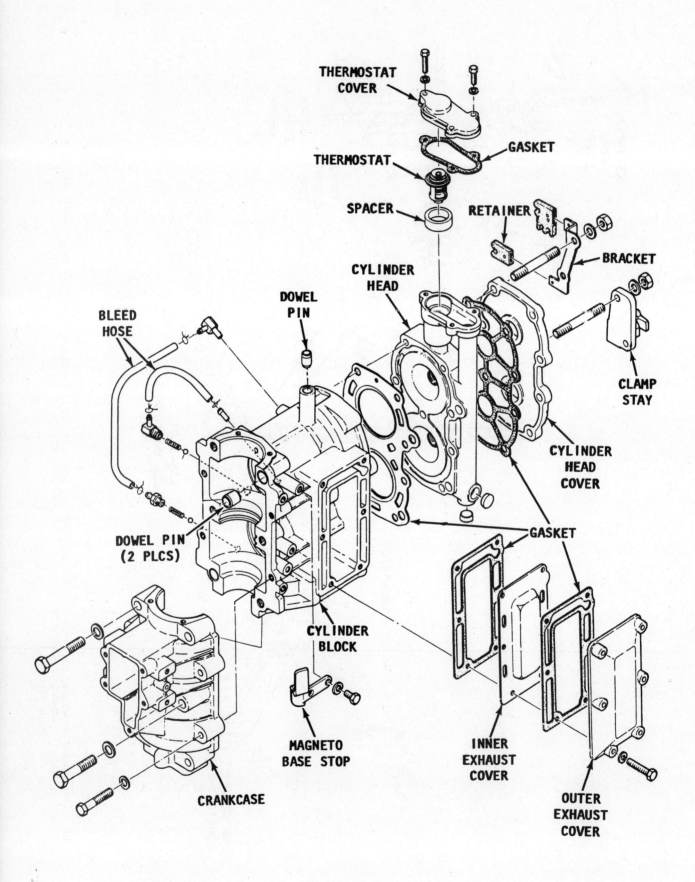

Exploded drawing of the 15hp powerhead prior to 1979. Major parts are identified.

ASSEMBLE 2-CYLINDER   7-47

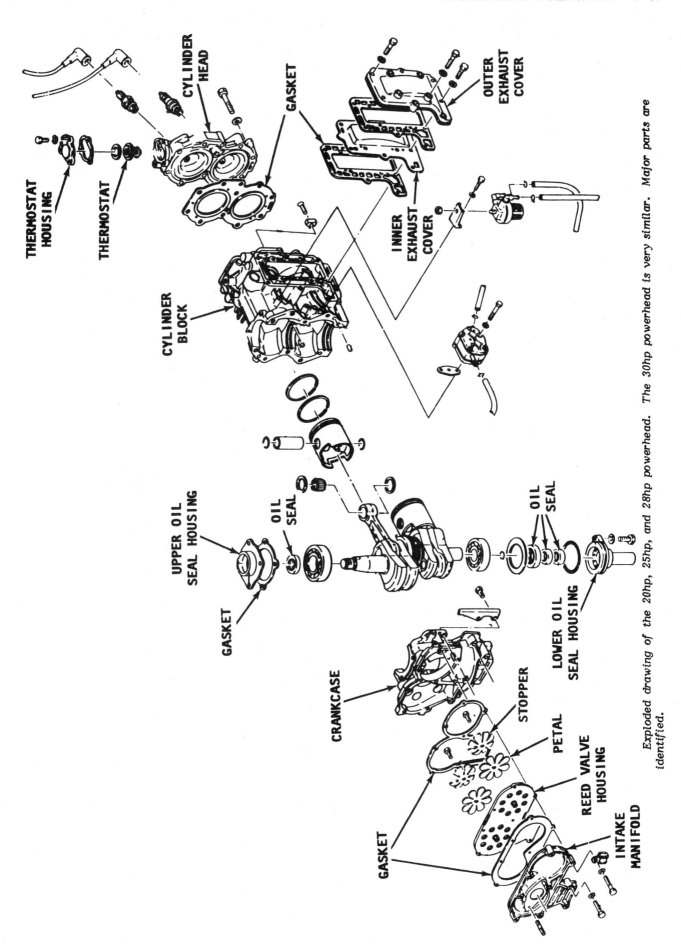

Exploded drawing of the 20hp, 25hp, and 28hp powerhead. The 30hp powerhead is very similar. Major parts are identified.

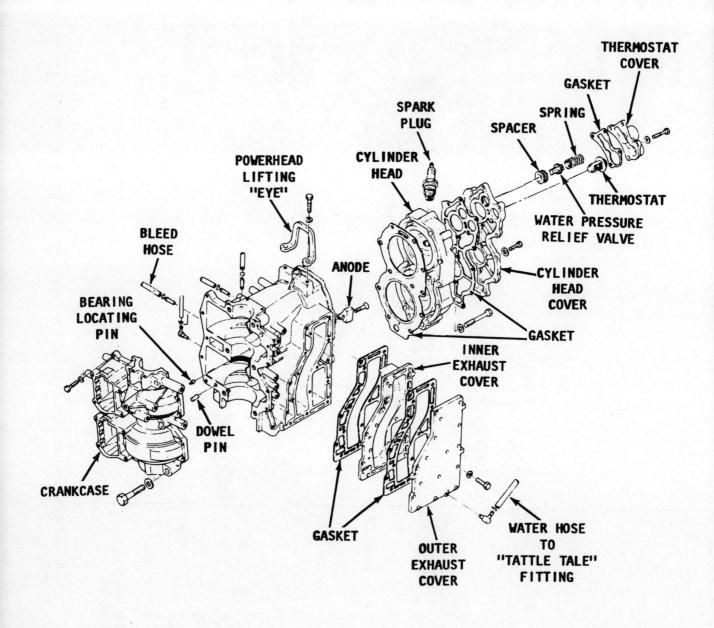

Exploded drawing of the 40hp cylinder block. Major parts are identified. An exploded drawing of the crankshaft assembly is on Page 7-50.

ASSEMBLE 2-CYLINDER 7-49

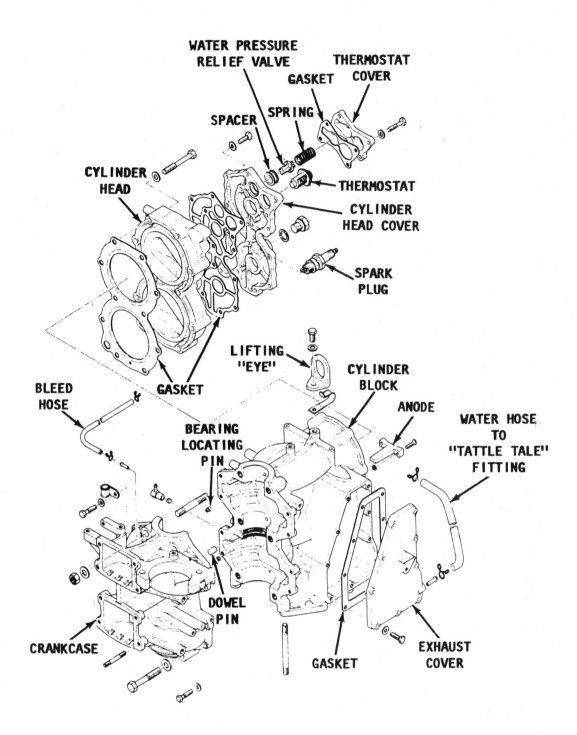

Exploded drawing of the 48hp, 55hp, and 60hp powerhead. Major parts are identified. An exploded drawing of the crankshaft assembly is on Page 7-50.

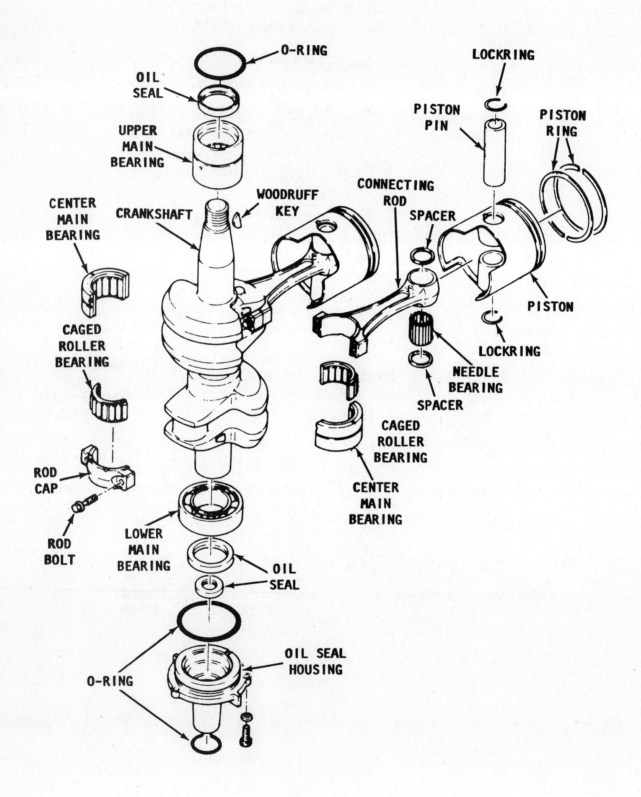

Exploded drawing of the crankshaft assembly for powerheads 40hp and larger. Major parts are identified.

ASSEMBLE 2-CYLINDER 7-51

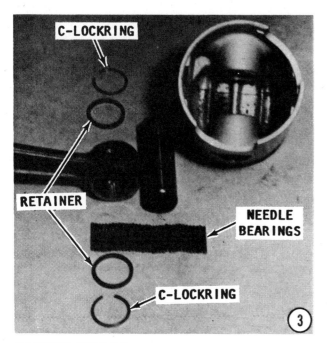

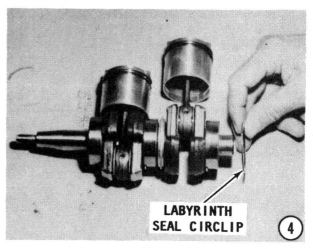

SAFETY WORDS

The piston pin lockrings are made of spring steel and may slip out of the pliers or pop out of the groove with considerable force. Therefore, **WEAR** eye protection glasses while installing the piston pin lockrings in the next step.

3- Select the set of needle bearings removed from the No. 1 piston or obtain a new set of bearings.

Coat the inner circumference of the small end of the No. 1 connecting rod with multi-purpose water resistant lubricant. Position the needle bearings one by one around the circumference. Dab some lubricant on the sides of the rod and "stick" the retainers in place.

Position the end of the rod with the needle bearings and retainers in place up into the piston. The word **UP** on the piston crown **MUST** face **TOWARD** the tapered (upper) end of the crankshaft. Slide the piston pin through the piston and connecting rod. Center the pin in the piston. Install the **C**-lockring at each end of the piston pin.

Perform the procedures in this step to install the No. 2 needle bearing set and the piston onto the No. 2 connecting rod.

4- Slide a labyrinth seal circlip on each end of the assembled crankshaft.

Model 40hp and Larger

The next three steps, No. 5, No. 6 and No. 7, apply only to the 40hp and larger powerheads. If the powerhead being serviced is smaller than 40hp, proceed directly to **Step 8**, on Page 7-53.

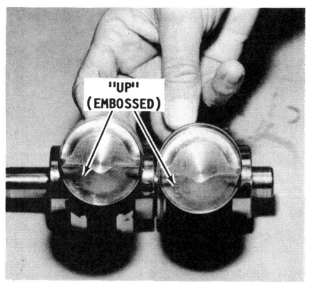

During installation, the word "UP" on the piston crown MUST face toward the tapered (upper), end of the crankshaft.

5- Coat the cylinder bores with a good grade of engine oil. **BEFORE** installing the the piston into the cylinder, make the following test.

Run your finger along the top rim of the cylinder. If the surface of the block and the surface of the bore have a sharp edge, the piston may be installed from the top. If the the slightest groove or ridge is felt on the rim, the piston **MUST** be installed from the lower end of the cylinder. Attempting to install the piston from the top will not be successful because the piston ring will bottom on the ridge. If force is used the ring may very well break.

Top Installation

Obtain special tool P/N M-91-84543M. Compress the aligned rings and, at the same time, use the end of a wooden mallet handle to gently tap the piston down into the cylinder bore. The word **UP** embossed on the piston crown **MUST** face toward the flywheel end of the block.

Bottom Installation

Installation of the piston from the lower end of the cylinder bore is an easy matter and can be accomplished without the use of special tools. Verify the ring end gaps are properly aligned and the word **UP**, embossed on the piston crown, faces toward the flywheel end of the block.

Three hands are better than two for this task. Simply compress each ring, one by one and, at the same time, push the piston up into the cylinder bore.

Repeat either procedure for the other piston.

After both pistons have been installed, slide each piston up and down in the cylinder bore several times. Check for binding. Listen for scratching noises. Scratching, or any other "spooky" noise) may indicate a ring was broken during installation.

SPECIAL BEARING WORDS

If the old roller bearings are to be installed for further service, each must be installed in the same location from which it was removed.

Place the lower half roller bearings in their original positions on the connecting rod caps. The locating pin hole for the first and third bearing outer race/cage should face **UPWARD**, toward the tapered end of the crankshaft, when installed.

The tang on the bearing will slide into the groove in the crankshaft.

6- Position the crankshaft into place in the block. Pack the lip of the upper oil seal with multi-purpose water resistant lubricant. Install the upper bearing, with the seal installed, with the lip of the seal facing **DOWNWARD**, towards the center of the crankshaft. Place the upper half of the roller bearings in their original positions, as recorded during disassembling, on the crankshaft journals. Place the connecting rod caps over the caged roller bearings in their original locations.

SPECIAL ROD CAP BOLT WORDS

The manufacturer does not recommend rod cap bolts be used a second time. Also, the manufacturer recommends a specific method of tightening the rod cap bolts, be followed as outlined in the next step.

Work on just one rod at a time.

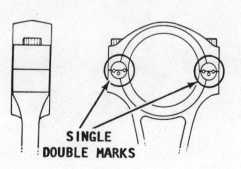

ASSEMBLE 2-CYLINDER 7-53

7- Thread in **NEW** connecting rod cap bolts a few turns each. Now, tighten the cap bolts to a torque value of 12 ft lbs (17Nm). Check to be sure the single mark on each rod cap end is centrally located between the two matching embossed marks on the rod end and on the cap, as shown in the accompanying illustration.

If the marks are aligned, as described and shown, loosen each bolt one half turn, and then tighten the bolt to a torque value of 25 ft lbs (35Nm).

If the marks are **NOT** aligned, as described and shown, remove the cap bolts, and then the upper caged roller bearing half. Install the upper caged roller bearing half again. Check the alignment. If the second attempt fails to align the marks as described and shown, the crankshaft **MUST** be removed and the lower caged roller bearing half checked for correct alignment.

Models Less Than 40hp

8- Coat the upper sides of each piston with 2-stroke engine oil. Hold the crankshaft at right angles to the cylinder bores and slowly lower one piston at a time into the appropriate cylinder. The upper edge of the each cylinder bore has a slight taper to squeeze in the rings around the locating pin and allow the piston to center the bore.

CRITICAL WORDS

If difficulty is experienced in fitting the piston into the cylinder, **DO NOT** force the piston. Such action might result in a broken piston ring. Raise the crankshaft and make

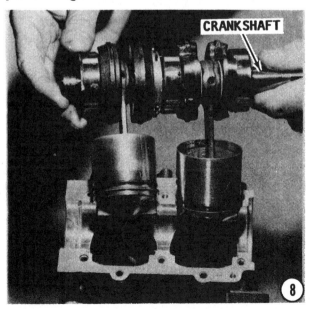

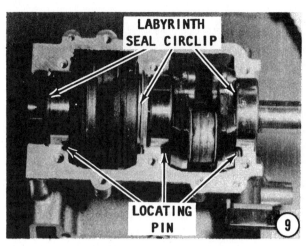

sure the ring end gap is aligned with the locating pin.

9- Push the crankshaft assembly down until it seats in the block. Fit the three labyrinth seal circlips into the grooves in the block. The center seal was never removed as it normally remains on the crankshaft. Rotate the upper, center, and lower bearings until their locating pins fit into the recesses in the block.

All Models

10- Apply a thin bead of Loctite 514 P/N C-92-75505-1 around both surfaces of the crankcase and block. Check to be sure the two dowel pins are in place and install the crankcase to the block.

11- Install and tighten the attaching bolts in the sequence shown in the accompanying illustration. Tighten the bolts in two rounds and to the torque value as follows:

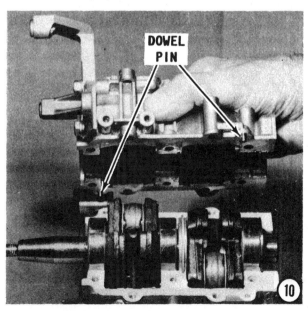

7-54 POWERHEAD

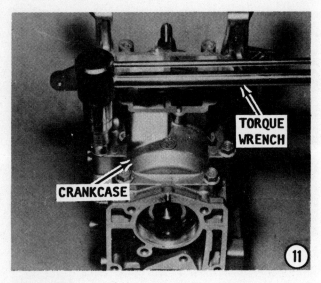

Model 8C
First round -- 4.3 ft lbs (6Nm). Second round -- 8.7 ft lbs (12Nm).

Models 9.9hp, 15hp and all Other 8hp
First round -- large bolts -- 11 ft lbs (15Nm). Small bolts 4.3 ft lbs (6Nm). Second round -- large bolts -- 22 ft lbs (30Nm). Small bolts -- 8.7 ft lbs (12Nm).

Models 20hp and Larger
First round -- 11 ft lbs (15Nm). Second round -- 19 ft lbs (27Nm).

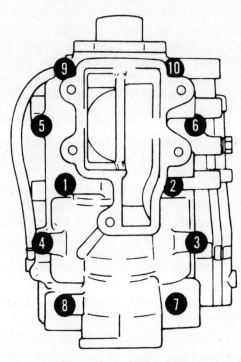

ALL OTHER MODEL 8HP SINCE 1979
(Except Model 8C)

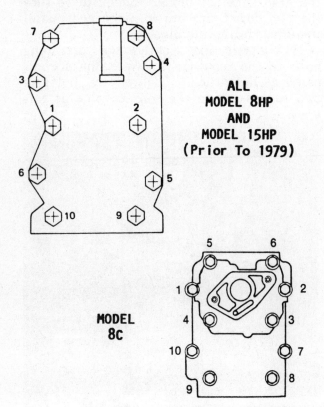

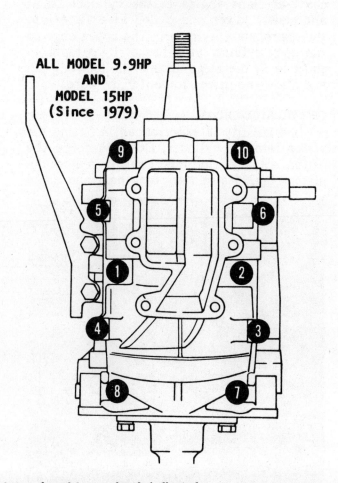

ALL MODEL 9.9HP AND MODEL 15HP (Since 1979)

Tightening sequence for the crankcase bolts of model powerheads indicated.

ASSEMBLE 2-CYLINDER 7-55

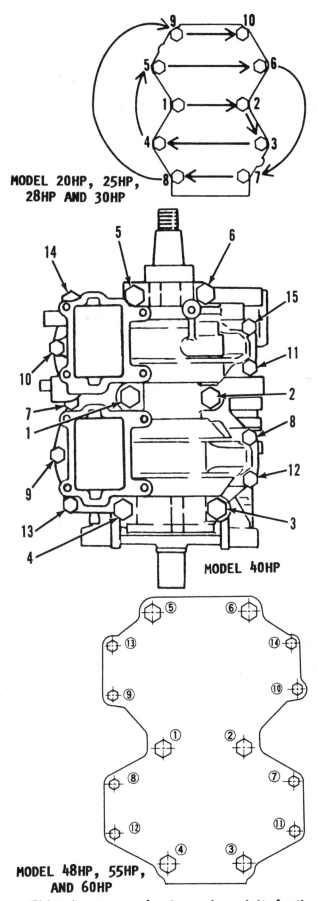

Tightening sequence for the crankcase bolts for the powerhead models indicated.

Rotate the crankshaft by hand to be sure the crankshaft does not bind.

BAD NEWS

If binding is felt, it wil be necessary to remove the crankcase and reseat the crankshaft and also to check the positioning of the labyrinth seal circlips and the bearing locating pins. If binding is still a problem after the crankcase has been installed a second time, the cause might very well be a broken piston ring.

PISTON RING TENSION

Check to be sure each piston ring has spring tension. This is accomplished by **CAREFULLY** pressing on each ring with a screwdriver extended through the exhaust ports, as shown in the accompanying illustration. If spring tension cannot be felt (the spring fails to return to its original position), the ring was probably broken during the piston and crankshaft installation process. **TAKE CARE** not to burr the piston rings while checking for spring tension.

12- Install the following parts in the order given over the exhaust ports: the gasket, the inner exhaust cover, a second gasket, and finally the outer exhaust cover.

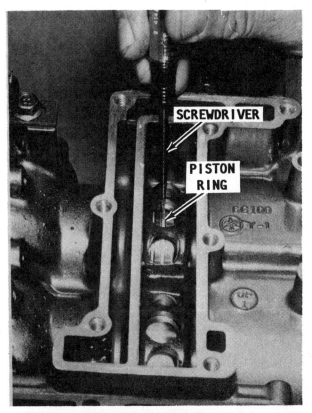

Checking piston ring spring tension with a small screwdriver through the exhaust port.

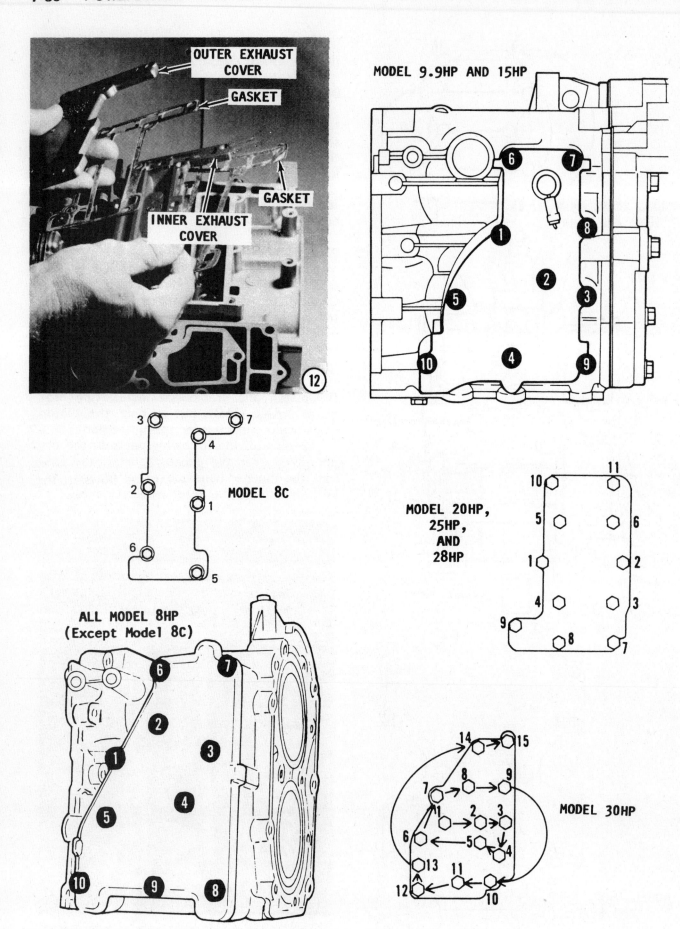

Tightening sequence for the exhaust cover of the powerheads indicated.

ASSEMBLE 2-CYLINDER 7-57

SPECIAL WORDS
The manufacturer recommends **NO** sealing agent be applied to any of the gasket sealing surfaces.

13- Install and tighten the exhaust cover attaching bolts in the sequence shown in the accompanying illustration in two rounds as listed.

Model 8C
First round -- 4.3 ft lbs (6Nm). Second round -- 7.1 ft lbs (10Nm).

Model 9.9hp, 15hp and All Other 8hp
First round -- 4.3 ft lbs (6Nm). Second round -- 8.7 ft lbs (12Nm).

Model 20hp and Larger
First round -- 2.9 ft lbs (4Nm). Second round -- 6.5 ft lbs (9Nm).

Model 8C
14- Install the following parts onto the intake port in the order given: the gasket, the reed valve housing, another gasket, and finally the intake manifold. Install and tighten the manifold attaching bolts in the sequence indicated in the accompanying illustration to a torque value of 8 ft lbs (11Nm).

Model 9.9hp, 15hp and All Other 8hp
15- Apply a light coating of Loctite to the upper portion **ONLY** of the reed valve housing gasket. Install the gasket and reed valve housing to the intake port. Install and

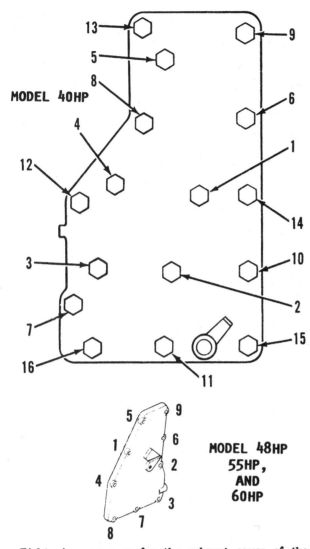

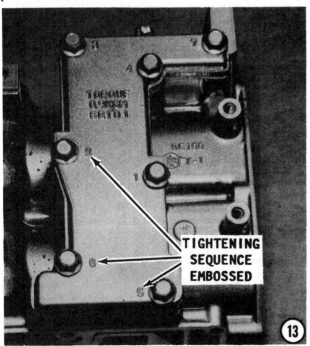

Tightening sequence for the exhaust cover of the powerhead models indicated.

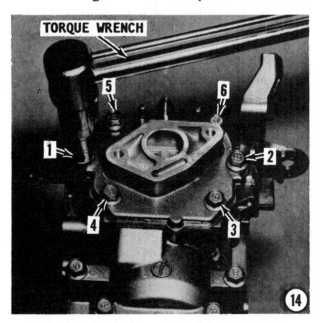

7-58 POWERHEAD

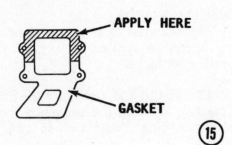

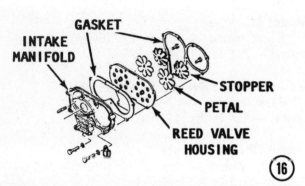

tighten the attaching bolts in a circular pattern to 4.3 ft lbs (6Nm) on the first round and to 8.7 ft lbs (12Nm) on the second round.

Model 20hp and Larger
16- Install the following parts onto the intake port in the order given: the gasket with the **NARROW** edges, the reed valve housing with the reeds facing **TOWARD** the powerhead, the gasket with the **WIDE** edges, and finally the intake manifold. Install the various length bolts in the locations noted

Reeds, reed petals, and stoppers for a 20hp thru 30hp powerhead.

during disassembling, Step 18. Tighten the bolts in a circular pattern to a torque value of 5.8 ft lbs (8Nm).

Model 8C
17- Insert the thermostat into the powerhead with the spring end going in **FIRST**. Install the thermostat collar. Position a new head gasket in place on the powerhead. The manufacturer recommends **NO** sealing agent whatsoever be used on either side of the gasket.

Model 8C
18- Start the seven attaching bolts. The spark plug lead stay is secured with the lower left bolt. Tighten the bolts in the sequence shown in the accompanying illustration to a torque value of 5.8 ft lbs (8Nm) on the first round and to 7.7 ft lbs (10Nm) on the second round.

Model 9.9hp, 15hp and All Other 8hp
19- Position a new head gasket in place on the cylinder head.

SPECIAL WORDS
The manufacturer recommends **NO** sealing agent be used on either side of the thermostat gasket and the head gasket.

Install the cylinder head cover and secure it with the attaching bolts. Tighten the bolts in the sequence shown in the accompanying illustration to a torque value of 5.8 ft lbs (8Nm) on the first round and to 12 ft lbs (17Nm) on the second round. In-

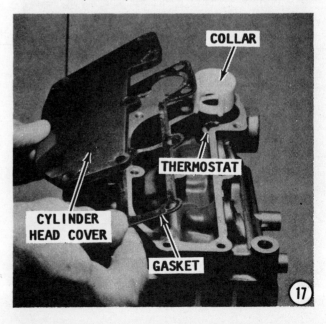

ASSEMBLE 2-CYLINDER 7-59

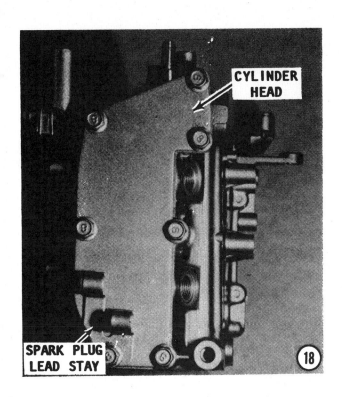

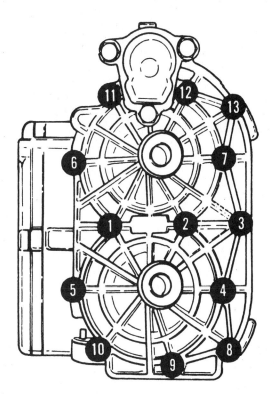

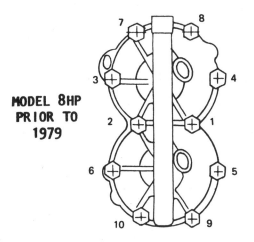

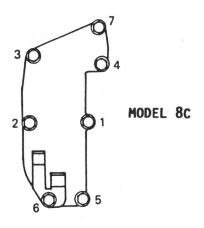

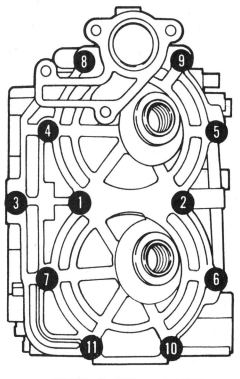

Tightening sequence for the cylinder head or cylinder head cover for the powerheads indicated.

7-60 POWERHEAD

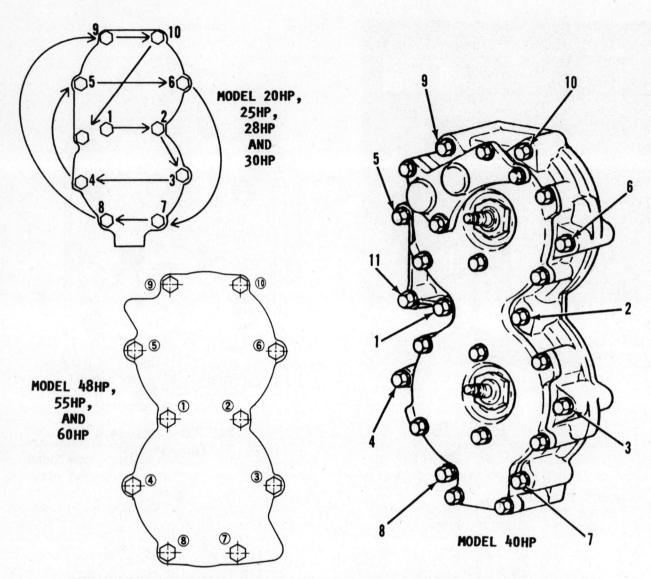

Tightening sequence for the cylinder head or cylinder head cover for the powerheads indicated.

stall the thermostat into the head cover with the spring end **TOWARD** the cover. Position a new gasket in place, and then install the thermostat cover. Secure the cover with the four attaching bolts. Tighten the bolts in a circular pattern to a torque value of 5.8 ft lbs (8Nm).

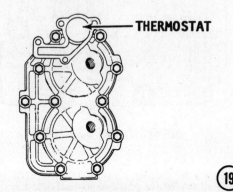

Model 20hp and Larger

20- Position a new head gasket in place on the powerhead.

SPECIAL WORDS

The manufacturer recommends **NO** sealing agent be used on either side of the thermostat gasket and the head gasket.

If the anode was removed from the cylinder block during disassembling, install a new anode at this time. Secure it in place with the screw.

Install the cylinder head onto the powerhead. Apply Loctite to the threads of the attaching bolts. Install the bolts and tighten them in the sequence shown in the accompanying illustration to a torque value of 11 ft lbs (15Nm) on the first round and to 19 ft lbs (27Nm) on the second round.

ASSEMBLE 2-CYLINDER   7-61

Insert the thermostat into the powerhead with the spring end **TOWARD** the powerhead. Position a new gasket in place, and then the thermostat cover. Secure the cover with the attaching bolts. Tighten the bolts in a circular pattern to a torque value of 3.8 ft lbs (8Nm).

SEALING SURFACE WORDS
CRITICAL READING

Because of the high temperatures and pressures developed, the sealing surfaces of the cylinder head and the block are the most prone to water leaks. No sealing agent is recommended **BECAUSE** it is almost impossible to apply an even coat of sealer. An even coat would be essential to ensure a air/water tight seal.

Some head gaskets are supplied with a "tacky" coating on both surfaces applied at the time of manufacture. This "tacky" substance will provide an even coating all around. Therefore, no further sealing agent is required.

HOWEVER, if a slight water leak should be noticed following completed assembly work and powerhead start up, **DO NOT** attempt to stop the leak by tightening the head bolts beyond the recommended torque value. Such action will only aggravate the problem and most likely distort the head.

FURTHERMORE, tightening the bolts, which are case hardened aluminum, may force the bolt beyond its elastic limit and cause the bolt to **FRACTURE. BAD NEWS,** very **BAD NEWS** indeed. A fractured bolt must usually be drilled out and the hole re- tapped to accommodate an oversize bolt, etc. Avoid such a situation.

Probable causes and remedies of a new head gasket leaking are:

a- Sealing surfaces not thoroughly cleaned of old gasket material. Disassemble and remove **ALL** traces of old gasket.

b- Damage to the machined surface of the head or the block. The remedy for this damage is the same as for the next case "c".

c- Permanently distorted head or block. Spray a light **EVEN** coat of any type metallic spray paint on both sides of a new head gasket. Use only metallic paint -- any color will do. Regular spray paint does not have the particle content required to provide the extra sealing properties this procedure requires.

Assemble the block and head with the gasket while the paint is till **TACKY.** Install the head bolts and tighten in the recommended sequence and to the proper torque value and **NO** more!

Allow the paint to set for at least **24** hours before starting the powerhead.

Consider this procedure as a temporary "band aid" type solution until a new head may be purchased or other permanent measures can be performed.

Under normal circumstances, if procedures have been followed to the letter, the head gasket will not leak.

End — Sealing Lecture

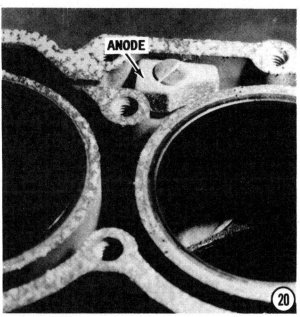

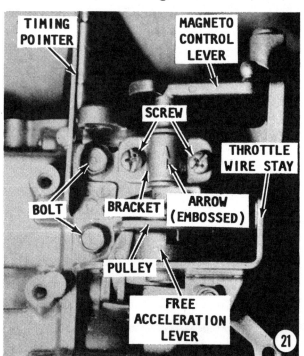

Model 8C

21- Install the throttle wire stay onto the block with the two bolts. Tighten the bolts to a torque value of 5.8 ft lbs (8Nm). Position the free acceleration lever and pulley onto the throttle wire stay. Insert the magneto control lever shaft into the free acceleration lever and pulley, and then install the magneto control lever between the bracket (installed with the embossed arrow pointing **UP**), and the throttle wire stay. Install the joint link to the free acceleration lever. Slide the bolt thru the limited throttle opening collar and arm, then thread it through the bolt hole next to the timing pointer. Tighten the bolt securely. Install the starter wire end (the end with the two locknuts), into the arm. Place the wire and two locknuts over the wire stay and tighten them temporarily to hold the wire in place. The procedure for adjusting the length of this wire is covered in Chapter 11 when the manual starter is installed and the "no start in-gear" system is adjusted.

All Models

22- If the unit being serviced is equipped with an oil seal housing: Install new oil seals in the housing using the appropriate driver and mandrel. Drive the seal in from the bottom of the housing with the lip facing **UP**. After the housing is installed, the seal lip will actually face down, toward the powerhead. Pack the seal lip with multi-purpose water resistant lubricant.

SPECIAL WORDS
FROM MARINER ENGINEERS

When two seals are installed, the lips of **BOTH** seals face in the **SAME** direction. Mariner engineers have concluded it is better to have both seals face the same direction rather than have them "back to back" as directed by other outboard manufacturers. In this position, with both lips facing toward the water after oil seal installation, the engineers feel the seals will be more effective in keeping water out of the oil seal housing. Any lubricant lost will be negligible.

If the unit being servied is not equipped with an oil seal housing: The upper crankshaft seal is installed in the stator plate. If this seal is damaged, it must be replaced,

Pack the lips of the seal with multi-purpose water resistant lubricant. Install the seal into the stator plate with the lips facing **DOWN** toward the crankshaft. Install the seal using the appropriate driver and handle.

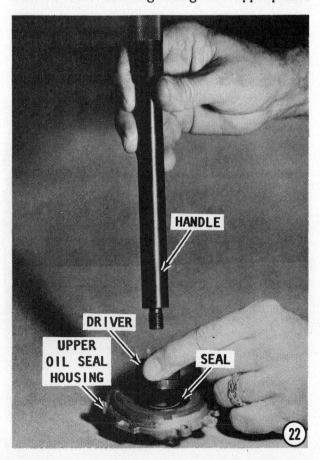

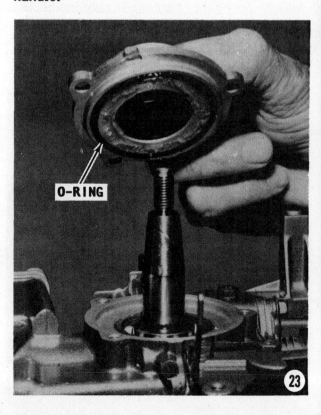

ASSEMBLE 2-CYLINDER 7-63

Model 8C

23- Coat the O-ring and sealing surfaces of the upper oil seal housing with multi-purpose water resistant lubricant. Install the O-ring into the recess of the housing and push the housing squarely into place over the crankshaft.

Model 20hp

Install a new gasket and the oil seal housing over the crankshaft. Seat the housing in the powerhead.

All Models

If the oil seals in the lower oil seal housing were removed during disassembly, Step 12, install new oil seals using a handle and appropriate driver.

24- Secure the lower oil seal housing in a vice equipped with soft jaws. Pack the lips with multi-purpose water resistant lubricant. Drive the seals in from the top of the housing with the lip of each seal facing **TOWARD** the lower unit.

Some units use a small oil seal, installed first, then one large oil seal.

Other units use two small oil seals, installed first, then one large oil seal. The lips of all three seals face in the same direction --toward the lower unit **AFTER** the housing is installed.

25- Coat the sealing surfaces of the lower oil seal housing and a new O-ring with multi-purpose water resistant lubricant. Install the new O-ring over the housing. Seat the housing into its recess in the powerhead.

Rotate the housing until the bolt hole/s align/s. Install and tighten the bolt/s to a torque value of 5.8 ft lbs (8Nm).

Model 8C

These units have an exhaust manifold installed next to the oil seal housing. Install the manifold and tighten the attaching bolts to a torque value of 5.8 ft lbs (8Nm).

All Models

26- Check to be sure the two dowel pins on the intermediate housing are in place forward and aft of the driveshaft area.

Model 9.9hp, 15hp and All Other 8hp

Apply Permatex, or equivalent material to the gasket area shown in the accompanying illustration.

All Other Models

No sealing agent is required on the gasket.

All Models

Position the gasket in place with the dowel pins passing up through the correct holes in the gasket.

Model 9.9hp, 15hp and All Other 8hp Except Model 8C

Position the three rubber seals into the intermediate housing and bring the stator plate into contact with the stopper on the full open side. Set the throttle grip to the wide open position.

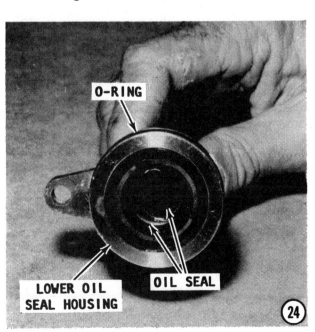

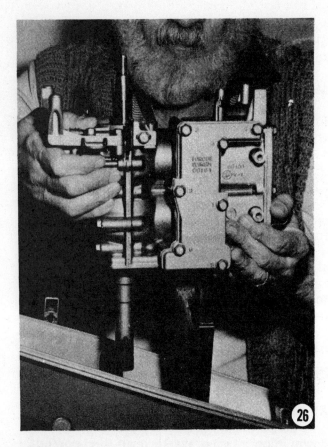

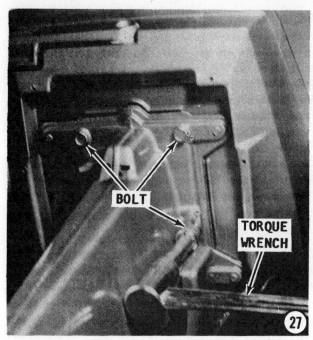

All Models

Apply a coating of multi-purpose water resistant lubricant to the lower end of the driveshaft. Install the powerhead onto the intermediate housing using the dowel pins to align the two surfaces. The propeller may have to be rotated slightly to permit the crankshaft to index with the driveshaft and allow the powerhead to seat properly.

27- Install the powerhead bolts. The outboard unit may have to be turned port or starboard to permit installation of the two front bolts. Tighten the bolts evenly and alternately to the following torque value for the models listed.

8C -- 5.8 ft lbs (8Nm)
9.9hp, 15hp
& All Other 8hp -- 13 ft lbs (18Nm)
20hp and Larger -- 13 ft lbs (18Nm)

Model 8C

28- Twist the throttle grip from **SLOW** to **FAST**. Fit the shorter pulled cable end into the outer slot - the one close to the cowling edge. Fit the longer slack cable end into the inner slot -- the one closest to the powerhead. Fit the wires down into the

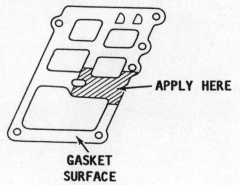

Apply Permatex, or equivalent, to the shaded portion of the powerhead to lower unit gasket.
Note: This illustration is valid for Models 9.9hp, 15hp, and all 8hp units after 1979, with the exception of the Model 8C.

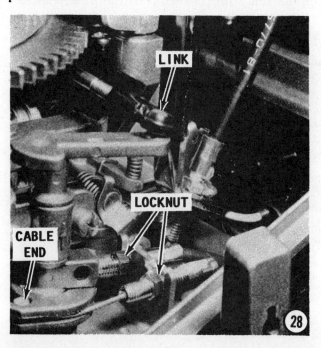

ASSEMBLE 2-CYLINDER 7-65

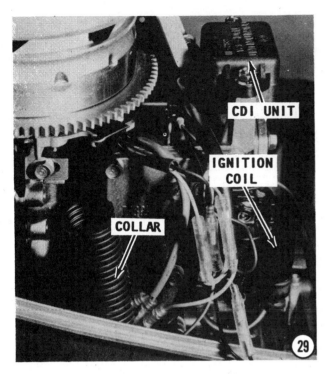

wire stay with the locknuts on both sides of the stay. Secure the locknuts in place. Snap the link over the ball joint.

Model 8C

29- Install the ignition coil and CDI unit and secure them in place with the attaching hardware. Match the colors and connect the wires at their quick disconnect fitting. Ground the Black wires with eyes at their original locations. Tidy the wire harness and wrap the collar around it to keep it together neatly.

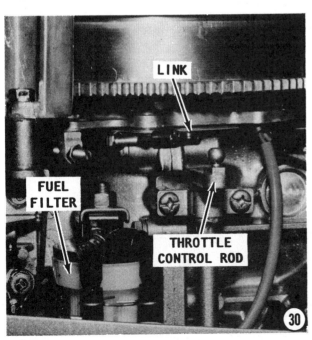

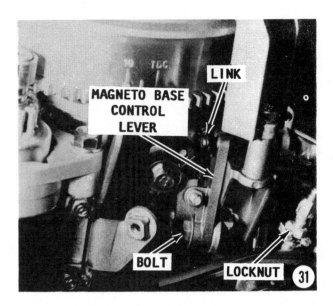

Model 9.9hp, 15hp and All Other 8hp

The following four Steps, No. 30, 31, 32, and 33, apply only to the models listed in this heading.

30- Install the fuel filter. Connect the fuel lines leading to and from the filter according to the arrows embossed on top of the fittings. Secure the wire type clamps over the hoses. Snap the link over the ball joint on the throttle control rod.

31- Install the starter cable end into the bracket. Push the cable into the wire stay with one locknut on each side of the stay. Tighten the nuts to secure the cable in place. Install the magneto base control lever and snap the link onto the ball joint.

32- Install the CDI unit bracket. Seat the bracket by pushing it away from the flywheel. Install the CDI unit in the bracket. Install the ignition coil and secure the

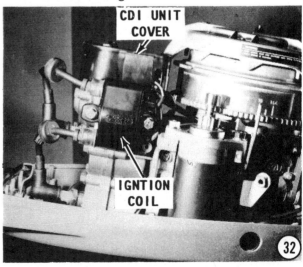

7-66 POWERHEAD

CDI unit and coil in place. Match colors and connect wires at their quick disconnect fittings. Ground any Black wires with eyes at their original locations. Install the cover over the CDI unit and tidy the wire harness.

33- Install the hose from the pilot hole to the fitting on the exhaust cover.

Models 20hp and Larger

The following two Steps, No. 34 and 35, apply only to the models listed in this heading.

34- Install the starter wire stay bracket and arm. Insert the starter cable end into the arm and push the cable into the stay with a locknut on both sides of the stay. Tighten the locknuts to secure the cable in place. Install the bushing into the center of the throttle cable stay and magneto control lever. Secure the cable and lever to the block using a washer and bolts. Install the other retaining bolt located on the upper right of the circular throttle cable end retainer.

Tighten both bolts to a torque value of 5.8 ft lbs (8Nm).

Install the throttle control and magneto control link rods.

SPECIAL WORDS

When installing the various link rods, check to be sure the adjustable end is fully snapped onto the ball joints on the levers.

Slide the throttle cable ends into their appropriate recesses in the circular throttle cable end retainer. Position the cables in the stays. Tighten the locknuts around the stays to secure the cables in place.

35- Install the CDI unit bracket with the three attaching bolts. Connect the Black ground lead to the bracket. Install the CDI unit and the ignition coil. Match colors and connect wires at their quick disconnect fittings. Tidy the wire harness into the wire retainers.

CLOSING TASKS

All Models

After the powerhead is secured to the intermediate housing, complete the work by first following the procedures listed in Chapter 5 beginning with the stator plate installation.

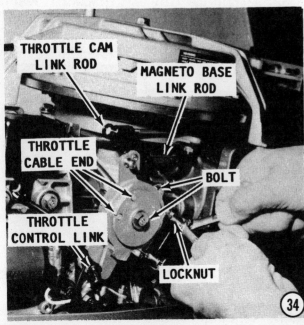

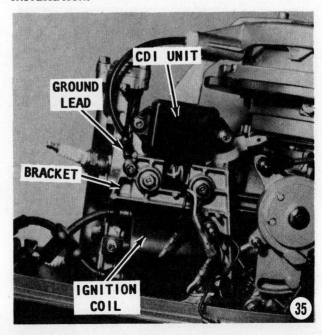

Install the flywheel.

Install the carburetor and fuel tank, see Chapter 4.

Install the hand rewind starter. After the starter is installed, check the action of the no-start-in-gear protection system. If adjustment is required, see the instructions at the end of the installation procedures in Chapter 11.

Mount the engine in a test tank, on a boat in a body of water, or connect a flush attachment to the lower unit.

REMEMBER, the powerhead will **NOT** start without the emergency tether in place behind the kill switch knob.

Start the engine and check the completed work.

CAUTION
Water must circulate through the lower unit to the powerhead anytime the powerhead is operating to prevent damage to the water pump in the lower unit. Just five seconds without water will damage the water pump impeller.

Attempt to start and run the engine without the cowling installed. This will provide the opportunity to check for fuel and oil leaks, without the cowling in place.

After the engine is operating properly, install the cowling. Follow the break-in procedures with the unit on a boat and with a load on the propeller.

Break-in Procedures

As soon as the engine starts, CHECK to be sure the water pump is operating. If the water pump is operating, a fine stream will be discharged from the exhaust relief hole at the rear of the drive shaft housing.

DO NOT operate the engine at full throttle except for **VERY** short periods, until after 10 hours of operation as follows:

a- Operate at 1/2 throttle, approximately 2500 to 3500 rpm, for 2 hours.

b- Operate at any speed after 2 hours **BUT NOT** at sustained full throttle until another 8 hours of operation.

c- Mix gasoline and oil during the break-in period, total of 10 hours, at a ratio of 25:1.

d- While the engine is operating during the initial period, check the fuel, exhaust, and water systems for leaks.

e- See Chapter 2 for tuning procedures.

7-7 CLEANING AND INSPECTING ALL POWERHEADS

The success of the overhaul work is largely dependent on how well the cleaning and inspecting tasks are completed. If some parts are not thoroughly cleaned, or if an unsatisfactory unit is allowed to be returned to service through negligent inspection, the time and expense involved in the work will not be justified with peak powerhead performance and long operating life. Therefore, the procedures in the following sections should be followed closely and the work performed with patience and attention to detail.

REED BLOCK SERVICE

Disassemble the reed block housing by first removing the screws securing the reed stoppers and reed petals to the housing. After the screws are removed, lift the stoppers and petals from the housing.

Clean the gasket surfaces of the housing. Check the surfaces for deep grooves, cracks or any distortion which could cause leakage.

Replace the reed block housing if it is damaged. The reed petals should fit flush against their seats, and not be preloaded against their seats or bend away from their seats. The following limits are recommended by the manufacturer.

Model	Petal Clearance
2hp --	0.012" (0.3mm)
3.5hp --	0.008" (0.2mm)
4hp & 5hp --	0.008" (0.2mm)
8hp --	0.024" (0.6mm)
9.9hp & 15hp --	0.008" (0.2mm)
20hp, 25hp & 28hp --	0.008" (0.2mm)
30hp --	0.022" (0.5mm)
40hp & Larger --	0.008" (0.2mm)

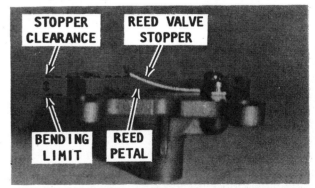

A 2hp reed valve housing with single reed petal. Critical measurements are explained in the text.

If the reed petal is distorted beyond its bending limit, rarely, if ever, can the petal be successfully straightened. **THEREFORE**, it must be replaced.

The valve stopper clearance is the distance between the bottom edge of the stopper and the top of the reed petal. This measurement **MUST** be within the following limits.

Model	Stopper Clearance
2hp --	0.24" (6mm)
3.5hp --	0.28" (7mm)
4hp & 5hp --	0.28" (7mm)
8hp Prior to 1979 --	0.26" (6.5mm)
8hp 1979 & Later --	0.18" (4.5mm)
9.9hp --	0.05" (1.3mm)
15hp --	0.16" (4mm)
20hp, 25hp, 28hp, 30hp & 40hp --	0.19" (5mm)
48hp --	0.12" (3mm)
55hp & 60hp --	0.39" (10mm)

The valve stopper can sometimes be successfully bent to achieve the required clearance, if not, it **MUST** be replaced.

Do not remove the reed valves unless they, or the stoppers, are to be replaced. If servicing a 9.9hp unit or larger, the reeds **MUST** be replaced in sets.

Apply Loctite to the threads of the reed retaining screws. Tighten each screw gradually, starting in the center and working outwards across the reed block. Tighten the screws to the following torque values:

Model	Torque Value
2hp --	0.6 ft lbs (0.8Nm)
3.5hp thru 15hp --	0.7 ft lbs (1Nm)
20hp & Larger --	3.5 ft lbs (5Nm)

EXHAUST COVER

The exhaust covers are one of the most neglected items on any outboard powerhead. Seldom are they checked and serviced. Many times a powerhead may be overhauled and returned to service without the exhaust covers ever having been removed.

The exhaust manifold is located on the port side of the powerheads covered in this manual.

One reason the exhaust covers are not removed is because the attaching bolts usually become corroded in place. This means they are very difficult to remove, but the work should be done. Heat applied to the bolt head and around the exhaust cover will help in removal. However, some bolts may still be broken. If the bolt is broken it must be drilled out and the hole tapped with new threads.

The exhaust covers are installed over the exhaust ports to allow the exhaust to leave the powerhead and be transferred to the exhaust housing. If the cover was the only item over the exhaust ports, they would become so hot from the exhaust gases they might cause a fire or a person would be severely burned if they came in contact with the cover.

Therefore, an inner plate is installed to help dissipate the exhaust heat. Two gaskets are installed -- one on either side of the inner plate. Water is channeled to circulate between the exhaust cover and the inner plate. This circulating water cools the exhaust cover and prevents it from becoming a hazard.

A thorough cleaning of the inner plate behind the exhaust covers should be performed during a major powerhead overhaul. If the integrity of the exhaust cover assembly is in doubt, replace the inner plate.

BLEED SYSTEM SERVICE
(If equipped)

Check the condition of the rubber bleed hose. Replace the hose if it shows signs of deterioration or leakage. Check the operation of the two check valves. The air/fuel mixture should be able to pass through the valve in only **ONE** direction. The check

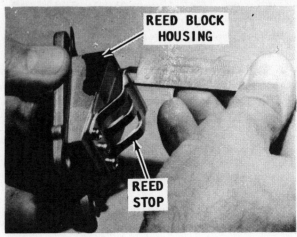

Taking measurements on a 4hp or 5hp reed valve housing with three reed petals.

valves and hoses connect the upper main bearing to the lower main bearing.

Defective check valves cannot be serviced. If defective, they **MUST** be replaced.

CRANKSHAFT SERVICE

Clean the crankshaft with solvent and wipe the journals dry with a lint free cloth. Inspect the main journals and connecting rod journals for cracks, scratches, grooves, or scores. Inspect the crankshaft oil seal surface for nicks, sharp edges or burrs which might damage the oil seal during installation or might cause premature seal wear. **ALWAYS** handle the crankshaft carefully to avoid damaging the highly finished journal surfaces. Blow out all oil passages with compressed air. The oil passageway leads from the rod to the main bearing journal. **TAKE CARE** not to blow dirt into the main bearing journal bore.

Inspect the internal splines at one end and threads at the other end for signs of abnormal wear. Check the crankshaft for runout by supporting it on two "V" blocks at the main bearing surfaces.

Install a dial indicator gauge above the main bearing journals. Rotate the crankshaft and measure the runout (or the out-of-round) and the taper at both ends (and in the center journal on all two-cylinder models).

Crankshaft runout and taper limits are as follows:

Model	Runout/Taper
2hp --	0.0008" (0.02mm)
All Others --	0.0010" (0.03mm)

If **"V"** blocks or a dial indicator are not available, a micrometer may be used to measure the diameter of the journal. Make a second measurement at right angles to the first. Check the difference between the first and second measurement for out-of-round condition. If the journals are tapered, ridged, or out-of-round by more than the specification allows, the journals should be reground, or the crankshaft replaced.

Any out-of-round or taper shortens bearing life. Good shop practice dictates new main bearings be installed with a new or reground crankshaft.

SPECIAL WORDS

Normally the connecting rod/s would **ONLY** be presssed from the crankshaft if either the crankshaft and/or the connecting rod/s were to be replaced. Therefore, the connecting rod axial "play" is checked at the piston end to determine the amount of wear at the crankshaft end of the connecting rod. Checking the "play" is described in the following paragraph.

Crankshaft setup with "V" blocks and dial indicator to measure main bearing journals on an assembled crankshaft for a 30hp and smaller powerhead. The ball bearings and connecting rods will not affect the readings.

Crankshaft setup with "V" blocks and dial indicator to measure main bearing journals of a crankshaft for a 40hp or larger powerhead.

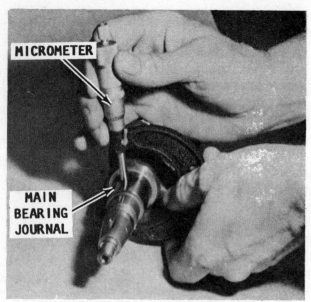

Using a micrometer to take the first measurement on a crankshaft main journal for out-of-round and taper.

Setup the crankshaft in the **"V"** blocks. Setup a dial indicator to touch the flat surface of the piston end of the rod. Now, hold the crankshaft steady in the **"V"** blocks and at the same time, rock the piston end of the rod along the same axis as the crankshaft. If the dial indicator needle moves through more than 0.08" (2mm) for all two-stroke models covered in this manual, the "play" is considered excessive. The rod must be pressed from the crankshaft, the

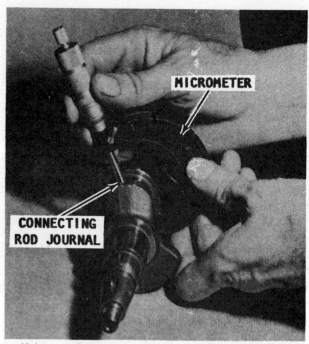

Using a micrometer to take the first measurement on a connecting rod journal for out-of-round and taper.

journal checked and a determination made as to whether the crankshaft and/or the rod must be replaced. A new rod may be purchased, **BUT** it must be pressed onto the crankshaft throw with a hydraulic press.

To check the connecting rod side clearance at the crankshaft, first insert a feeler gauge between the connecting rod and the counterweight of the crankshaft. Acceptable clearances are as follows:

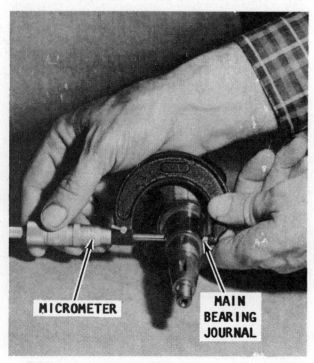

Taking the second measurement on a crankshaft main bearing journal at $90°$ to the first measurement.

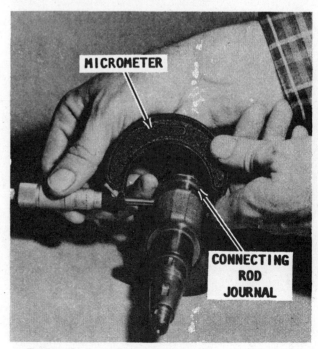

Taking the second measurement on a connecting rod journal at $90°$ to the first measurement.

CLEANING & INSPECTING   7-71

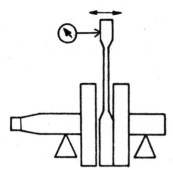

Crankshaft and connecting rod setup to check for side clearance, as explained in the text.

Model	Clearance
2hp --	0.012 - 0.024" (0.30 - 0.60mm)
3.5hp thru 15hp --	0.008 - 0.028" (0.20 - 0.70mm)
20hp and Larger --	0.008 - 0.012" (0.20 - 0.30mm)

If the measurement is not within the limits listed, measure the distance from the outside edge of one counterweight to the outside edge of the other counterweight for the throw of the connecting rod involved. This measurement will give an indication if the counterweight is "walking" on the crankpin or if the clearance is due to worn parts. The connecting rod would wear before the edge of the counterweight because the rod is manufactured from a much softer material. The distance measured -- from outside edge to outside edge -- should be as follows:

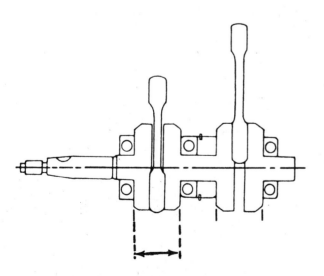

Crankshaft counterweight width measurement, as explained in the text.

Model	Width
2hp --	1.098 - 1.104" (27.90 - 28.05mm)
3.5hp -- (Air-Cooled)	No info. available
5hp -- (Air-Cooled)	1.689 - 1.695" (42.90 - 43.05mm)
4hp & 5hp -- (Water-Cooled)	1.564 - 1.574" (39.90 - 40.05mm)
8hp -- (Prior to 1979)	1.492 - 1.500" (37.90 - 38.10mm)
8C --	1.564 - 1.574" (39.90 - 40.05mm)
8hp -- (All Others)	1.571 - 1.573" (39.90 - 39.95mm)
9.9hp & 15hp --	1.846 - 1.852" (46.90 - 47.05mm)
20hp, 25hp, & 28hp --	2.122 - 2.124" (53.90 - 53.95mm)
30hp --	2.240 - 2.242" (56.90 - 56.95mm)

All models 40hp and larger have a conventional crankshaft with removable connecting rods. Therefore, this measurement does not apply.

Inspect the crankshaft oil seal surfaces to be sure they are not grooved, pitted, or scratched. Replace the crankshaft if it is severely damaged or worn. Check all crankshaft bearing surfaces for rust, water marks, chatter marks, uneven wear or overheating. Clean the crankshaft surfaces with 320-grit carborundum cloth. **NEVER** spin-dry a crankshaft ball bearing with compressed air.

Clean the crankshaft and crankshaft ball bearing with solvent. Dry the parts, but not the ball bearing, with compressed air. Check the crankshaft surfaces a second time. Replace the crankshaft if the surfaces cannot be cleaned properly for satisfactory service. If the crankshaft is to be installed for service, lubricate the surfaces with light oil. **DO NOT** lubricate the crankshaft ball bearing at this time.

7-72 POWERHEAD

After the crankshaft has been cleaned, grasp the outer race of the crankshaft ball bearing installed on the lower end of the crankshaft, and attempt to work the race back-and-forth. There should not be excessive "play". A very slight amount of side "play" is acceptable because there is only about 0.001" (0.025mm) clearance in the bearing.

Lubricate the ball bearing with light oil. Check the action of the bearing by rotating the outer bearing race. The bearing should have a smooth action and no rust stains. If the ball bearing sounds or feels rough or catches, the bearing should be removed and discarded.

CONNECTING ROD SERVICE

Inspect the connecting rod bearings for rust or signs of bearing failure. **NEVER** intermix new and used bearings. If even one bearing in a set needs to be replaced, all bearings at that location **MUST** be replaced.

Clean the inside diameter of the piston pin end of the connecting rod with crocus cloth.

Clean the connecting rod **ONLY** enough to remove marks. **DO NOT** continue, once the marks have disappeared.

Assemble the piston end of the connecting rod with loose needle bearings, caged needle bearings, or no needle bearing, depending on the model being serviced. Insert the piston pin and check for vertical "play". The piston pin should have **NO** noticeable vertical "play".

If the pin is loose or there is vertical "play" check for and replace the worn part/s.

Inspect the piston pin and matching rod end for signs of heat discoloration. Overheating is identified as a bluish bearing surface color and is caused by inadequate lubrication or by operating the powerhead at excessive high rpm.

Next Four Paragraphs
Model 40hp and Larger

Check the bearing surface of the rod and rod cap for signs of chatter marks. This condition is identified by a rough bearing surface resembling a tiny washboard. The condition is caused by a combination of low-speed low-load operation in cold water. The condition is aggravated by inadequate lubrication and improper fuel.

Under these conditions, the crankshaft journal is hammered by the connecting rod. As ignition occurs in the cylinder, the piston pushes the connecting rod with tremendous force. This force is transferred to the connecting rod journal. Since there is little or no load on the crankshaft, it bounces away from the connecting rod.

The crankshaft then remains immobile for a split second, until the piston travel causes the connecting rod to catch up to the waiting crankshaft journal, then hammers it. In some instances, the connecting rod crankpin bore becomes highly polished.

While the powerhead is running, a "whir" and/or "chirp" sound may be heard when the powerhead is accelerated rapidly -- say from idle speed to about 1500 rpm, then quickly returned to idle. If chatter marks are discovered, the crankshaft and the connecting rods should be replaced.

Inspect the bearing surface of the rod and rod cap for signs of uneven wear and possible overheating. Overheating is identi-

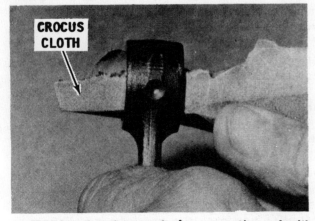

Cleaning the piston end of a connecting rod with crocus cloth.

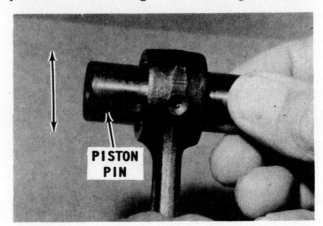

Checking the piston end of a connecting rod for vertical free "play" using a piston pin.

CLEANING & INSPECTING

fied as a bluish bearing surface color and is caused by inadequate lubrication or by operating the powerhead at excessive high rpm.

PISTON SERVICE

Inspect each piston for evidence of scoring, cracks, metal damage, cracked piston pin boss, or worn pin boss. Be especially critical during inspection if the outboard unit has been submerged. If the piston pin is bent, the piston and pin **MUST** be replaced as a set for two reasons. First, a bent pin will damage the boss when it is removed. Secondly, a piston pin is not sold as a separate item.

Check the piston ring grooves for wear, burns distortion, or loose locating pins. During an overhaul, the rings should be replaced to ensure lasting repair and proper powerhead performance after the work is completed. Clean the piston dome, ring grooves and the piston skirt. Clean carbon deposits from the ring grooves using the recessed end of a broken piston ring.

NEVER use a rectangular ring to clean the groove for a tapered ring, or use a tapered ring to clean the groove for a rectangular ring.

NEVER use an automotive-type ring

Using part of a broken piston ring to clean the ring groove. Exercise care not to disturb the pin in each groove.

groove cleaner, because such a tool may loosen the piston ring locating pins.

Clean carbon deposits from the top of the piston using a soft wire brush, carbon removal solution or by sand blasting. If a wire brush is used, **TAKE CARE** not to burr or round machined edges. Clean the piston skirt with crocus cloth.

Install the piston pin through the first boss only. Check for vertical free "play". There should be **NO** vertical free "play". The presence of "play" is an indication the piston boss is worn. The piston is manufactured from a softer material than the piston pin. Therefore, the piston boss will wear more quickly than the pin.

Excessive piston skirt wear **CANNOT** be visually detected. Therefore, good shop practice dictates, the piston skirt diameter be measured with a micrometer.

Piston skirt diameter should be measured at right angles to the piston pin axis at a point 0.2" (5mm) for the 2hp model and 0.4" (10mm) for all other models, above the bottom edge of the piston.

Piston diameters are as follows:

It is believed, this crown siezed with the cylinder wall when the unit was operated at high rpm and the timing was not adjusted properly. At the same instant, the rod apparently pulled the lower part of the piston downward, severing it from the crown.

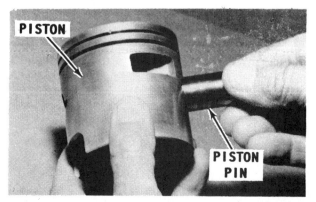

Checking free "play" between the piston pin and the piston boss. There should be NO "play".

Model	Diameter
2hp --	1.535" (39.00mm)
3.5hp -- (Air-Cooled)	1.772" (45.00mm)
4hp --	1.969" (50.00mm)
5hp -- (1977-79)	1.969" (50.00mm)
5hp -- (1980 & on)	2.126" (54.00mm)
8hp -- (Prior to 1988)	1.969" (50.00mm)
8hp -- (1988 & on)	2.126" (54.00mm)
9.9hp -- (Prior to 1986)	2.205" (56.00mm)
9.9hp -- (1986 & on)	2.126" (54.00mm)
15hp -- (Prior to 1989)	2.205" (56.00mm)
15hp -- (1989 & on)	2.363" (60.00mm)
20hp -- (1977-79)	2.520" (64.00mm)
20hp -- (1980-85)	2.638" (67.00mm)
20hp & 25hp -- (1986 & on)	2.560" (65.00mm)
25hp -- (Prior to 1986)	2.638" (67.00mm)
28hp --	2.638" (67.00mm)
30hp --	2.835" (72.00mm)
40hp --	2.953" (75.00mm)
48hp, 55hp, & 60hp --	3.228" (82.00mm)

RING END GAP CLEARANCE

FIRST, THESE IMPORTANT WORDS

Before the piston rings are installed onto the piston, the ring end gap clearance for each ring must be determined. The purpose of the piston rings is to prevent the blowby of gases in the combustion chamber. This cannot be achieved unless the correct oil film thickness is left on the cylinder wall.

This thin coating of oil acts as a seal between the cylinder wall and the face of the piston ring. An excessive end gap will allow blowby and the cylinder will lose compression. An inadequate end gap will

The pitted damage to this piston crown was probably caused by a broken piston ring working its way into the combustion chamber. The little "hills" then became "hot" spots on the crown, contributing to "dieseling" after the powerhead was shut down.

scrape too much oil from the cylinder wall and limit lubrication. Lack of adequate lubrication will cause excessive heat and wear.

IDEALLY the ring end gap measurement should be taken **AFTER** the cylinder bore has been measured for wear and taper **AND** after any corrective work, such as boring or honing, has been completed.

IF the ring end gap is measured with a taper to the cylinder wall, the diameter at

The rings on this piston became stuck due to lack of adequate lubrication, incorrect timing, or overheating.

CLEANING & INSPECTING 7-75

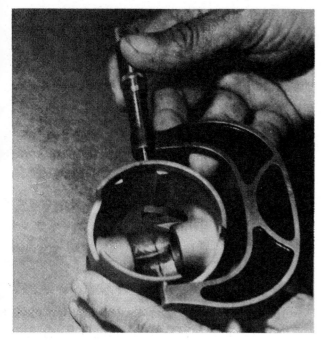

A micrometer must be used to check the diameter of the piston.

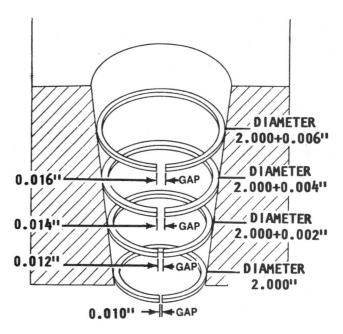

The cylinder taper drastically affects ring and end gap, as shown in this cross-section line drawing.

the lower limit of ring travel will be smaller than the diameter at the top of the cylinder.

IF the ring is fitted to the upper part of a cylinder with a taper, the ring end gap will not be great enough at the lower limit of ring travel. Such a condition could result in a broken ring and/or damage to the cylinder wall and/or damage to the piston and/or damage to the cylinder head.

IF the cylinder is to be only honed, not bored, **OR** if only cleaned, not honed, the ring end gap should be measured at the lower limit of ring travel.

The manufacturer actually gives the precise depth the ring should be inserted into the cylinder -- usually just above the ports -- and assumes the cylinder walls are **PARALLEL** with no taper.

Insert the piston ring from the top or the bottom, as listed in the following table.

The rings on this piston were broken, possibly during installation, and then caused extensive damage to the piston. Ring parts found their way into the combustion chamber and caused damage to the piston crown.

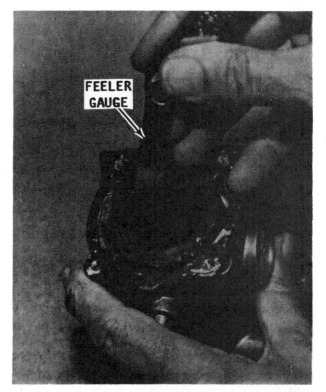

Measuring the piston ring end gap with a feeler gauge. The text states some rings are inserted from the top and others from the bottom of the cylinder.

Push the ring down, or up, the specified distance from the surface using the piston crown. Once the ring is in the proper position, measure the ring end gap with a feeler gauge.

If the end gap is greater than the amount listed, replace the entire ring set.

If the end gap is less than the amount listed, carefully file the ends of the ring -- just a little at a time -- until the correct end gap is obtained.

Model	Insert From	Depth	End Gap
2hp	Top	0.8" (20mm)	0.004 - 0.012" (0.1 - 0.3mm)
3.5hp to 8hp	Bottom	2.7" (70mm)	0.006 - 0.014" (0.15 - 0.35mm)
9.9hp & 15hp	Top	0.8" (20mm)	0.006 - 0.014" (0.15 - 0.35mm)
20hp, 25hp, 28hp, 40hp, & Larger	Top	0.8" (20mm)	0.012 - 0.020" (0.30 - 0.50mm)
30hp	Top	0.8" (20mm)	0.008 - 0.016" (0.20 - 0.40mm)

SPECIAL WORDS

Inspect the piston ring locating pins to be sure they are tight. There is one locating pin in each ring groove. If the locating pins are loose, the piston MUST be replaced.

PISTON RING SIDE CLEARANCE

After the rings are installed on the piston, the clearance in the grooves needs to be checked with a feeler gauge. Check the clearance between the ring and the upper "land" and compare your measurement with the Specifications listed in the following table. Ring wear forms a step at the inner portion of the upper "land". If the piston grooves have worn to the extent to cause high steps on the upper "land", the step will interfere with the operation of new rings and the ring clearance will be too much. Therefore, if steps exist in any of the upper "lands", the piston should be replaced.

On a plain ring this clearance may be measured either above or below the ring. On a Keystone ring -- one with a taper at the top -- the clearance MUST be measured below the ring.

Model	Ring	Side Clearance
2hp thru 5hp	Both	0.0010 - 0.0030" (0.03 - 0.07mm)
8hp (Prior to 1979)	Both	0.0012 - 0.0028" (0.03 - 0.07mm)
8hp (After 1979)	Top	0.0008 - 0.0024" (0.02 - 0.06mm)
	Bottom	0.0016 - 0.0030" (0.04 - 0.08mm)
9.9hp & 15hp	Both	0.0014 - 0.0040" (0.04 - 0.10mm)
20hp & Larger	Top	0.0008 - 0.0024" (0.02 - 0.06mm)
	Bottom	0.0016 - 0.0030" (0.04 - 0.08mm)

OVERSIZE PISTONS AND OVERSIZE RINGS

Scored cylinder blocks can be saved for further service by reboring and installing oversize pistons and piston rings. **HOWEVER**, if the scoring is over 0.0075" (0.13mm) deep, the block cannot be effectively rebored for continued use.

Check with the parts department at your local Mariner dealer for the model being serviced.

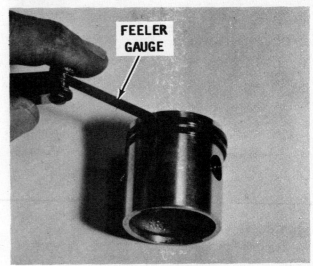

Measuring piston ring side clearance with a feeler gauge, as explained in the text.

The following table indicates the oversize pistons with matching ring sets available, at press time, for the units listed.

MODEL	PISTON SIZE
2hp --	1.545 and 1.555" (39.25 and 39.50mm)
3.5hp --	See note below
4hp & 5hp --	See note below
8hp -- (Prior to 1988)	1.978 and 1.989" (50.25 and 50.50mm)
8hp -- (1989 & on)	No info. avail.
9.9hp & 15hp -- (2.205" bore)	2.215 and 2.224" (56.25 and 56.50mm)
20hp thru 28hp -- (2.638" bore)	2.648 and 2.657" (67.25 and 67.50mm)
30hp --	2.845 and 2.854" (72.25 and 72.50mm)
40hp --	2.963 and 2.972" (75.25 and 75.50mm)
48hp & Larger --	See note below

NOTE
Oversize pistons are available. Check with your local Mariner dealer.
If oversize pistons are not readily available, the local marine shop may have the facilities to "knurl" the piston, making it larger.

CYLINDER BLOCK SERVICE

Inspect the cylinder block and cylinder bores for cracks or other damage. Remove carbon with a fine wire brush on a shaft attached to an electric drill or use a carbon remover solution.
STOP: If the cylinder block is to be submerged in a carbon removal solution, the crankcase bleed system **MUST** be removed from the block to prevent damage to hoses and check valves.
Use an inside micrometer or telescopic gauge and micrometer to check the cylin-

The walls of this cylinder were damaged beyond repair when a piston ring broke and worked its way into the combustion chamber.

ders for wear. Check the bore for out-of-round and/or oversize bore. If the bore is tapered, out-of-round or worn more than the wear limit specified by the manufacturer, the cylinder/s should be rebored -- provided oversize pistons and rings are available.
Check with the Mariner dealer **PRIOR** to reboring. If oversize pistons and matching rings are not available, the block **MUST** be replaced.

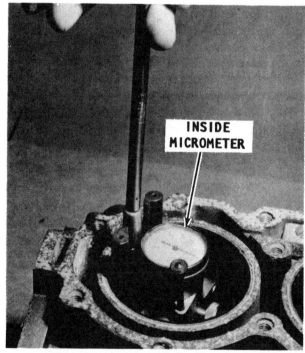

Checking the cylinder for taper using an inside micrometer. One measurement should be taken near the top and another near the bottom. The difference between the two measurements is the amount of taper.

GOOD WORDS

Oversize piston weight is approximately the same as a standard size piston. Therefore, it is **NOT** necessary to rebore all cylinders in a block just because one cylinder requires reboring.

Cylinder sleeves are an integral part of the die cast cylinder block and **CANNOT** be replaced. In other words, the cylinder cannot be "resleeved".

Four inside cylinder bore measurements must be taken for each cylinder to determine an out-of-round condition, the maximum taper, and the maximum bore diameter.

In the accompanying illustration, measurements D1 and D2 are diameters measured at 0.8" (20mm) from the top of the cylinder at right angles to each other. Measurements D3 and D4 are diameters measured at 2.4" (60mm) from the top of the cylinder at right angles to each other.

Out-of-Round

Measure the cylinder diameter at D1 and D2. The manufacturer requires the difference between the two measurements should be less than 0.002" (0.050mm) for **ALL** models.

Maximum Taper

Measure the cylinder diameter at D1, D2, D3, and D4. Take the largest of the D1 or D2 measurements and subtract the smallest measurement at D3 or D4. The answer to the subtraction -- the cylinder taper -- should be less than 0.003" (0.08mm) for **ALL** models.

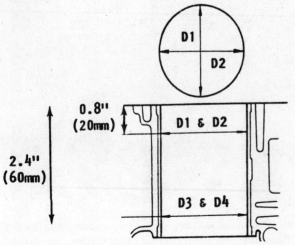

Top view diagram and cross section of a typical cylinder to indicate where measurements are to be taken for wear limit, taper, and out-of-round limits.

Bore Wear Limit

The maximum cylinder diameter D1, D2, D3, and D4 must not exceed the bore wear limits indicated in the following table **BEFORE** the bore is rebored for the **FIRST** time. These limits are only imposed on original parts because the sealing ability of the rings would be lost, resulting in power loss, increased powerhead noise, unnecessary vibration, piston slap, and excessive oil consumption.

The limits indicated are usually 0.003 to 0.005" (0.83 to 0.127mm) above the standard bore. Therefore, if the bore is resized, it may sustain another 0.003 to 0.005" (0.83 to 0.127mm) wear before a second reboring is required -- provided the oversize pistons and matching rings are available.

Model	Std. Cyl. Bore	Wear Lmt.
2hp	1.535" (39.0mm)	1.555" (39.5mm)
3.5hp (Air-Cooled)	1.772" (45.0mm)	1.775" (45.1mm)
4hp	1.969" (50.0mm)	1.972" (50.1mm)
5hp (Air-Cooled)	1.969" (50.0mm)	1.972" (50.1mm)
5hp (Water-Cooled)	2.126" (54.0mm)	2.129" (54.1mm)
8hp (Prior to 1988)	1.969" (50.0mm)	1.972" (50.1mm)
8hp (1988 & on)	2.126" (54.0mm)	2.130" (54.1mm)
9.9hp (Prior to 1986)	2.205" (56.0mm)	2.209" (56.1mm)
9.9hp (1986 & on)	2.126" (54.0mm)	2.130" (54.1mm)
15hp (Prior to 1989)	2.205" (56.0mm)	2.209" (56.1mm)
15hp (1989 & on)	2.363" (60.0mm)	2.367" (60.1mm)
20hp (1977-97)	2.520" (64.0mm)	2.524" (64.1mm)
20hp (1980-85)	2.638" (67.0mm)	2.642" (67.1mm)
20hp & 25hp (1986 & on)	2.560" (65.0mm)	2.564" (65.1mm)
25hp (Prior to 1986)	2.638" (67.0mm)	2.642" (67.1mm)
30hp	2.834" (72.0mm)	2.838" (72.1mm)
40hp	2.953" (75.0mm)	2.957" (75.1mm)
48hp, 55hp & 60hp	3.228" (82.0mm)	3.232" (82.1mm)

CLEANING & INSPECTING

Piston Clearance

Piston clearance is the difference between a maximum piston diameter and a minimum cylinder bore diameter. If this clearance is excessive, the powerhead will develop the same symptoms as for excessive cylinder bore wear — loss of ring sealing ability, loss of power, increased powerhead noise, unnecessary vibration, and excessive oil consumption.

Maximum piston diameter was described earlier in this section. Minimum cylinder bore diameter is usually determined by measurement D3 or D4 also described earlier in this section.

If the piston clearance exceeds the limits outlined in the following table, either the piston or the cylinder block **MUST** be replaced.

Calculate the piston clearance by subtracting the maximum piston skirt diameter from the maximum cylinder bore measurement and compare the results for the model being serviced.

Model	Piston Clearance
2hp --	0.0012 - 0.0016" (0.03 - 0.04mm)
3.5hp -- (Air-Cooled)	0.0012 - 0.0016" (0.03 - 0.04mm)
4hp & 5hp -- (Water-Cooled)	0.0012 - 0.0016" (0.03 - 0.04mm)
5hp -- (Air-Cooled)	0.0014 - 0.0016" (0.035 - 0.04mm)
8hp -- (Prior to 1979)	0.0012 - 0.0014" (0.03 - 0.035mm)
8C --	0.0020 - 0.0040" (0.05 - 0.10mm)
9.9hp, 15hp & All Other 8hp --	0.0014 - 0.0016" (0.035 - 0.040mm)
20hp, 25hp, 28hp, 30hp & 40hp --	0.0020 - 0.0040" (0.05 - 0.10mm)
48hp, 55hp, & 60hp --	0.0020 - 0.0021" (0.050 - 0.055mm)

HONING CYLINDER WALLS

Hone the cylinder walls lightly to seat the new piston rings, as outlined in this section. If the cylinders have been scored, but are not out-of-round or the bore is rough, clean the surface of the cylinder with a cylinder hone as described in the following procedures.

SPECIAL WORDS

If overheating has occurred, check and resurface the spark plug end of the cylinder

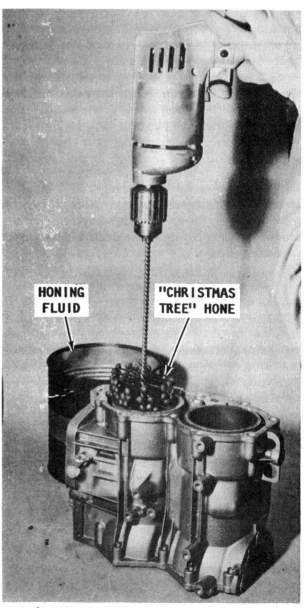

*Refinishing the cylinder wall using an electric drill and "Christmas tree" hone. **ALWAYS** keep the tool moving in long even strokes the entire cylinder depth. Use a continuous liberal amount of honing fluid.*

block, if necessary. This can be accomplished with 240-grit sandpaper and a small flat block of wood.

To ensure satisfactory powerhead performance and long life following the overhaul work, the honing work should be performed with patience, skill, and in the following sequence:

a- Follow the hone manufacturer's recommendations for use of the hone and for cleaning and lubricating during the honing operation. A "Christmas tree" hone may also be used.

b- Pump a continuous flow of honing oil into the work area. If pumping is not practical, use an oil can. Apply the oil generously and frequently on both the stones and work surface.

c- Begin the stroking at the smallest diameter. Maintain a firm stone pressure against the cylinder wall to assure fast stock removal and accurate results.

d- Expand the stones as necessary to compensate for stock removal and stone wear. The best cross-hatch pattern is obtained using a stroke rate of 30 complete cycles per minute. Again, use the honing oil generously.

e- Hone the cylinder walls **ONLY** enough to de-glaze the walls.

f- After the honing operation has been completed, clean the cylinder bores with hot water and detergent. Scrub the walls with a stiff bristle brush and rinse thoroughly with hot water. The cylinders **MUST** be thoroughly cleaned to prevent any abrasive material from remaining in the cylinder bore. Such material will cause rapid wear of new piston rings, the cylinder bore, and the bearings.

g- After cleaning, swab the bores several times with engine oil and a clean cloth, and then wipe them dry with a clean cloth. **NEVER** use kerosene or gasoline to clean the cylinders.

h- Clean the remainder of the cylinder block to remove any excess material spread during the honing operation.

BLOCK AND CYLINDER HEAD WARPAGE

First, check to be sure all old gasket material has been removed from the contact surfaces of the block and the cylinder head. Clean both surfaces down to shiny metal, to ensure a true measurement.

Next, place a straight edge across the gasket surface. Check under the straight edge with a 0.004" (0.1mm) feeler gauge. Move the straight edge to at least eight different locations. If the feeler gauge can pass under the straight edge -- anywhere contact with the other is made -- the surface will have to be resurfaced.

The block or the cylinder head may be resurfaced by placing the warped surface on 400-600 grit **WET** sandpaper, with the sandpaper resting on a **FLAT MACHINED** surface. If a machined surface is not available a large piece of glass or mirror may be used. **DO NOT** attempt to use a workbench or similar surface for this task. A workbench is never perfectly flat and the block or cylinder head will pickup the imperfections of the surface and the warpage will be made worse.

Sand -- work -- the warped surface on the wet sandpaper using large figure "8" motions. Rotate the block, or head, through $180°$ (turn it end-for-end) and spend an equal amount of time in each position to avoid removing too much material from one side.

If a suitable flat surface is not available, the next best method is to wrap 400-600 grit wet sandpaper around a **LARGE** file. Draw the file as evenly as possible in one sweep across the surface. Do not file in one place. Draw the file in many, many directions to get as even a finish as possible.

As the work moves along, check the progress with the straight edge and feeler gauge. Once the 0.004" (0.1mm) feeler gauge will no longer slide under the straight edge, consider the work completed.

If the warpage cannot be reduced using one of the described methods, the block or cylinder head should be replaced.

ONE LAST CHANCE

If the warpage cannot be reduced and it is not possible to obtain new items -- and the warped part must be assembled for further use -- there is a strong possibility of a water leak at the head gasket. In an effort to prevent a water leak, follow the instructions outlined in Assembling for powerhead "D" on Page 7-61.

8
ELECTRICAL

8-1 INTRODUCTION

The battery, charging system, and the cranking system are considered subsystems of the electrical system. Each of these units will be covered in detail in this chapter beginning with the battery.

8-2 BATTERIES

The battery is one of the most important parts of the electrical system. In addition to providing electrical power to start the engine, it also provides power for operation of the running lights, radio, electrical accessories, and possibly the pump for a bait tank.

Because of its job and the consequences, (failure to perform in an emergency) the best advice is to purchase a well-known brand, with an extended warranty period, from a reputable dealer.

The usual warranty covers a prorated replacement policy, which means the purchaser would be entitled to a consideration for the time left on the warranty period if the battery should prove defective before its time.

Do not consider a battery of less than **70-ampere hour** or **100-minute** reserve capacity. If in doubt as to how large the boat requires, make a liberal estimate and then purchase the one with the next higher ampere rating.

MARINE BATTERIES

Because marine batteries are required to perform under much more rigorous conditions than automotive batteries, they are constructed much differently than those used in automobiles or trucks. Therefore, a marine battery should always be the No. 1 unit for the boat and other types of batteries used only in an emergency.

Marine batteries have a much heavier exterior case to withstand the violent pounding and shocks imposed on it as the boat moves through rough water and in extremely tight turns.

The plates in marine batteries are thicker than in automotive batteries and each plate is securely anchored within the battery case to ensure extended life.

The caps of marine batteries are "spill proof" to prevent acid from spilling into the bilges when the boat heels to one side in a tight turn, or is moving through rough water.

Because of these features, the marine battery will recover from a low charge condition and give satisfactory service over

A fully charged battery, filled to the proper level with electrolyte, is the heart of the ignition and electrical systems. Engine cranking and efficient performance of electrical items depend on a full-rated battery.

a much longer period of time than any type intended for automotive use.

NEVER use a "Maintenance Free" type battery with an outboard unit. The charging system is not regulated as with automotive installations and the battery may be quickly damaged.

BATTERY CONSTRUCTION

A battery consists of a number of positive and negative plates immersed in a solution of diluted sulfuric acid. The plates contain dissimilar active materials and are kept apart by separators. The plates are grouped into what are termed elements. Plate straps on top of each element connect all of the positive plates and all of the negative plates into groups.

The battery is divided into cells which hold a number of the elements apart from the others. The entire arrangement is contained within a hard-rubber case. The top is a one-piece cover and contains the filler caps for each cell. The terminal posts protrude through the top where the battery connections for the boat are made. Each of the cells is connected to its neighbor in a positive-to-negative manner with a heavy strap called the cell connector.

BATTERY RATINGS

Four different methods are used to measure and indicate battery electrical capacity:

1- Ampere-hour rating
2- Cold cranking performance
3- Reserve capacity
4- Watt hour rating

The **ampere-hour** rating of a battery refers to the battery's ability to provide a set amount of amperes for a given amount of time under test conditions at a constant temperature of $80°$ ($27°C$). Amperes x hours equals ampere-hour rating. Therefore, if the battery is capable of suppling 4 amperes of current for 20 consecutive hours, the battery is rated as an 80 ampere-hour battery.

The ampere-hour rating is useful for some service operations, such as slow charging or battery testing.

Cold cranking performance is measured by cooling a fully charged battery to $0°F$ ($-17°C$) and then testing it for 30 seconds to determine the maximum current flow. In this manner the cold cranking amperes rating is the number of amperes available to be drawn from the battery before the voltage drops below 7.2 volts.

The accompanying graphic illustration depicts the amount of power in watts available from a battery at different temperatures and the amount of power in watts required of the engine at the same temperature. It becomes quite obvious --the colder the climate -- the more necessary for the battery to be **FULLY** charged.

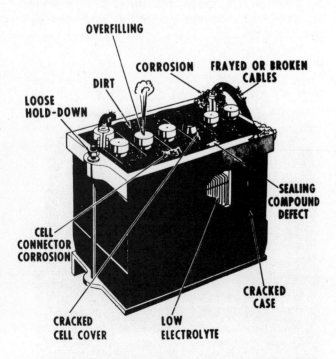

A visual inspection of the battery should be made each time the boat is used. Such a quick check may reveal a potential problem in its early stages. A dead battery in a busy waterway or far from assistance could have serious consequences.

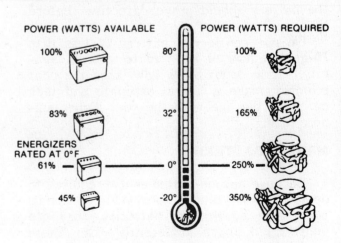

Comparison of battery efficiency and engine demands at various temperatures.

Reserve capacity of a battery is considered the length of time -- in minutes -- at 80°F (27°C) a 25 ampere current can be maintained before the voltage drops below 10.5 volts. This test is intended to provide an approximation of how long the engine, including electrical accessories such as bilge pump, running lights, etc., could operate satisfactorily if the stator assembly or lighting coil did not produce sufficient current. A typical rating is 100 minutes.

Watt-hour is a very useful rating of battery power. It is determined by multiplying the number of ampere hours times the voltage. Therefore, a 12-volt battery rated at 80 ampere-hours would be rated at 960 watt-hours (80 x 12 = 960).

If possible, the new battery should have a power rating equal to or higher than the unit it is replacing.

BATTERY LOCATION

Every battery installed in a boat must be secured in a well-protected ventilated area. If the battery area lacks adequate ventilation, hydrogen gas which is given off during charging could become very explosive. This is especially true if the gas is concentrated and confined.

BATTERY SERVICE

The battery requires periodic servicing and a definite maintenance program will ensure extended life. If the battery should test satisfactorily, but still fails to perform properly, one of five problems could be the cause.

1- An accessory might have accidentally been left on overnight or for a long period during the day. Such an oversight would result in a discharged battery.

2- Slow speed engine operation for long periods of time resulting in an undercharged condition.

3- Using more electrical power than the stator assembly or lighting coil can replace would result in an undercharged condition.

4- A defect in the charging system. A faulty stator assembly or lighting coil, defective rectifier, or high resistance somewhere in the system could cause the battery to become undercharged.

5- Failure to maintain the battery in good order. This might include a low level of electrolyte in the cells; loose or dirty cable connections at the battery terminals; or possibly an excessively dirty battery top.

Electrolyte Level

The most common practice of checking the electrolyte level in a battery is to remove the cell cap and visually observe the level in the vent well. The bottom of each vent well has a split vent which will cause the surface of the electrolyte to appear distorted when it makes contact. When the distortion first appears at the bottom of the split vent, the electrolyte level is correct.

Some late-model batteries have an electrolyte-level indicator installed which operates in the following manner:

A transparent rod extends through the center of one of the cell caps. The lower tip of the rod is immersed in the electrolyte when the level is correct. If the level should drop below normal, the lower tip of the rod is exposed and the upper end glows as a warning to add water. Such a device is only necessary on one cell cap because if the electrolyte is low in one cell it is also low in the other cells. **BE SURE** to replace the cap with the indicator onto the second cell from the positive terminal.

During hot weather and periods of heavy use, the electrolyte level should be checked more often than during normal operation. Add potable (drinking) water to bring the level of electrolyte in each cell to the proper level. **TAKE CARE** not to overfill, because adding an excessive amount of water will cause loss of electrolyte and any loss will result in poor performance, short

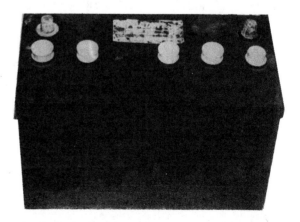

An explosive hydrogen gas is normally released from the cells under a wide range of circumstances. This battery exploded when the gas ignited from someone smoking in the area when the caps were removed. Such an explosion could also be caused by a spark from the battery terminals igniting the gas.

8-4 ELECTRICAL

battery life, and will contribute quickly to corrosion. **NEVER** add electrolyte from another battery. Use only clean pure water.

Battery Testing

A hydrometer is a device to measure the percentage of sulfuric acid in the battery electrolyte in terms of specific gravity. When the condition of the battery drops from fully charged to discharged, the acid leaves the solution and enters the plates, causing the specific gravity of the electrolyte to drop.

It may not be common knowledge, but hydrometer floats are calibrated for use at 80°F (27°C). If the hydrometer is used at any other temperature, hotter or colder, a correction factor must be applied. (Remember, a liquid will expand if it is heated and will contract if cooled. Such expansion and contraction will cause a definite change in the specific gravity of the liquid, in this case the electrolyte.)

A quality hydrometer will have a thermometer/temperature correction table in the lower portion, as shown in the accompanying illustration. By knowing the air temperature around the battery and from the table, a correction factor may be applied to the specific gravity reading of the hydrometer float. In this manner, an accurate determination may be made as to the condition of the battery.

The following six points should be observed when using a hydrometer.

1- **NEVER** attempt to take a reading immediately after adding water to the battery. Allow at least 1/4 hour of charging at a high rate to thoroughly mix the electrolyte with the new water. This time will also allow for the necessary gasses to be created.

2- **ALWAYS** be sure the hydrometer is clean inside and out as a precaution against contaminating the electrolyte.

3- If a thermometer is an integral part of the hydrometer, draw liquid into it several times to ensure the correct temperature before taking a reading.

4- **BE SURE** to hold the hydrometer vertically and suck up liquid only until the float is free and floating.

5- **ALWAYS** hold the hydrometer at eye level and take the reading at the surface of the liquid with the float free and floating.

Disregard the light curvature appearing where the liquid rises against the float stem. This phenomenon is due to surface tension.

6- **DO NOT** drop any of the battery fluid on the boat or on your clothing, because it is extremely caustic. Use water and baking soda to neutralize any battery liquid that does accidentally drop.

After withdrawing electrolyte from the battery cell until the float is barely free, note the level of the liquid inside the hydrometer. If the level is within the green band range for all cells, the condition of the

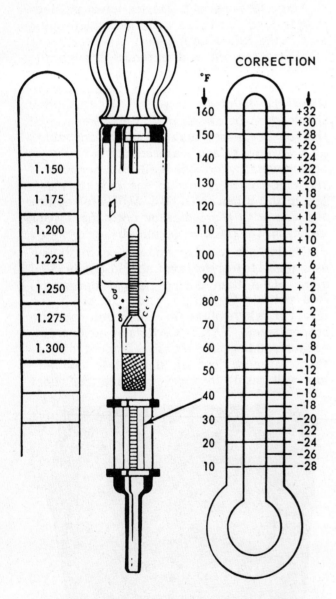

A check of the electrolyte in the battery should be on the maintenance schedule for any boat. A hydrometer reading of 1.300, or in the green band, indicates the battery is in satisfactory condition. If the reading is 1.150 or in the red band, the battery must be charged. Observe the six safety points listed in the text when using a hydrometer.

battery is satisfactory. If the level is within the white band for all cells, the battery is in fair condition.

If the level is within the green or white band for all cells except one, which registers in the red, the cell is shorted internally. No amount of charging will bring the battery back to satisfactory condition.

If the level in all cells is about the same, even if it falls in the red band, the battery may be recharged and returned to service. If the level fails to rise above the red band after charging, the only solution is to replace the battery.

Cleaning

Dirt and corrosion should be cleaned from the battery just as soon as it is discovered. Any accumulation of acid film or dirt will permit current to flow between the terminals. Such a current flow will drain the battery over a period of time.

Clean the exterior of the battery with a solution of diluted ammonia or a soda solution to neutralize any acid which may be present. Flush the cleaning solution off with clean water. **TAKE CARE** to prevent any of the neutralizing solution from entering the cells, by keeping the caps tight.

A poor contact at the terminals will add resistance to the charging circuit. This resistance will cause the voltage regulator to register a fully charged battery, and thus cut down on the stator assembly or lighting coil output adding to the low battery charge problem.

Scrape the battery posts clean with a suitable tool or with a stiff wire brush. Clean the inside of the cable clamps to be sure they do not cause any resistance in the circuit.

JUMPER CABLES

If booster batteries are used for starting an engine the jumper cables must be connected correctly and in the proper sequence to prevent damage to either battery, or to the alternator diodes.

ALWAYS connect a cable from the positive terminal of the dead battery to the positive terminal of the good battery **FIRST**. **NEXT,** connect one end of the other cable to the negative terminal of the good battery and the other end to a good ground on the powerhead. **DO NOT** connect the negative jumper from the good battery to the negative terminal of the low battery. Such action will almost always cause a spark which could ignite gases escaping through the vent holes in the battery filler caps. Igniting the gases may result in an explosion destroying the battery and causing severe personal **INJURY**.

By making the negative (ground) connection on the powerhead, if an arc is created, it will not be near the battery.

The second part of the tool shown at left is used to clean the battery lead terminals.

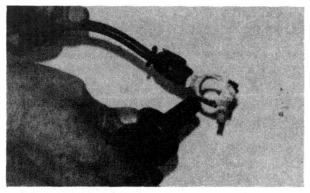

An inexpensive two-part tool will do an excellent job of cleaning the battery terminals.

DISCONNECT the battery ground cable before replacing a stator assembly or lighting coil, or before connecting any type of meter to the ignition system.

If it is necessary to use a fast-charger on a dead battery, **ALWAYS** disconnect one of the boat cables from the battery **FIRST**, to prevent burning out the diodes in the rectifier.

NEVER use a fast-charger as a booster to start the engine because the diodes in the alternator will be **DAMAGED**.

STORAGE

If the boat is to be laid up for the winter or for more than a few weeks, special attention must be given to the battery to prevent complete discharge or possible damage to the terminals and wiring. Before putting the boat in storage, disconnect and remove the batteries. Clean them thoroughly of any dirt or corrosion, and then charge them to full specific gravity reading. After they are fully charged, store them in a clean cool dry place where they will not be damaged or knocked over, preferably on a couple blocks of wood. Storing the battery up off the deck, will permit air to circulate freely around and under the battery and will help to prevent condensation.

NEVER store the battery with anything on top of it or cover the battery in such a manner as to prevent air from circulating around the fillercaps. All batteries, both new and old, will discharge during periods of storage, more so if they are hot than if they remain cool. Therefore, the electrolyte level and the specific gravity should be checked at regular intervals. A drop in the specific gravity reading is cause to charge them back to a full reading.

In cold climates, care should be exercised in selecting the battery storage area. A fully-charged battery will freeze at about 60 degress below zero. A discharged battery, almost dead, will have ice forming at about 19 degrees above zero.

8-3 THERMOMELT STICKS

Thermomelt sticks are an easy method of determining if the powerhead is running at the proper temperature. Thermomelt sticks are not expensive and are available at your local marine dealer.

Start the engine with the propeller in the water and run it for about 5 minutes at roughly 3000 rpm.

CAUTION

Water must circulate through the lower unit to the powerhead anytime the powerhead is operating to prevent damage to the water pump in the lower unit. Just five seconds without water will damage the water pump impeller.

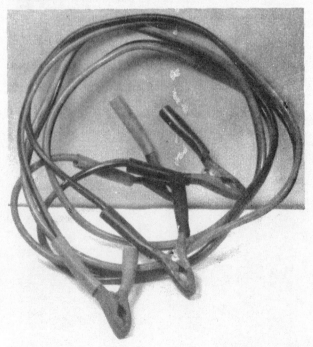

A common set of heavy-duty jumper cables. The booster battery must be connected correctly and in the proper sequence, as outlined in the text. Proper procedure will prevent damage to the battery, the diodes, or the rectifier.

Thermomelt sticks, available at a welding supply shop, may be used to fairly accurately determine powerhead operating temperature and overheat warning sensors, as explained in the text.

TACHOMETERS 8-7

The 140 degree stick should melt when you touch it to the lower thermostat housing or on the top cylinder. If it does not melt, the thermostat is stuck in the open position and the engine temperature is too low.

Touch the 170 degree stick to the same spot on the lower thermostat housing or on the top cylinder. The stick should not melt. If it does, the thermostat is stuck in the closed position or the water pump is not operating properly because the engine is running too hot.

If the powerhead is not equipped with a thermostat, the problem may be solved by reverse flushing to clean out the cooling system and/or servicing the water pump. For service procedures for the thermostat, see Chapter 7. For service procedures for the water pump, see Chapter 9.

8-4 TACHOMETER

An accurate tachometer can be installed on any engine. Such an instrument provides an indication of engine speed in revolutions per minute (rpm). This is accomplished by measuring the number of electrical pulses per minute generated in the primary circuit of the ignition system.

The meter readings range from 0 to 6,000 rpm, in increments of 100. Tachometers have solid-state electronic circuits which eliminates the need for relays or batteries and contributes to their accuracy. The electronic parts of the tachometer susceptible to moisture are coated to prolong their life.

Maximum engine performance can only be obtained through proper tuning using a tachometer.

SPECIAL WORDS ON TACHOMETERS AND CONNECTIONS

The 8 to 30hp powerheads use a CDI system firing a twin lead ignition coil twice for each crankshaft revolution. If an induction tachometer is installed to measure powerhead speed, the tachometer will probably indicate **DOUBLE** the actual crankshaft rotation. Check the instructions with the tachometer to be used. Some tachometer manufacturers have allowed for the double reading and others have not.

For all models **EXCEPT** units with Type I ignition (magneto type -- points under the flywheel), connect the two tachometer leads to the two Green leads from the stator. Either tachometer lead may be connected to either Green lead.

For the Models with the Type I ignition system, connect one tachometer lead (may be White, Red, or Yellow, depending on the manufacturer), to the primary negative terminal of the coil (usually a small black lead), and the other tachometer lead, Black, to a suitable ground.

8-5 ELECTRICAL SYSTEM GENERAL INFORMATION

In the early days, all outboard engines were started by simply pulling on a rope wound around the flywheel. As time passed and owners were reluctant to use muscle power, it was necessary to replace the rope starter with some form of power cranking system. Today, many small engines are still started by pulling on a rope, but others have a powered cranking motor installed.

The system utilized to replace the rope method was an electric cranking motor coupled with a mechanical gear mesh between the cranking motor and the powerhead flywheel, similar to the method used to crank an automobile engine.

The electrical system consists of three circuits:
    a- Charging circuit
    b- Cranking motor circuit
    c- Ignition circuit

Charging Circuit

The charging circuit consists of permanent magnets and a stator located within the flywheel; a lighting coil installed on the stator plate; a rectifier located elsewhere

8-8 ELECTRICAL

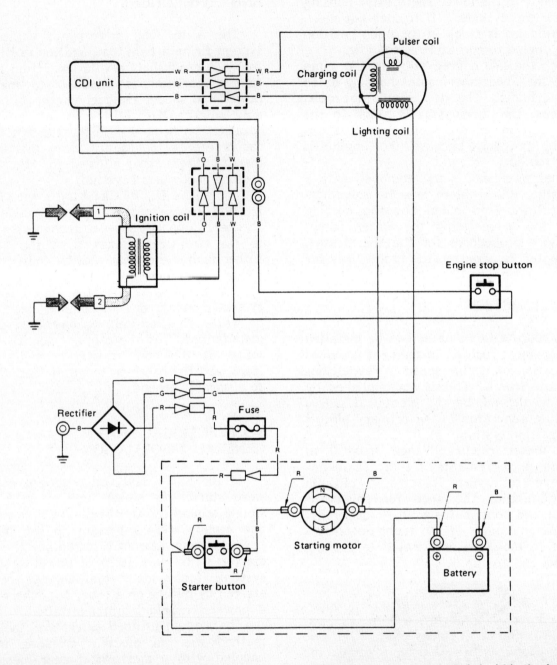

Simplified wiring diagram of a typical charging system. The cranking motor circuit is included within the dashed line at the lower portion of the diagram.

on the powerhead; an external battery; and the necessary wiring to connect the units. The negative side of the rectifier is grounded. The positive side of the rectifier passes through the internal harness plug to the battery. The negative side of the battery is connected, through the connector, to a good ground on the engine.

The alternating current generated in the stator windings passes to the rectifier. The rectifier changes the alternating current (AC) to direct current (DC) to charge the 12-volt battery.

Cranking Motor Circuit

The cranking motor circuit consists of a cranking motor and a starter-engaging mechanism. A starter relay is used as a heavy-duty switch to carry the heavy current from the battery to the cranking motor. On most models, the starter relay is actuated by depressing the **START** button.

CRANKING MOTOR CIRCUIT 8-9

On boats equipped with a remote control shift box, the ignition key is turned to the **START** position.

Ignition Circuit

The ignition circuit is covered extensively in Chapter 5.

8-6 CHARGING CIRCUIT SERVICE

The stator is located under, and protected by, the flywheel. Therefore, the stator, including the lighting coil, seldom causes problems in the charging circuit. Most problems in the charging circuit can be traced to the rectifier or to the battery. If either the stator or the rectifier fails the troubleshooting tests, the defective unit cannot be repaired, it **MUST** be replaced.

8-7 CRANKING MOTOR CIRCUIT SERVICE

DESCRIPTION

As the name implies, the sole purpose of the cranking motor circuit is to control operation of the cranking motor to crank the powerhead until the engine is operating. The circuit includes a solenoid or magnetic switch to connect or disconnect the motor from the battery. The operator controls the switch with a push button or key switch.

A neutral start switch is installed into the circuit to permit operation of the cranking motor **ONLY** if the shift control lever is in **NEUTRAL**. This switch is a safety device to prevent accidental engine start when the engine is in gear.

The cranking motor is a series wound electric motor which draws a heavy current from the battery. It is designed to be used only for short periods of time to crank the engine for starting. To prevent overheating the motor, cranking should not be continued for more than 30-seconds without allowing the motor to cool for at least three minutes. Actually, this time can be spent in making preliminary checks to determine why the engine fails to start.

Theory of Operation

Power is transmitted from the cranking motor to the powerhead flywheel through a Bendix drive. This drive has a pinion gear mounted on screw threads. When the motor is operated, the pinion gear moves upward and meshes with the teeth on the flywheel ring gear.

When the powerhead starts, the pinion gear is driven faster than the shaft, and as a result, it screws out of mesh with the flywheel. A rubber cushion is built into the Bendix drive to absorb the shock when the pinion meshes with the flywheel ring gear. The parts of the drive **MUST** be properly assembled for efficient operation. If the drive is removed for cleaning, **TAKE CARE** to assemble the parts as shown in the accompanying illustrations in this section. If the screw shaft assembly is reversed, it will strike the splines and the rubber cushion will not absorb the shock.

The sound of the motor during cranking is a good indication of whether the cranking motor is operating properly or not. Naturally, temperature conditions will affect the speed at which the cranking motor is able to crank the engine. The speed of cranking a cold engine will be much slower than when cranking a warm engine. An experienced operator will learn to recognize the favorable sounds of the powerhead cranking under various conditions.

Faulty Symptoms

If the cranking motor spins, but fails to crank the engine, the cause is usually a corroded or gummy Bendix drive. The drive should be removed, cleaned, and given an inspection.

If the cranking motor cranks the engine too slowly, the following are possible causes and the corrective actions that may be taken:

a- Battery charge is low. Charge the battery to full capacity.

b- High resistance connections at the battery, solenoid, or motor. Clean and tighten all connections.

c- Undersize battery cables. Replace cables with sufficient size.

d- Battery cables too long. Relocate the battery to shorten the run to the solenoid.

Maintenance

The cranking motor does not require periodic maintenance or lubrication. If the motor fails to perform properly, the checks outlined in the previous paragraph should be performed.

The frequency of starts governs how often the motor should be removed and

8-10 ELECTRICAL

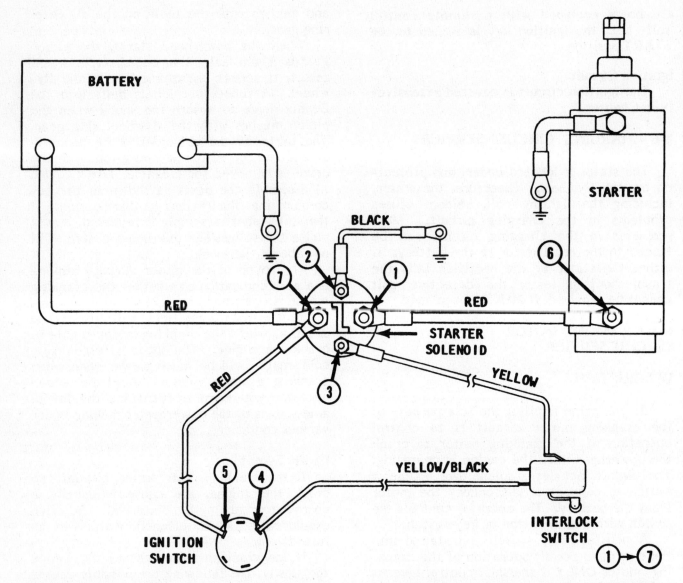

reconditioned. The manufacturer recommends removal and reconditioning every 1000 hours.

Naturally, the motor will have to be removed if the corrective actions outlined under **Faulty Symptoms** above, does not restore the motor to satisfactory operation.

CRANKING MOTOR TROUBLESHOOTING

Before wasting too much time troubleshooting the cranking motor circuit, the following checks should be made. Many times, the problem will be corrected.

a- Battery fully charged.
b- Shift control lever in NEUTRAL.
c- Main 20-amp fuse located on top of the fuse cover is "good" (not blown).
d- All electrical connections clean and tight.

e- Wiring in good condition, insulation not worn or frayed.

Two more areas may cause the powerhead to crank slowly even though the cranking motor circuit is in excellent condition: a tight or "frozen" powerhead and water in the lower unit. The following troubleshooting procedures are presented in a logical sequence, with the most common and easily corrected areas listed first in each problem area. The connection number refers to the numbered positions in the accompanying illustrations.

Perform the following quick checks and corrective actions for following problems:

1- Cranking Motor Rotates Slowly
a- Battery charge is low. Charge the battery to full capacity.
b- Electrical connections corroded or loose. Clean and tighten.

c- Defective cranking motor. Perform an amp draw test. Lay an amp draw-gauge on the cable leading to the cranking motor. Turn the key on and attempt to crank the engine. If the gauge indicates an excessive amperage draw, the cranking motor **MUST** be replaced or rebuilt.

2- Cranking Motor Fails to Crank Powerhead

Test Motor

a- Disconnect the cranking motor lead from the solenoid to prevent the powerhead from starting during the testing process.

NOTE: This lead is to remain disconnected from the solenoid during tests No. 2 thru No. 6.

b- Disconnect the black ground wire from the No. 2.

c- Connect a voltmeter between the No. 2 and a common engine ground.

d- Turn the key switch to the **START** position.

e- Observe the voltmeter reading. If there is the slightest amount of reading, check the black ground wire connection or check for an open circuit. Connect the ground wire back to the No. 2 and move to Step 6. If there is no voltmeter reading, proceed with Step 3.

3- Test Cranking Motor Solenoid

a- Connect a voltmeter between the engine common ground and the No. 3.

b- Turn the ignition key switch to the **START** position, or depress the start button.

c- Observe the voltmeter reading. If there is the slightest indication of a reading, the solenoid is defective and must be replaced. If there is no reading, proceed with Step 4.

4- Test Neutral Start Switch

a- Connect a voltmeter between the common engine ground and the No. 4. Turn the ignition key switch to the **START** position, or depress the start button.

b- Observe the voltmeter. If there is any indication of a reading, the neutral start switch is open or the brown wire lead is open between the No. 3 and No. 4. If there is no voltmeter reading, proceed to Step 5.

5- Test for Open Wire

a- Connect a voltmeter between the common engine ground and No. 5.

b- The voltmeter should indicate 12-volts. If the meter needle flickers (fails to hold steady), check the circuit between No. 5 and common engine ground. If meter fails to indicate voltage, replace the positive battery cable.

6- Further Tests for Solenoid

a- Connect the voltmeter between the common engine ground and No. 1.

b- Turn the ignition key switch to the **START** position.

c- Observe the voltmeter. If there is no reading, the cranking motor solenoid is defective and must be replaced. If a reading is indicated and a click sound is heard, proceed to Step 7.

7- Test Large Red Cable

a- Connect the red cable to the cranking motor solenoid.

b- Connect the voltmeter between the engine common ground and No. 6.

c- Turn the ignition key switch to the **START** position, or depress the start button.

d- Observe the voltmeter. If there is no reading, check the red cable for a poor connection or an open circuit. If there is any indication of a reading, and the cranking motor does not rotate, the cranking motor must be replaced.

CRANKING MOTOR SOLENOID TROUBLESHOOTING

Description

The cranking motor solenoid is actually a switch between the battery and the cranking motor. The solenoid cannot be serviced. Therefore, if troubleshooting indicates the solenoid to be faulty, it **MUST** be replaced.

Solenoid Testing
Model 20hp, 25hp, 30hp, and 40hp

The solenoid used for these model powerheads is actually an electrical "switch". The solenoid on the Model 40hp is mounted on the starter frame. Other models listed in the heading have the solenoid mounted elsewhere on the powerhead.

The following test must be conducted with the solenoid removed from the powerhead.

8-12 ELECTRICAL

a- Connect one test lead of an ohmmeter to each of the large solenoid terminals.

b- Connect the positive (+) lead from a fully charged 12-volt battery to the small solenoid terminal where the brown wire was attached.

c- Momentarily make contact with the ground lead from the battery to the other small solenoid terminal where the black wire was attached. If a loud "click" sound is heard, and the ohmmeter indicates continuity, the solenoid is in serviceable condition. If, however, a "click" sound is not heard, and/or the ohmmeter does not indicate continuity, the solenoid is defective and must be replaced **ONLY** with a **MARINE** type solenoid.

Cranking Motor Solenoid
Model 28hp, 48hp, 55hp, and 60hp

The solenoid on these powerheads is an electrical/mechanical unit to physically activate the pinion gear. Cranking motors with this type solenoid also have a magnetic starter switch. Testing the switch is outlined after the solenoid test.

This test may be performed with the starter/solenoid assembly removed from the powerhead or with just the solenoid removed.

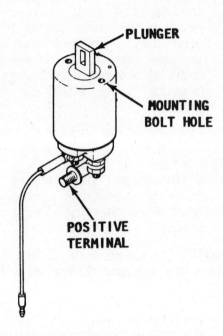

Line drawing of a cranking motor solenoid for a 28hp, 48hp, 55hp, or 60hp powerhead.

a- Obtain a fully charged 12-volt battery.

b- Using a pair of jumper leads, connect the negative battery terminal to the solenoid housing. (One of the mounting bolts may have to be temporarily installed to make this connection.)

c- Momentarily make contact with the lead from the positive battery terminal to the large positive solenoid terminal. If the solenoid has been removed from the starter, the plunger should "retract", move inward. If the solenoid is still in place on the starter, the pinion gear will be forced upward.

Starter Switch Testing
Model 28hp, 48hp, 55hp, and 60hp

As mentioned in the previous short section, these model powerheads are equipped with a magnetic starter switch in the cranking motor circuit. The switch is usually mounted on top of the exhaust cover.

a- Obtain an ohmmeter. Select the Rx1000 scale. Check for continuity across the Blue and Red leads at the center of the switch.

b- Connect the Blue and Red leads from the center of the switch across the 12-volt battery, as indicated in the accompanying illustration. Connect the ohmmeter across the two larger outer switch

Testing a cranking motor solenoid removed from a 20hp, 25hp, 30hp, or 40hp powerhead, as explained in the text.

CRANKING MOTOR SERVICE **8-13**

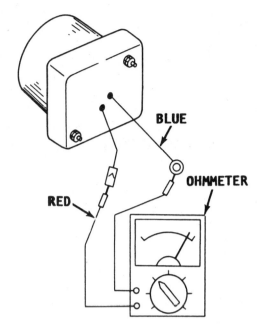

Hookup for testing the starter switch for a Model 28hp, 48hp, 55hp, or 60hp powerhead.

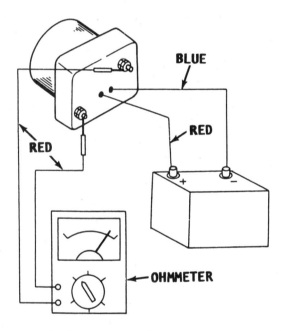

Hookup for further tests of the same switch, shown above, as outlined in the text.

terminals. The meter should indicate continuity.

8-8 CRANKING MOTOR SERVICE

Description

One type cranking motor is used on the powerheads covered in this manual.

Marine cranking motors are very similar in construction and operation to the units used in the automotive industry.

All marine cranking motors use the inertia-type drive assembly. This type assembly is mounted on an armature shaft with external spiral splines which mate with the internal splines of the drive assembly.

NEVER operate a cranking motor for more than 30-seconds without allowing it to cool for at least three minutes. Continuous operation without the cooling period can cause serious damage to the cranking motor.

CRANKING MOTOR REMOVAL MODEL 20HP, 25HP, 30HP, AND 40HP

Before beginning any work on the cranking motor, disconnect the positive (+) lead from the battery terminal. Remove the cowling from the powerhead.

1- Disconnect the cranking motor lead by removing the Phillips screw from the relay. The terminal on the cranking motor is not accessible. Therefore, the lead will come off with the motor.

8-14 ELECTRICAL

2- Remove the mounting bolts and washers from the cranking motor housing. One of the bolts is partially under the flywheel. However, by rotating the flywheel until a cutout is directly over the bolt head, the bolt may be "wiggled" out.

3- Remove the cranking motor from the powerhead. Disconnect the large red lead from the motor.

CRANKING MOTOR REMOVAL MODEL 28HP, 48HP, 55HP, AND 60HP

Disconnect both positive and negative battery cables. Disconnect the large Black negative cable from the solenoid ground terminal. Disconnect the large Red positive cable, the small Red wire, and the small Orange wire (all on one eyelet connector) from the positive terminal on the solenoid. Disconnect the other small Red wire from the ignition terminal on the solenoid.

Remove the four cranking motor securing bolts and lift the motor free of the powerhead.

CRANKING MOTOR DISASSEMBLING

The exploded drawings on Page 8-15 and 8-16 will be most helpful while performing the following procedures.

1- Slide the "boot" clear of the solenoid terminal. Loosen the terminal nut on the solenoid and slide the cranking motor lead free of the terminal.

2- Remove the two bolts and washers securing the solenoid to the cranking motor nose cone.

3- "Wiggle" the solenoid plunger free of the yoke arm. Once the plunger is free, remove the torsion spring. Separate the solenoid from the nose cone. Remove and **SAVE** the spacers atop the solenoid.

4- With the cranking motor on a work surface, pry the snap ring out of its groove and free of the armature shaft.

5- After the snap ring is free, slide the collar, spring, and pinion gear free of the armature. The spring is only used on cranking motors installed on Model 20hp, 25hp, 30hp and 40hp powerheads.

6- Make a mark on upper end cap/nose cone, and another on the lower end cap and matching marks on the field frame assembly as an aid to alignment and installation of the thru bolts.

SPECIAL WORDS

Without some type of alignment marks, the task of inserting the thru bolts and starting the threads into the end cap is almost impossible.

Model 20hp, 25hp, and 30hp

Remove the thru bolts, and then the upper end cap, armature assembly and lower end cap.

Note the number of washers on each end of the armature. Incorrect placement of these washers will cause the armature to be misaligned with the brushes and the field magnets during assembling. Two washers are usually used at the upper end, and three washers at the lower end. Remove and discard the O-ring around each end cap.

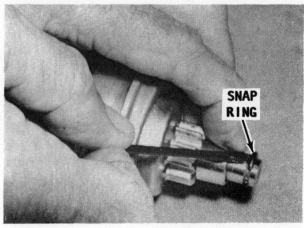

The collar is held back while the snap ring is pried free from the top end of the armature shaft.

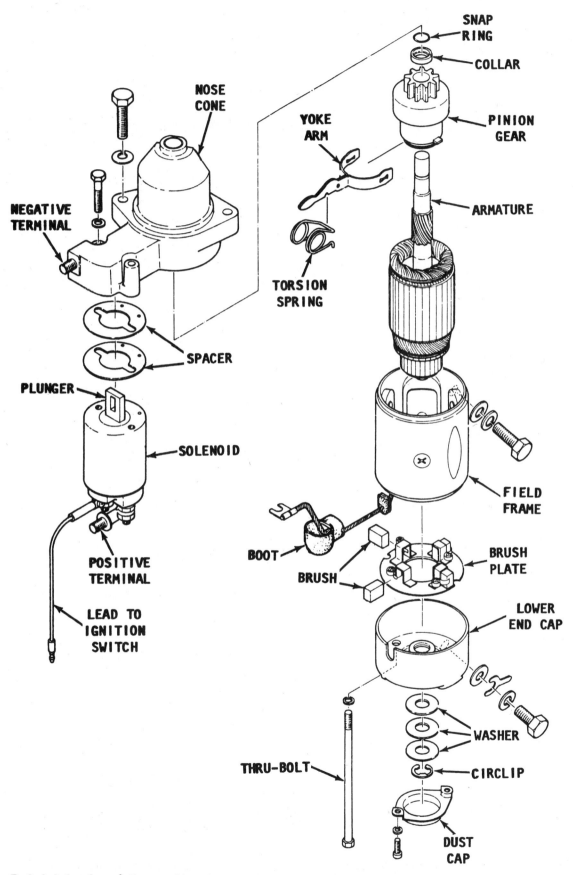

Exploded drawing of the cranking motor and solenoid used on the Model 28hp, 48hp, 55hp, and 60hp powerhead. Major parts are identified.

8-16 ELECTRICAL

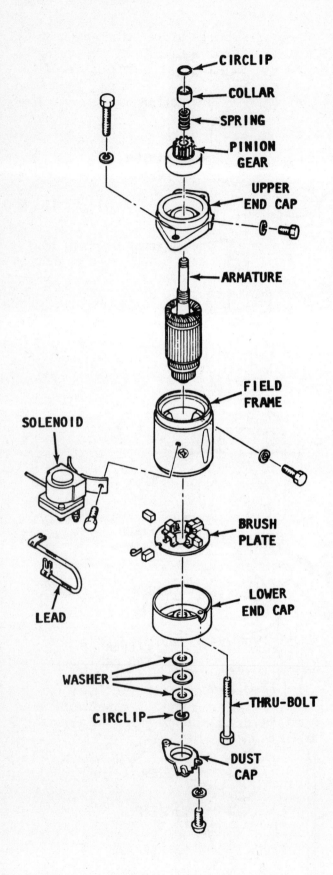

Exploded drawing of a cranking motor used on a 40hp powerhead, with major parts identified. Note how the solenoid is mounted on the field frame.

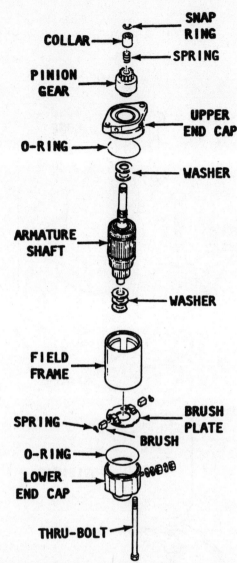

Exploded drawing of the cranking motor used on Model 20hp, 25hp, and 30hp powerheads. Major parts are identified.

Model 28hp, 40hp, 48hp, 55hp, and 60hp

Remove the two screws and washers from the dust cover at the base of the lower end cap. Use a slotted screwdriver and pry the circlip from the lower end of the armature. Remove and **SAVE** the washers from under the circlip.

The pinion gear and associated parts disassembled from the armature shaft of a cranking motor used on Model 20hp, 25hp, 30hp, and 40hp powerheads.

Remove the thru bolts, and then the upper end cap/nose cone, armature assembly and lower end cap.

CLEANING AND INSPECTING

Pinion Gear Assembly

Inspect the pinion gear teeth for chips, cracks, or a broken tooth. Check the splines inside the pinion gear for burrs and to be sure the gear moves freely on the armature shaft.

BAD NEWS

This type of cranking motor pinion gear assembly cannot be repaired if the unit is defective. Replacement of the pinion gear assembly as a unit is the only answer.

Check to be sure the return spring is flexible and has not become distorted. Check both ends of the spring for signs of damage.

Clean the armature shaft and check to be sure the shaft is free of any burrs. If burrs are discovered, they may be removed with crocus cloth.

Frame Assembly

Clean the field coils, armature, commutator, armature shaft, brush-end plate and drive-end housing with a brush or compressed air. Wash all other parts in solvent and blow them dry with compressed air.

Inspect the insulation and the unsoldered connections of the armature windings for breaks or burns.

Perform tests on any suspected defective part, according to the procedures outlined beginning in the next column.

Check the commutator for runout. Inspect the armature shaft and both bearings for scoring.

Turn the commutator in a lathe if it is out-of-round by more than 0.005" (0.13mm).

Check the springs in the brush holder to be sure none are broken. Check the spring tension and replace if the tension is not 32-40 ounces (900-1135gm). Check the insulated brush holders for shorts to ground. If the brushes are worn down to 0.4" (10mm) or less, they must be replaced.

The armature, fields, and brush holders must be checked before assembling the starter motor. Procedures to test cranking motor parts begin towards the end of this column.

Both the positive and negative brushes are mounted in the lower cap. The positive brush is attached to the positive terminal. The negative brush is attached to the lower end cap with a screw.

The negative brush is removed first, by simply removing the screw attaching the brush lead to the lower cap. Take care not to lose the small spring behing the brush. The positive brush is removed after the negative brush by removing the screw from the brush plate. Remove the brush plate with the positive brush still attached to the plate. Slide the positive brush out of the slot in the end plate. Take care not to lose the small spring behind the brush.

Installation of the new positive brush is accomplished by first inserting the small spring into the brush retainer, and then sliding the new positive brush into the slot of the end cap. Install the negative brush by first inserting the small spring into the brush retainer, and then slide the new negative brush into the slot of the end cap. Secure each lead with the attaching screw.

TESTING CRANKING MOTOR PARTS

SPECIAL WORDS

Most marine shops and all electrical motor rebuild shops will test an armature for a modest charge. If the armature has a short, it **MUST** be replaced.

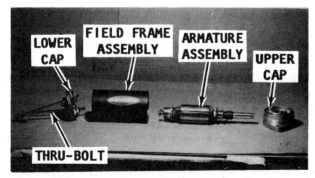

Cranking motor diassembled ready for cleaning and inspecting.

The internal splines can only be inspected after the pinion gear has been removed from the cranking motor.

ELECTRICAL

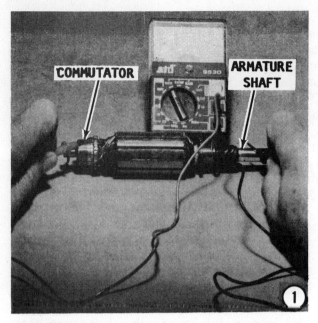

Check the armature for a short circuit by placing it on a growler and holding a hack saw blade over the armature core while the armature is rotated. If the saw blade vibrates, the armature is shorted. Clean between the armature bars, and then check again on the growler. If the saw blade still vibrates, the armature must be replaced. Occasionally carbon dust from the brushes will short the armature. Therefore, blow the slots in the armature clean with compressed air.

1- Make contact with one probe of the test light on the armature core or shaft. Make contact with the other probe on the commutator. If the light comes on, the armature is grounded and must be replaced.

Turning the Commutator

2- True the commutator, if necessary, in a lathe. **NEVER** undercut the mica because the brushes are harder than the insulation. Undercut the insulation between the commutator bars 1/32" (0.80mm) to the full width of the insulation and flat at the bottom. A triangular groove is not satisfactory. After the undercutting work is completed, clean out the slots carefully to remove dirt and copper dust. Sand the commutator lightly with No. 500 or No. 600 sandpaper to remove any burrs left from the undercutting.

3- Test light probes, placed on any two commutator bars, should light and indicate continuity.

4- Check the armature a second time on the growler for possible short circuits.

Positive Brush

5- Obtain an ohmmeter. Make contact with one lead of the ohmmeter to the positive brush. Make contact with the other test lead to the case. The ohmmeter should indicate **NO** continuity. If the meter indicates continuity, the positive lead is shorted to the case and must be replaced.

Negative Brush

6- Make contact with one lead of the ohmmeter to the negative brush. Make contact with the other test lead to the case.

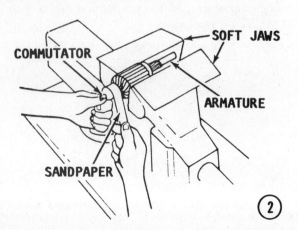

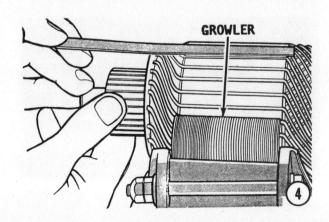

ASSEMBLE CRANKING MOTOR

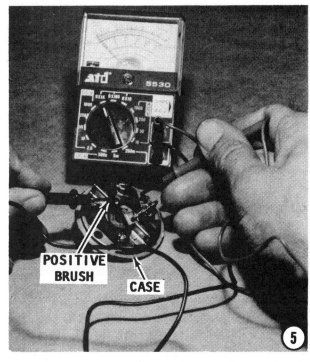

back and insert the small end of the armature into the end cap bushing. **TAKE CARE** not to distort the brush springs. Once assembled, rotate the shaft inside the bushing and check to be sure there is no binding and the brushes sweep across the commutator smoothly.

2- Slide the free end of the armature into the field frame end with the magnets recessed. **TAKE CARE** not to lose the brushes from the commutator. Seat the end cap into the field frame with the marks made prior to disassembly, aligned.

3- Stretch a new O-ring around the circumference of the upper end cap. Install the washers over the armature shaft. Seat the upper end cap with the marks made during disassembly, aligned.

Now, hold it all together and slide the two thru bolts through the lower end cap, through the field assembly, and then thread them into the upper end cap.

SPECIAL WORDS

Installing the two thru bolts is not as simple a task as it sounds. The reason: as the bolts pass the magnets inside the field frame, the magnets will attract the bolts, "pulling" them out of line. With a bit of patience, the bolts can be made to align with the holes in the upper end cap and the threads started.

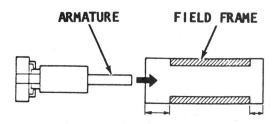

The ohmmeter should indicate continuity. If the meter indicates no continuity, there is an open in the negative lead or the lead is not grounded properly to the end cap.

ASSEMBLING

Models 20hp, 25hp, and 30hp

1- Stretch a new O-ring around the circumference of the lower end cap. Slide the three washers over the commutator end of the armature shaft. Apply a thin coating of Quicksilver All Purpose Lubricant, or equivalent waterproof anti-sieze lubricant, onto the bushing in the end cap. Hold the brushes

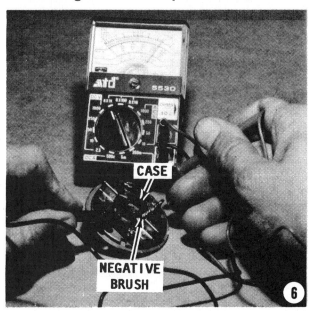

8-20 ELECTRICAL

2- Slide the three washers onto the armature shaft. Snap the circlip into place on the end of the shaft. Install the dust cover and secure it in place with the two screws and washers.

3- Slide the free end of the armature into the field frame end with the magnets recessed. **TAKE CARE** not to lose the brushes from the commutator. Seat the end cap into the field frame with the marks made prior to disassembly, aligned.

4- Seat the upper end cap with the marks made during disassembly, aligned.

SPECIAL WORDS

Installing the two thru bolts is not as simple a task as it sounds. The reason: as the bolts pass the magnets inside the field frame, the magnets will attract the bolts, "pulling" them out of line. With a bit of patience, the bolts can be made to align with the holes in the upper end cap and the threads started.

After the bolts are successfully threaded into the end cap, tighten them securely.

4- Install the following components onto the armature shaft in the sequence given: first, the pinion gear, followed by the spring, the collar, and finally the snap ring. It will be necessary to hold the collar down tight against the spring while the snap ring is installed. The ring must be crimped around the groove in the shaft with a pair of pliers. When properly installed the collar will completely hide the snap ring.

Model 28hp, 40hp, 48hp, 55hp, and 60hp

The exploded drawings on Page 8-15 and 8-16 will be most helpful during assembly of this cranking motor.

1- Apply a thin coating of Quicksilver All Purpose Lubricant, or equivalent waterproof anti-sieze lubricant, onto the bushing in the end cap. Hold the brushes back and insert the small end of the armature into the end cap bushing. **TAKE CARE** not to distort the brush springs. Once assembled, rotate the shaft inside the bushing and check to be sure there is no binding and the brushes sweep across the commutator smoothly.

After the bolts are successfully threaded into the end cap, tighten them securely.

5- Install the following components onto the armature shaft in the sequence given: first, the pinion gear, the spring (Model 40hp only), followed by the collar, and finally the snap ring. The ring must be crimped around the groove in the shaft with a pair of pliers. When properly installed the collar will completely hide the snap ring.

Model 28hp, 48hp, 55hp, and 60hp

6- Slide the two spacers onto the top of the solenoid with the pair of small holes in the spacers aligned with the holes in the end of the solenoid.

SPECIAL WORDS

The spacers can only be installed properly one way. The holes **MUST** be aligned to permit proper installation of the torsion spring.

7- Position the torsion spring in place atop the solenoid stradling the plunger and with the spring ends indexed into the two holes in the spacers and end of the solenoid.

8- Bring the solenoid and end cap/nose cone together with the flat portion of the spring facing toward the cranking motor, as indicated in the drawing. Now, work the end of the shaft to index over the yoke arm and the flat of the spring to index into the

TEST MISCELLANEOUS PARTS

small notch on the underside of the yoke arm.

9- Install and tighten the two bolts and washer to secure the solenoid to the end cap/nose cone together.

10- Connect the field frame lead to the solenoid terminal. Slide the protective "boot" over the terminal.

INSTALLATION

Model 20hp, 25hp, 30hp, and 40hp

1- Move the cranking motor close to position on the powerhead. Connect the Red lead to the cranking motor terminal. Now, position the cranking motor against the mounting bracket. Secure the motor in place with the two bolts and washers. One of the bolts goes into place partially under the flywheel. However, by rotating the flywheel, a cutout in the flywheel can be positioned directly over the bolt hole allowing the bolt to be "wiggled" into place and the threads started. Tighten the mounting bolts to a torque value of 13 ft lbs (18Nm).

2- Connect the cranking motor lead to the relay with the Phillips head screw.

Model 28hp, 48hp, 55hp, and 60hp

Install the cranking motor and solenoid assembly to the powerhead with the four securing bolts. Tighten the bolts to a torque value of 13 ft lb (18Nm).

Connect the small Red wire from the starter switch to the ignition terminal on the solenoid. Connect the large Red cable, the small Orange wire and the other small Red wire (these are all grouped together on a single eyelet connector), to the positive terminal on the solenoid. Connect the large Black cable to the negative terminal on the solenoid. Make a final check of all connections before connecting the positive and negative cables to the battery.

Check Completed Work

Mount the outboard unit in a test tank, on the boat in a body of water, or connect a flush attachment and hose to the lower unit.

CAUTION

Water must circulate through the lower unit to the powerhead anytime the powerhead is operating to prevent damage to the water pump in the lower unit. Just five seconds without water will damage the water pump impeller.

NEVER, AGAIN, NEVER operate the engine at high speed with a flush device attached. The engine, operating at high speed with such a device attached, would **RUNAWAY** from lack of a load on the propeller, causing extensive damage.

Crank the powerhead with the cranking motor and start the unit. Shut the powerhead down and start it several times to check operation of the cranking motor.

8-9 TESTING OTHER ELECTRICAL COMPONENTS

This short section provides testing procedures for other electrical parts installed on the powerhead. If a unit fails the testing, the faulty part must be replaced. In most cases, removal and installation is through attaching hardware.

Start Button Test

Trace the start button harness containing two wires from the switch to their nearest quick disconnect fitting. The colors

Using an ohmmeter to test the start button.

8-22 ELECTRICAL

may vary for different models. Refer to the wiring diagram in the Appendix for proper color identification.

Disconnect the two leads and connect an ohmmeter across the disconnected leads. Depress the start button. The meter should register continuity. Release the button and the meter should now register no continuity. Both tests **MUST** be successful. If the tests are not successful, the start button **MUST** be replaced. The start button is a one piece sealed unit and cannot be serviced.

"Kill" Switch Test

Trace the "Kill" switch button harness containing two wires from the switch to their nearest quick disconnect fitting. The colors may vary for different models, but the "hot" wire is usually White/Black. Refer to the wiring diagram in the Appendix for proper color identification.

Disconnect the two leads and connect an ohmmeter across the disconnected leads. Verify the emergency tether is in place behind the "Kill" switch button. Select the Rx1000 scale on the meter. Depress the kill button. The meter should register continuity. Release the button. The meter should now register no continuity. Both tests must be successful. If the switch fails either test, the switch is defective and **MUST** be replaced. The switch is a one piece sealed unit and cannot be serviced.

Choke Solenoid Test

Disconnect the Blue lead from the choke solenoid and the eyelet connector to ground. Connect an ohmmeter across the two disconnected leads. Select the Rx1000 scale on the meter. With the choke in the closed position, the meter should register continuity.

Manually open the choke valve until the gap between the solenoid coil stay grommet and the plunger is about 5/16" (7.9mm). With the plunger is this position, the meter should register no continuity. If the solenoid fails either test, the solenoid **MUST** be replaced.

Neutral Safety Switch Test

Trace the neutral safety switch leads from the switch to their nearest quick disconnect fitting. Both of these leads are usually Brown, but may vary for different models. Refer to the wiring diagram in the Appendix for the proper color identification.

Disconnect the two leads and connect an ohmmeter across the two disconnected leads. Select the Rx1000 scale on the meter. When the shift lever is in the **NEUTRAL** position, the meter should register continuity.

When the lower unit is shifted to either **FORWARD** or **REVERSE,** the meter should register no continuity.

The switch must pass all three tests to indicate the safety switch is functioning properly. If the switch fails any one or more of the tests, the switch **MUST** be replaced.

REMEMBER this is a safety switch. A faulty switch may allow the powerhead to be started with the lower unit in gear -- an extremely dangerous situation for persons aboard and the boat.

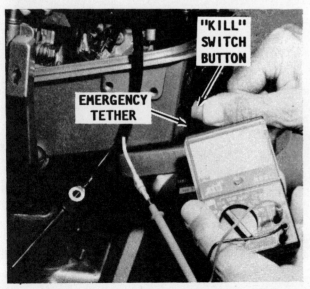

Using an ohmmeter to test the "kill" switch.

The neutral safety switch is "hidden", but is located on the axis of the shift lever just inside the lower cowling pan. The wire harness leads to the switch.

9
LOWER UNIT

9-1 DESCRIPTION

The lower unit is considered as that part of the outboard below the intermediate housing. The unit contains the propeller shaft, the driven and pinion gears, the drive shaft from the powerhead and the water pump. The lower unit of all 2-cylinder units and the 4 and 5hp 1-cylinder unit are equipped with shifting capabilities.

The lower unit matched with 3.5hp powerheads has shifting capabilities and therefore, its internal components are similar to the larger horsepower units. However, this lower unit has only a forward gear and a neutral position, no reverse gear. The operator must swing the outboard 180° to move the boat sternward.

The lower unit matched with the 2hp powerhead has no shifting capabities and therefore, no neutral position. The forward gear in the gearcase is in constant mesh with the pinion gear, and therefore, as soon as the powerhead is started, the propeller rotates.

The forward and reverse gears together with the clutch, shift assembly, and related linkage are all housed within the lower unit.

Shifting on units equipped with remote controls is accomplished through a cable arrangement from a shift box, installed near the helm, to the engine. The cable hookup involves two individual cables, one for shifting and the other for throttle control.

The lower unit may be removed and serviced without disturbing the remainder of the outboard motor.

CHAPTER COVERAGE

Three distinctly different lower units, identified as Type A, Type B, and Type C are covered in this chapter with separate sections for each. Each section is presented with complete detailed instructions for removal, disassembly, cleaning and inspecting, assembling, adjusting, and installation of only one unit. There are no cross-references. Each section is complete from removal of the first item to final testing.

Gear backlash tolerances are given in the text under the appropriate step.

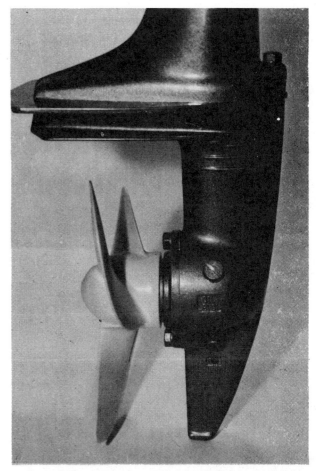

Type "A" lower unit used with the 2hp powerhead -- no reverse capability.

BE SURE to check the section heading for the model unit being serviced to ensure the correct procedures are followed.

The sections and the type unit covered are as follows:

9-5 Type A -- No shifting capability -- constant mesh in forward gear -- operator swings the outboard 180° to move boat sternward -- on all 2hp units.

9-6 Type B -- Shifting capability -- F-N or F-N-R -- one piece gearcase, on all other units covered in this manual, except those listed in Type A and C.

9-7 Type C -- Shifting capability -- F-N-R -- split (two piece) gearcase, on early 20, 25, and 40hp units.

ILLUSTRATIONS

Because this chapter covers such a wide range of models, the illustrations included with the procedural steps are those of the most popular lower units. In some cases, the unit being serviced may not appear to be identical with the unit illustrated. However, the step-by-step work sequence will be valid in all cases. If there is a special procedure for a unique lower unit, the differences will be clearly indicated in the step.

SPECIAL WORDS

All threaded parts are right-hand unless otherwise indicated.

If there is any water in the lower unit or metal particles are discovered in the gear lubricant, the lower unit should be completely disassembled, cleaned and inspected.

9-2 TROUBLESHOOTING

Troubleshooting MUST be done BEFORE the unit is removed from the powerhead to permit isolating the problem to one area. Always attempt to proceed with troubleshooting in an orderly manner. The shot-in-the-dark approach will only result in wasted time, incorrect diagnosis, frustration, and replacement of unnecessary parts.

The following procedures are presented in a logical sequence with the most prevalent, easiest, and less costly items to be checked listed first.

1- Check the propeller and the rubber hub. See if the hub is shredded. If the propeller has been subjected to many strikes against underwater objects, it could slip on its hub. If the hub appears to be damaged, replace it with a NEW hub. Replacement of the hub must be done by a propeller rebuilding shop equipped with the proper tools and experience for such work.

2- **Shift mechanism check:** Verify that the ignition switch is OFF, to prevent possible personal injury, should the engine start. Shift the unit into REVERSE gear (if so equipped), and at the same time have an assistant turn the propeller shaft to ensure the clutch is fully engaged. If the shift

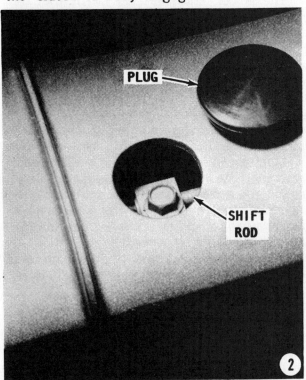

PROPELLER REMOVAL 9-3

handle is hard to move, the trouble may be in the lower unit shift rod or the shift box.

3- Isolate the problem: Disconnect the remote-control cable at the engine and then lift off the remote-control shift cable. Operate the shift lever. If shifting is still hard, the problem is in the shift cable or control box, see Chapter 10. If the shifting feels normal with the remote-control cable disconnected, the problem must be in the lower unit. To verify the problem is in the lower unit, have an assistant turn the propeller and at the same time move the shift cable back-and-forth. Determine if the clutch engages properly.

9-3 PROPELLER REMOVAL

Propeller Secured with Cotter Pin

Disconnect the high tension lead from the spark plug to prevent the powerhead from starting accidently. Straighten the cotter pin, and then pull it free of the propeller with a pair of pliers or a cotter pin removal tool. Remove the propeller, and then push the shear pin free of the propeller shaft. If the propeller is "frozen" to the shaft, follow the directions given under the heading **"Frozen Propeller"**, on Page 9-4, starting with Step 2.

Propeller Secured with "Nose Cone" Type Propeller Nut

1- Disconnect the high tension leads from the spark plugs to prevent the powerhead from starting accidently. Pull the

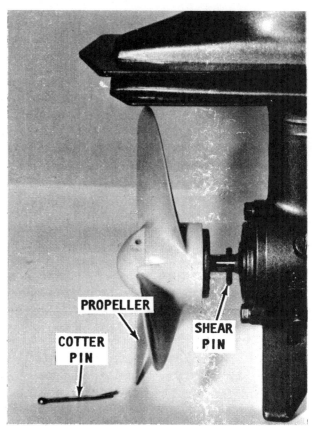

This type propeller indexes with the shear pin and is secured with a cotter pin.

cotter pin free of the propeller nut with a pair of pliers. Place a block of wood between one blade of the propeller and the anti-cavitation plate to prevent the shaft from rotating. Remove the propeller nut.

2- Remove the propeller lock pin, and then slide the propeller free of the propeller shaft.

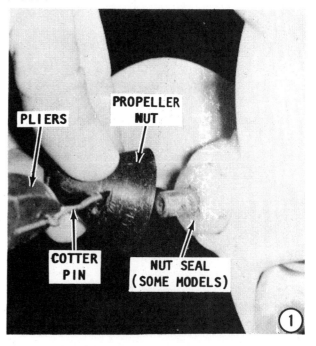

9-4 LOWER UNIT

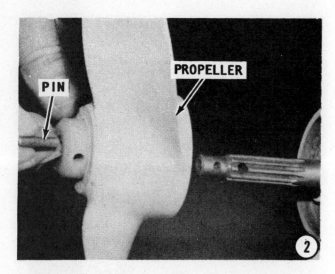

If the propeller is "frozen" to the shaft, follow the directions given under the heading **"Frozen Propeller"**, on this page, starting with Step 2.

Propeller Secured with Castellated Nut

1- Disconnect the high tension leads from the spark plugs to prevent the powerhead from starting accidently. Straighten the cotter pin and pull it free of the propeller shaft. Remove the castellated nut and washer. **NEVER** pry on the edge of the propeller. Any small distortion will affect propeller performance.

If the nut is "frozen", place a block of wood between one blade of the propeller and the anti-cavitation plate to keep the shaft

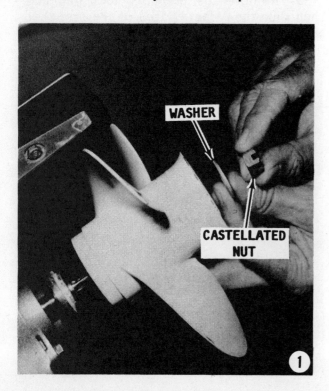

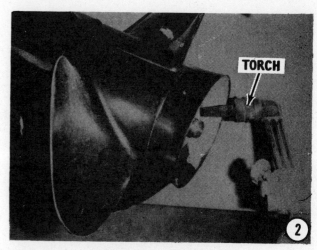

from turning. Use a socket and breaker bar to loosen the castellated nut. Remove the nut, washer, and then the propeller.

Frozen Propeller

2- If the propeller is frozen to the shaft, heat must be applied to the shaft to melt out the rubber inside the hub. Using heat will destroy the hub, but there is no other way. As heat is applied, the rubber will expand and the propeller will actually be blown from the shaft. Therefore, **STAND CLEAR** to avoid personal injury.

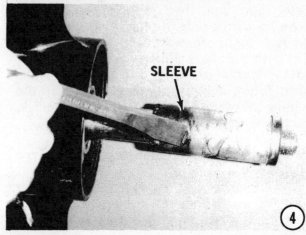

3- Use a knife and cut the hub off the inner sleeve.

4- The sleeve can be removed by cutting it with a hacksaw, or it can be removed with a puller. Again, if the sleeve is frozen, it may be necessary to apply heat. Remove the thrust hub from the propeller shaft.

9-4 PROPELLER INSTALLATION

SPECIAL WORDS
ALL MODELS

Apply Perfect Seal anti-seizing compound to the propeller shaft. The compound will prevent the propeller from "freezing" to the shaft and permit the propeller to be removed, without difficulty, the next time removal is required.

Propeller Secured with Cotter Pin

Install a new shear pin through the propeller shaft. Guide the propeller onto the pin. Insert a new cotter pin into the hole and bend the ends of the cotter pin in opposite directions.

Propeller Secured with "Nose Cone" Type Propeller Nut

1- Install a new shear pin through the propeller shaft. Slide the propeller onto the shaft with the internal splines of the propeller indexing with the splines on the shaft. Wedge a block of wood between one of the propeller blades and the anti-cavitation plate to prevent the propeller from rotating.

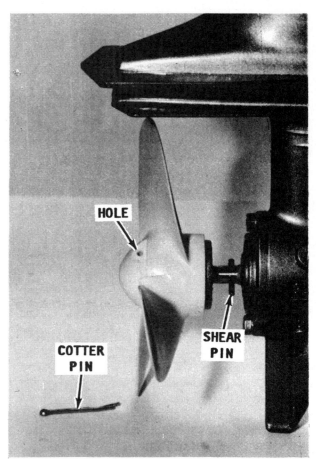

This type propeller is installed onto the propeller shaft until it indexes over the shear pin, and then is secured with a cotter pin.

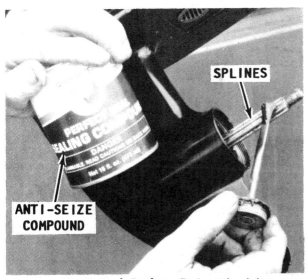

An application of Perfect Seal anti-seizing compound will prevent the propeller from "freezing" on the shaft making the next removal much easier.

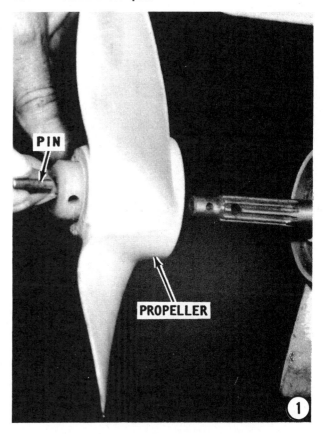

9-6 LOWER UNIT

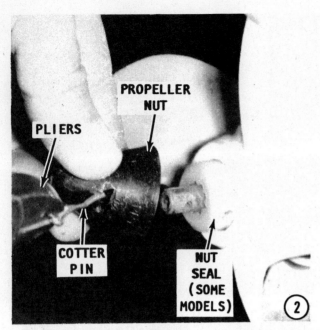

2- Install the propeller nut. Tighten the nut securely using a pair of large channel lock pliers or an adjustable wrench, and then check to determine if the holes in the nut are aligned with the hole in the propeller shaft. If not, tighten the nut just a "whisker" more until the holes are aligned. Insert and secure the cotter pin.

Propeller Secured with Castellated Nut

Install the spacer onto the propeller shaft. Install the propeller, the washer and the castellated nut. Place a block of wood between one of the propeller blades and the

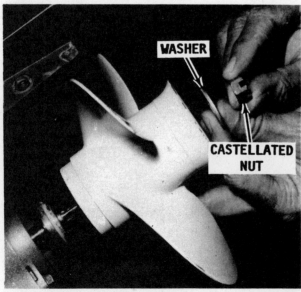

This type propeller is used on lower units matched with the larger powerheads covered in this manual. The propeller is secured with a washer, castellated nut and a cotter pin.

anti-cavitation plate to prevent the propeller from rotating. Tighten the nut to a torque value of 13 ft lbs (18Nm) for models 8, 9.9, and 15hp, or 22 ft lbs (30Nm) for models 20, 25, 28, and 30hp, or 25 ft lbs (34Nm) for models 40hp and above.

If necessary, back off the nut until the cotter pin may be inserted through the nut and the hole in the propeller shaft. Bend the arms of the cotter pin around the nut to secure it in place.

9-5 LOWER UNIT TYPE "A" MATCHED WITH 2HP POWERHEAD

DESCRIPTION

The Type "A" lower unit is a direct drive unit -- the pinion gear on the lower end of the driveshaft is in constant mesh with the forward gear. Reverse action of the propeller is accomplished by the boat operator swinging the engine with the tiller handle 180° and holding it in this position while the boat moves sternward. When the operator is ready to move forward again, he/she simply swings the tiller handle back to the normal forward position.

WORDS OF WISDOM

Before beginning work on the lower unit, take time to **READ** and **UNDERSTAND** the information presented in Section 9-1, this chapter.

Disconnect the high tension spark plug lead and remove the spark plug before working on the lower unit.

LOWER UNIT REMOVAL

The following procedures present complete instructions to remove, disassemble, assemble, and adjust the lower unit of the 2 hp unit.

SPECIAL WORDS

In order to remove the water pump impeller, the driveshaft **MUST** be removed from the lower unit. The impeller and pump housing can only be removed from the lower end of the driveshaft. Therefore, the circlip securing the pinion gear to the driveshaft must first be removed. This can only be accomplished by removing the lower unit bearing carrier cap and removing the propeller shaft.

DISASSEMBLING TYPE "A"

Before purchasing a lower unit gasket replacement kit, take time to establish the year of manufacture for the unit being serviced. Since 1986, the cap gasket has been replaced with an "O"-ring.

DISASSEMBLING

1- Position a suitable container under the lower unit, and then remove the **OIL** screw and the **OIL LEVEL** screw. Allow the gear lubricant to drain into the container. As the lubricant drains, catch some with your fingers from time to time, and rub it between your thumb and finger to determine if any metal particles are present. If metal is detected in the lubricant, the unit must be completely disassembled, inspected, and the damaged parts replaced. Check the color of the lubricant as it drains. A whitish or creamy color indicates the presence of water in the lubricant. Check the drain pan for signs of water separation from the lubricant. The presence of any water in the gear lubricant is bad news. The unit must be completely disassembled, inspected, the cause of the problem determined, and then corrected.

After the lubricant has drained, temporarily install both the drain and oil level screws.

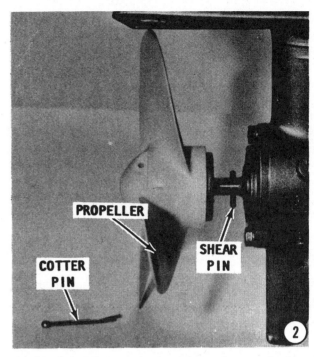

2- Straighten the cotter pin, and then pull it free of the propeller with a pair of pliers or a cotter pin removal tool. Remove the propeller, and then push the shear pin out of the propeller shaft.

3- Remove the two bolts securing the bearing carrier cap.

Units Prior to 1986

4- Use two small screwdrivers, one on each side, and simultaneously pry the cap from the lower unit. **TAKE CARE** not to mar the gasket sealing surfaces.

Units 1986 and On

Rotate the cap through 90°, and then gently tap the bearing carrier cap to break it free from the lower unit.

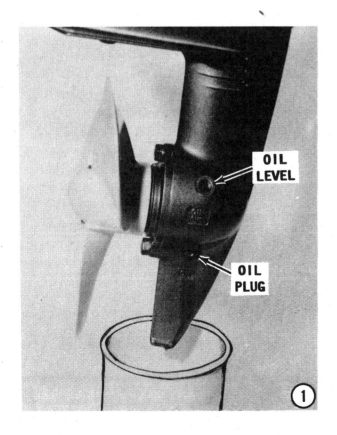

9-8 LOWER UNIT

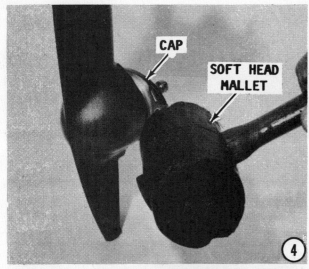

5- Remove the cap from the lower unit. Remove and discard the gasket from units prior to 1986. Remove and discard the "O"-ring from units since 1986.

CRITICAL WORDS
Perform Step 6 only if the seal has been damaged and is no longer fit for service.

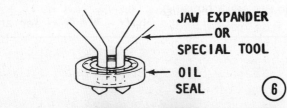

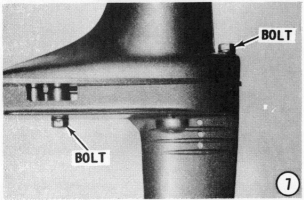

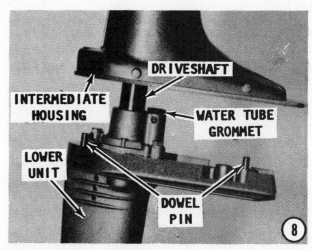

Removal of the seal destroys its sealing qualities and it cannot be installed a second time. Therefore, be absolutely sure a new seal is available before removing the old seal in the next step.

6- Obtain a slide hammer with jaw expander attachment. Use the slide hammer to remove the oil seal from the cap.

7- Remove the two attaching bolts securing the lower unit to the intermediate housing.

8- Separate the lower unit from the intermediate housing. **WATCH FOR** and **SAVE** the two dowel pins when the two units are separated. The water tube will come out of the grommet and remain with the intermediate housing. The driveshaft will remain with the lower unit.

DISASSEMBLING TYPE "A" 9-9

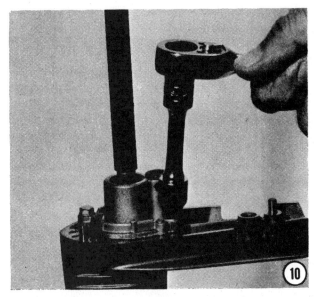

9- Pull the driveshaft tube free of the driveshaft.

10- Remove the two bolts securing the water pump housing cover to the lower unit.

11- Raise the water pump cover a bit to clear the indexing pin. Leave the water pump cover in this position at this time.

12- Pry the circlip free from the end of the driveshaft with a thin screwdriver. This clip holds the pinion gear onto the driveshaft. The clip may not come free on the first try, but have patience and it will come free. With one hand, remove driveshaft up and out of the lower unit housing and at the same time, with the other hand, catch the pinion gear and **SAVE** any shim material from behind the gear. The same thickness of shim material will be used upon later installation.

13- With the driveshaft on the workbench, remove the oil seal protector (a white plastic cap), the insert cartridge (a metal cup), and the water pump impeller.

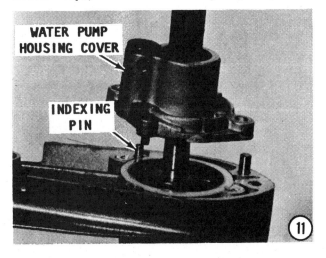

Check the condition of the impeller carefully and replace with a new one if there is any question as to its condition for satisfactory service. **NEVER** turn the impeller over in an attempt to gain further life from the impeller.

Remove the pin from the driveshaft and slide the water pump housing cover and cartridge outer plate free of the shaft.

Remove the O-ring from the water pump housing.

SPECIAL WORDS

Removal of the seal, as described in the next paragraph, destroys its sealing qualities and it cannot be installed a second time. Therefore, remove the seal **ONLY** if it is unfit for further service **AND** be absolutely sure a new seal is available before removing the old seal.

Obtain a slide hammer with jaw expander attachment. Use the tool to remove the oil seal from the housing cover.

14- Remove the propeller shaft from the lower unit housing. The forward gear will come out with the shaft. **WATCH** for and **SAVE** any shim material from the back side of the forward gear. The shim material is critical to obtaining the original backlash during installation.

15- Inspect the condition of the upper seal in the lower unit housing. If replace-

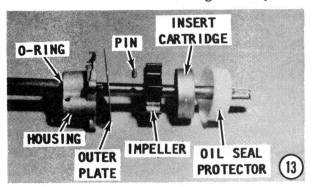

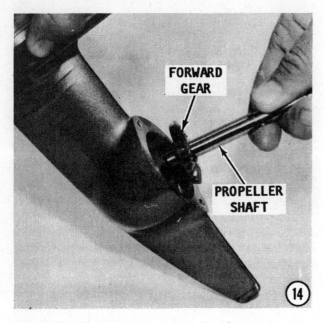

ment is required, use the same slide hammer and jaw attachment as used in Step No. 13 to remove the seal.

16- Three driveshaft bushings are used in the lower unit housing. These bushings need not be removed unless the determination is made they are no longer fit for service. Obtain Driver rod P/N M-91-83168M and drive all three down the driveshaft bore in one operation.

The bushings may also be removed using a suitably sized socket and extension.

CRITICAL WORDS

Removal of the forward propeller shaft bearing in the next step will destroy the bearing. Therefore, remove the bearing **ONLY** if it is no longer fit for service **AND** be absolutely sure a replacement bearing is available prior to removal.

Rotate the forward propeller shaft bearing. If any roughness is felt, the bearing must be removed and a new one installed.

Bearing Removal
Using Special Tools

17- Removal of the bearing is accomplished using a Bearing Puller Assembly Tool P/N M-91-83165M and jaw "B" attachment P/N M-91-84528M, or a slide hammer with jaw expander attachment.

If these special tools are not available, see Bearing Removal Without Special Tools, which follows and it may be possible to remove this bearing in the manner described.

Bearing Removal
Without Special Tools

The following two steps are to be performed only if the forward propeller shaft bearing must be removed and a suitable bearing puller is not available. The procedure will change the temperature between the bearing retainer and the housing substantially, hopefully about $80°F$ ($50°C$). This change will contract one metal -- the bearing retainer, and expand the other metal -- the housing, giving perhaps as much as .003" (.08mm) clearance to allow the bearing to fall free. Read the complete steps before commencing the work because three things are necessary: a freezer, refrigerator, or ice chest; some ice cubes or crushed ice; and a container large enough to immerse about 1 1/2" (3.8cm) of the forward part of the lower gear housing in boiling water.

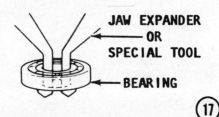

CLEAN & INSPECT TYPE "A"   9-11

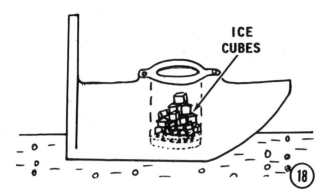

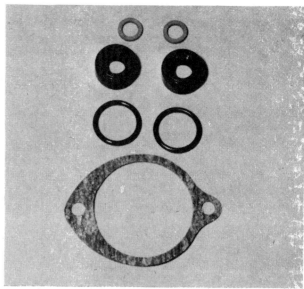

Parts included in a lower unit gasket replacement kit for Lower Unit "A" prior to 1986.

After all parts, including all seals, grommets, etc., have been removed from the lower gear housing, place the gear housing in a freezer, preferably overnight. If a freezer is not available try an electric refrigerator or ice chest. The next morning, obtain a container of suitable size to hold about 1 1/2" (3.8cm) of the forward part of the lower gear housing. Fill the container with water and bring to a rapid boil. While the water is coming to a boil, place a folded towel on a flat surface for padding.

18- After the water is boiling, remove the lower gear housing from its cold storage area. Fill the propeller shaft cavity with ice cubes or crushed ice. Hold the lower gear housing by the trim tab end and the lower end of the housing. Now, immerse the lower unit in the boiling water for about 20 or 30 seconds.

19- Quickly remove the housing from the boiling water; dump the ice; and with the open end of the housing facing downward, **SLAM** the housing onto the padded surface. **PRESTO**, the bearing should fall out.

If the bearing fails to come free, try the complete procedure a second time. Failure on the second attempt will require the use of a bearing puller.

CLEANING AND INSPECTING

Clean all water pump parts with solvent, and then dry them with compressed air. Inspect the water pump cover and base for cracks and distortion. If possible, **ALWAYS** install a new water pump impeller while the lower unit is disassembled. A new impeller will ensure extended satisfactory service and give "peace of mind" to the owner. If the old impeller must be returned to ser-

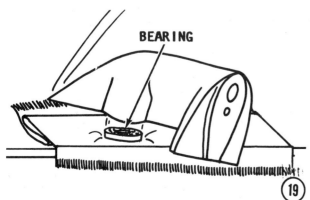

Parts included in a water pump repair kit for a 2hp unit.

9-12 LOWER UNIT

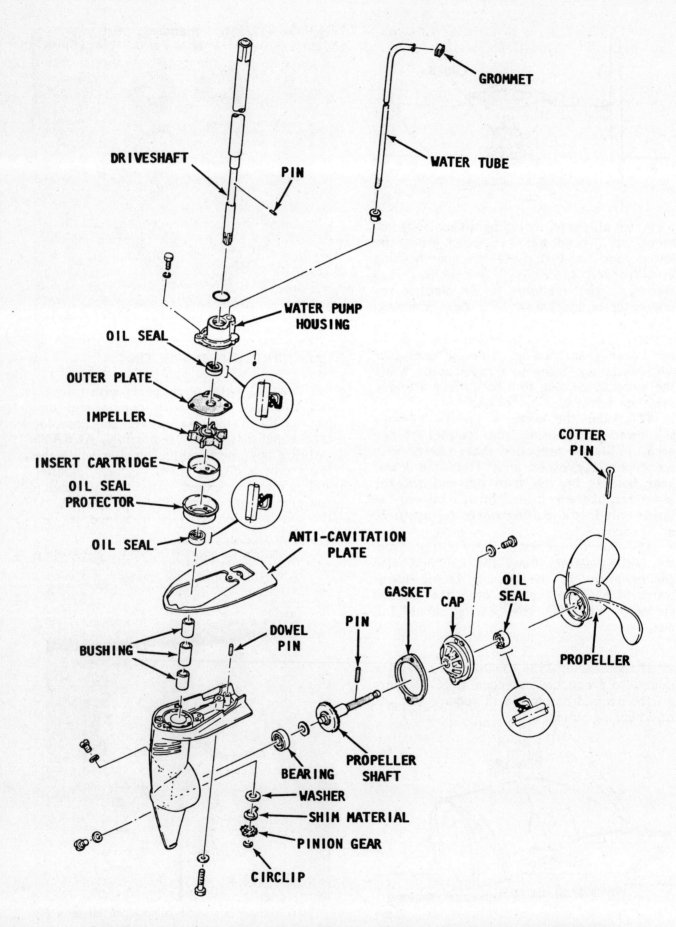

Exploded drawing of the Type "A" lower unit used on the Model 2hp, with major parts identified.

vice, **NEVER** install it in reverse to the original direction of rotation. Installation in reverse will cause premature impeller failure.

Inspect the ends of the impeller blades for cracks, tears, and wear. Check for a glazed or melted appearance, caused from operating without sufficient water. If any question exists, as previously stated, install a **NEW** impeller if at all possible.

GOOD WORDS

If an old impeller is installed be **SURE** the impeller is installed in the same manner from which it was removed -- the blades will rotate in the same direction. **NEVER** turn the impeller over thinking it will extend its life. On the contrary, the blades would crack and break after just a short time of operation.

Inspect the bearing surface of the propeller shaft. Check the shaft surface for pitting, scoring, grooving, imbedded particles, uneven wear and discoloration.

Check the straightness of the propeller shaft with a set of **V**-blocks. Rotate the propeller on the blocks

Good shop practice dictates installation of new O-rings and oil seals **REGARDLESS** of their appearance.

Clean the pinion gear and the propeller shaft with solvent. Dry the cleaned parts with compressed air.

Check the pinion gear and the drive gear for abnormal wear.

ASSEMBLING LOWER UNIT "A"

FIRST, THESE WORDS

The first three assembling steps apply **ONLY** if the forward propeller shaft bearing was removed from the housing. Therefore, the assumption is made disassembling Step 17 or Steps 18 and 19 were followed in order to remove the bearing. The paragraphs prior to Step 18 explain in detail the theory behind the procedure, without the use of the special tool. If this bearing was not removed, proceed directly to Step 4.

Bearing Installation
Using a Driver Tool

1- Lubricate the ball bearing with Formula 50 oil. Place the propeller shaft forward bearing squarely into the housing with the side embossed with the bearing size facing **OUTWARD**. Obtain Driver rod P/N M-91-84529M, mandrel P/N/ M-91-84530M and pilot P/N M-91-84534M. Slide the mandrel over the rod and insert both pieces into the gearcase and into the bearing to be installed. Slide the pilot over the end of the driver and allow the pilot to rest against the face of the gearcase. The pilot keeps the driver rod square in the bore. Tap the end of the driver rod to install and seat the bearing in place. A change in "tone" will be heard when the bearing finally seats. Remove the special tools.

If these tools are not available: Lubricate the ball bearing with Formula 50 oil. Obtain a suitably sized socket which will rest on the **OUTER** bearing race, and a long extension. Place the propeller shaft forward bearing squarely into the housing with the side embossed with the bearing size facing **OUTWARD**. Drive the bearing into the housing until it is fully seated. A change in "tone" will be heard when the bearing finally seats.

Bearing Installation
Without a Driver Tool

2- Place a **NEW** forward propeller shaft bearing in a freezer, refrigerator, or ice chest, preferably overnight. The next morning, boil water in a container of sufficient size to allow about 1-1/2" (3mm) of the

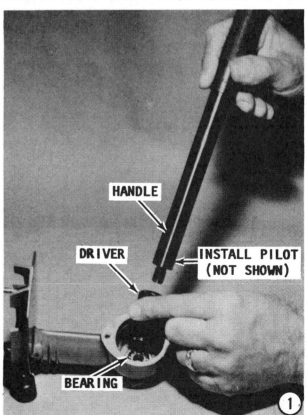

9-14 LOWER UNIT

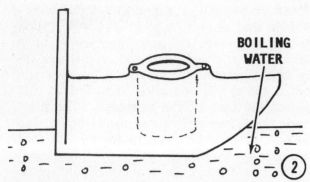

forward part of the lower gear housing to be immersed. While the water is coming to a boil, place a folded towel on a flat surface for padding. Immerse the forward part of the gear housing in the boiling water for about a minute.

3- Quickly remove the lower gear housing from the boiling water and place it on the padded surface with the open end facing upward. At the same time, have an assistant bring the bearing from the cold storage area. Continue working rapidly. Carefully place the bearing squarely into the housing as far as possible, with the embossed side facing **OUTWARD**. Push the bearing into place. Obtain a blunt punch or piece of tubing to bear on the complete circumference of the retainer. **CAREFULLY** tap the bearing retainer all the way into its forward position -- until it "bottoms-out". Tap evenly around the outer perimeter of the retainer, shifting from one side to the other to **ENSURE** the bearing is going squarely into place. The bearing must be properly installed to receive the forward end of the propeller shaft.

Driveshaft Bushing Installation

The following step is to be performed if any one or all three of the driveshaft bushings were removed. Three short bushings are used instead of one long bushing to facilitate installation and still provide the required amount of bearing surface. If none of the bushings were removed, proceed directly to Step 5.

4- Obtain special tool Bushing Installer Assembly P/N M-91-83170M. Position the special tool plate on top of the lower unit. Lower the long threaded bushing installer into the lower unit from the top until the threads protrude into the lower cavity of the lower unit, as shown. Now, thread the nut provided with the tool down onto the installer tool until the nut is snug against the plate. Slide the first bushing over the threads at the lower end of the driveshaft, and then install the bushing retainer onto the threads. Tighten the bushing installer from the upper end of the tool using the proper size wrench. Continue rotating the bushing installer until the bushing is drawn up into the lower unit and seats properly. Install the second and third bushing in the same manner.

Driveshaft Oil Seal Installation

5- Obtain seal installer special tool P/N M-91-83178M and driver rod P/N M-91-84529M. Thread the seal installer onto the handle. Place the oil seal over the end of the installer with the lip facing **UPWARD**.

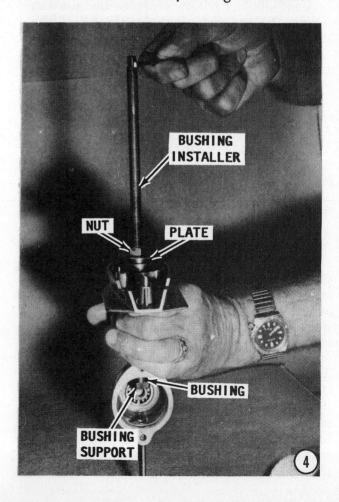

ASSEMBLING TYPE "A" 9-15

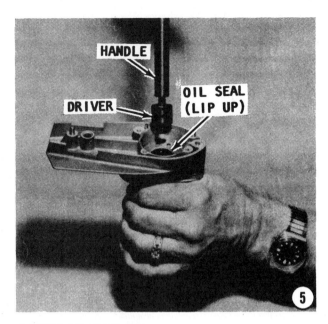

CRITICAL WORDS

The seal must be installed with the lip facing upward to prevent water from entering and contaminating the lubricant in the lower unit.

Lower the seal installer and the seal **SQUARELY** into the seal recess. Tap the end of the handle with a hammer until the seal is fully seated. After installation, pack the seal with multi-purpose water resistant lubricant.

Propeller Shaft Installation

6- Place the shim material saved during disassembling, Step 13, onto the propeller shaft ahead of the forward gear. The shim material should give the same amount of

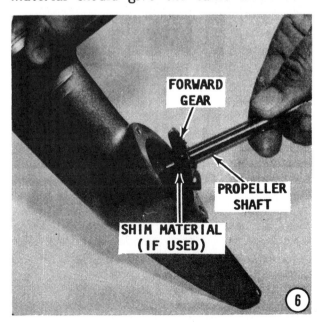

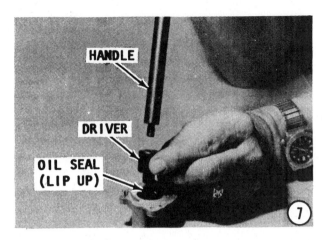

backlash between the pinion and forward gear as before disassembling. Insert the propeller shaft into the lower unit housing. Push the propeller shaft into the forward propeller shaft bearing as far as possible.

Water Pump Assembling

7- Obtain seal installer tool P/N M-91-83178M and driver P/N M-91-84529M.

CRITICAL WORDS

After installation of the water pump housing cover, the oil seal lip faces **DOWNWARD** to prevent water in the water pump from contaminating the lubricant around the driveshaft. **HOWEVER,** to install the seal, the water pump cover is turned upside down on the work surface. In this position, the lip of the seal must face **UPWARD**.

Using the seal installer and handle, install the seal. After installation, pack the seal with multi-purpose water resistant lubricant.

Install a new O-ring around the water pump housing.

8- Slide the water pump housing cover over the splined end of the driveshaft. Apply a coating of Loctite Type "A" to both sides of the water pump plate. Next, install

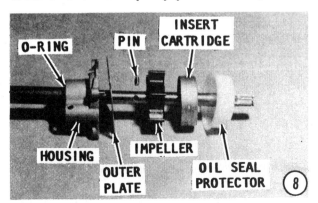

9-16 LOWER UNIT

the plate with the hole in the plate indexed over the pin on the water pump housing. Dab some multi-purpose water resistant lubricant, onto the dowel pin and insert it into the hole in the driveshaft. Slide the water pump impeller onto the driveshaft with the notch in the impeller indexed over the dowel pin. Slide the metal cup over the impeller, and then the plastic cup over the metal cup.

9- Lower the assembled driveshaft into the lower unit but do **NOT** mate the cover with the lower unit surface at this time. Leave some space, as shown. The splines on the lower end of the driveshaft will protrude into the lower unit cavity.

10- Slide the thrust washer and any shim material saved during disassembly, Step 12, onto the lower end of the driveshaft. Slide the pinion gear up onto the end of the driveshaft. The splines of the pinion gear will index with the splines of the driveshaft and the gear teeth will mesh with the teeth of the forward gear. Rotating the pinion gear slightly will permit the splines to index and the gears to mesh.

Now, comes the hard part. Snap the circlip into the groove on the end of the driveshaft to secure the pinion gear in place. If the first attempt is not successful, try again. Take a break, have a cup of coffee, tea, whatever, then give it another go. With patience, the task can be accomplished.

11- Apply some Loctite, or equivalent, to the two water pump cover retaining bolts. Tighten the bolts alternately and evenly to a torque value of 3.2 ft lbs (4.4 Nm).

Pinion Gear Depth
SPECIAL WORDS

The proper amount of pinion gear depth (pinion gear engagement with the forward gear), is critical for proper operation of the lower unit and long life of the internal parts.

12- Grasp the driveshaft and **PULL UPWARD**. At the same time, check the pinion gear tooth engagement with the forward gear teeth to be sure contact is made the full length of the tooth. This can be accomplished by using a flashlight and looking into the lower unit opening. If the pinion gear depth is **NOT** correct, the amount of shim material behind the pinion gear must be adjusted.

Bearing Carrier Installation

13- Place the bearing carrier cap on the work surface with the painted side facing **UPWARD**. Install the oil seal with the lip facing **DOWNWARD**. In this position the

ASSEMBLING TYPE "A" 9-17

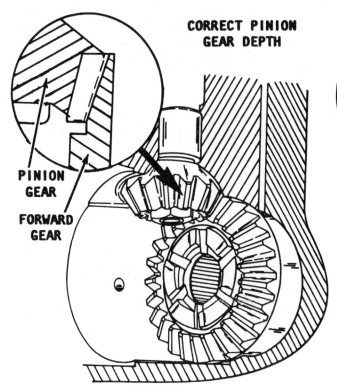

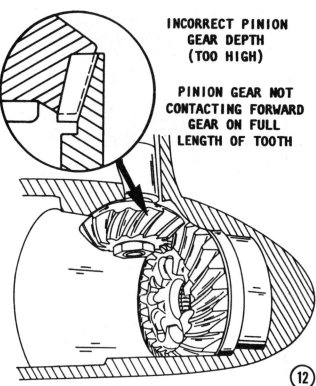

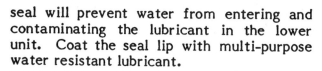

seal will prevent water from entering and contaminating the lubricant in the lower unit. Coat the seal lip with multi-purpose water resistant lubricant.

SPECIAL WORDS

On models through 1985, a gasket is installed between the bearing carrier cap and the lower unit housing. Since 1986, the gasket has been replaced with an O-ring. On all units, apply a coating of Loctite Type "A" to the gasket or O-ring and to its mating surface.

14- Install the bearing carrier cap to the lower unit and make sure the cap is seated evenly all the way around.

15- Apply Loctite, or equivalent, to the threads of the bearing carrier attaching bolts. Install and tighten the bolts alternately and evenly to a torque value of 5.8 ft lbs (8 Nm).

Backlash Education

Backlash is the acceptable clearance between two meshing gears, in order to take into account possible errors in machining, deformation due to load, expansion due to heat generated in the lower unit, and center-to-center distance tolerances. A "no backlash" condition is unacceptable, as such a condition would mean the gears are locked together or are too tight against each other

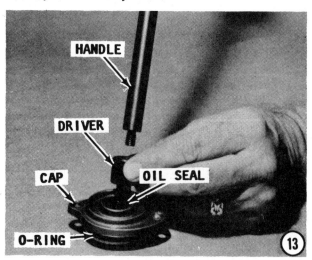

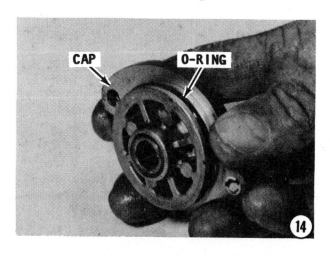

9-18 LOWER UNIT

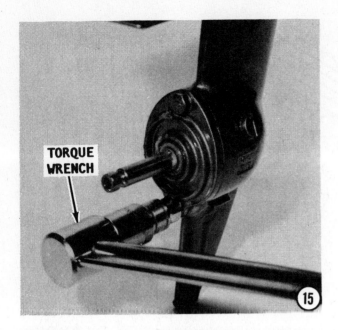

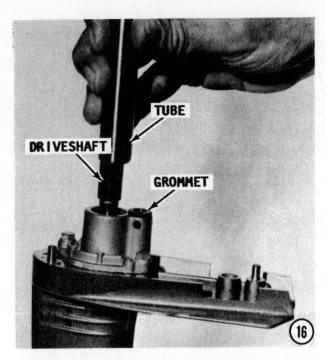

which would cause phenomenal wear and generate excessive heat from the resulting friction.

Excessive backlash which cannot be corrected with shim material adjustment indicates worn gears. Such worn gears must be replaced. Excessive backlash is usually accompanied by a loud whine when the lower unit is operating at idle speeds.

Checking Gear Backlash

Prevent the driveshaft from rotating with one hand, and at the same time pull **UPWARD** on the shaft. With the other hand, **PUSH IN** on the propeller shaft and at the same time, attempt to rock the shaft back and forth. The propeller shaft should be free to move "just a whisker" to have an acceptable gear backlash. The propeller shaft should definately **NOT** be "locked" in place.

If the propeller shaft is too "tight", remove shim material from behind the forward gear. If the propeller shaft is too "sloppy", add shim material behind the forward gear.

Repeat the procedure to check reverse gear backlash, but **PULL OUT** on the propeller shaft, while attempting to rock the shaft back and forth.

If the propeller shaft is too "tight", remove shim material from behind the reverse gear. If the propeller shaft is too "sloppy", add shim material behind the reverse gear.

16- Apply a small amount of multi-purpose water resistant lubricant to the squared section at the upper end of the driveshaft.

BAD NEWS

An excessive amount of lubricant on top of the driveshaft to crankshaft splines will be trapped in the clearance space. This trapped lubricant will not allow the driveshaft to fully engage with the crankshaft.

Slide the tube over the driveshaft and push a new grommet in place where the water tube enters the water pump housing cover.

INSTALLATION

17- Apply just a dab of multi-purpose water resistant lubricant, to the indexing pin on the mating surface of the lower unit.

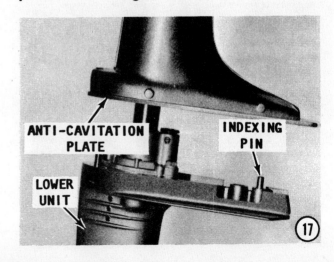

INSTALLING TYPE "A"

Install the anticavitation plate onto the lower surface of the intermediate housing. Begin to bring the intermediate housing and the lower gear housing together. As the two units come closer, rotate the propeller shaft slightly to index the upper end of the lower driveshaft tube with the upper rectangular driveshaft. At the same time, feed the water tube into the water tube seal. Push the lower gear housing and the intermediate housing together. The pin on the upper surface of the lower unit will index with a matching hole in the lower surface of the anticavitation plate.

18- Apply Loctite Type "A" to the threads of the two bolts used to secure the lower unit to the intermediate housing. Install and tighten the two bolts to a torque value of 5.8 ft lbs (8Nm).

19- Apply Perfect Seal anti-seizing compound to the propeller shaft.

SPECIAL WORDS

The compound will prevent the propeller from "freezing" to the shaft and permit the propeller to be removed, without difficulty, the next time removal is required.

Install a new shear pin through the propeller shaft. Guide the propeller onto the pin. Insert a new cotter pin into the hole and bend the ends of the cotter pin in opposite directions.

20- Remove the oil level plug and the oil plug. Fill the lower unit with Quicksilver

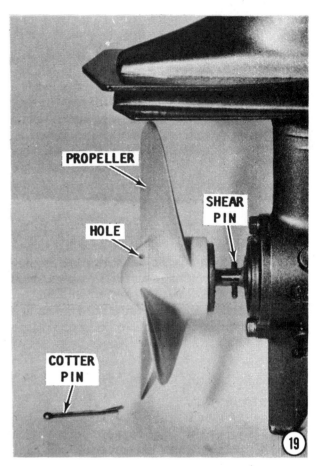

gearcase lubricant until the lubricant flows from the top hole. The capacity of the lower unit is 1.5 U.S. oz (45cc). Install both plugs and clean any excess fluid from the lower unit.

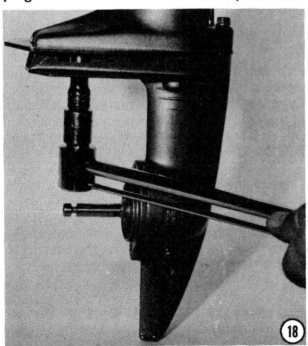

9-6 LOWER UNIT TYPE "B"

Type "B" lower units are one piece units, matched with **ALL** 3.5, 4, 5, 8, 9.9, 15, 28, 30, 48, 55, and 60hp and **MOST** (late model) 20, 25, and 40hp outboards. The rest (early model) of the 20, 25, and 40hp units have two piece lower units and are identified as Type "C" units, even though their shifting capabilites are identical to those in Type "B".

DESCRIPTION

In addition to the normal forward capability, most Type "B" lower units are equipped with a clutch "dog" permitting operation in **NEUTRAL, FORWARD,** and **REVERSE.**

The 3.5hp air cooled outboard is equipped with a clutch "dog" permitting operation in **NEUTRAL,** and **FORWARD** gear only.

Shifting Operation
All Units Equipped with F-N-R Capability

The pinion gear is in constant mesh with both the forward and reverse gear. These three gears constantly rotate anytime the powerhead is operating.

A sliding clutch "dog" is mounted on the propeller shaft. The clutch "dog" is operated manually through linkage from a shift lever located on the starboard side of the powerhead. When the clutch "dog" is moved forward, it engages with the forward gear. The clutch "dog" then picks up the rotation of the forward gear. Because the clutch is secured to the propeller shaft with a pin, the shaft rotates at the same speed as the clutch. The propeller is thereby rotated to move the boat forward.

When the clutch "dog" is moved aft, the the clutch engages only the reverse gear. The propeller shaft and the propeller are thus moved in the opposite direction to move the boat sternward.

When the clutch "dog" is in the neutral position, neither the forward nor reverse gear is engaged with the clutch and the propeller shaft does not rotate.

From this explanation, an understanding of wear characteristics can be appreciated. The pinion gear and the clutch "dog" receive the most wear, followed by the forward gear, with the reverse gear receiving the least wear.

Shifting Operation
All Units Equipped with F-N Capability

The pinion gear is in constant mesh with the forward gear. These two gears constantly rotate anytime the powerhead is operating.

A sliding clutch "dog" is mounted on the propeller shaft. The clutch "dog" is operated manually through linkage from a shift lever located on the starboard side of the powerhead. When the clutch "dog" is moved forward, it engages with the forward gear. The clutch "dog" then picks up the rotation of the forward gear. Because the clutch is secured to the propeller shaft with a pin, the shaft rotates at the same speed as the clutch. The propeller is thereby rotated to move the boat forward.

When the clutch "dog" is moved aft, the the clutch disengages from the forward gear into the neutral position and the propeller shaft does not rotate. In the neutral position the clutch "dog" rotates freely.

Exhaust Gases

At low engine speed, the exhaust gases from the powerhead exit from an idle hole in the intermediate housing. As powerhead rpm increases, these gases, for smaller models, are partly channeled through the exhaust passage in the intermediate housing and partly through the idle hole.

For larger horsepower models, the gases are forced out with the cycled water and stream exit through the propeller at normal cruising rpm and faster.

Water for cooling is pumped to the powerhead by the water pump and is expelled with the exhaust gases. The water pump impeller is installed on the driveshaft. Therefore, the water output of the pump is directly proportional to powerhead rpm.

WORDS OF WISDOM

Procedural steps are given to remove all items in the lower unit. However, do **NOT** remove bearings, bushings, or seals, if a determination can be made the item is fit for further service. As the work progresses, simply skip the steps involving these parts and continue with required work.

Before beginning work on the lower unit, take time to **READ** and **UNDERSTAND** the information presented in Section 9-1, this chapter.

REMOVING TYPE "A" 9-21

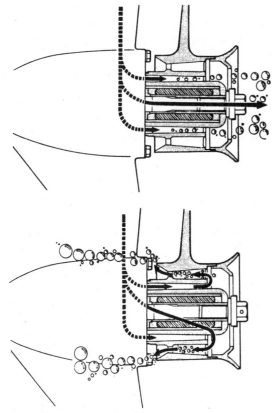

Cross-section drawing of the lower unit showing route of the exhaust gases with the unit in forward gear (top), and in reverse gear (bottom).

Disconnect the high tension spark plug lead and remove the spark plug before working on the lower unit.

LOWER UNIT REMOVAL

The following procedures present complete instructions to remove, disassemble, assemble, and adjust the lower unit matched with **ALL** 3.5, 4, 5, 8, 9.9, 15, 28, 30, 48, 55, and 60hp and **MOST** (late model) 20, 25, and 40hp powerheads.

AUTHORS' WORDS

Because so many different models are covered in this manual, it would not be feasible to provide an illustration of each and every unit.

Therefore, the accompanying illustrations are of a "typical" unit. The unit being serviced may differ slightly in appearance due to engineering or cosmetic changes but the procedures are valid. If a difference should occur, the models affected will be clearly identified.

1- Position a suitable container under the lower unit, and then remove the **OIL** screw and the **OIL LEVEL** screw. Allow the gear lubricant to drain into the container. As the lubricant drains, catch some with your fingers from time to time, and rub it between your thumb and finger to determine if any metal particles are present. If metal is detected in the lubricant, the unit must be completely disassembled, inspected, and the damaged parts replaced.

Check the color of the lubricant as it drains. A whitish or creamy color indicates the presence of water in the lubricant. Check the drain pan for signs of water separation from the lubricant. The presence of any water in the gear lubricant is bad news. The unit must be completely disassembled, inspected, the cause of the problem determined, and then corrected.

After the lubricant has drained, temporarily install both the drain and oil level screws.

Propeller Secured with "Nose Cone" Type Propeller Nut

2- Straighten the cotter pin, and then pull it free of the propeller nut with a pair of pliers or a cotter pin removal tool. Remove the propeller nut using a large pair of channel lock pliers or an adjustable wrench. Push the shear pin free of the propeller shaft.

Propeller Secured with Castellated Nut

Straighten the cotter pin, and then pull it free of the castellated nut with a pair of

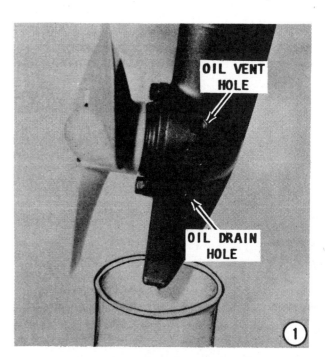

pliers or a cotter pin removal tool. Remove the castellated propeller nut by first placing a block of wood between one of the propeller blades and the anti-cavitation plate to prevent the propeller from rotating, and then remove the nut. Remove the washer. Remove the outer thrust hub from the propeller shaft. If the thrust hub is stubborn and refuses to budge, use two **PADDED** pry bars on oppostie sides of the hub and work the hub loose. **TAKE CARE** not to damage the lower unit.

Remove the propeller. On some Model 3.5hp lower units two spacers are used between the bearing carrier and the propeller. Slide these spacers free of the propeller shaft, and then remove the propeller and the rubber damper.

On other Model 3.5hp lower units, the cotter pin secures the propeller directly to the propeller shaft. Refer to the exploded diagram on Page 9-35.

If the propeller is "frozen" to the shaft, see Section 9-3 for special procedures to "break" it loose.

Remove the inner thrust hub. If this hub is also stubborn, use padded pry bars and work the hub loose. Again, **TAKE CARE** not to damage the lower unit.

3- Tilt and lock the outboard unit in the raised position. For models 3.5, 4, 5, and some early 8hp: If the unit has a black grommet on the intermediate housing, pry it free of the housing. All other 8hp models have the shift rod connector outside the intermediate housing, as shown in the accompanying illustration.

Models 4hp, 5hp, and 8hp
Shift the lower unit into reverse gear. If shifting is difficult, rotate the flywheel **COUNTERCLOCKWISE** as an aid in shifting.

Loosen but **DO NOT** remove the shift rod locknut on the two piece shift rod connector.

Models 3.5hp, 9.9hp, and Larger
Check to be sure the lower unit is in **NEUTRAL**.

All Models Except 3.5hp, 4hp, 5hp, and 8hp
Loosen the locknut. Separate the lower and upper shift rods by rotating the shift rod threaded connector until the lower rod is free.

4- Remove the bolts securing the lower unit to the intermediate housing.

5- Separate the lower unit from the intermediate housing. **WATCH FOR** and **SAVE** the two dowel pins when the two units are separated. The water tube will come out of the grommet and remain with the intermediate housing. The driveshaft will remain with the lower unit.

SPECIAL WORDS
3.5HP AND 5HP AIR-COOLED MODELS
These two units have a gasket between the lower unit and the intermediate housing. Also, the anti-cavitation plate will come free of the gearcase.

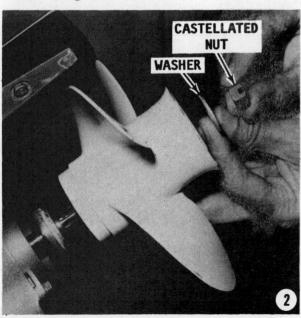

DISASSEMBLING TYPE "B" 9-23

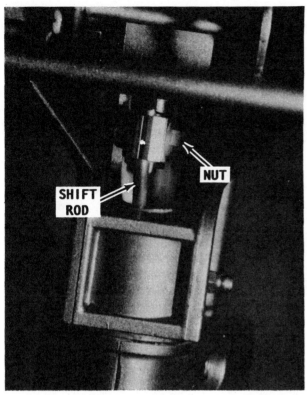

Shift rod connection arrangement on the lower unit matched with the 8hp powerhead.

Because the units are air-cooled, there is no water pump. Cooling is accomplished by fins around the powerhead casting.

A long thin water tube is attached to the upper face of the anti-cavitation plate, inside the trim tab. Water pressure from the propeller forces water up this tube, not for cooling purposes, but to mix with the exhaust gases to reduce noise level.

Model 5hp Only

Remove the shift rod limit screw on the starboard side of the gearcase.

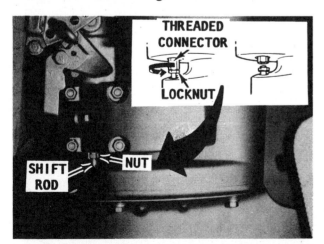

Shift rod connection arrangement on the lower unit matched with the Model 9.9hp and larger powerheads.

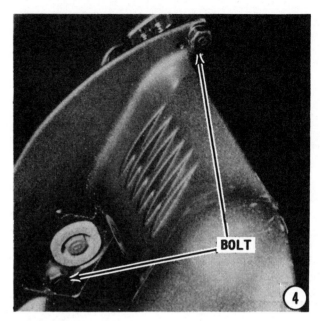

Model 3.5hp and 5hp

Remove any attaching hardware and boots and pull the shift rod up clear of the gearcase.

Skip the following steps related to the water pump and proceed directly to Step 11.

DISASSEMBLING

Water Pump Removal
All Models Except 3.5hp and 5hp
Air-Cooled Models

6- Remove the four bolts and washers securing the water pump cover to the oil

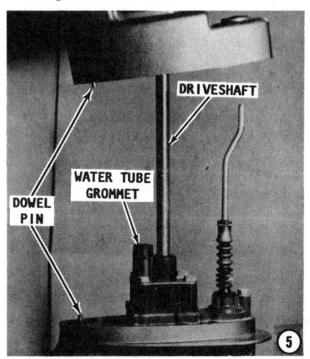

9-24 LOWER UNIT

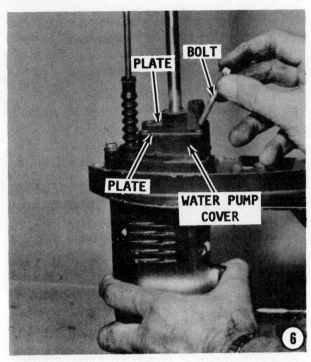

seal housing. Models 4, 5, 9.9, 15hp, and some 8hp have a plate on each side of the cover secured by the same bolts.

Some Models 8, 9.9, and 15hp have an extra retaining plate on top of the water pump housing. This plate is secured in place with two separate Phillips head screws.

GOOD WORDS

Move the water pump cover **SLOWLY** past the splines on the driveshaft. This action will help prevent the splines on the end of the shaft from damaging the water pump cover oil seal. Splines on the end of a shaft or threads on the end of a rod are one of the greatest "enemies" of an oil seal. The damage to the seal is caused as the seal passes over the splines or threads during removal and installation.

7- Slide the following components upward and free of the driveshaft: First, the water pump housing. On Models 4hp, and 5hp, the indexing pin may now be removed from the driveshaft. Next, slide the inner plate up and free of the driveshaft on Models 20hp and larger. Now, slide the water pump impeller and the insert cartridge free of the driveshaft. Remove and **SAVE** the small Woodruff key on all models except the 4hp, and 5hp. These models have the indexing pin. Remove the water pump gasket. Pull out the water tube grommet from the water pump housing.

Models 4hp, 5hp, and 8hp

8- Lift up the outer plate and the gasket beneath it. The 8hp units have an oil seal housing and a gasket under the housing. The shift rod and boot will come away with the outer plate. **TAKE CARE** not to lose the two locating pins in the bearing housing.

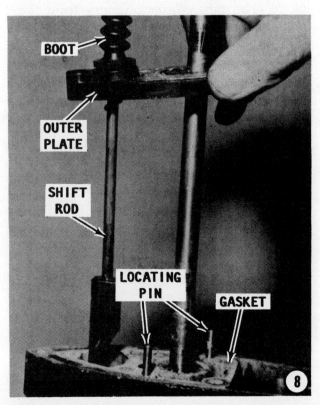

DISASSEMBLING TYPE "B" 9-25

Models 9.9hp, 15hp, 40hp, 48hp, 55hp, and 60hp

9- Remove the outer plate and the gasket under the plate. Remove the oil seal housing and the gasket under the housing.

Special Instructions
Some Early 8hp and 15hp Models

Loosen and remove the shift rod detent screw and washer on the starboard side of the gearcase.

Some 8hp models have a spring loaded detent ball pressing against the back of the shift cam at the end of the shift rod. After removal, if the shift rod has a series of three notches cut in the back of the shift cam, remove the detent plug, spring and ball bearing at the forward "nose" of the gearcase. The purpose of this arrangement is to provide a positive shift action. Therefore, this particular design 8hp lower unit will be refered to as the "Positive Shift" 8hp model, in the remainder of the procedures.

All Models

Loosen but **DO NOT** remove the small Phillips screw on the white plastic shift rod retainer. Pull out the shift rod and boot assembly.

Models 48hp, 55hp, and 60hp

Observe the end of the shift rod. If the rod has a splined end, a shift cam was left behind in the lower unit. This cam will be retrieved at the end of Step 32.

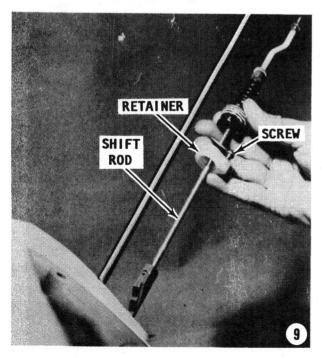

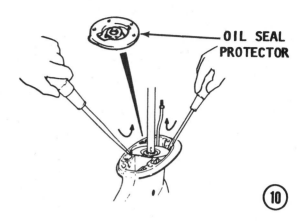

Model 40hp

Remove the shift limit screw and washer on the starboard side of the gearcase.

Models 20hp, 25hp, 28hp, and 30hp

10- Use two screwdrivers and lift out the oil seal protector, and then remove the oil seal housing and O-ring in the housing.

Loosen but **DO NOT** remove the small Phillips screw on the white plastic shift rod retainer. Pull out the shift rod and boot assembly.

All Models

SPECIAL WORDS

Removal of the seal, as described in the next paragraph, destroys its sealing qualities. As a result, it cannot be installed a second time. Therefore, remove the seal **ONLY** if it is unfit for further service **AND** be absolutely sure a **NEW** seal is available before removing the old seal.

11- Obtain a slide hammer with jaw expander attachment.

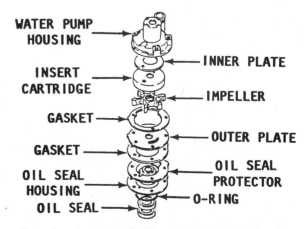

Exploded drawing of the water pump used on the Model 20hp, 25hp, 28hp and 30hp. This pump is of a different design than others in this manual.

9-26 LOWER UNIT

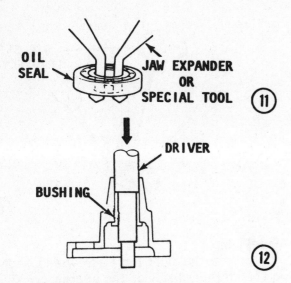

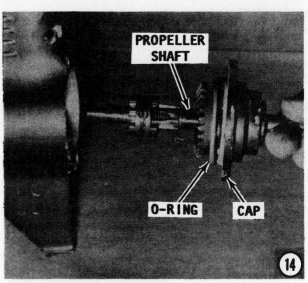

Air-Cooled Models 3.5hp and 5hp

Use the slide hammer and remove the oil seal from the anti-cavitation plate. Remove and discard the O-ring around the driveshaft bore.

Water-Cooled Models 4hp, and 5hp

Use the tool to remove the oil seals from the lower unit housing.

All Other Models

Use the slide hammer to remove the oil seals from the oil seal housing.

OIL SEAL HOUSING BUSHING

Models 8hp, 9.9hp, and 15hp

12- Obtain Driver Rod P/N M-91-83169M, or a suitable size socket and extension and drive the bushing free of the oil seal housing.

Place the housing down on its large flat face and drive the bushing down toward the large face.

VERY SPECIAL WORDS

If the only work to be performed on the lower unit is servicing the water pump, proceed directly to Step 26, Assembling.

BEARING CARRIER CAP REMOVAL

Models 3.5hp, 4hp and 5hp

13- Remove the two bolts securing the bearing carrier cap.

Rotate the cap 90° (1/4 turn), and then gently tap the bearing carrier cap to "break" it free from the lower unit. Remove the cap from the lower unit. Remove and discard the O-ring, or gasket. Pull out the propeller shaft.

Model 8hp, 9.9hp and 15 hp

14- Remove the two bolts securing the bearing carrier cap.

Early W15 and 15hp models have three bolts securing the bearing carrier to the gearcase.

Use two small screwdrivers, one on each side, and simultaneously pry the cap from the lower unit. TAKE CARE not to mar the gasket sealing surfaces. Pull out the propeller shaft, and then remove the cap from the shaft. Remove and discard the O-ring.

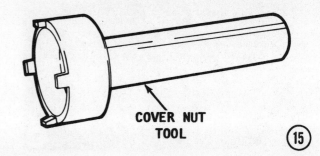

Model 20hp and Larger

15- Straighten the tabs on the lockwasher bent over the ring nut. Obtain cover nut tool P/N:

C-91-78729 for Models 20, 25, and 30hp
M-91-84546M, or C-91-74241 for Model 40hp

Check with the local Mariner dealer for the correct part number if the model being serviced is a 28hp, 48hp, 55hp or 60hp.

Note the word "OFF" embossed on the ring nut. Install the tool into the ring and rotate the ring in the direction indicated. Rotate the ring nut **COUNTERCLOCKWISE** until the nut is free. Remove the lockwasher. Attempt to remove the propeller shaft. If the shaft and the bearing carrier cannot be removed easily, they must be "pulled" as described in the next step.

SAVE the bearing carrier locating key on the keyway at the bottom of the gearcase.

Stubborn Carriers
Models 20hp and Larger

16- Remove the securing bolts and washers from the front of the bearing carrier. Obtain a universal gear puller. Hook the ends of the J-bolts into the bearing carrier ribs across from each other. Insert the threaded ends through the puller, and then install the washers and nuts to take up the slack. Rotate the center threaded shaft **CLOCKWISE** to separate the bearing carrier from the lower unit.

Remove the tool and withdraw the propeller shaft with the bearing carrier, reverse gear, and washer still on the shaft.

TAKE CARE not to lose the small key fitted into the bottom of the bearing carrier.

All Models

17- Remove the bearing carrier from the propeller shaft. Lift off the washer between the reverse gear and the clutch dog. Air-cooled Models 3.5hp and 5hp do not have this washer. Some early 8hp and 15hp Models have an additional washer between the forward gear and the clutch "dog".

Models 4hp thru 15hp

Remove the reverse gear from the bearing carrier. **WATCH** for and **SAVE** any shim material from the back side of the reverse gear. The shim material is critical in obtaining the correct backlash during assembling. Using the old shim material will save considerable time, especially starting with no shim material.

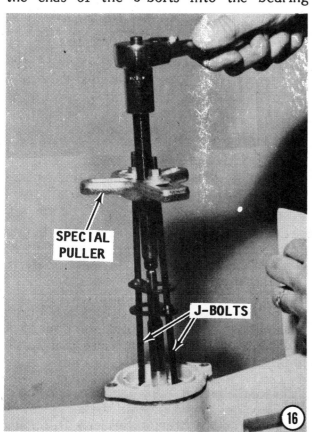

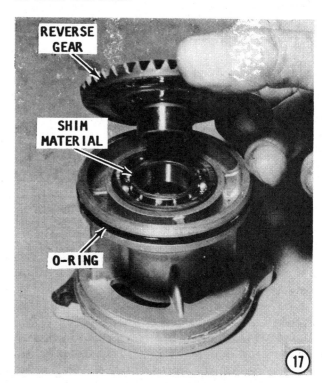

9-28 LOWER UNIT

Models 20hp and Larger

The reverse gear and ball bearing assembly is pressed into the bearing carrier on these models. This same ball bearing is pressed onto the back of the reverse gear. If necessary, after the assembly has been removed from the carrier, the bearing may be separated from the gear using a bearing separator. This procedure is outlined in Step 19.

All Models

The ball bearing or the ball bearing and reverse gear together must be "pulled" from the bearing carrier.

Obtain a slide hammer with jaw expander attachment and "pull" the bearing or the bearing and reverse gear from the bearing carrier. When using the tool check to be **SURE** the jaws are hooked onto the inner race.

SPECIAL WORDS
AIR-COOLED MODEL 3.5HP

Even though this model has no reverse gear, a ball bearing supporting the propeller shaft is located inside the bearing carrier.

CRITICAL WORDS

Perform Step 18 only if the seals have been damaged and are no longer fit for service. Removal of the seals destroys their sealing qualities. Therefore, they cannot be installed a second time. Be absolutely sure a new seal is available **BEFORE** removing the old seal in the next step.

SPECIAL WORDS

Perform Steps 18, and 19 **ONLY** if the bearing in question is no longer fit for service. Perform Step 20 **ONLY** if the bushing is no longer fit for service.

18- Inspect the condition of the oil seal/s in the bearing carrier cap for the 3.5hp, 4hp, and 5hp models, or in the bearing carrier for all other models. If the seals appear to be damaged and replacement is required, use the same slide hammer and jaw attachments as used in Step 11 to remove the seals.

Models 20hp and Larger

19- The ball bearing is pressed onto the back of the reverse gear. To remove this bearing, first obtain a universal bearing sep-

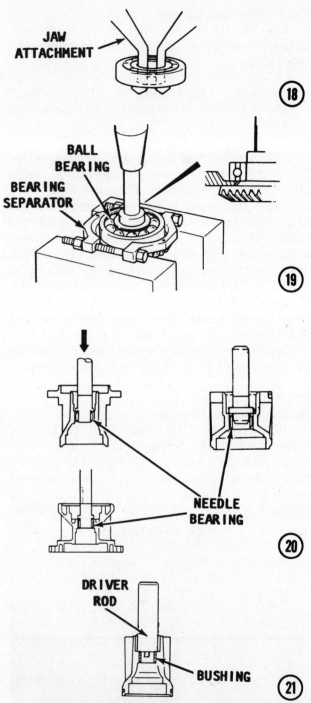

arator tool, and then press on the inner race to separate the bearing from the gear.

WATCH for and **SAVE** any shim material and thrust washer from the back side of the reverse gear. The shim material is critical in obtaining the correct backlash during assembling. Using the old shim material will save considerable time, especially starting with no shim material.

All Models Except 3.5hp 4hp, 5hp, and 8hp

20- The caged needle bearing set is pressed into the bearing carrier. To remove the bearing set, obtain the following special tools, P/N:

Models 9.9hp and 15hp - Driver Rod P/N M-91-37323M, and Mandrel P/N M-91-83184M

Models 20hp, 25hp, and 30hp - Driver Rod P/N M-91-84529M, and Mandrel P/N M-91-84570M

Model 40hp - Driver Rod P/N M-91-84562M, and Mandrel P/N M-91-84575M

For all other models not listed above, check with the local Mariner dealer for the correct part numbers of special tools for the models being serviced, or obtain a suitable size socket and extension. The socket **MUST** rest against the shoulder of the bearing.

Drive the needle bearing set free of the bearing carrier. Note the direction to drive out the needle bearing. The direction is different for different model lower units.

Models 9.9hp to 30hp

Drive the bearing from the propeller end of the carrier toward the forward end.

Models 40hp and Larger

Drive the bearing from the forward end of the carrier toward the propeller end.

Model 8hp Only

21- Obtain Driver Rod P/N M-91-83169M or a suitable size socket and extension. The socket **MUST** rest against the shoulder of the bushing. Drive the bushing free of the bearing carrier with the special tool positioned at the propeller end of the carrier. Some early model 8hp units have a "short" carrier which does not have a bushing.

DRIVESHAFT REMOVAL

Before the driveshaft can be removed, the pinion gear at the lower end of the shaft must be removed. The pinion gear may be secured to the driveshaft with a simple circlip, or with a nut.

Models with Circlip

22- Pry the circlip free from the end of the driveshaft with a thin screwdriver. This clip holds the pinion gear onto the lower end of the driveshaft. The clip may not come free on the first try, but have patience and it will come free. With one hand, lift the driveshaft up and out of the lower unit housing and at the same time, with the other hand, catch the pinion gear and **SAVE** any shim material from behind the gear. The shim material is critical to obtaining the correct backlash during installation. Using the old shim material will save considerable time, especially starting with no shim material.

Models with Nut

SPECIAL WORDS

In most cases, when working with tools, a nut is rotated to remove or install it to a particular bolt, shaft, etc. In the next two steps, the reverse is required because there is no room to move a wrench inside the lower unit cavity. The nut on the lower end of the driveshaft is held steady and the shaft is rotated until the nut is free.

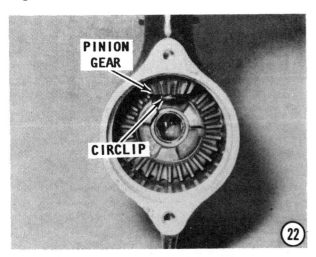

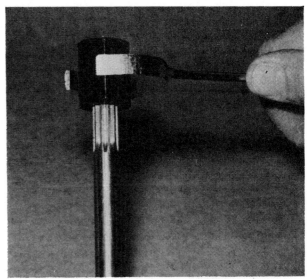

The driveshaft is rotated CLOCKWISE using the special tool and wrench while the pinion nut is being held, see illustration No. 23.

9-30 LOWER UNIT

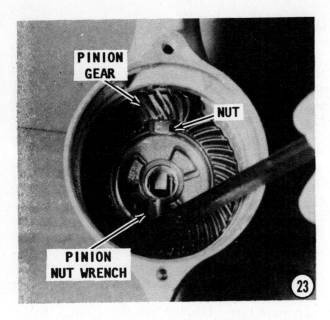

23- Obtain a wrench suitable for holding the pinion nut steady. This wrench is used to prevent the nut from rotating while the driveshaft is rotated.

Obtain Driveshaft Holding Tool:

P/N M-91-43073M for Models 9.9hp and 15hp

P/N M-91-83180M for Models 20hp, 25hp, 30hp, and 40hp

For all other models not listed above, check with the local Mariner dealer for the correct part number of special tool, for the model being serviced.

Now, hold the pinion nut steady with the special tool, and at the same time install the proper driveshaft tool, for the model being serviced, on top of the driveshaft. With both tools in place, one holding the nut and the other on the driveshaft, rotate the driveshaft **COUNTERCLOCKWISE** to "break" the nut free.

24- Remove the pinion nut, and gently pull up on the driveshaft and at the same time rotate the driveshaft. The pinion gear, followed by the shim material, washer, thrust bearing (not on all models), and a washer will come free from the lower end of the driveshaft. Pull the driveshaft up out of the lower unit housing. Keep the parts from the lower end of the driveshaft in **ORDER**, as an aid during assembling.

SAVE any shim material from behind the pinion gear. The shim material is critical to obtaining the correct backlash during installation. Using the old shim material will save considerable time, especially starting with no shim material.

If the driveshaft has a tapered roller bearing, and the bearing is unfit for further service, press the bearing free using the proper size mandrel. **TAKE CARE** not to bend or distort the driveshaft because of its length.

DRIVESHAFT BUSHING OR NEEDLE BEARING REMOVAL

If the driveshaft bushing, upper or lower, on some models, or the needle bearing on

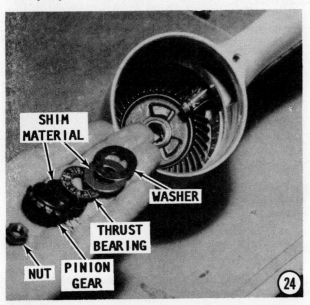

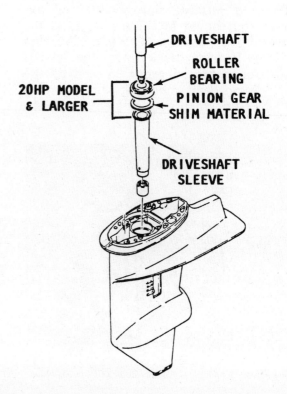

DISASSEMBLING TYPE "B"   9-31

other models, is unfit for service, perform the following steps, depending on the model being serviced.

Models 4hp thru 15hp
25- Lift out the driveshaft sleeve.

Models 20hp, and Larger
Before the driveshaft sleeve may be removed the upper driveshaft tapered roller bearing race and pinion gear shim material must be removed. The bearing race may be removed using a slide hammer and jaw attachment.

WATCH for and **SAVE** any shim material from behind the bearing. The shim material is critical to obtaining the correct backlash during installation. Using the old shim material will save considerable time, especially starting with no shim material.

After the bearing race and shim material has been removed, lift out the driveshaft sleeve.

Model 4hp and 5hp
26- One bushing is located in the lower unit just above the pinion gear and another at the upper end of the lower unit housing. Obtain a suitable size socket and extension. Insert the tools from the top of the lower unit and drive the lower bushing downward and free. Remove the upper bushing out the top of the housing, using a slide hammer and jaw puller attachment.

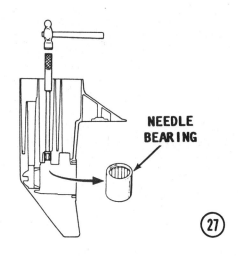

Model 8hp
This model has only one bushing in the lower unit just above the pinion gear. Obtain Driver Rod P/N M-91-83169M or a suitable size socket and extension. Insert the tool from the top of the lower unit and drive the bushing downward and free.

Model 9.9hp and Larger
27- Two small caged needle bearing sets or a single bushing are used at the lower end of the driveshaft, on the model 9.9hp and 15hp. One small needle bearing set is used at the lower end of the driveshaft on the model 20hp and larger. Obtain the following special tools:

Models 9.9hp and 15hp - Driver Rod P/N M-91-83169M

Models 20hp, 25hp, and 30hp - Driver Rod P/N M-91-84529M and Mandrel P/N M-91-83184M

Model 40hp - Driver Rod P/N M-91-84529M and Mandrel P/N M-91-83185M

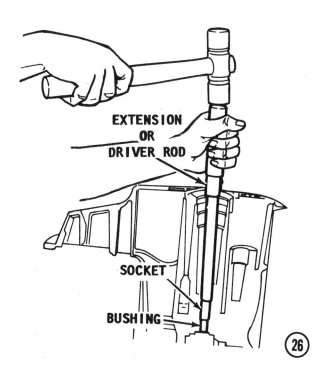

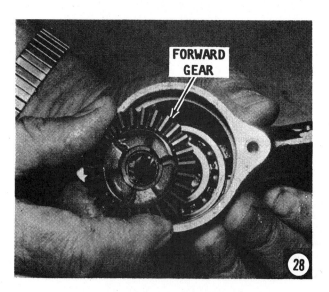

On some models, a driver rod and mandrel are used, while on other models, only a driver rod is used.

For all other models not listed above, check with the local Mariner dealer for the correct part numbers of special tools for the model being serviced, or obtain a suitable size socket and extension. The socket **MUST** rest against the outer bearing cage.

Insert the tools from the top of the lower unit and drive the needle bearing/s downward and free.

FORWARD GEAR REMOVAL

Model 3.5hp, 4hp and 5hp

28- Remove the forward gear from the lower unit.

29- If the forward ball bearing is no longer fit for service, obtain a slide hammer with a jaw attachment, and pull the ball bearing set from the lower unit.

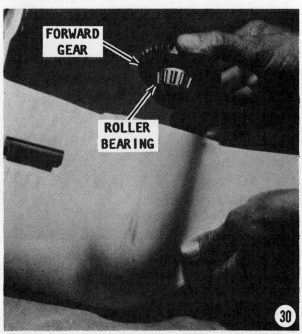

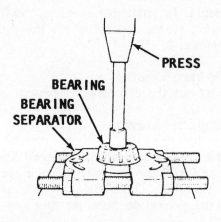

All Models Except 3.5hp, 4hp and 5hp

30- Lift out the forward gear and the tapered roller bearing.

CRITICAL WORDS

If a two piece bearing is to be replaced, the bearing and the race **MUST** be replaced as a matched set. Remove the bearing only if it is unfit for further service.

All Models Except 3.5hp, 4hp and 5hp

31- Position a bearing separator between the forward gear and the tapered roller bearing. Using a hydraulic press, separate the gear from the bearing.

32- Obtain a slide hammer and jaw attachment. Use the slide hammer and special tool to pull the bearing race from the lower unit housing. Be sure to hold the slide hammer at right angles to the driveshaft while working the slide hammer. **WATCH** for and **SAVE** any shim material found behind the forward gear. The shim material is

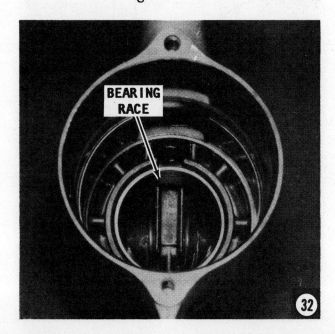

CLEAN & INSPECT TYPE "B" 9-33

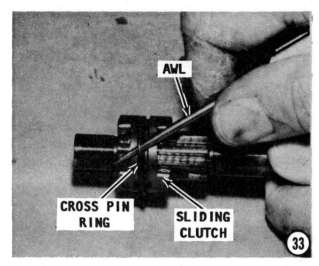

critical to obtaining the correct backlash during installation. Using the old shim material will save considerable time, especially starting with no shim material.

SPECIAL WORDS
MODELS 48HP, 55HP, AND 60HP

If the shift rod, removed in Step 9, was found to have a splined end, reach inside the lower unit and retrieve the shift cam.

CLUTCH "DOG" REMOVAL
ALL MODELS EXCEPT AIR-COOLED 3.5HP

33- Insert an awl under the end loop of the cross pin ring and pry the ring free of the clutch "dog".

34- Use a long pointed punch and press out the cross pin. Remove the plunger,

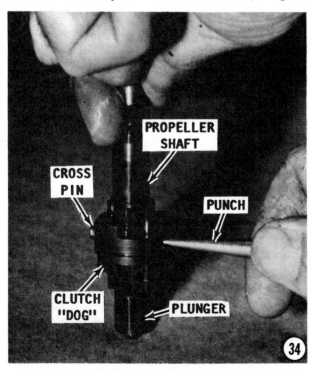

spring, and spring guide from the end of the propeller shaft. Slide the clutch "dog" from the shaft. Observe how the clutch "dog" was installed. It **MUST** be installed in the same direction.

CLEANING AND INSPECTING

Good shop practice requires installation of new O-rings and oil seals **REGARDLESS** of their appearance.

Clean all water pump parts with solvent, and then dry them with compressed air. Inspect the water pump housing and oil seal housing for cracks and distortion, possibly caused from overheating. Inspect the inner and outer plates and water pump cartridge for grooves and/or rough surfaces.

If possible, **ALWAYS** install a new water pump impeller while the lower unit is disassembled. A new impeller will ensure extended satisfactory service and give "peace of mind" to the owner. If the old impeller must be returned to service, **NEVER** install it in reverse to the original direction of rotation. Installation in reverse will cause premature impeller failure.

If installation of a new impeller is not possible, check the seal surfaces. All must be in good condition to ensure proper pump operation. Check the upper, lower, and ends of the impeller vanes for grooves, cracking, and wear. Check to be sure the indexing notch of the impeller hub is intact and will not allow the impeller to slip.

Clean around the Woodruff key or impeller pin. Clean all bearings with solvent, dry them with compressed air, and inspect them carefully. Be sure there is no water in the air line. Direct the air stream through the bearing. **NEVER** spin a bearing with compressed air. Such action is highly dangerous and may cause the bearing to score from lack of lubrication. After the bearings are

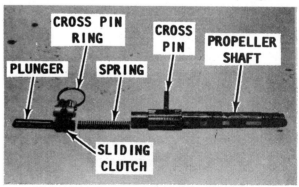

Arrangement of shift parts on the propeller shaft.

clean and dry, lubricate them with Formula 50 oil, or equivalent. Do not lubricate tapered bearing cups until after they have been inspected.

Inspect all ball bearings for roughness, scratches and bearing race side wear. Hold the outer race, and work the inner bearing race in-and-out, to check for side wear.

Determine the condition of tapered bearing rollers and inner bearing race, by inspecting the bearing cup for pitting, scoring, grooves, uneven wear, imbedded particles, and discoloration caused from overheating. **ALWAYS** replace tapered roller bearings as a set.

Clean the forward gear with solvent, and then dry it with compressed air. Inspect the gear teeth for wear. Under normal conditions the gear will show signs of wear but it will be smooth and even.

Clean the bearing carrier or cap with solvent, and then dry it with compressed air. **NEVER** spin bearings with compressed air. Such action is highly dangerous and may cause the bearing to score from lack of lubrication. Check the gear teeth of the reverse gear for wear. The wear should be smooth and even.

Check the clutch "dogs" to be sure they are not rounded-off, or chipped. Such damage is usually the result of poor operator habits and is caused by shifting too slowly or shifting while the engine is operating at high rpm. Such damage might also be caused by improper shift rod adjustments.

Rotate the reverse gear and check for catches and roughness. Check the bearing for side wear of the bearing races.

Inspect the roller bearing surface of the propeller shaft. Check the shaft surface for pitting, scoring, grooving, embedded particles, uneven wear and discoloration caused from overheating.

Clean the driveshaft with solvent, and then dry it with compressed air. **NEVER** spin bearings with compressed air. Such action is dangerous and could damage the bearing. Inspect the bearing for roughness, scratches, or side wear. If the bearing shows signs of such damage, it should be replaced. If the bearing is satisfactory for further service coat it with oil.

Inspect the driveshaft splines for excessive wear. Check the oil seal surfaces above and below the water pump drive pin or Woodruff key area for grooves. Replace the shaft if grooves are discovered.

Inspect the driveshaft bearing surface above the pinion gear splines for pitting, grooves, scoring, uneven wear, embedded metal particles and discoloration caused by overheating.

Inspect the propeller shaft oil seal surface to be sure it is not pitted, grooved, or scratched. Inspect the roller bearing contact surface on the propeller shaft for pitting, grooves, scoring, uneven wear, embedded metal particles, and discoloration caused from overheating.

Inspect the propeller shaft splines for wear and corrosion damage. Check the propeller shaft for straightness.

Inspect the following parts for wear, corrosion, or other signs of damage:

Shift shaft boot
Shift shaft retainer
Shift cam

On Models 15hp and smaller, check the pinion retaining circlip to be sure it is not bent or stretched. If the clip is deformed, it must be replaced.

Clean all parts with solvent, and then dry them with compressed air.
Inspect:
All bearing bores for loose fitting bearings.
Gear housing for impact damage.
Cover nut threads on Models 20hp and above for cross-threading and corrosion damage.
Check the pinion nut corners for wear or damage. This nut is a special locknut. Therefore, do **NOT** attempt to replace it with a standard nut. Obtain the correct nut from an authorized Mariner dealer.

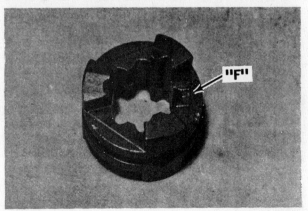

During installation onto the propeller shaft, the "F" embossed on the clutch dog MUST face the forward gear.

CLEAN & INSPECT TYPE "B" 9-35

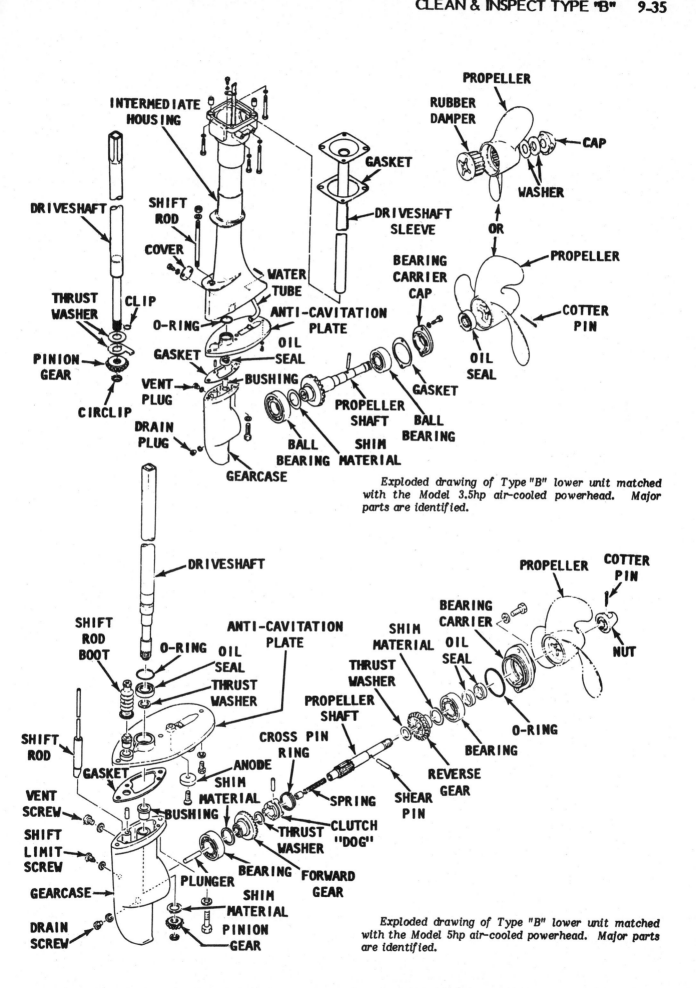

Exploded drawing of Type "B" lower unit matched with the Model 3.5hp air-cooled powerhead. Major parts are identified.

Exploded drawing of Type "B" lower unit matched with the Model 5hp air-cooled powerhead. Major parts are identified.

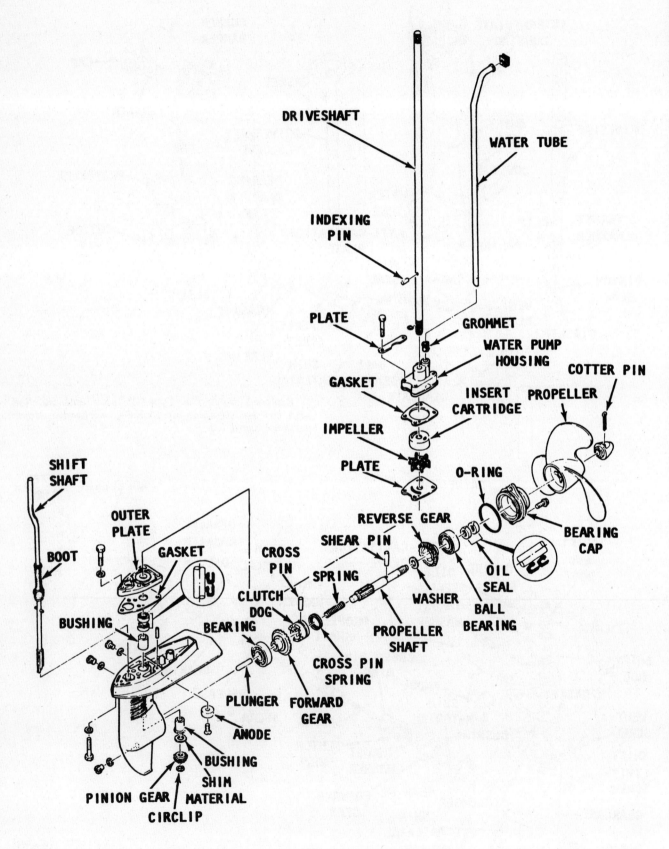

Exploded drawing of the Type "B" lower unit matched with the Model 4hp and 5hp powerheads. Major parts are identified.

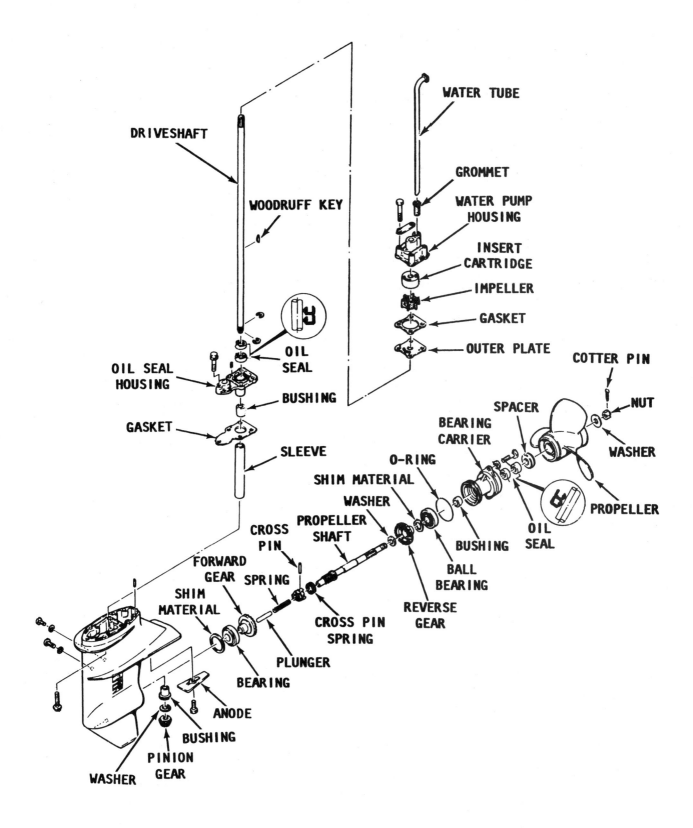

Exploded drawing of the Type "B" lower unit matched with the Model 8C powerhead. Major parts are identified.

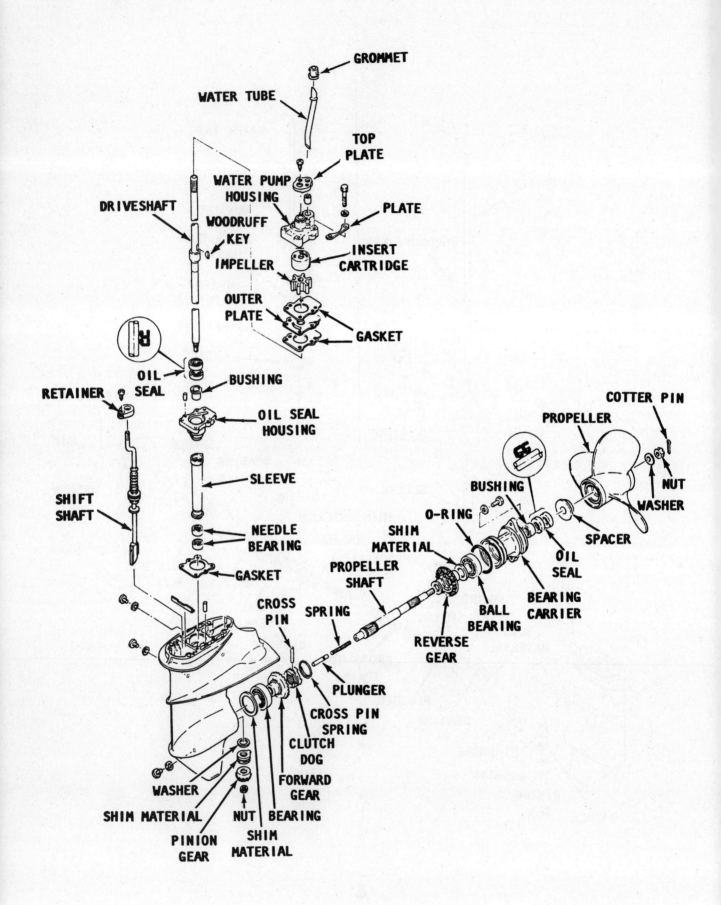

Exploded drawing of the Type "B" lower unit matched with the Model 9.9hp, other 8hp besides those indicated on the previous page, and the 15hp powerheads. Major parts are identified.

CLEAN & INSPECT TYPE "B" 9-39

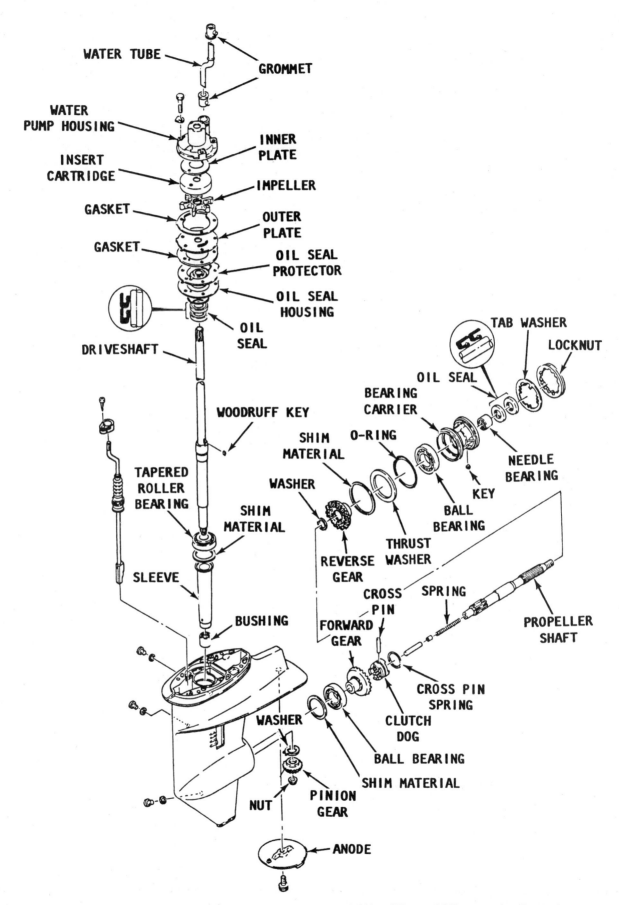

Exploded drawing of the Type "B" lower unit matched with the Model 20hp, 25hp, and 30hp powerheads. Major parts are identified.

9-40 LOWER UNIT

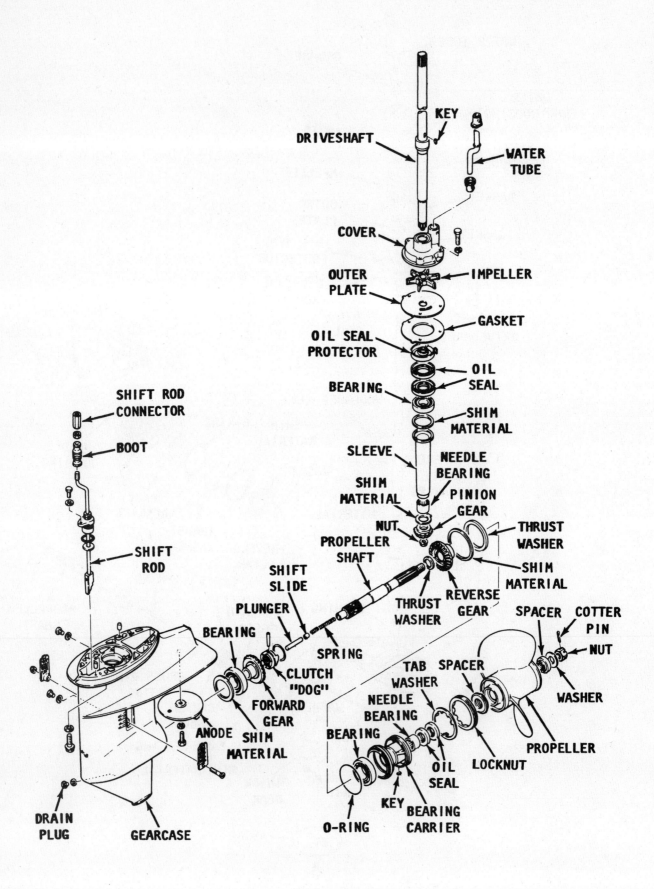

Exploded drawing of the Type "B" lower unit matched with the Model 28hp powerhead. Major parts are identified.

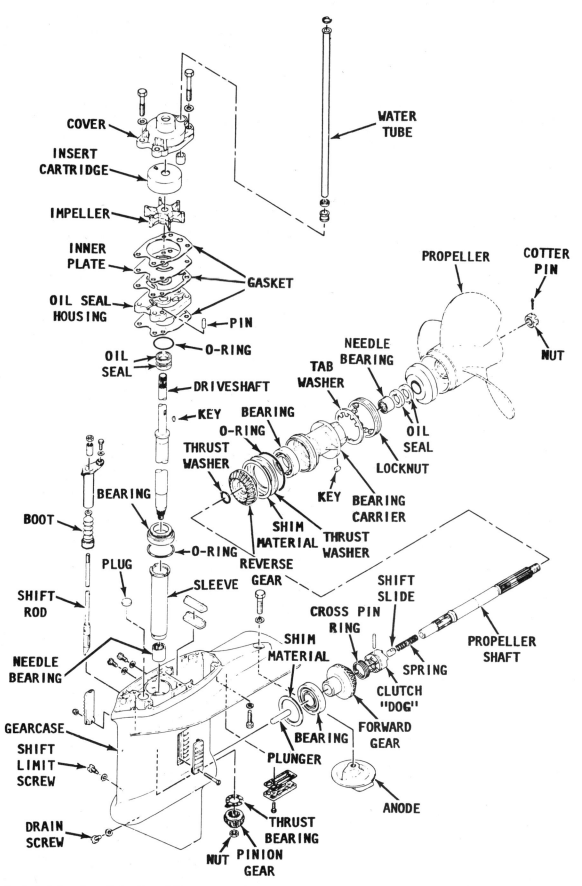

Exploded drawing of the Type "B" lower unit matched with the Model 40hp powerhead. Major parts are identified.

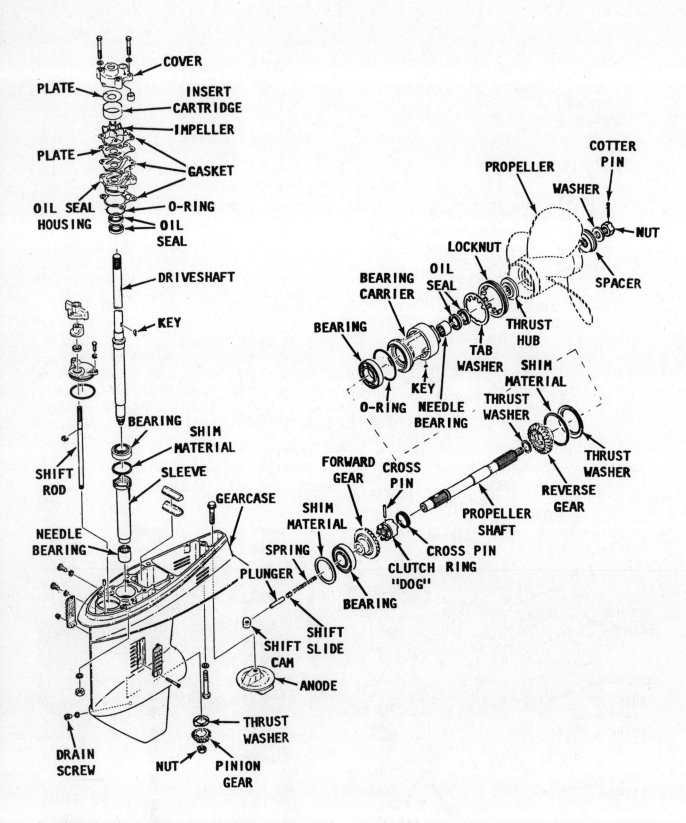

Exploded drawing of the Type "B" lower unit matched with the Model 48hp, 55hp, and 60hp powerheads. Major parts are identified.

ASSEMBLING TYPE "B" 9-43

Models 48hp, 55hp, and 60hp
Clean the "L" shaped oil passageway at the lower end of the driveshaft.

Model 9.9hp and 15hp
Observe the cut of the gears. Notice if the gears are a straight bevel cut or a spiral bevel cut. This difference does affect service procedures, but when filling the lower unit after the work is complete, the gearcase with the straight cut bevel gears needs 1.5 U.S. oz more to fill than the spiral cut bevel gear gearcase!

ASSEMBLING

FIRST, THESE WORDS
Procedural steps are given to assemble and install virtually all items in the lower unit. However, if certain items, i.e. bearings, bushings, seals, etc. were found fit for further service and were not removed, simply skip the assembly steps involved. Proceed with the required tasks to assemble and install the necessary components.

CLUTCH "DOG" INSTALLATION
ALL MODELS EXCEPT AIR-COOLED 3.5HP

1- Slide the spring down into the propeller shaft, followed by the spring guide. Insert a narrow screwdriver into the slot in the shaft. Compress the spring and guide until approximately 1/2" (12mm) is obtained between the top of the slot and the screwdriver.

Hold the spring compressed, and at the same time, slide the clutch "dog" over the splines of the propeller shaft with the hole in the "dog" aligned with the slot in the shaft. The letter "F" embossed on the "dog" **MUST** face toward the forward gear.

Insert the cross pin into the clutch "dog" and through the space held open by the screwdriver. Center the pin and then remove the screwdriver allowing the spring to pop back into place.

Fit the cross pin ring into the groove around the clutch "dog" to retain the cross-pin in place.

Insert the flat end of the plunger into the propeller shaft, with the rounded end protruding to permit the plunger to slide along the cam of the shift rod.

FORWARD GEAR
BALL BEARING INSTALLATION

The next three steps apply **ONLY** if the forward gear ball bearing was removed. If the bearing was not disturbed, proceed directly to Step 5.

Model 3.5hp, 4hp, and 5hp
2- For air-cooled models 3.5 and 5hp: Obtain a suitable size socket which contacts the outer bearing race **ONLY**, and an extension. For water-cooled Models 4hp and 5hp: Obtain Driver Rod P/N M-91-84529M and Mandrel P/N M-91-84533M. Place the propeller shaft forward bearing squarely into the housing with the side embossed with the bearing size facing **OUTWARD**. Drive the bearing into the housing until it is fully seated.

SPECIAL WORDS
MODELS 48HP, 55HP, AND 60HP
If the shift rod, removed in Step 9 of disassembly, has a splined end, and later if a

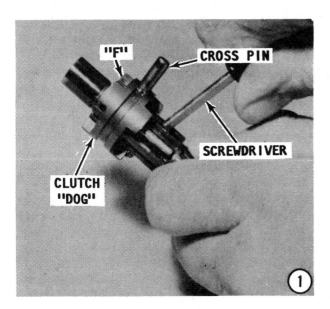

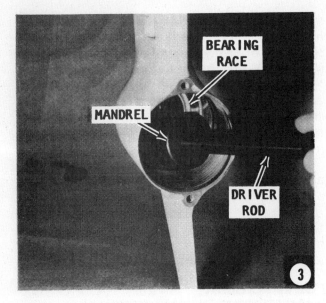

shift cam was removed from the gearcase in Step 32, **NOW** is the time to install these two parts, before the gearcase is built up.

Hold the shift cam inside the gearcase, with any embossed numbers or letters facing **UPWARD**, directly under the shift rod bore. Slide the shift rod down into the bore to index with the cam.

If the shift rod, removed in Step 9 of disassembly, has a tapered cam end, do not install the rod at this time. This type rod will be installed in a later step.

All Models Except 3.5hp, 4hp, and 5hp

3- Insert the shim material saved during disassembling, Step 31, into the lower unit bearing race cavity. The shim material should give the same amount of backlash between the pinion gear and the forward gear as before disassembling. Insert the bearing race squarely into the lower unit housing. Obtain the following special tools:

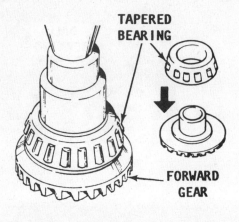

Model 8B, W8, Marathon 8hp, 9.9hp and 15hp - Driver Rod P/N M-91-84529M, Mandrel P/N M-91-84532M and Pilot P/N M-91-84535M
Model 8C - Driver Rod P/N M-91-84529M and Mandrel P/N M-91-84719M
Models 20hp, 25hp, and 30hp - Driver Rod P/N M-91-84529M and Mandrel P/N M-83182M
Model 40hp - Driver Rod P/N M-91-84529M and Mandrel P/N M-91-84584M

For all other models not listed above, check with the local Mariner dealer for the correct part numbers of special tools for the model being serviced, or obtain a suitable size socket and extension. The socket **MUST** rest against the outer bearing cage.

4- Position the forward gear tapered bearing over the forward gear. Use a suitable mandrel and press the bearing flush against the shoulder of the gear. **ALWAYS** press on the **INNER** race, never on the cage or the rollers.

PLAN AHEAD
ALL MODELS

Obtain a suitable substance which can be used to indicate a wear pattern on the forward and pinion gears as they mesh. Machine dye may be used and if this material is not available, Desenex Foot Powder (obtainable at the local Drug Store/Pharmacy), or equivalent may be substituted. Desenex is a white powder available in an aerosol container. Before assembling either gear, apply a light film of the dye, Desenex,

or equivalent, to the driven side of the gear. After the gears are assembled and rotated several times, they will be disassembled and the wear pattern can be examined. The substance will be removed from the gears prior to final assembly.

5- Position the forward gear assembly into the forward bearing race, or bearing.

DRIVESHAFT BUSHING AND BEARING INSTALLATION

Perform the next two or three steps **ONLY** if the bushings and/or bearings were removed during diassembling. If these items were not disturbed, proceed directly to Step 9.

Air-cooled Models 3.5hp and 5hp

These models have a single driveshaft bushing at the top of the lower unit housing.

Obtain a suitable size socket and extension and drive the upper bushing into place from the top until it seats.

Water-cooled Models 4hp and 5hp
Upper Bushing

6- These models have two bushings for the driveshaft. One is located at the top of the lower unit housing and the other just above the pinion gear at the lower end.

Obtain a suitable size socket and extension and drive the upper bushing into place from the top until it seats.

Model 4hp and 5hp Lower Bushing
Model 8hp, 9.9hp, and 15hp
Single Lower Bushing

For Models 4hp and 5hp: Check with the local Mariner dealer for the correct part numbers of the Bushing Installer, driver and plate kit, required for this installation procedure. For Models 8hp, 9.9hp, and 15hp: Obtain Bushing Puller Assembly P/N M-91-83170M. Position the plate on top of the lower unit. Lower the long threaded bushing installer into the lower unit from the top until the threads protrude into the lower unit cavity, as shown.

Now, thread the nut provided with the tool down onto the installer tool until the nut is snug against the plate. Slide the first bushing over the threads at the lower end of the tool, and then install the bushing retainer onto the threads. Tighten the bushing installer from the upper end of the tool, using the proper size wrench. Continue rotating the bushing installer until the bushing is drawn up into the lower unit and seats properly.

Model 9.9hp and 15hp

7- These models use two needle bearings at the lower end of the driveshaft just above the pinion gear.

Position the first needle bearing with the stamp mark on the bearing facing **UPWARD** squarely in the driveshaft bore. Obtain Driver Rod P/N M-91-83169M or a suitable size socket which contacts the outer bearing race **ONLY**, and an extension. Be sure the socket does **NOT** hang over the edge of the bearing to be installed, or it will wedge

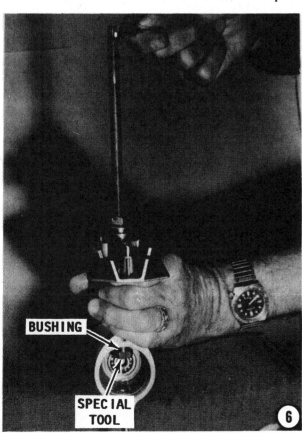

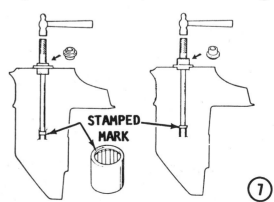

9-46 LOWER UNIT

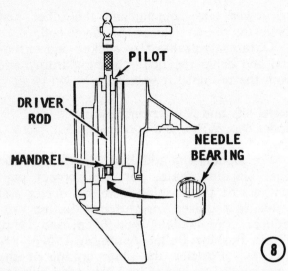

itself in the bore before the bearing is seated. Tap the bearing into place. The "tone" of the hammer striking the tool will change noteceably as soon as the bearing seats.

Install the second bearing in a similar manner, with the stamp mark on the bearing facing **UPWARD**.

Model 20hp and Larger

8- These models have a single needle bearing at the lower end of the driveshaft just above the pinion gear.

Position the needle bearing with the stamp mark on the bearing facing **UPWARD** squarely in the driveshaft bore. Obtain the following special tools:

Model 20hp, 25hp, and 30hp -- Driver Rod P/N M-91-84529M and Mandrel P/N M-91-83184M

Model 40hp -- Driver Rod P/N M-91-84529M, Mandrel P/N M-91-83185M, and Pilot P/N M-91-84573M

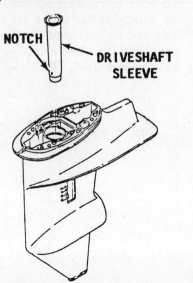

For all other models not listed above, check with the local Mariner dealer for the correct part numbers of special tools for the model being serviced. An alternate method is to obtain a suitable size socket and long extension. The socket **MUST** rest against the outer bearing cage. Check to be sure the socket does **NOT** hang over the edge of the bearing to be installed, or it will wedge itself in the bore before the bearing is seated. Tap the bearing into place. The "tone" of the hammer striking the tool will change noticeably as soon as the bearing seats.

All Models Except 3.5hp, 4hp and 5hp

9- Slide the driveshaft sleeve into the upper end of the lower unit with the notch in the lower flange facing **AFT**.

OIL SEAL HOUSING BUSHING

Models 8hp, 9.9hp, and 15hp

10- Obtain the following special tools:
Model 8C - Driver Rod P/N M-91-84529M, Mandrel P/N M-91-83178M
Model 9.9hp and 15hp - Driver Rod P/N M-91-83169M

For all other models not listed above, check with the local Mariner dealer for the correct part numbers of special tools for the model being serviced. An alternate method

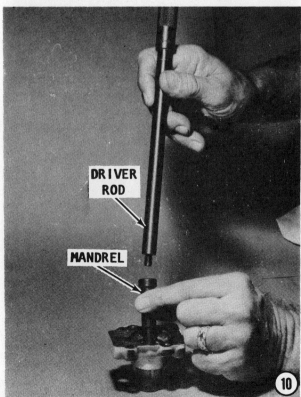

is to obtain a suitable size socket and extension.

Align the notch in the bushing shoulder with the hole in the oil seal housing. Drive the bushing into place until the bushing shoulder contacts the housing.

Model 20hp, 25hp, 28hp, and 30hp

11- Press a new tapered roller bearing onto the driveshaft with the pinion end facing **TOWARD** the tapered end of the bearing. Because of the driveshaft length, **TAKE CARE** not to bend or distort the driveshaft while the bearing is being pressed into place.

Install the shim material saved during disassembly, Step 24. Obtain the following special tool: Driver Rod P/N M-91-84529M and Mandrel P/N M-91-84719M or a suitable size socket which will contact the outer bearing race **ONLY**, and an extension. Check to be sure the socket does **NOT** hang over the edge of the race to be installed, or it will wedge itself in the bore before the race is seated. Install the tapered roller bearing race over the shim material. Drive the race in squarely. The "tone" of the hammer striking the tool will change noticeably as soon as the bearing seats.

DRIVESHAFT OIL SEALS

SPECIAL WORDS
FROM MARINER ENGINEERS

Two seals are installed into the top of the lower unit housing on the Model 4hp and 5hp units. On all other units, the oil seals are installed into the oil seal housing. These seals on all models are installed with the lip of both seals facing **UPWARD**. Mariner engineers have concluded it is better to have both seals face the same direction rather than have them "back to back" as directed by other outboard manufacturers.

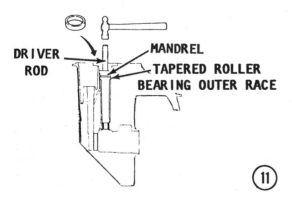

Model 3.5hp, 4hp and 5hp

SPECIAL WORDS FOR
AIR-COOLED MODELS 3.5HP AND 5HP

These two models have a single oil seal and an O-ring installed in the anti-cavitation plate. Perform the following step, installing the oul seal with the lip facing **DOWNWARD** toward the lower unit, until the seal is flush with the lower surface of the anti-cavitation plate.

Air-Cooled Models 3.5hp and 5hp

12- Obtain a suitable size socket, just a "whisker" smaller then the outer diameter of the oil seal to be installed, and an extension.

Water-Cooled Models 4hp and 5hp

Obtain Driver Rod P/N M-91-84529M and Mandrel P/N M-91-84530M.

Lower the first seal **SQUARELY** into the seal recess on top of the lower unit, with the lip of the seal facing **UPWARD**, toward the water pump. Tap the end of extension with a hammer until the seal is fully seated. After installation, pack the seal with multi-purpose water resistant lubricant.

Install the second seal in the same manner, with the seal lip facing **UPWARD**, toward the water pump. Pack the second seal with lubricant.

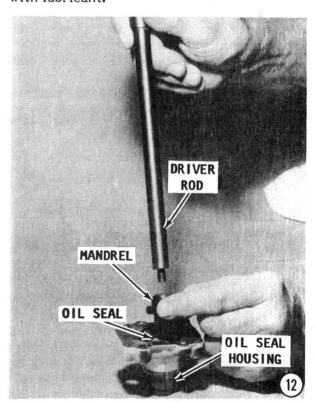

SPECIAL WORDS FOR ALL
MODELS EXCEPT 3.5HP, 4HP, AND 5HP

The manufacturer recommends the driveshaft oil seals, in the oil seal housing, **NOT** be installed at this time, but later, after the forward and reverse gear backlash procedures have been completed. Therefore, the oil seal housing must be temporarily installed over the driveshaft and removed later for seal installation.

Models 8hp, 9.9hp, and 15hp

Install the gasket and oil seal housing into the top of the lower unit over the driveshaft sleeve, to index with the dowel pin. Install two more dowel pins on the top surface of the oil seal housing.

PINION GEAR
AND DRIVESHAFT INSTALLATION

All Units

13- Lower the driveshaft down through the sleeve in the upper end of the lower unit. Coat the pinion gear with a fine spray of Desenex, as instructed in the "Plan Ahead" paragraph prior to assembling Step 5. Handle the gear carefully to prevent disturbing the powder.

Hold the driveshaft in place with one hand, and at the same time assemble the parts onto the lower end of the driveshaft with the other hand in the same order as they were removed. Hopefully they were kept in order on the work surface.

Correct order is first, the washer followed by the thrust bearing (on some units only). Next, install the same amount of shim material removed during disassembling. After the shim material is in place, slide on the washer, and then the pinion gear. It may be necessary to rotate the driveshaft slightly to allow the splines on the driveshaft to index with the internal splines of the pinion gear. The teeth of the pinion gear will index with the teeth of the forward gear.

Models with Circlip

Secure the pinion gear in place with the circlip. Slipping the circlip into the groove on the end of the driveshaft is not an easy task, but have patience and success will be the reward.

Models with Nut

Once the pinion gear is in place, then start the threads of the pinion gear nut. Tighten the nut as much as possible by rotating the driveshaft with one hand and holding the nut with the other hand. The next step tightens the nut.

14- Tighten the pinion gear nut as follows: Obtain the correct size driveshaft holding tool and a suitable socket or wrench to hold the pinion nut steady inside the gearcase:

Models 9.9hp and 15hp - P/N M-91-43073M

Models 20hp, 25hp, 30hp, and 40hp - P/N M-91-83180M

For all other models not listed above, check with the local Mariner dealer for the correct part numbers, of the special tool, for the model being serviced. It is **NOT**

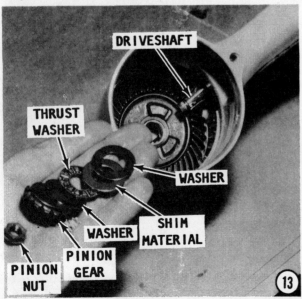

ASSEMBLING TYPE "B"   9-49

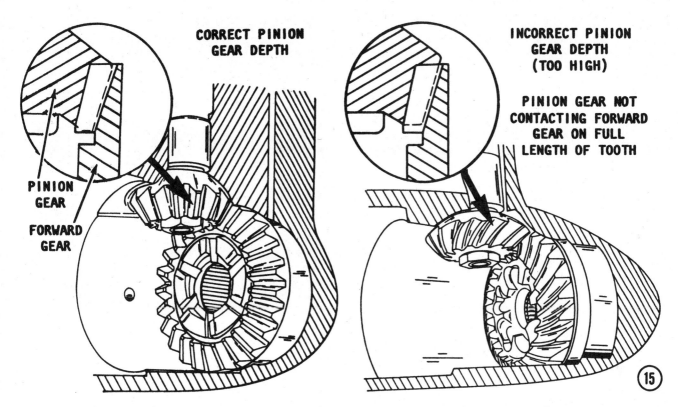

possible to tighten the pinion nut to the correct torque value **WITHOUT** the use of this special tool.

Now, place the proper tool over the splines on the upper end of the driveshaft. Position a socket and torque wrench on the tool. Reach into the lower unit cavity and hold the pinion gear nut steady with the wrench or socket. Rotate the driveshaft **CLOCKWISE** until the torque wrench indicates the following values for the models listed:

Model 15hp and smaller -- 19 ft lbs (26Nm)

Model 20hp, 25hp, 28hp, and 30hp -- 25 ft lbs (35Nm)

Model 40hp and larger -- 54 ft lbs (73Nm)

PINION GEAR DEPTH
ALL MODELS EXCEPT 3.5HP
SPECIAL WORDS

The correct amount of pinion gear depth (pinion gear engagement with the forward gear), is critical for proper operation of the lower unit and long life of internal parts.

All Models
Except Model 3.5hp

15- Grasp the driveshaft and **PULL UP-WARD**. At the same time, check the pinion gear tooth engagement with the forward gear teeth to be sure contact is made the full length of the tooth. This can be accomplished by using a flashlight and looking into the lower unit opening. If the pinion gear depth is **NOT** correct, the amount of shim material behind the pinion gear must be adjusted. The addition or removal of shim material behind the pinion gear will not affect the forward gear backlash procedures, which begin on Page 9-54.

BEARING CARRIER CAP

SPECIAL WORDS
FROM MARINER ENGINEERS

Two seals are installed into the bearing carrier cap. These seals on 4hp and 5hp models are installed with the lip of both seals facing **DOWNWARD**, during installation. Mariner engineers have concluded it is better to have both seals face the same direction rather than have them "back to back" as directed by other outboard manufacturers. In this position, with both lips facing outward toward the water, after cap installation, the engineers feel the seals will be more effective in keeping water out of the lower unit. Any lubricant lost will be negligible. If the unit being serviced has only one oil seal in the bearing carrier cap, install the seal with the seal lips facing **TOWARD** the propeller, following installation.

Model 3.5hp, 4hp, and 5hp

16- For air-cooled Models 3.5 and 5hp: Obtain a suitable size socket and extension. For water-cooled Models 4hp and 5hp: Obtain Driver Rod P/N M-91-84529M and Mandrel P/N M-91-83176M. Place the bearing carrier cap on the work surface with the **PAINTED** side facing **UPWARD**. Install the first oil seal, using the driver and handle, with the lip facing **DOWNWARD**. After the seal is installed, coat the seal lip with multi-purpose water resistant lubricant from the propeller side of the cap.

Coat the lip of the second seal with grease. Install the second seal in the same manner, with the lip of the seal facing **DOWNWARD**.

REVERSE GEAR BALL BEARING

GOOD WORDS

Even though the small air-cooled Model 3.5hp has no reverse gear, a ball bearing supporting the propeller shaft is located in the bearing carrier cap. Therefore, the following step also applies to this model.

Model 3.5hp, 4hp, and 5hp

17- For air-cooled models 3.5hp and 5hp: Obtain a suitable size socket or other driv-

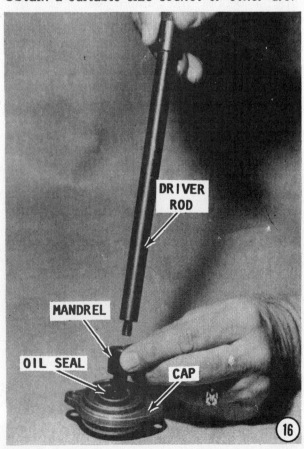

er, which will rest against the outer bearing race, and an extension. For water-cooled Models 4hp and 5hp: Obtain Driver Rod P/N M-91-84529M and Mandrel P/N M-91-84532M. Install the reverse gear ball bearing squarely into the carrier cap with the embossed letters facing **OUTWARD**. Stretch a new outer O-ring around the cap or install a new gasket. Coat the outside perimeter of the cap with multi-purpose water resistant lubricant.

BEARING CARRIER BUSHING

Model 8hp

18- Obtain Driver Rod P/N M-91-84529M and Mandrel P/N M-91-83174M or a suitable size socket and extension. Install the bushing into the bearing carrier from the oil seal side until the bushing seats within the carrier.

BEARING CARRIER NEEDLE BEARING

19- Check the paragraph headings to ensure the proper procedures are being performed for the model being serviced.

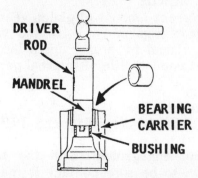

ASSEMBLING TYPE "B"

Model 9.9hp and 15hp

Obtain Driver Rod P/N M-91-84529M and Mandrel P/N M-91-83184M. Lower the needle bearing into the bearing carrier from the **PROPELLER** end of the carrier.

Model 20hp and Larger

Obtain the following special tools:
Model 20hp, 25hp, and 30hp -- Driver Rod P/N M-91-84529M and Mandrel P/N M-91-84570M
Model 40hp -- Driver Rod P/N M-91-84529M and Mandrel P/N M-91-84574M

For all other models not listed above, check with the local Mariner dealer for the correct part numbers of special tools for the model being serviced, or obtain a suitable size socket and extension. The socket **MUST** rest against the outer bearing cage. Check to be sure the socket does **NOT** hang over the edge of the cage to be installed, or it will wedge itself in the bore before the bearing is properly seated.

Models 20hp, 25hp, 28hp, and 30hp

Lower the needle bearing into the bearing carrier from the **LOWER UNIT** end of the carrier.

Models 40hp and Larger

Lower the needle bearing into the bearing carrier from the **PROPELLER** end of the carrier.

In all cases, install the bearing with the stamped mark on the bearing facing **UPWARD** toward the driver. Drive the bearing in squarely. The "tone" of the hammer striking the tool will change noticeably when the bearing seats.

BEARING CARRIER OIL SEAL

20- Pack the lips of both seals with multi-purpose water resistant lubricant.

SPECIAL WORDS
FROM MARINER ENGINEERS

Two seals are installed into bearing carrier on all models except the 4hp and 5hp units. After installation of the bearing carrier to the lower unit, the lips of **BOTH** seals face toward the propeller. Mariner engineers have concluded it is better to have both seals face the same direction rather than have them "back to back" as directed by other outboard manufacturers. In this position, with both lips facing outward toward the water after bearing carrier installation, the engineers feel the seals will be more effective in keeping water out of the lower unit. Any lubricant lost will be negligible.

The direction of installation into the bearing carrier varies depending on the model being serviced.

On some models the seals are installed from the propeller end of the carrier. On other models the seals are installed from the lower unit end of the carrier. In all cases, after the seals are properly installed, the lips will face toward the propeller.

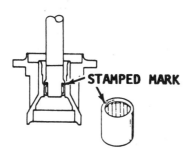

9.9hp, 15hp, and 40hp

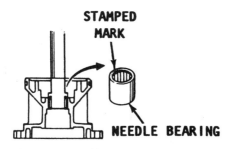

20hp, 25hp, 28hp, and 30hp

Model 8hp, 9.9hp, and 15hp

Both seals are driven into the bearing carrier from the propeller end of the carrier with the lips of the seals facing **UPWARD** toward the driver.

Model 20hp, 25hp, 28hp, and 30hp

Both seals are driven into the bearing carrier from the lower unit end of the carrier with the lips of the seals facing **DOWNWARD** away from the driver.

Model 40hp and Larger

Both seals are driven into the bearing carrier from the propeller end of the carrier with the lips of the seals facing **TOWARD** the driver.

Obtain the following special tools:
Model 8hp -- Driver Rod P/N M-91-84529M and Mandel P/N M-91-84533M
Models 9.9hp and 15hp -- Driver Rod P/N M-91-84529M and Mandrel P/N M-91-83175M
Models 20hp, 25hp, and 30hp -- Drive Rod P/N M-91-84529M and Mandrel P/N M-91-84718M
Model 40hp - Mandrel P/N M-91-84576M and appropriate Driver Rod

For all other models not listed above, check with the local Mariner dealer for the correct part numbers of special tools for the model being serviced, or obtain a suitable size socket and extension.

Install both seals in the direction indicated.

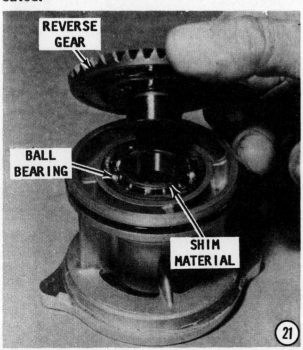

REVERSE GEAR BALL BEARING

If the reverse gear ball bearing is pressed into the bearing carrier, perform Step 21.

If the reverse gear ball bearing is pressed onto the reverse gear, perform Step 22.

If the reverse gear is a press fit **DO NOT** forget to install any shim material removed from behind the reverse gear during disassembling. This shim material and, in some cases, a thrust washer, **MUST** be installed between the reverse gear and the ball bearing assembly in order to ensure proper mesh between the reverse gear and the pinion gear.

21- Place the ball bearing squarely into the bearing carrier with the marks embossed on the bearing facing **UPWARD** toward the driver.

Obtain a suitable size mandrel which will rest against the outer bearing cage. Press the bearing into position using an arbor press.

CRITICAL WORDS

Take care to ensure the mandrel is pressing on the inner race and not on the outer race or the ball bearings. Such action would destroy the bearing.

Install the same amount of shim material saved during Step 17, disassembling. Install the reverse gear.

22- Position the reverse gear on a press with the gear teeth facing **DOWN**. Place the same amount of shim material saved from Step 19, disassembling, on the back of the gear. If servicing a Model 20hp or larger, install the thrust washer on top of the shim material.

Position the ball bearing assembly on top of the shim material, or the thrust washer,

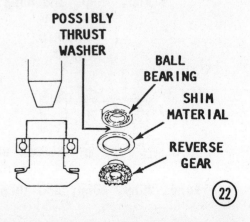

ASSEMBLING TYPE "B" 9-53

as the case may be, with the embossed marks on the bearing facing **UPWARD** toward the press shaft.

Now, use a suitable mandrel and press against the inner bearing race.

CRITICAL WORDS
Take care to ensure the mandrel is pressing on the inner race and not on the outer race or the ball bearings. Such action would destroy the bearing.

Continue to press the bearing into place until the bearing, thrust washer, shim material, and back of the reverse gear are all seated against each other.

Leave the assembled unit in place on the press and install the bearing carrier over the ball bearing. This is accomplished by using a suitable mandrel which will contact the inner hub of the bearing carrier and does not apply pressure on the ribs.

Continue to press until the outer race of the bearing seats against the bearing carrier surface.

All Models
23- Install a new O-ring into the outer groove of the bearing carrier. Apply a coating of Loctite Type "A" to the O-ring and to its mating surface.

Coat the teeth of the reverse gear with a fine spray of Desenex, as instructed in the "Plan Ahead" paragraph prior to assembling, Step 5. Handle the gear carefully to prevent disturbing the powder.

Some early 8hp (those with positive shift action) and early 15hp models have an additional thrust washer between the clutch "dog" and the forward gear.

Place the washer over the reverse gear. Slide the threaded end of the assembled propeller shaft into the bearing carrier.

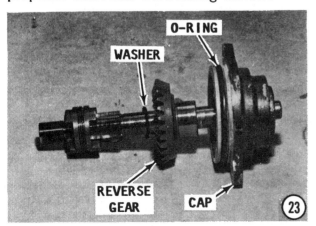

All Models
24- Install the propeller shaft and bearing carrier into the gearcase.

Models 15hp and smaller
Install the bolts and lockwashers securing the bearing carrier to the gearcase. Tighten the bolts to a torque value of 5.8 ft lbs (8Nm).

VERY SPECIAL WORDS
Assembling of parts at this time is **NOT** to be considered **FINAL**. The three gears are coated with the Desenex powder, or equivalent, to determine a gear pattern. Therefore, the assemblies will be separated to check the pattern. During final installation the two mounting bolts will be coated with Loctite, or equivalent.

If the assembler has omitted the application of the Desenex powder and does not have plans to check the gear pattern, then this step may be considered as the final assembly of the bearing carrier. Loctite, or equivalent, should be applied to the threads of the bearing carrier attaching bolts.

Model 20hp and Larger
Align the keyway in the lower unit housing with the keyway in the bearing carrier.

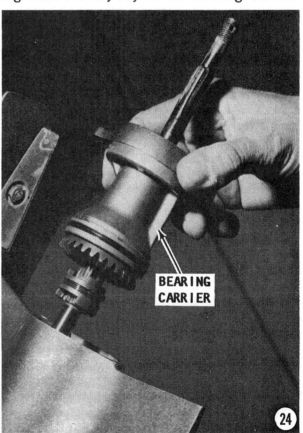

Insert the key into both grooves, and then push the bearing carrier into place in the lower unit housing. Install the ring nut with the embossed marks facing **OUTWARD**, away from the bearing carrier.

Obtain the correct size Ring Nut tool:

Models 20hp, 25hp, and 30hp -- P/N C-91-78729

Model 40hp -- P/N C-91-74241 or M-91-84546M

For all other models not listed above, check with the local Mariner dealer for the correct part number of the special tool for the model being serviced. It is **NOT** possible to tighten the ring nut to the correct torque value **WITHOUT** the use of this special tool.

Tighten the locknut, using the tool and a torque wrench, to a torque value of 65 ft lbs (90Nm) for Models 20hp, 25hp, 28hp, 30hp, 48hp, 55hp, and 60hp or to a torque value of 95 ft lbs (130Nm), for the Model 40hp. Tighten the locknut in the direction indicated by the embossed **ON** mark. Remove the special tool.

VERY SPECIAL WORDS

Assembling of parts at this time is **NOT** to be considered as final. The three gears are coated with the Desenex powder, or equivalent, to determine a gear pattern. Therefore, the assemblies will be separated to check the pattern. During final installation, one or more of the lockwasher tabs will be bent down over the locknut.

If the assembler has omitted the application of Desenex powder and does not plan to check the gear pattern, then this step may be considered as the final assembly of the bearing carrier. If such is the case, bend one or more of the lockwasher tabs down over the locknut to secure it in place.

All Models

The propeller will be installed after the gear backlash measurements have been made; the water pump installed; and the lower unit attached to the intermediate housing.

Backlash Education

Backlash is the acceptable clearance between two meshing gears to compensate for: possible errors in machining; deformation due to load; expansion due to heat generated in the lower unit; and center-to-center distance tolerances. A "no backlash" condition is unacceptable. Such a condition would indicate the gears are locked together or are too tight against each other resulting in phenomenal wear and the generation of excessive heat from the friction.

Excessive backlash, which cannot be corrected with shim material adjustment, indicates worn gears. Therefore, worn gears must be replaced. Excessive backlash is usually accompanied by a loud "whine" when the lower unit is operating in **NEUTRAL** gear.

As a general rule, if the lower unit was merely disassembled, cleaned and then assembled with only a new water pump impeller, new gaskets, seals and O-rings, there is no reason to believe the backlash would have changed. Therefore, it is safe to say this next section may be skipped.

HOWEVER, if any one or more of the following components were replaced, the gear backlash should be checked for possible shim adjustment:

New lower unit housing -- check forward and reverse gear shim material and pinion gear depth.

New forward gear tapered roller bearing -- check forward gear backlash.

New reverse gear ball bearing -- check reverse gear backlash.

New pinion gear -- check pinion gear depth.

New forward gear -- check forward gear backlash.

New reverse gear -- check reverse gear backlash.

New bearing carrier -- check reverse gear backlash.

New thrust washer on the 20hp and larger models -- check reverse gear backlash.

The backlash is measured at this time before the water pump is installed. If the amount of backlash needs to be adjusted, the lower unit must be disassembled to change the amount of shim material behind one of the gears.

Checking Gear Backlash
Models 3.5 hp, 4hp, and 5hp
Models 28hp, 48hp, 55hp, and 60hp

To check forward gear backlash: With one hand, hold the driveshaft steady to prevent it from rotating and at the same time, pull **UPWARD** on the shaft. With the other hand, **PUSH IN** on the propeller shaft and at the same time, attempt to rock the propeller shaft back and forth. The propeller shaft should be free to move "just a whisker" to have an acceptable gear backlash. (The manufacturer does not give an actual measurement for "acceptable" backlash for the models listed in the heading.)

The propeller shaft should definately **NOT** be "locked" in place.

If the propeller shaft is too "tight", remove shim material from behind the forward gear. If the propeller shaft is too "sloppy", add shim material behind the forward gear.

Repeat the procedure to check reverse gear backlash, but **PULL OUT** on the propeller shaft, while attempting to rock the shaft back and forth.

If the propeller shaft is too "tight", remove shim material from behind the reverse gear. If the propeller shaft is too "sloppy", add shim material behind the reverse gear.

Water-Cooled Models
Listed in Heading
Proceed directly to "Water Pump Installation" on Page 9-58.

Air-Cooled Models
Proceed directly to either Step 30 or Step 31, on Page 9-60, depending on the model being serviced.

Forward Gear Backlash
Models 8hp, 9.9hp, 15hp, 20hp,
25hp, 30hp, and 40hp
Install the bearing carrier puller and "J" bolts onto the ribs of the carrier. With the bearing carrier attaching bolts or the bearing carrier locknut secured, the puller locks the shaft and prevents rotation in any direction. The puller will also lock the lower unit in **FORWARD** gear.

Obtain backlash adjusting plate P/N M-91-84563M. Secure the plate over the top of the lower unit with the nut and bolt provided with the tool.

Secure a dial indicator to the plate.

Obtain and install a backlash indicator gauge onto the driveshaft, P/N M-91-85725M for 8hp, 9.9hp, 15hp, 20hp, 25hp, and 30hp units, or P/N M-91-84551M for 40hp units. Adjust the end of the dial indicator to rest on the gauge as follows:

On the "30A" mark for the 8hp, 9.9hp, 15hp, 20hp, 25, and 30hp models.
On the "40A" mark for the 40hp models.

Backlash Measurement
Push down on the driveshaft, and at the same time slowly rotate the shaft, "rocking" the shaft back and forth through about a 25° to 30° arc.

As soon as a "click" is felt, stop all motion and observe the maximum deflection of the dial indicator needle. Acceptable backlash is as follows for the models listed:

8C hp	0.010-0.030"
	(0.25-0.75mm)
8B, W8, & Marthn 8	0.004-0.012"
	(0.10-0.30mm)
8*hp, 9.9hp, & 15*hp	0.009-0.027"
	(0.20-0.69mm)
W15, & 15**hp	0.004-0.012"
	(0.10-0.30mm)
20hp, 25hp & 30hp	0.008-0.020"
	(0.20-0.23mm)
40hp	0.004-0.007"
	(0.10-0.23mm)

* All other models not listed.
** Units with three bolts securing the bearing carrier.

Remove or add shim material behind the forward gear to bring the backlash within specifications.

Forward gear -- adding shim material **DECREASES** backlash.

Forward gear -- removing shim material **INCREASES** backlash.

HELPFUL WORDS
To determine how much shim material must be moved to obtain the desired backlash, the manufacturer gives a very simple formula:

A constant (a number depending on the unit being serviced) minus the backlash measurement, and then the answer multiplied by two. This may sound complicated but it is **NOT**.

Formula: (Constant - measurement) x 2.

Model 8hp (all), 9.9hp, and (all) 15hp
Constant is 0.5 if measuring in millimeters, and 0.020 if measuring in inches. Therefore: Assuming the measurement was 0.008" and we needed 0.010". (0.020 - 0.008) x 2 = 0.024.

To increase backlash remove 0.024" shim material.

To decrease backlash add 0.024" shim material.

If measuring in millimeters use the same formula with the millimeter constant, then add or remove shim material as required.

Model 20hp, 25hp, and 30hp

The constant for these units is 0.35 if measuring in millimeters and 0.014 if measuring in inches.

Model 40hp

The constant for these units is 0.43 if measuring in millimeters and 0.017 if measuring in inches.

CRITICAL WORDS

If the backlash specification cannot be reached by adding and removing shim material, the gears may have to be replaced.

Remove the bearing carrier puller and "J" bolts from the propeller shaft.

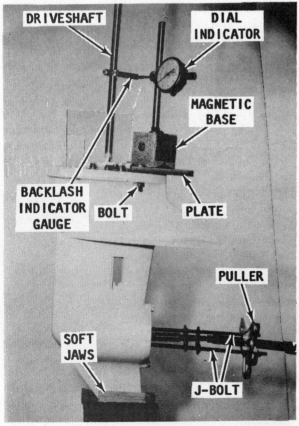

Setup for measuring backlash for lower units matched with 8hp thru 40hp powerheads, as explained in the text.

REVERSE GEAR BACKLASH

All Models

Temporarily install the shift rod and the propeller. Push the shift rod into the lower unit a "notch" to shift the unit into **NEUTRAL**. Move the dial indicator out of the way temporarily and check to be sure the propeller shaft does not rotate, when the driveshaft is rotated in either direction. Now, push the shift rod into the lower unit one more "notch" to shift the lower unit into **REVERSE** gear. Rotate the driveshaft **CLOCKWISE** and check to be sure the propeller shaft rotates **COUNTERCLOCKWISE**.

This is a test to see if the clutch mechanism was properly assembled and functions correctly.

Move the shift rod out from the lower unit one "notch" to shift the lower unit back into **NEUTRAL**. Rotate the driveshaft **CLOCKWISE** $20°$ to $30°$ with the other hand. This action will pre-load the reverse gear. Push the shift rod into the lower unit one "notch" to shift the unit into **REVERSE** gear. Hold the propeller with one hand to prevent it from rotating and at the same time push down on the driveshaft with the other hand. Continue to hold the propeller immobile and while still pushing down on the driveshaft, rotate the driveshaft back and forth through about $20°$ to $30°$. When a "click" is felt, stop all motion and observe the maximum deflection on the dial indicator.

Acceptable reverse gear backlash is as follows for the models listed.

Model	Backlash
8C hp	0.010-0.030" (0.25-0.75mm)
8B, W8, & Marthn 8	0.004-0.012" (0.10-0.30mm)
8*hp, 9.9hp, & 15*hp	0.030-0.045" (0.76-1.15mm)
W15, & 15**hp	0.004-0.012" (0.10-0.30mm)
20hp, 25hp & 30hp	0.028-0.039" (0.71-1.00mm)
40hp	0.016-0.021" (0.40-0.54mm)

* All other models not listed.
** Units with three bolts securing the bearing carrier.

Remove or add shim material behind the reverse gear to bring the backlash within specifications.

Reverse gear -- adding shim material **INCREASES** backlash.

Reverse gear -- removing shim material **DECREASES** backlash.

HELPFUL WORDS

To determine how much shim material must be moved to obtain the desired backlash, the manufacturer gives a very simple formula:

A constant (a number depending on the unit being serviced) minus the backlash measurement, and then the answer multiplied by two. This may sound complicated but it is **NOT**.

Formula: (Constant - measurement) x 2.

Model 8C hp

The constant for these units is 1.00 if measuring in millimeters, and 0.039 if measuring in inches. Therefore: Assuming the measurement was 0.020" and we needed 0.031". (0.039 - 0.020) x 2 = 0.038.

To increase backlash add 0.016" shim material.

To decrease backlash remove 0.016" shim material.

If measuring in millimeters use the same formula with the millimeter constant, then add or remove shim material as required.

All Other 8hp and 15*hp Models

*Units with three bolts securing the bearing carrier.

The constant for these units is 0.50 if measuring in millimeters and 0.020 if measuring in inches.

Model 9.9hp, and all other 15hp

The constant for these units is 0.98 if measuring in millimeters and 0.038 if measuring in inches.

Model 20hp, 25hp, and 30hp

The constant for these units is 0.85 if measuring in millimeters and 0.034 if measuring in inches.

Model 40hp

The constant for these units is 0.47 if measuring in millimeters and 0.018 if measuring in inches.

CRITICAL WORDS

If the backlash specification cannot be obtained by adding and removing shim material, the pinion gear and/or the reverse gear may have to be replaced.

Remove the shift rod and the propeller.

SPECIAL WORDS
MODEL 20HP, 25HP, 28HP, 30HP, AND 40HP

These five units have a thrust washer between the shim material and the reverse gear. The thrust washer absorbs backlash. Therefore, it may be possible to replace the thrust washer and again check the backlash. The cost of a new thrust washer is minimal compared to the cost of a pinion gear and a reverse gear.

After the proper amount of backlash and pinion gear depth has been obtained, check the gear mesh pattern.

GEAR MESH PATTERN

This step is only necessary if Desenex, or similar material, was applied to the three gears prior to assembling.

All models to be held with the lower unit in the upright (normal) position.

25- Grasp the driveshaft and pull **UPWARD**. At the same time, rotate the propeller shaft **COUNTERCLOCKWISE** through about six or eight complete revolutions. This action will establish a wear pattern on the gears with the Desenex powder.

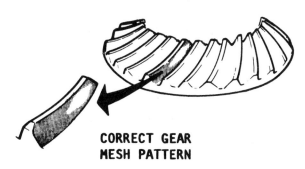

CORRECT GEAR MESH PATTERN

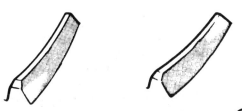

BOTH INCORRECT PATTERNS

9-58 LOWER UNIT

Now, disassemble the unit and compare the pattern made on the gear teeth with the accompanying illustrations. The pattern should almost be oval on the drive side and be positioned about halfway up the gear teeth.

If the pattern appears to be satisfactory, clean the dye or powder from the gear teeth and assemble the unit one final time.

If the pattern does **NOT** appear to be satisfactory, add or remove shim material, as required. Adding or removing shim material will move the gear pattern towards or away from the center of the teeth.

After the gear mesh pattern is determined to be satisfactory, assemble the bearing carrier one final time. For all units **EXCEPT** the 20hp model and larger, apply Loctite Type "A" to the threads of the attaching bolts. Tighten the bolts to a torque value of 5.8 ft lbs (8Nm).

If servicing a 20hp or larger unit, install the bearing carrier using the special instruction outlined in Step 23 and then bend down one or more lockwasher tabs over the locknut.

All Models Except 3.5hp, 4hp and 5hp

Remove the temporarily installed oil seal housing. Obtain the following special tools:

Model 8hp - Driver Rod - P/N M-91-84529M, Mandrel P/N M-91-83177M

Models 9.9hp and 15hp - Driver Rod P/N M-91-84529M and Mandrel P/N M-91-83175M

Models 20hp, 25hp, and 30hp - Driver Rod P/N M-91-84529M and Mandrel P/N M-91-84718M

Model 40hp - Driver Rod P/N M-91-84529M and Mandrel P/N M-91-84676M

For all other models not listed above, check with the local Mariner dealer for the correct part numbers of special tools for the model being serviced, or obtain a suitable size socket and extension. The socket should be just a "whisker" smaller than the outer diameter of the oil seal to be installed.

Lower the first seal **SQUARELY** into the oil seal housing recess, with the lip of the seal facing **UPWARD**. Tap the end of extension with a hammer until the seal is fully seated. After installation, pack the seal with multi-purpose water resistant lubricant.

WATER PUMP INSTALLATION

Installation of the water pump varies for the different models covered in this manual. Follow only the numbered steps with the heading for the model being serviced.

Model 4hp and 5hp

26- Install the gasket and the base plate with the shift rod over the driveshaft. Index the gasket and plate over the two dowel pins. Secure the plate to the lower unit with the one bolt next to the shift rod. Seat the boot well onto the plate.

Model 20hp and Larger

27- Install a new O-ring into the oil seal housing. Slide the housing over the driveshaft and into the lower unit. Check to be sure the projection on the housing fits into the recess in the lower unit.

Install the two dowel pins on the top surface of the oil seal housing. Install the following parts over the driveshaft and index each over the two dowel pins: the oil seal protector, the gasket, and then the outer plate cartridge.

Fit the Woodruff key into the driveshaft. Just a dab of grease will help to hold the

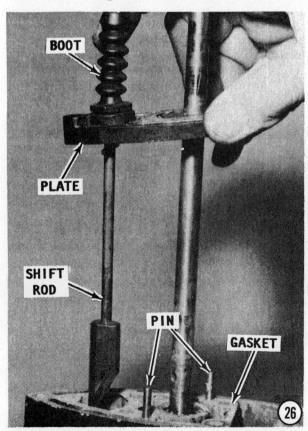

key in place. Slide the water pump impeller over the driveshaft with the rubber membrane on the top side and the keyway in the impeller indexed over the Woodruff key. **TAKE CARE** not to damage the membrane. Coat the impeller blades with multi-purpose water resistant lubricant.

Install the insert cartridge, the inner plate, and finally the water pump housing over the driveshaft. Rotate the insert cartridge **COUNTERCLOCKWISE** over the impeller to tuck in the blades. Seat all parts over the two dowel pins and secure the water pump housing with the three washers and bolts. Tighten the bolts to a torque value of 5.8 ft lbs (8Nm).

All Models

28- Install a new water tube grommet into the water pump housing. Slide the gasket and the outer plate over the driveshaft. Fit the Woodruff key into place in the driveshaft. Just a dab of grease will help to hold the key in place. The model 4hp and 5hp have an indexing pin.

Slide the water pump impeller onto the driveshaft with the keyway in the impeller indexed over the Woodruff key, or the indexing pin. Coat the impeller blades with multi-purpose water resistant lubricant.

Install the insert cartridge over the driveshaft. Rotate the cartridge **COUNTERCLOCKWISE** over the impeller to tuck in the blades. Install another gasket and finally the water pump housing over the driveshaft. Check to be sure all parts are indexed over the two dowel pins.

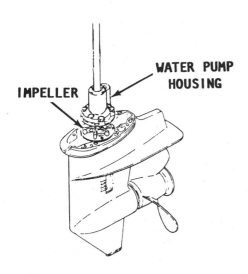

Model 15hp and Smaller

29- Position the left and right side plates over the water pump housing and secure them in place with the four washers and bolts. Tighten the bolts to a torque value of 5.8 ft lbs (8Nm). Some early model 8hp and 15hp units are not equipped with these plates.

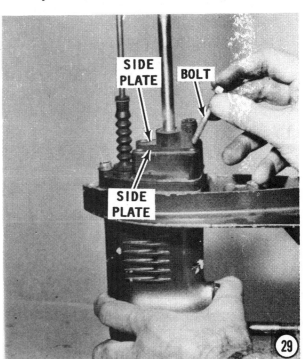

Model 9.9hp, and 15hp

These units have an extra retaining plate on top of the water pump housing. This extra plate is secured in place with two Phillips head screws.

All Models EXCEPT:
Air-Cooled 5hp
Some Early 8hp and 15hp Models
and Model 40hp

30- Install the shift rod and retaining bracket into the lower unit with the cam lobe facing toward the propeller shaft. Secure the rod in place by tightening the Phillips head screws into the retaining bracket. Fit the boot over the bracket.

Air-Cooled 5hp and Some Early Model 8hp
Model 15hp and 40hp

Insert the shift rod and boot into the lower unit with the cam on the rod facing toward the propeller. Secure the rod in place by installing and **BARELY** tightening the shift limit screw on the starboard side of the lower unit housing. This screw will be tightened later in Step 33, when the shift linkage is adjusted.

The 8hp with positive shift action also has a shift detent screw installed at the forward "nose" of the housing. After the shift shaft is installed, insert the detent ball, spring, washer and plug into the threaded hole at the forward "nose". Tighten the plug securely.

INSTALLATION — LOWER UNIT TO INTERMEDIATE HOUSING

The following steps apply to all Lower Unit "B" models covered in this section.

31- Apply just a **"DAB"** of multi-purpose water resistant lubricant to the splines at the upper end of the driveshaft.

BAD NEWS

An excessive amount of lubricant on top of the driveshaft to crankshaft splines will be trapped in the clearance space. This trapped lubricant will not allow the driveshaft to fully engage with the crankshaft.

Apply some of the same lubricant to the end of the water tube in the intermediate housing.

Apply just a **"DAB"** of lubricant, to the indexing pin on the mating surface of the lower unit.

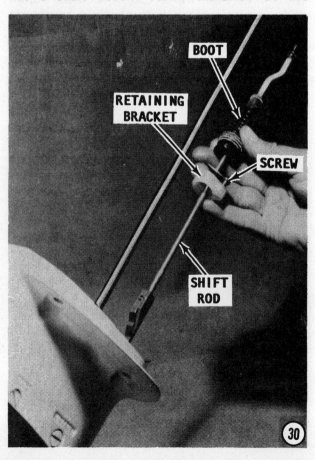

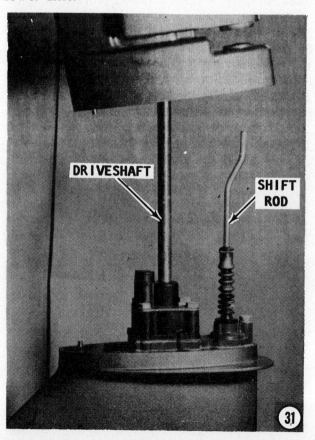

ASSEMBLING TYPE "B"

For units with a stepped cam at the end of the shift rod:
Models 4hp thru 8hp: Push the shift rod into the **REVERSE** gear position, and then push the shift lever on the powerhead into the reverse gear position.
Models 3.5hp, and 9.9hp thru 40hp: Pull the shift rod **UP** slightly to engage **NEUTRAL**, and then position the shift lever in the neutral postion.

For units with splines at the end of the shift rod:
Models 48hp, 55hp, and 60hp: Rotate the shift rod to engage neutral gear. The rod will rotate through approximately $60°$. Therefore, neutral gear will be $30°$ from either stop.

Special Instructions
Air-Cooled Models 3.5hp and 5hp

Position the gasket, followed by the anti-cavitation plate, over the dowel pin on the lower unit. Slide the water tube into the grommet on the anti-cavitation plate as the two halves come together.

Begin to bring the intermediate housing and lower gear housing together.

SPECIAL WORDS

The remainder of this step takes time and patience. Success will probably not be achieved on the first attempt. **THREE** items must mate at the same time before the lower unit can be seated against the intermediate housing.

If servicing an air-cooled unit, only part **a-** and part **c-** apply.

a- The top of the driveshaft on the lower unit indexes with the lower end of the crankshaft.
b- The water tube in the intermediate housing slides into the grommet on the water pump housing.
c- The top of the lower shift rod in the lower unit slides into the connector and is secured to the upper shift rod in the intermediate housing.

32- As the two units come closer, rotate the propeller shaft slightly to index the upper end of the lower driveshaft with the crankshaft. At the same time, feed the water tube into the water tube grommet, and feed the lower shift rod into the threaded connector on the upper shift rod.

Push the lower unit housing and the intermediate housing together. **TAKE CARE** not to bend the long narrow water tube. The pin on the upper surface of the lower unit will index with a matching hole in the lower surface of the anti-cavitation plate.

Tighten the bolt on the connector to secure the two shift rods together or secure the threaded connector onto the lower shift rod at least **FIVE** turns and lock it in place with the locknut.

WORDS FROM EXPERIENCE

If all three items appear to mate properly, but the lower unit seems locked in posi-

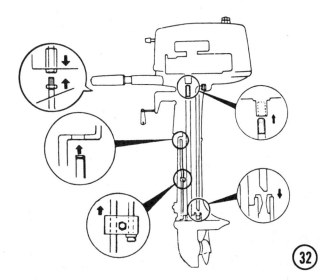

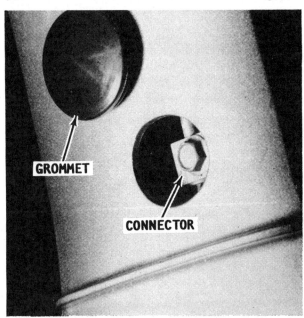

Shift rod connection arrangement on the Model 4hp powerhead.

9-62 LOWER UNIT

Shift rod connection arrangement for lower units matched with the 8hp powerhead.

tion about 6 inches (15cm) away from the intermediate housing and it is not possible, with ease, to bring the two housings closer, the driveshaft has missed the cylindrical oil seal housing leading to the crankshaft.

Move the lower unit out of the way. Shine a flashlight up into the intermediate housing and find the oil seal housing. Now, on the next attempt, "find" the edges of the oil seal housing with the driveshaft before trying to mate anything else -- the water tube and the top of the shift rod. If the driveshaft can be made to enter the oil seal housing, the driveshaft can then be easily indexed with the crankshaft.

33- Apply Loctite Type "A" to the threads of the bolts used to secure the lower unit to the intermediate housing. Install and tighten the bolts to the following torque value, depending on the unit being serviced:

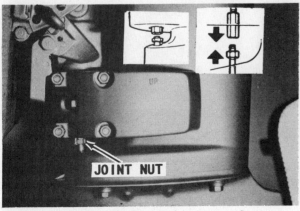

Shift rod connection arrangement for lower units matched with the 9.9hp and larger powerheads.

15hp and smaller -- 5.8 ft lbs (8Nm).
20hp, 25hp, 28hp, 30hp, 48hp, 55hp, and 60hp -- 26 ft lbs (36Nm).
40hp -- 16 ft lbs (22Nm)

Install the anode, if it was removed.

Check the Shift Action
Operate the shift lever through all gears. The shifting should be smooth and the propeller should rotate in the proper direction when the flywheel is rotated by hand in a **CLOCKWISE** direction. Naturally the propeller should not rotate when the unit is in neutral.

SHIFT LINKAGE ADJUSTMENT
This task is very much a "trial and error" procedure.

On models with a clamping connector: Loosen, but do not remove, the connector bolt and adjust the lower shift rod within the connector until the shift positions are correct.

On models with a threaded connector: Back off the locknut, and adjust the threaded connector to raise or lower, the lower shift rod until the shift positions are correct. Make sure there are at least five threads engaging the lower shift rod after the adjustment is completed. Tighten the locknut against the threaded connector to hold the adjustment.

On models with a shift limit screw on the starboard side of the lower unit housing: Tighten the screw and check the shift ac-

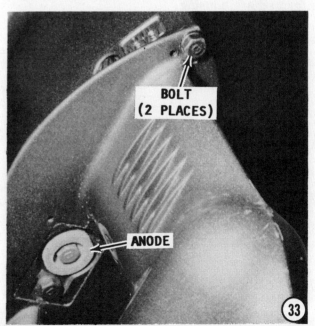

ASSEMBLING TYPE "B"

tion. If tightening the screw caused the shift rod to "freeze", the end of the screw did not index into the slot in the shift rod, in which case, the rod must be removed and repositioned.

PROPELLER INSTALLATION

GOOD WORDS

An anti-seizing compound, such as Perfect Seal, will prevent the propeller from "freezing" to the shaft and permit propeller removal, without difficulty, the next time the propeller needs to be "pulled".

All Models

Apply Perfect Seal or equivalent anti-seizing compound, to the propeller shaft.

Model 3.5hp

34- Install a new shear pin through the propeller shaft. Some units have two spacers and a rubber damper between the bearing carrier and the propeller. Slide the propeller onto the shaft with the internal splines of the propeller indexing with the splines on the shaft. Wedge a block of wood between one of the propeller blades and the anti-cavitation plate to prevent the propeller from rotating. Install the propeller nut.

Tighten the nut securely using a pair of large channel lock pliers or an adjustable wrench, and then check to determine if the holes in the nut are aligned with the hole in the propeller shaft. If not, tighten the nut just a "whisker" more until the holes are aligned. Insert and secure the cotter pin.

Some 3.5hp units do not have a propeller nut. If servicing this type unit: Guide the propeller onto the pin. Insert a new cotter pin into the hole and bend the ends of the cotter pin in opposite directions.

Propeller secured with "Nose Cone" Type Propeller Nut

35- Install a new shear pin through the propeller shaft. Slide the propeller onto the shaft with the internal splines of the propeller indexing with the splines on the shaft. Wedge a block of wood between one of the propeller blades and the anti-cavitation plate to prevent the propeller from rotating.

Install the propeller nut. Tighten the nut securely using a pair of large channel lock pliers or an adjustable wrench, and then check to determine if the holes in the nut are aligned with the hole in the propeller shaft. If not, tighten the nut just a "whisker" more until the holes are aligned. Insert and secure the cotter pin.

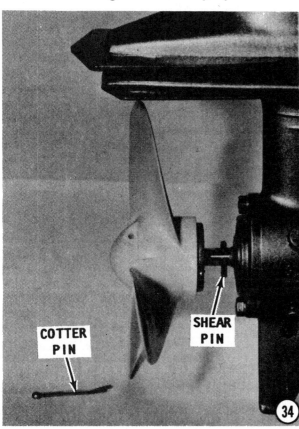

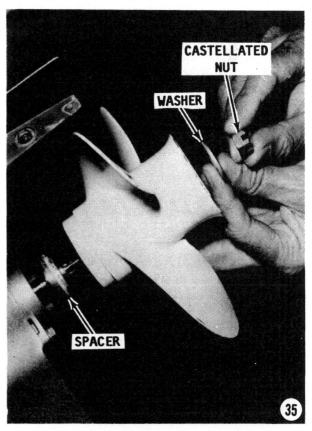

9-64 LOWER UNIT

Propeller secured with Castellated Nut

Install the spacer onto the propeller shaft. Install the propeller, the washer and the castellated nut. Place a block of wood between one of the propeller blades and the anti-cavitation plate to prevent the propeller from rotating. Tighten the nut as follows:

Models 8hp, 9.9hp, and 15hp
Torque value of 13 ft lbs (18Nm).

Models 20hp, 25hp, 28hp and 30hp
Torque value of 22 ft lbs (30Nm).

Models 40hp and Larger
Torque value of 25 ft lbs (34Nm).

If necessary, back off the nut until the cotter pin may be inserted through the nut and the hole in the propeller shaft. Bend the arms of the cotter pin around the nut to secure it in place.

Remove the block of wood. Connect the spark plug wires to the spark plugs. Connect the electrical lead to the battery terminal.

TRIM TAB ADJUSTMENT

The trim tab should be positioned to enable the helmsperson to handle the boat with equal ease to starboard and port at normal cruising speed. If the boat seems to turn more easily to starboard, loosen the socket head screw and move the trim tab trailing edge to the right. Move the trailing edge of the trim tab to the left if the boat tends to turn more easily to port.

CLOSING TASKS

Release the tilt lock lever and lower the outboard to the normal operating position.

All Models

36- Remove the oil level plug and the drain plug.

SPECIAL WORDS

Temporarily, remove the spark plug lead/s to prevent the powerhead from starting. Shift the unit into forward gear and rotate the propeller clockwise, from time to time, while filling the gearcase. This action will help dislodge any air pockets trapped inside and make the task go faster.

Fill the lower unit with gearcase lubricant or Hypoid gear oil 90 weight until the lubricant begins to escape from the top hole. Install both plugs and clean any excess lubricant from the lower unit.

Use the following table as a guide to lower unit capacity for the model indicated.

Model	Capacity	
3.5hp	1.7 U.S. oz	80cc
4hp	3.6 U.S. oz	105cc
5hp (air cooled)	1.7 U.S. oz	80cc
5hp (water cooled)	3.6 U.S. oz	105cc
8C hp	5.4 U.S. oz	160cc
8hp (all other Models)	7.8 U.S. oz	230cc
9.9hp (with straight cut gears)	7.6 U.S. oz	225cc
9.9hp (with spiral cut gears)	6.2 U.S. oz	185cc
15hp (with straight cut gears)	7.6 U.S. oz	225cc
15hp (with spiral cut gears)	6.2 U.S. oz	185cc
20hp	6.1 U.S. oz	180cc
25hp	6.1 U.S. oz	180cc
28hp	6.1 U.S. oz	180cc
30hp	6.1 U.S. oz	180cc
40hp	11.2 U.S. oz	330cc
48hp	17.2 U.S. oz	500cc
55hp	17.2 U.S. oz	500cc
60hp	17.2 U.S. oz	500cc

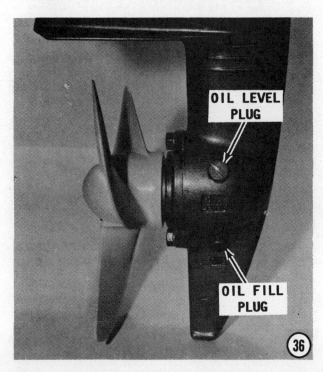

Mount the outboard unit in a test tank or body of water.

CAUTION

Water must circulate through the lower unit to the powerhead anytime the powerhead is operating to prevent damage to the water pump in the lower unit. Just five seconds without water will damage the water pump impeller.

Start the engine and check the completed work for satisfactory operation, shifting, and **NO** leaks.

9-7 TYPE "C" LOWER UNIT

Type "C" lower units are two piece units, matched with early Model 20hp, 25hp, and 40hp powerheads.

DESCRIPTION

In addition to the normal forward capability, most Type "C" lower units are equipped with a clutch "dog" permitting operation in **NEUTRAL, FORWARD,** and **REVERSE.**

Shifting Operation

The pinion gear remains in constant mesh with both the forward and reverse gears. These three gears constantly rotate anytime the powerhead is operating.

A sliding clutch "dog" is mounted on the propeller shaft. The clutch "dog" is operated manually through linkage from a shift lever located on the starboard side of the powerhead. When the clutch "dog" is moved forward, it engages with the forward gear. The clutch "dog" then picks up the rotation of the forward gear. Because the clutch is secured to the propeller shaft with a pin, the shaft rotates at the same speed as the clutch. The propeller is thereby rotated to move the boat forward.

When the clutch "dog" is moved aft, the the clutch engages only the reverse gear. The propeller shaft and the propeller are thus moved in the opposite direction to move the boat sternward.

When the clutch "dog" is in the neutral position, neither the forward nor reverse gear is engaged with the clutch and the propeller shaft does not rotate.

From this explanation, an understanding of wear characteristics can be appreciated. The pinion gear and the clutch "dog" receive the most wear, followed by the forward gear, with the reverse gear receiving the least wear.

WORDS OF WISDOM

Procedural steps are given to remove all items in the lower unit. However, do **NOT** remove bearings, bushings, or seals, if a determination can be made the item is fit for further service. As the work progresses, simply skip the steps involving these parts and continue with required work.

Before beginning work on the lower unit, take time to **READ** and **UNDERSTAND** the information presented in Section 9-1, this chapter.

Disconnect the high tension spark plug leads and remove the spark plugs before working on the lower unit.

LOWER UNIT REMOVAL

AUTHORS' WORDS

Because so many different models are covered in this section, it would not be feasible or practical to provide an illustration of each and every unit.

Therefore, the accompanying illustrations are of a "typical" lower unit "C" with split gearcase. The appearance of the unit being serviced may differ slightly due to engineering or cosmetic changes but the procedures are valid. If a difference should occur, the models affected will be clearly identified.

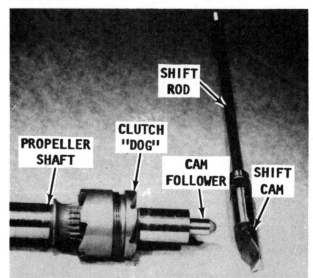

The shift cam ramp moves the cam follower inward and allows spring action to move the follower outward. This action causes the clutch "dog" to engage either the forward gear or the reverse gear.

LOWER UNIT REMOVAL

1- Tilt and lock the outboard unit in the raised position. Check to be sure the lower unit is in neutral gear.

Separate the lower and upper shift rod by backing off the locknut and rotating the long threaded connecting nut upward.

2- Remove the six bolts securing the lower unit to the intermediate housing.

Separate the lower unit from the intermediate housing. **WATCH FOR** and **SAVE** the two dowel pins when the two units are separated. The water tube will come out of the grommet and remain with the intermediate housing. The driveshaft will remain with the lower unit.

DISASSEMBLING

Propeller Removal

3- Straighten the cotter pin, and then pull it free of the propeller nut with a pair of pliers or a cotter pin removal tool. Remove the propeller nut using a large pair of channel lock pliers or an adjustable wrench. Push the shear pin out of the propeller shaft.

Draining the Gearcase

4- Position a suitable container under the lower unit, and then remove the **OIL** screw and the **OIL LEVEL** screw. Allow the gear lubricant to drain into the container. As the lubricant drains, catch some with your fingers from time to time, and rub it between your thumb and finger to determine if any metal particles are present. If metal is detected in the lubricant, the unit must be completely disassembled, inspected, and the damaged parts replaced.

Check the color of the lubricant as it drains. A whitish or creamy color indicates the presence of water in the lubricant. Check the drain pan for signs of water separation from the lubricant. The presence of any water in the gear lubricant is **BAD NEWS**. The unit must be completely disassembled, inspected, the cause of the problem determined, and then corrected.

After the lubricant has drained, temporarily install both the drain and oil level screws.

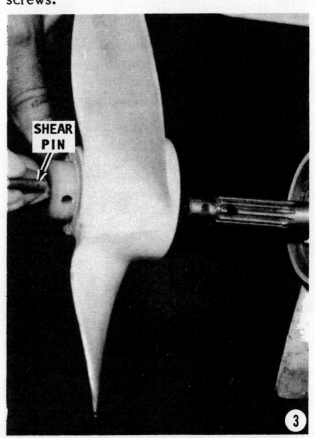

DISASSEMBLING TYPE "C" 9-67

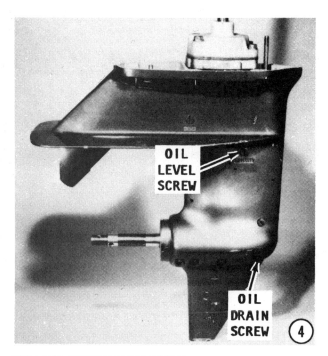

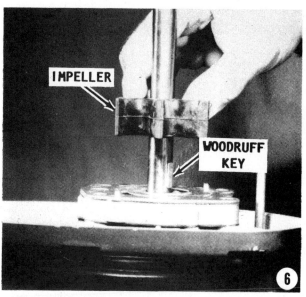

action will help prevent the splines on the end of the shaft from damaging the water pump cover oil seal. Splines on the end of a shaft or threads on the end of a rod are one of the greatest "enemies" of an oil seal. The damage to the seal is caused as the seal passes over the splines or threads during removal and installation.

WATER PUMP REMOVAL

5- Clamp the skeg in a vise equipped with soft jaws or padded with a couple of shop towels. Remove the four bolts and washers securing the water pump cover. Lift the water pump cover and insert cartridge up and free of the driveshaft. Remove and discard the gasket.

6- Work the impeller up and free of the Woodruff key. Remove the key.

GOOD WORDS

Move the water pump cover **SLOWLY** past the splines on the driveshaft. This

7- Lift the water pump inner plate from the driveshaft. Remove and discard the gasket.

Use a gasket scraper and remove all traces of gasket material from the water pump body. A clean surface will provide the best seal for a new gasket.

If the only work to be performed is service of the water pump, proceed directly to Step 17 of the Assembling instructions on Page 9-79.

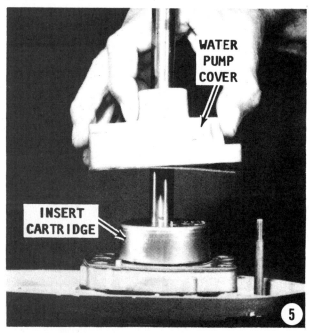

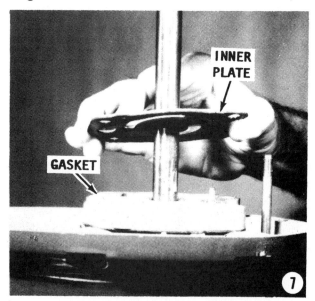

9-68 LOWER UNIT

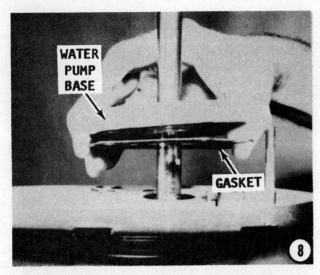

8- Remove the water pump base. If necessary, place a couple of shop towels on the gearcase surface around the water pump base to protect the machined surface. Now, using a couple of screwdrivers, pry the base from the gearcase. Lift the base up and free of the driveshaft. Do not lose the two dowel pins which may remain in either mating surface.

GEARCASE SEPARATION

CRITICAL WORDS

DO NOT clamp the driveshaft in a vise without any protection. Cut notches in a couple of pieces of wood to fit around the shaft, or use a thick pad of shop towels when securing the driveshaft in the vise.

Remove the gearcase from the vise and clamp the driveshaft in the vise.

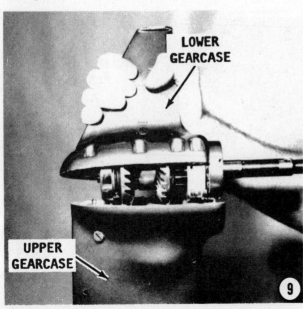

VERY CRITICAL WORDS

STOP, and observe the bearing carrier. The notches cut in the bearing carrier are deceiving! They are **NOT** there for the purpose of engaging a special tool for removal, **NOR** for rotating the carrier using the "hammer and punch" method.

In fact, the bearing carrier does **NOT** unscrew from the gearcase. The two halves of the case must be separated before the carrier can be removed.

9- Remove the six or nine bolts, depending on the model being serviced, securing the lower gearcase to the upper gearcase, and then separate the two halves. If necessary, tap the skeg on both sides using a soft head mallet to jar the halves free.

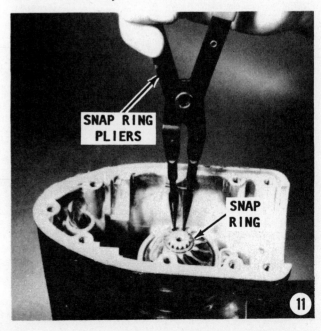

DISASSEMBLING TYPE "C" 9-69

10- Lift off the upper gearcase with propeller shaft still installed and set it aside for later disassembly. Remove and discard the rubber seal.

PINION GEAR AND DRIVESHAFT REMOVAL

11- Obtain a pair of external snap ring pliers. Remove the snap ring around the end of the driveshaft.

12- **CAREFULLY** lift out the pinion gear, taking care not to disturb the large roller bearings around the driveshaft. The inner roller bearing cage is pressed onto the pinion gear shaft. **WATCH** for and **SAVE** any shim material under the inner bearing race. The shim material is critical in obtaining the correct backlash during assembling. Using the old shim material will save considerable time, especially starting with no shim material.

13- Lift the gearcase up and free of the driveshaft. Exercise **CARE** not to disturb the driveshaft roller bearings. If the gearcase will not come free, perform Step 14. If the gearcase can be removed without problems, skip the following step and proceed directly to Step 15.

14- Unclamp the driveshaft from the vise. Open up the vise a couple of inches. Insert the driveshaft into the open vise and place some shop towels between the top of the vise and the machined surface of the gearcase and rest the gearcase on the padded vise. Using a soft head mallet, tap the end of the driveshaft down and out of the gearcase.

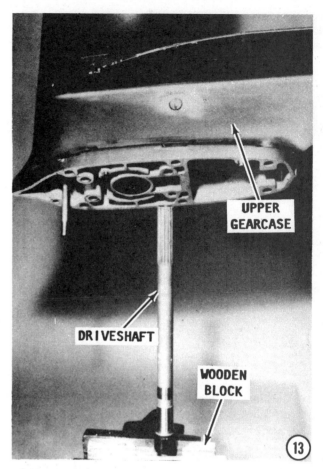

UPPER GEARCASE DISASSEMBLING

15- If the two oil seals at the top of the upper gearcase are no longer fit for service, obtain a slide hammer with a puller jaw attachment and remove the seals one at a time. Remove the driveshaft thrust washer under the lower seal.

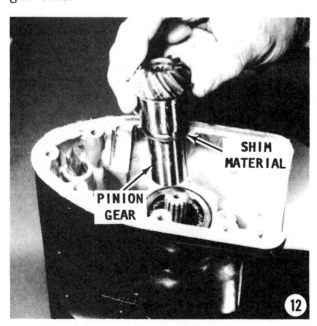

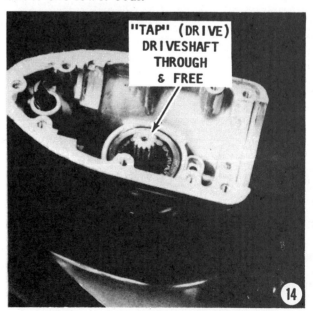

9-70 LOWER UNIT

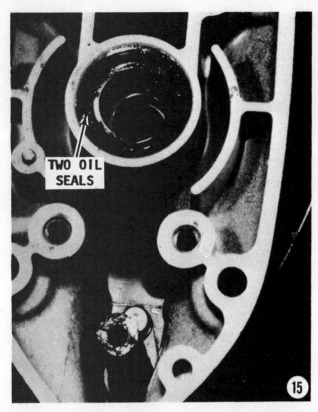

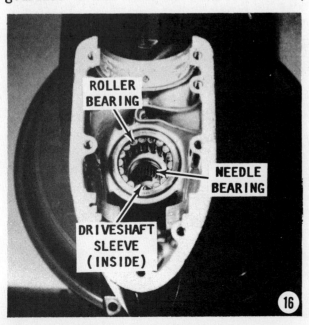

16- If the needle bearing at the top of the upper driveshaft is no longer fit for service, use the slide hammer with puller jaw attachment to remove the needle bearing. Do **NOT** remove this bearing unless a replacement is at hand, because this manner of removal distorts and destroys the bearing cage.

Lift out the long plastic driveshaft sleeve, and the thrust bearing.

If the large roller bearing in the upper gearcase is also unfit for further service,

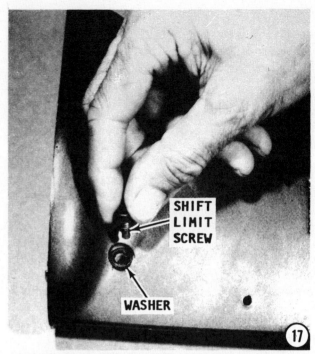

turn the gearcase upside down, and remove the bearing using the same tool.

17- Remove the shift limit screw and washer from the side of the upper gearcase.

18- Remove the bolt securing the shift rod retainer and boot to the gearcase. Pull out the shift rod.

LOWER GEARCASE AND PROPELLER SHAFT DISASSEMBLING

19- Lift out the propeller shaft assembly and bearing carrier from the lower gearcase.

20- Remove the cam follower, tapered bearing race, the forward gear and bearing

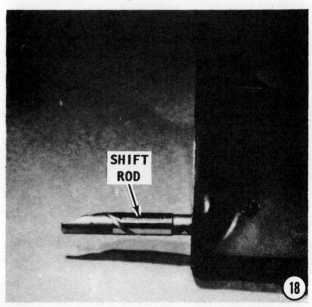

DISASSEMBLING TYPE "C" 9-71

assembly from the forward end of the shaft. Remove the bearing carrier, shim material, reverse gear and thrust washer from the aft end of the shaft. **SAVE** any shim material between the gear and the bearing race. The shim material is critical in obtaining the correct backlash during assembling. Using the old shim material will save considerable time, especially starting with no shim material.

21- The tapered forward bearing is pressed onto the forward gear shaft. If the bearing is no londer fit for service, obtain a universal bearing separator tool and separate the bearing from the gear. **WATCH** for and **SAVE** any shim material between the gear and the bearing race. The shim mater-

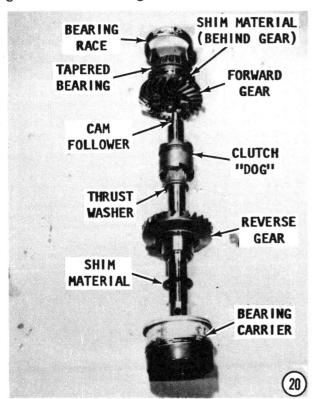

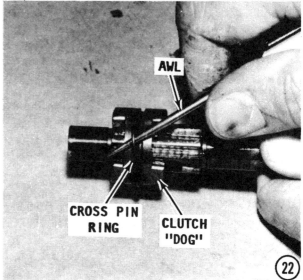

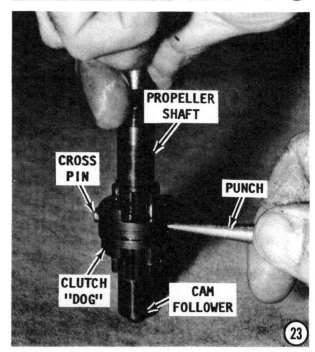

9-72 LOWER UNIT

ial is critical in obtaining the correct backlash during assembling. Using the old shim material will save considerable time, especially starting with no shim material.

22- Insert an awl under the end loop of the cross pin ring and unwind the ring free of the clutch "dog". Take **CARE** not to overstretch the spring.

23- Temporarily install the cam follower and push against a solid surface. At the same time, push the cross pin out of the sliding clutch "dog".

SLOWLY release pressure on the cam follower and slide the clutch forward off the end of the shaft. Tip the shaft and allow the cam follwer, shift slide, and spring to slide out the end of the shaft.

BEARING CARRIER DISASSEMBLING

24- Remove and discard the O-ring around the bearing carrier.

25- If the ball bearing at the aft end of the bearing carrier is no longer fit for service, use a slide hammer with puller jaw attachment to remove the bearing.

If the two oil seals behind the ball bearing are no longer fit for service, use a slide hammer with puller jaw attachment to remove the bearing and then use the same tool to pull the seals from the carrier.

CLEANING AND INSPECTING

Good shop practice requires installation of new O-rings and oil seals **REGARDLESS** of their appearance.

Clean all water pump parts with solvent, and then dry them with compressed air. Inspect the water pump housing and oil seal housing for cracks and distortion, possibly caused from overheating. Inspect the inner and outer plates and water pump cartridge for grooves and/or rough surfaces.

If possible, **ALWAYS** install a new water pump impeller while the lower unit is disassembled. A new impeller will ensure extended satisfactory service and give "peace of mind" to the owner. If the old impeller must be returned to service, **NEVER** install it in reverse to the original direction of rotation. Installation in reverse will cause premature impeller failure.

If installation of a new impeller is not possible, check the seal surfaces. All must be in good condition to ensure proper pump operation. Check the upper, lower, and ends of the impeller vanes for grooves, cracking, and wear. Check to be sure the indexing notch of the impeller hub is intact and will not allow the impeller to slip.

Clean around the Woodruff key or impeller pin. Clean all bearings with solvent, dry them with compressed air, and inspect them carefully. Be sure there is no water in the

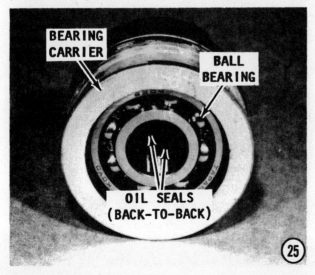

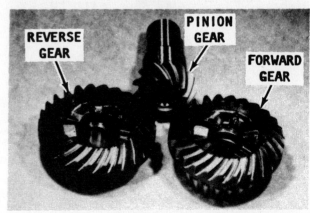

The reverse gear, pinion gear, and forward gear, cleaned, inspected, and ready for installation into the lower unit.

CLEAN & INSPECT TYPE "C" 9-73

air line. Direct the air stream through the bearing. **NEVER** spin a bearing with compressed air. Such action is highly dangerous and may cause the bearing to score from lack of lubrication. After the bearings are clean and dry, lubricate them with Formula 50 oil, or equivalent. Do not lubricate tapered bearing cups until after they have been inspected.

Inspect all ball bearings for roughness, scratches and bearing race side wear. Hold the outer race, and work the inner bearing race in-and-out, to check for side wear.

Determine the condition of tapered bearing rollers and inner bearing race, by inspecting the bearing cup for pitting, scoring, grooves, uneven wear, imbedded particles, and discoloration caused from overheating. **ALWAYS** replace tapered roller bearings as a set.

Clean the forward gear with solvent, and then dry it with compressed air. Inspect the gear teeth for wear. Under normal conditions the gear will show signs of wear but it will be smooth and even.

Clean the bearing carrier or cap with solvent, and then dry it with compressed air. **NEVER** spin bearings with compressed air. Such action is highly dangerous and may cause the bearing to score from lack of lubrication. Check the gear teeth of the reverse gear for wear. The wear should be smooth and even.

Check the clutch "dog" surfaces to be sure they are not rounded-off, or chipped. Such damage is usually the result of poor operator habits and is caused by shifting too slowly or shifting while the engine is operating at high rpm. Such damage might also be caused by improper shift rod adjustments.

Rotate the reverse gear and check for catches and roughness. Check the bearing for side wear of the bearing races.

Inspect the roller bearing surface of the propeller shaft. Check the shaft surface for pitting, scoring, grooving, embedded particles, uneven wear and discoloration caused from overheating.

Clean the driveshaft with solvent, and then dry it with compressed air. **NEVER** spin bearings with compressed air. Such action is dangerous and could damage the bearing. Inspect the bearing for roughness, scratches, or side wear. If the bearing shows signs of such damage, it should be replaced. If the bearing is satisfactory for further service coat it with oil.

After 60 seconds at 1500 rpm.

After 90 seconds at 1500 rpm.

After 30 seconds at 2000 rpm.

After 45 seconds at 2000 rpm.

After 60 seconds at 2000 rpm

Cautions throughout this manual point out the danger of operating the powerhead without water passing through the water pump. The above photographs are self evident.

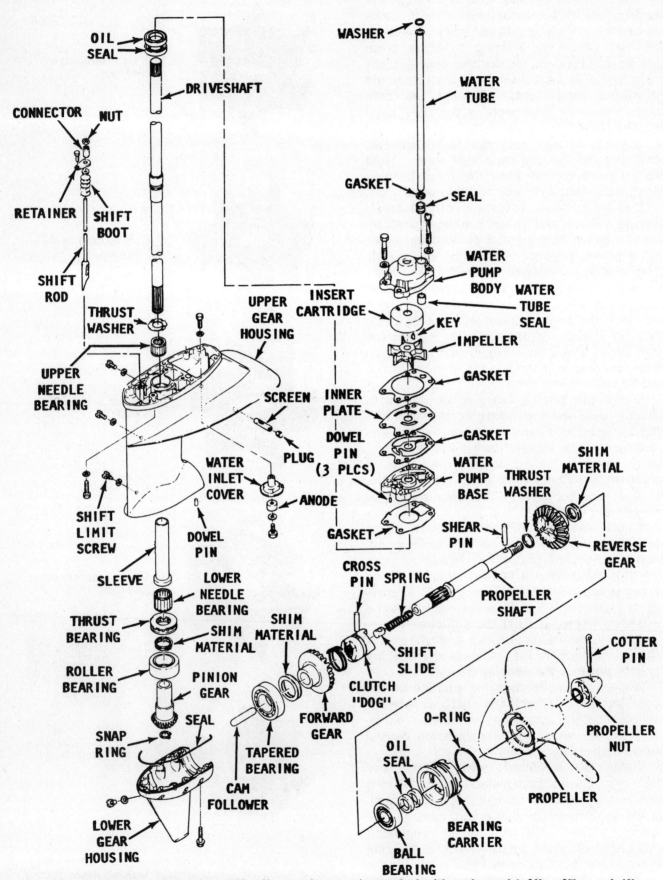

Exploded drawing of the Type "C" split case lower unit matched with early model 20hp, 25hp, and 40hp powerheads. Major parts are identified.

Inspect the driveshaft splines for excessive wear. Check the oil seal surfaces above and below the water pump drive pin or Woodruff key area for grooves. Replace the shaft if grooves are discovered.

Inspect the driveshaft bearing surface above the pinion gear splines for pitting, grooves, scoring, uneven wear, embedded metal particles and discoloration caused by overheating.

Inspect the propeller shaft oil seal surface to be sure it is not pitted, grooved, or scratched. Inspect the roller bearing contact surface on the propeller shaft for pitting, grooves, scoring, uneven wear, embedded metal particles, and discoloration caused from overheating.

Inspect the propeller shaft splines for wear and corrosion damage. Check the propeller shaft for straightness.

Check the driveshaft circlip to be sure it is not bent or stretched. If the clip is deformed, it must be replaced.

Clean all parts with solvent, and then dry them with compressed air.

Inspect all bearing bores for loose fitting bearings and the gearcase for impact damage.

Inspect the condition of the locating pins around the forward gear bearing race and especially those in the bearing carrier. If these locating pins are sheared away, someone, somewhere, sometime, attempted to remove the bearing carrier using the "hammer and chizel" method. The carrier **MUST** be replaced, as it will spin with the propeller shaft and allow water to enter the lower unit.

ASSEMBLING

FIRST, THESE WORDS

Procedural steps are given to assemble and install virtually all items in the lower unit. However, if certain items, i.e. bearings, bushings, seals, etc. were found fit for further service and were not removed, simply skip the assembly steps involved. Proceed with the required tasks to assemble and install the necessary components.

CLUTCH "DOG" INSTALLATION

1- Slide the spring into the propeller shaft, followed by the shift slide, with the flat end facing **AFT**. Insert a narrow screwdriver into the slot in the shaft. Compress the spring and shift slide together, until approximately 1/2" (12mm) distance is obtained between the top of the slot and the screwdriver.

Hold the spring compressed, and at the same time, slide the clutch "dog" over the splines of the propeller shaft with the hole in the "dog" aligned with the slot in the shaft. The letter **"F"** embossed on the "dog" **MUST** face toward the forward gear.

Insert the cross pin into the clutch "dog" and through the space held open by the screwdriver. Center the pin and then remove the screwdriver allowing the spring to pop back into place.

Fit the cross pin ring into the groove around the clutch "dog" to retain the cross-pin in place.

Insert the flat end of the plunger into the propeller shaft, with the rounded end protruding to permit the plunger to slide along the cam of the shift rod. Set the assembled propeller shaft aside for later installation.

FORWARD GEAR AND BEARING ASSEMBLING

2- Obtain a suitable support which contacts the clutch "dog" teeth on the forward gear, but does not rest on the main gear teeth. Position the support and gear on a press. Install the shim material, removed in Step 21 around the gear shaft. Position the tapered roller bearing over the gear shaft, with the taper facing **UPWARD**. Now, use a suitable mandrel and press against the inner race to seat the bearing over the gear.

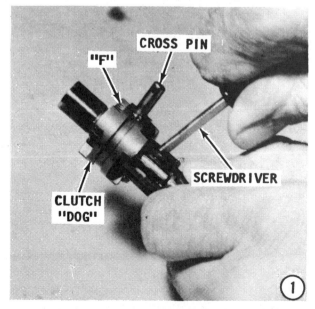

9-76 LOWER UNIT

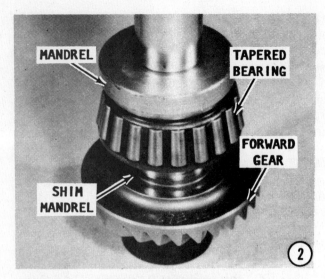

BEARING CARRIER ASSEMBLING

3- Pack the lips of both oil seals with multi-purpose water resistant lubricant. Install the seals one after the other, with both seal lips facing **DOWN**, using a driver and mandrel P/N M91-83175M. After installation, the seal lips will face **AFT --TOWARD** the propeller.

Obtain mandrel P/N M-91-84531M, and then using an arbor press, install the reverse gear ball bearing into the bearing carrier, with the embossed numbers on the bearing facing **UPWARD**.

4- Install a new O-ring around the bearing carrier.

PROPELLER SHAFT ASSEMBLING AND INSTALLATION

5- Slide first the thrust washer, followed by the assembled reverse gear, then the

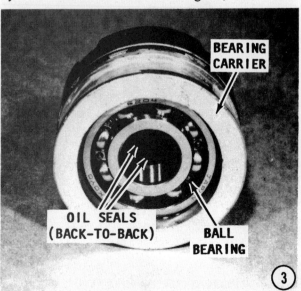

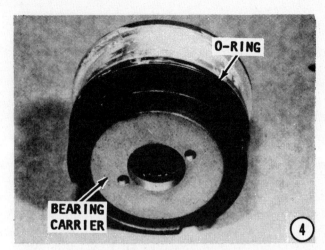

shim material (removed in Step 20), and finally the assembled bearing carrier onto the aft end of the propeller shaft. Slide the assembled forward gear, followed by the tapered bearing race onto the forward end of the propeller shaft. Check to be sure the cam follower is still in place in the forward end recess.

Keep the end of the shaft tilted upward to prevent loss of the cam follower.

Set the propeller shaft aside for later installation.

UPPER GEARCASE ASSEMBLING

6- Observe the long vertical slot in the shift shaft. Insert the shift shaft into the

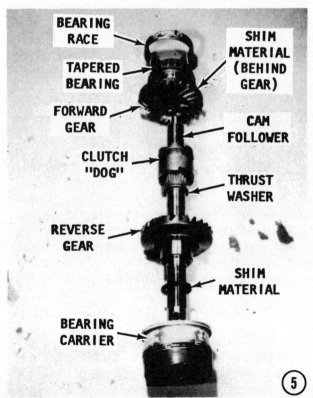

ASSEMBLING TYPE "C" 9-77

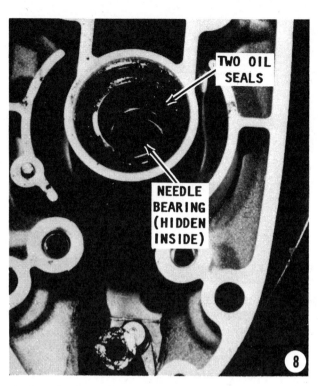

upper gearcase with the shift cam surface facing **AFT** and with the shaft slot aligned with the hole for the shift shaft limit screw. Install the bolt securing the shaft boot and retainer to the gearcase.

7- Install and **BARELY** tighten the shift shaft limit screw with a new washer. Move the shaft up and down a few times to verify the screw threads have indexed properly into the shift shaft slot.

8- With the gearcase in the upright position, install the driveshaft needle bearing using mandrel P/N M-91-M8437M and a suitable driver. The needle bearing is installed with the embossed letters **FACING** the man-

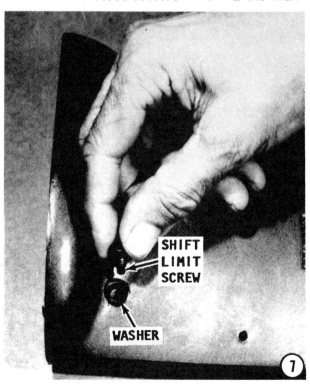

drel. Drive the bearing in until it seats. Install the driveshaft thrust washer over the needle bearing. Check to be sure the tabs on the washer index into the grooves in the housing. Pack the lips of both oil seals with multi-purpose water resistant lubricant.

Install the seals one after the other, with both seal lips facing **UPWARD** (toward the powerhead), using a driver and mandrel P/N M91-84533M.

9- Invert the upper gearcase and install the long plastic driveshaft sleeve, followed by the thrust bearing. Obtain mandrel P/N M-91-84532M and a suitable driver. Install the large roller bearing into the gear-

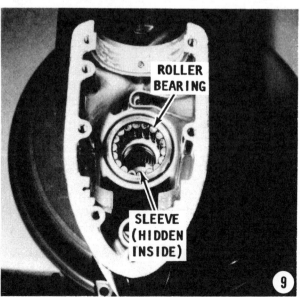

9-78 LOWER UNIT

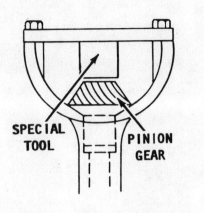

case with the embossed numbers on the bearing **FACING** the mandrel. Apply a liberal coating of multi-purpose water resistant lubricant to the roller bearings, to ensure they remain in the proper position during further assembly procedures.

DRIVESHAFT AND PINION GEAR INSTALLATION

PINION GEAR DEPTH

10- **WITHOUT** installing the driveshaft, slide the shim material, removed in Step 12

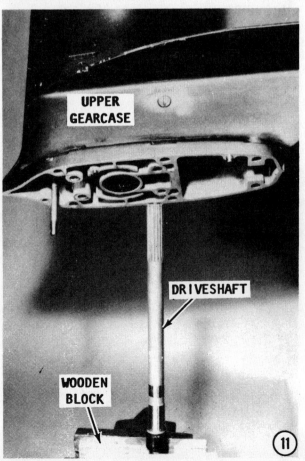

of disassembling, over the pinion gear shaft, and then place the gear into the gearcase, without disturbing the roller bearings. Obtain and secure Pinion Gear Depth Gauge P/N M-91-84544M over the pinion gear, using the two bolts. Insert a feeler gauge between the pinion gear and the depth gauge to determine the clearance. The measured distance should be 0.020" (0.5mm). Add or subtract pinion gear shim material to obtain the proper clearance. Remove the special tools and the pinion gear.

CRITICAL WORDS

DO NOT clamp the driveshaft in a vise without any protection. Cut notches in a couple pieces of wood to fit around the shaft, or use a thick pad of shop towels when securing the driveshaft in the vise.

11- Clamp the drive shaft in a vise as directed. Lower the upper gearcase over the driveshaft.

12- Slide the same amount of shim material, as removed in the previous step, over the pinion gear shaft. Install the pinion gear over the driveshaft. Exercise **CARE** to prevent disturbing the roller bearings.

13- Install the snap ring around the driveshaft groove, using a pair of external snap ring pliers.

FORWARD AND REVERSE GEAR BACKLASH CHECK

14- Install the assembled propeller shaft into the upper gearcase, and at the same time guide one of the posts on the bearing carrier into the hole in the gearcase. The

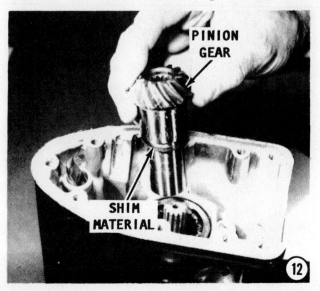

propeller shaft will not be correctly seated, if the post is not indexed **IN** the hole. Rotate the forward gear bearing race until the post on the race indexes into the notch in the gearcase.

Install a dial indicator onto a suitable mounting stand, and then secure the indicator over the propeller shaft.

Measuring Forward Gear Backlash

Adjust the position of the indicator to make contact with the flat side of the forward gear tooth.

Pull the shift shaft **DOWNWARD** (out) of the gearcase to engage the forward gear. The sliding clutch should now mesh with the forward gear. Grasp the forward gear and rock the gear back and forth, but not enough to rotate the gearcase on the driveshaft. Note the deflection of the dial indicator. This deflection indicates the backlash between the pinion gear and the forward gear. The manufacturer recommends a backlash tolerance as follows:

```
20hp  -- 0.004-0.012" (0.1-0.3mm)
25hp  -- 0.004-0.012" (0.1-0.3mm)
40hp  -- 0.002-0.006" (0.05-0.15mm)
```

If the forward gear backlash is not as specified, add or subtract shim material behind the forward gear to bring the measured dial indicator reading to the required value.

Measuring Reverse Gear Backlash

Adjust the position of the indicator to make contact with the flat side of the reverse gear tooth.

Push the shift shaft **UPWARD** (into) the gearcase to engage reverse gear. The sliding clutch should now mesh with the reverse gear. Grasp the reverse gear and rock the gear back and forth, but not enough to rotate the gearcase on the driveshaft. Note the deflection of the dial indicator. This deflection indicates the backlash between the pinion gear and the reverse gear. The manufacturer recommends a backlash tolerance as follows:

```
20hp  -- 0.004-0.012" (0.1-0.3mm)
25hp  -- 0.004-0.012" (0.1-0.3mm)
40hp  -- 0.002-0.006" (0.05-0.15mm)
```

If the reverse gear backlash is not as specified, add or subtract shim material behind the reverse gear to bring the measured dial indicator reading to the required value.

15- Apply a coating of Quicksilver water proof lubricant to the gearcase sealing ring. Install the ring in the groove of the lower gearcase half. Trim the sealing ring to leave 0.003" (0.8mm) overhanging on each side of the bearing carrier bore, as indicated in the accompanying line drawing. Install the lower gearcase over the upper gearcase with the locating pin on the forward bearing race and the pin on the bearing carrier indexed into the proper notches.

16- Install and tighten the attaching bolts securely.

WATER PUMP INSTALLATION

17- Slide a new gasket down the driveshaft, followed by the water pump base.

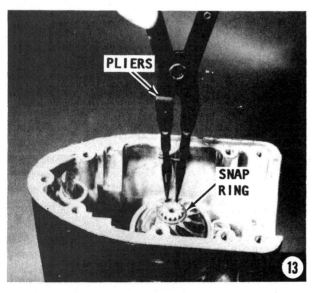

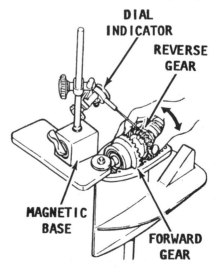

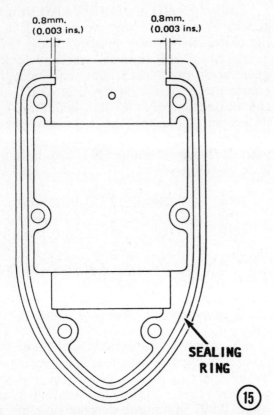

18- Install the two dowel pins on the top surface of the water pump base. Install the gasket, and then the outer plate down the driveshaft and index them over the two dowel pins.

19- Fit the Woodruff key into the driveshaft. Just a "dab" of grease on the key will help hold the key in place. Slide the water pump impeller down the driveshaft with the rubber membrane on the top side and the keyway in the impeller indexed over the Woodruff key. **TAKE CARE** not to damage the membrane. Coat the impeller blades with multi-purpose water resistant lubricant.

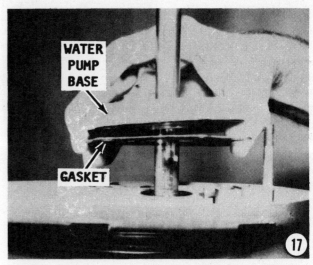

20- Slide the insert cartridge, another gasket, and finally the water pump housing down the driveshaft. Rotate the insert cartridge **COUNTERCLOCKWISE** over the impeller to tuck in the blades. Seat all parts over the two dowel pins and secure the water pump housing with the three washers and bolts. Tighten the bolts to a torque value of 5.8 ft lbs (8Nm).

INSTALLATION — LOWER UNIT TO INTERMEDIATE HOUSING

The following steps apply to all Lower Unit "C" models covered in this section.

Apply just a **"dab"** of multi-purpose water resistant lubricant to the splines at the upper end of the driveshaft.

BAD NEWS

An excessive amount of lubricant on top of the driveshaft to crankshaft splines will be trapped in the clearance space. This trapped lubricant will not allow the driveshaft to fully engage with the crankshaft.

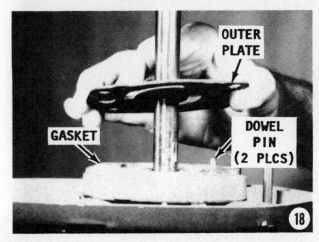

ASSEMBLING TYPE "C"

Apply some of the same lubricant to the end of the water tube in the intermediate housing.

Apply just a "dab" of lubricant, to the indexing pin on the mating surface of the lower unit.

Shift the lower unit into reverse gear by pushing down on the shift shaft. Check to be sure the shift lever is also in the reverse position.

Begin to bring the intermediate housing and lower gear housing together.

SPECIAL WORDS

The remainder of this step takes time and **PATIENCE**. Success will probably not be achieved on the first attempt. **THREE** items must mate at the same time before the lower unit can be seated against the intermediate housing.

a- The top of the driveshaft on the lower unit must index with the lower end of the crankshaft.

b- The water tube in the intermediate housing must slide into the grommet on the water pump housing.

c- The top of the lower shift rod in the lower unit must slide into the shift connector.

As the two units come closer, rotate the propeller shaft slightly to index the upper end of the driveshaft with the crankshaft. At the same time, feed the water tube into the water tube grommet, and feed the lower shift rod into the threaded connector on the upper shift rod.

Push the lower unit housing and the intermediate housing together. **TAKE CARE** not to bend the long narrow water tube. The pin on the upper surface of the lower unit will index with a matching hole in the lower surface of the anti-cavitation plate.

Secure the threaded connector onto the lower shift rod at least **FIVE** turns, and then lock it in place with the locknut.

Apply Loctite Type "A" to the threads of the bolts used to secure the lower unit to the intermediate housing. Install and tighten the bolts to a torque value of 16 ft lbs (22Nm).

Install the anode, if it was removed.

Check the Shifting

Operate the shift lever through all gears. The shifting should be smooth and the propeller should rotate in the proper direction when the flywheel is rotated by hand in a **CLOCKWISE** direction. Naturally the propeller should not rotate when the unit is in neutral.

SHIFT LINKAGE ADJUSTMENT

This task is very much a "trial and error" procedure.

Back off the locknut, and adjust the threaded connector to raise or lower, the lower shift rod until the shift positions are correct. Make sure there are at least five threads engaging the lower shift rod after the adjustment is completed. Tighten the locknut against the threaded connector to hold the adjustment.

These early 20hp, 25hp, and 40hp models are equipped with a shift limit screw on the starboard side of the lower unit housing. Tighten the screw and check the shift ac-

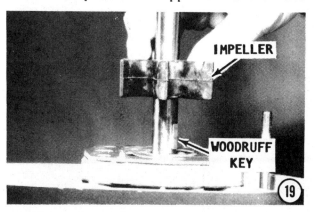

tion. If tightening the screw causes the shift rod to "freeze", the end of the screw did not index into the slot in the shift rod, in which case, the rod must be removed and repositioned.

CLOSING TASKS

Release the tilt lock lever and lower the outboard to the normal operating position.

FILLING THE LOWER UNIT

21- Remove the oil level plug and the drain plug.

SPECIAL WORDS

Temporarily, remove the spark plug lead/s to prevent the powerhead from starting. Shift the unit into **FORWARD** gear and rotate the propeller clockwise, from time to time, while filling the lower unit. This action will help dislodge any air pockets trapped inside and will speed the filling task.

Fill the lower unit with gearcase lubricant or Hypoid gear oil 90 weight until lubricant escapes from the top hole. Install both plugs and clean any excess lubricant from the lower unit.

PROPELLER INSTALLATION

22- Install a new shear pin through the propeller shaft. Slide the propeller onto the shaft with the internal splines of the propeller indexing with the splines on the shaft. Wedge a block of wood between one of the propeller blades and the anti-cavitation plate to prevent the propeller from rotating. Install the propeller nut. Tighten the nut securely using a pair of large channel lock pliers or an adjustable wrench. Check to determine if the holes in the nut are aligned with the hole in the propeller shaft. If not, tighten the nut just a "whisker" more until the holes are aligned. Insert and secure the cotter pin.

TRIM TAB ADJUSTMENT

The trim tab should be positioned to enable the helmsperson to handle the boat with equal ease to starboard and port at normal cruising speed. If the boat seems to turn more easily to starboard, loosen the socket head screw and move the trim tab trailing edge to the right. Move the trailing edge of the trim tab to the left if the boat tends to turn more easily to port.

Mount the engine in a test tank or body of water.

CAUTION

Water must circulate through the lower unit to the powerhead anytime the powerhead is operating to prevent damage to the water pump in the lower unit. Just five seconds without water will damage the water pump impeller.

Start the engine and check the completed work for satisfactory operation, shifting, and **NO** leaks.

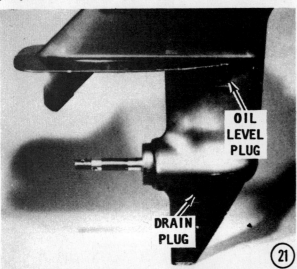

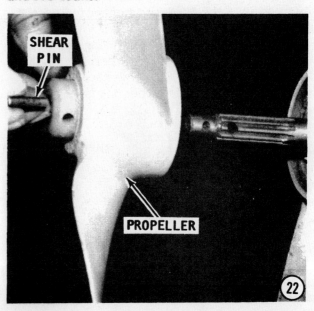

10
REMOTE CONTROLS

10-1 INTRODUCTION

Remote controls are seldom obtained from the original outboard manufacturer. Shift boxes and steering arrangements may be added by the boat manufacturer. Because of the wide assortment, styles, and price ranges of remote controls, the boat manufacturer, or customer, has a wide selection from which to draw, when outfitting the boat.

Therefore, the procedures and suggestions in this chapter are for the "Commander" shift controls widely used with the outboard units covered in this manual. The procedures are specific and in enough detail to allow troubleshooting, repair, and adjustment of the "Commander" shift unit for maximum comfort, performance and safety.

WOULD YOU BELIEVE

Probably 90% of steering cable problems are directly caused by the system not being operated, just sitting idle during the off-season. Without movement, all steering cables have a tendency to "freeze". **Would you also believe:** Service shops report almost 50% of the boat cables replaced every year are due to lack of movement. Therefore, during off-season when the boat is laid up in a yard, or on a trailer alongside the house, take time to go aboard and operate the steering wheel from hard-over to hard-over several times.

These sections provide step-by-step detailed instructions for the complete disassembly, cleaning and inspection, and assembly of the "Commander" shift box. Disassembly may be stopped at any point desired and the assembly process begun at that point. However, for best results and maximum performance, the entire system should be serviced if any one part is disassembled for repair.

An exploded drawing of the "Commander" shift box is presented between the assembling and disassembling procedures. This diagram will be most helpful in gaining an appreciation of how the shift box functions and the relationship of individual parts to one another.

If at all possible, keep the parts in order as they are removed. Make an effort to keep the work area clean and keep disassembled parts covered with a shop cloth to prevent contamination.

GOOD WORDS

If the control cable has a "Zerk" fitting at the engine end, the cable **MUST** be retracted, then the fitting lubricated with Quicksilver Multi-Purpose lubricant or Quicksilver 2-4C Lubricant.

STEERING CHECKS

The steering system may be checked by performing a few very simple tests. First, move the steering wheel from hard-over to hard-over, port and starboard several times. The outboard unit should move without any sign of stiffness. If binding or stiffness is encountered, the cause may be a defect in the swivel bearing.

Next, remove the steering bolt at the outboard unit, and again turn the steering wheel back-and-forth from hard-over to hard-over, port and starboard several times. If there is any sign of stiffness, it is proof the problem is with the cables. They may be corroded or there may be a defect in the steering mechanism.

10-2 REMOTE CONTROLS

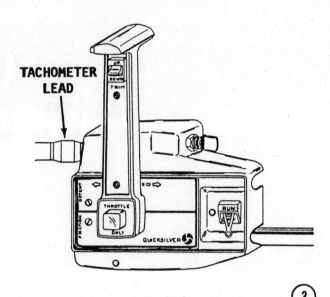

10-2 COMMANDER CONTROL SHIFT BOX REMOVAL AND DISASSEMBLING

The following detailed instructions cover removal and disassembly of the "Commander" control shift box from the mounting panel in the boat.

1- Turn the ignition key to the **OFF** position. Disconnect the high tension leads from the spark plugs, with a twisting motion.

2- Disconnect the remote control wiring harness plug from the outboard trim/tilt motor and pump assembly.

3- Disconnect the tachometer wiring plug from the forward end of the control housing.

4- Remove the three locknuts, flat washers, and bolts securing the control housing to the mounting panel. One is located next to the **RUN** button (the ignition safety stop switch), and the second is beneath the control handle on the lower portion of the plastic case. The third is located behind the control handle when the handle is in the **NEUTRAL** position. Shift the handle into **FORWARD** or **REVERSE** position to remove the bolt, then shift it back into the **NEUTRAL** position for the following steps.

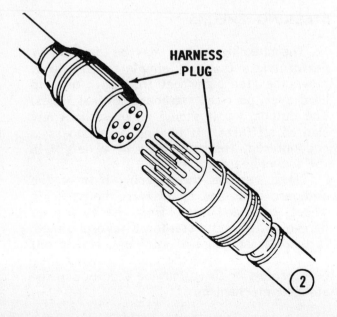

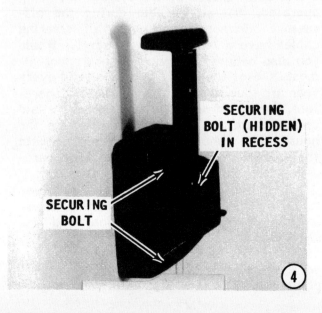

DISASSEMBLING SHIFT BOX 10-3

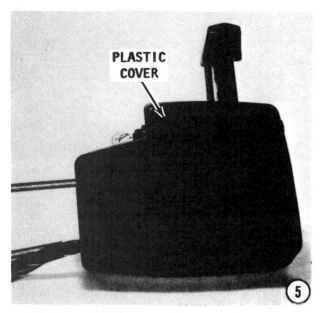

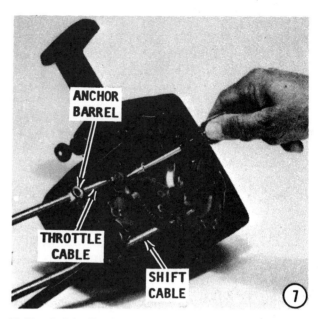

5- Pull the remote control housing away and free of the mounting panel. Remove the plastic cover from the back of the housing. Lift off the access cover from the housing. (Some "Commander" remote control units do not have an access cover.)

6- Remove the two screws securing the cable retainer over the throttle cable, wiring harness, and shift cable. Unscrew the two Phillips-head screws securing the back cover to the control module, and then lift off the cover.

Throttle Cable Removal

7- Loosen the cable retaining nut and raise the cable fastener enough to free the throttle cable from the pin. Lift the cable from the anchor barrel recess.

8- Remove the grommet.

Shift Cable Removal

9- Shift the outboard unit into **REVERSE** gear by depressing the neutral lock bar on the control handle and moving the control handle into the **REVERSE** position. **LOOSEN**, but do not remove, the shift cable retainer nut with a 3/8" deep socket as far as it will go without removing it. Raise the shift cable fastener enough to free the shift cable from the pin.

DO NOT attempt to shift into **REVERSE** while the cable fastener is loose. An attempt to shift may cause the cable fastener to strike the neutral safety microswitch and cause it damage.

Lift the wiring harness out of the cable anchor barrel recess and remove the shift cable from the control housing.

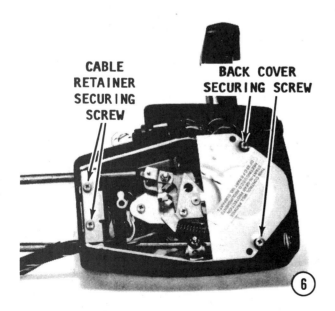

10-4 REMOTE CONTROLS

Control Handle Removal For Power Trim/Tilt With Toggle Trim Switch Or Push-Button Trim Switch

GOOD WORDS

For non-power trim/tilt units, it is not necessary to remove the cover of the control handle. If servicing one of these units, proceed directly to Step 13. All others perform Steps 11 and 12.

10- Depress the **NEUTRAL** lock bar on the control handle and shift the control handle back to the **NEUTRAL** position. Remove the two Phillips head screws which secure the cover to the handle, and then lift off the cover. The push button trim switch will come free with the cover, the toggle trim switch will stay in the handle body.

Unsnap and then remove the wire retainer. Carefully unplug the trim wires and straighten them out from the control panel hub for ease of removal later.

11- Back-off the set screw at the base of the control handle to allow the handle to be removed from the splined control shaft.

12- Grasp the "throttle only" button and pull it off the shaft.

SPECIAL WORDS

Take care not to damage the trim wires when removing the control handle, on power trim models.

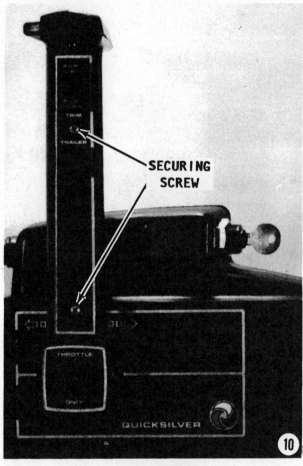

DISASSEMBLING SHIFT BOX   10-5

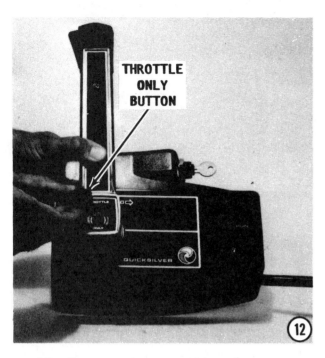

13- Remove the control handle.

14- Lift the neutral lockring from the control housing.

TAKE CARE to support the weight of the control housing to avoid placing any unnecessary stress on the control shaft during the following disassembling steps.

15- Remove the three Phillips-head screws securing the control module to the plastic case. Two are located on either side of the bearing plate and one is in the recess where the throttle cable enters the control housing.

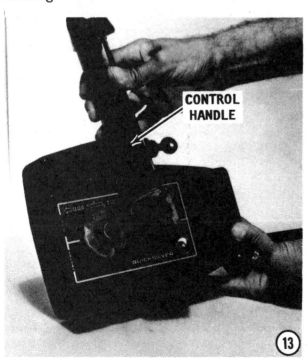

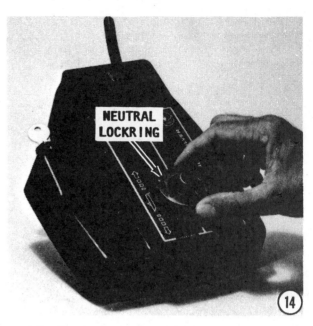

16- Back-out the detent adjustment screw and the control handle friction screw until their heads are flush with the control module casing. This action will reduce the pre-load from the two springs on the detent ball for later removal.

GOOD WORDS

As this next step is performed, count the number of turns for each screw as they are backed-out and record the figure somewhere. This will be a tremendous aid during assembling.

17- Remove the two locknuts securing the neutral safety switch to the plate assembly and lift out the micro-switch from the recess in the assembly.

18- Remove the Phillips-head screw securing the retaining clip to the control module.

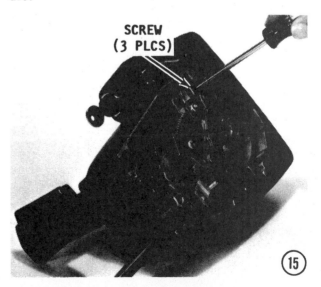

10-6 REMOTE CONTROLS

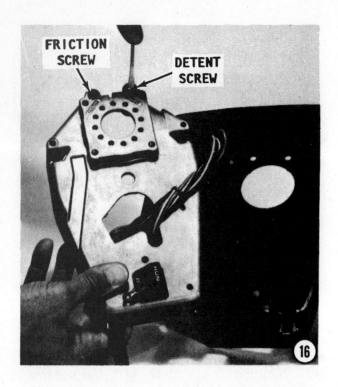

19- Support the module in your hand and tilt it until the shift gear spring, shift nylon pin (earlier models have a ball), shift gear pin, another ball the shift gear ball (inner), fall out from their recess. If the parts do not fall out into your hand, attach the control handle and ensure the unit is in the **NEUTRAL** position. The parts should come free when the handle is in the **NEUTRAL** position.

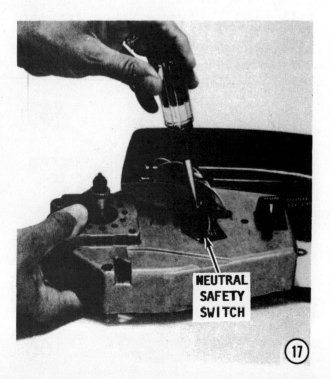

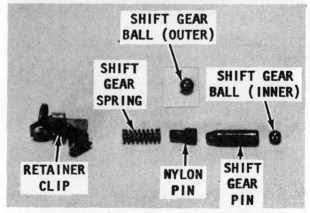

Arrangement of parts from the control module recess. As the parts are removed and cleaned, keep them in order, ready for installation.

DISASSEMBLING SHIFT BOX 10-7

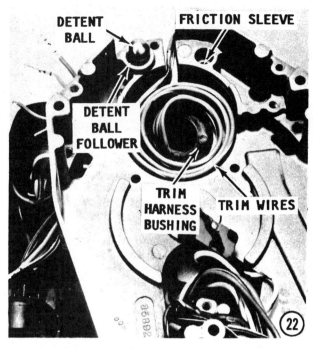

20- Remove the three Phillips-head screws securing the bearing plate assembly to the control module housing.

21- Lift out the bearing plate assembly from the control module housing.

Power Trim/Tilt Units Only

22- Uncoil the trim wires from the recess in the remote control module housing and lift them away with the trim harness bushing attached.

All Units

23- Remove the detent ball, the detent ball follower, and the two compression springs (located under the follower), from their recess in the control module housing.

24- If it is not part of the friction pad, remove the control handle friction sleeve from the recess in the control module housing.

25- Pull the throttle link assembly from the module. Remove the compression spring from the throttle lever. It is not necessary to remove this spring unless there is cause to replace it. At this point, there is the

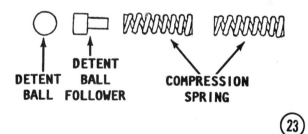

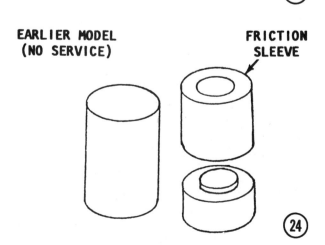

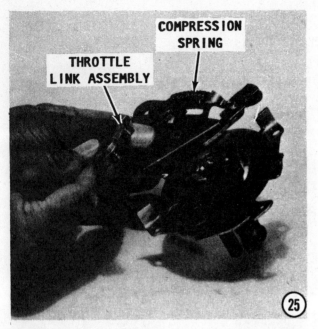

least amount of tension on the compression spring, therefore, now would be the time to replace it, if required.

26- Lift the shift pinion gear (with attached shift lever), off the pin on the bearing plate. The nylon bushing may come away with the shift lever or stay on the pin. Remove the shift lever and shift pinion gear as an assembly. **DO NOT** attempt to separate them. Both are replaced if one is worn.

Non-Power Trim/Tilt Units Only

27- Remove the trim harness bushing and wiring harness retainer from the control shaft. (On non-power trim/tilt units these two items act as spacers.)

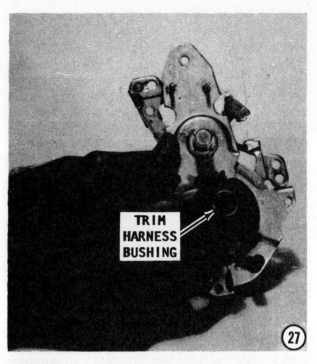

All Units

28- Remove the shift gear retaining ring from its groove with a pair of Circlip pliers.

SPECIAL WORDS

If the Circlip slipped out of its groove, this would allow the shift gear to ride up on the shaft and cause damage to the small parts contained in its recess. The shift gear ball (inner), the shift gear pin, the shift gear ball (outer), or the nylon pin and **PARTICULARLY** the shift gear spring **MUST** be inspected closely.

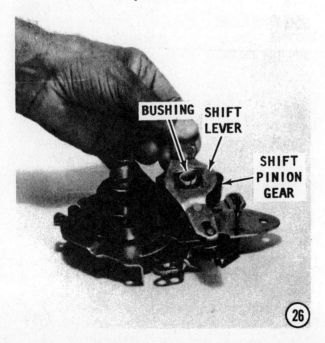

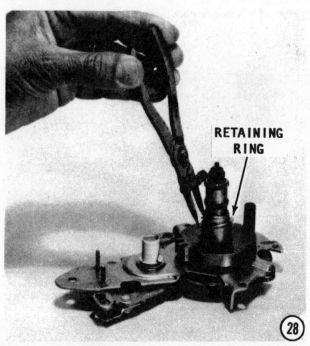

CLEANING & INSPECTING 10-9

29- Lift the gear from the control shaft.
30- Remove the "throttle only" shaft pin and "throttle only" shaft from the control shaft.
31- Remove the step washer from the base of the bearing plate.

CLEANING AND INSPECTING

Clean all metal parts with solvent, and then blow them dry with compressed air.

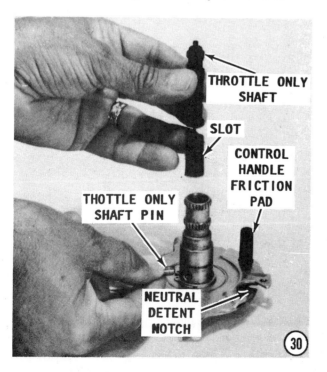

NEVER allow nylon bushings, plastic washers, nylon pins, wiring harness retainers, and the like, to remain submerged in solvent more than just a few moments. The solvent will cause these type parts to expand slightly. They are already considered a "tight fit" and even the slightest amount of expansion would make them very difficult to install. If force is used, the part is most likely to be distorted.

Inspect the control housing plastic case for cracks or other damage allowing moisture to enter and cause problems with the mechanism.

Carefully check the teeth on the shift gear and shift lever for signs of wear. Inspect all ball bearings for nicks or grooves which would CAUSE them to bind and fail to move freely.

Closely inspect the condition of all wires and their protective insulation. Look for exposed wires caused by the insulation rubbing on a moving part, cuts and nicks in the insulation and severe kinking which could cause internal breakage of the wires.

Inspect the surface area above the groove in which the Circlip is positioned for signs of the Circlip rising out of the groove. This would occur if the clip had lost its "spring" or worn away the top surface of the groove as mentioned previously in Step 29. If the Circlip slipped out of its groove, this would allow the shift gear to ride up on the shaft and cause damage to the small parts contained in its recess. The shift gear ball (inner), the shift gear pin, the shift gear ball (outer), or the nylon pin and **PARTICULARLY** the shift gear spring **MUST** be inspected closely.

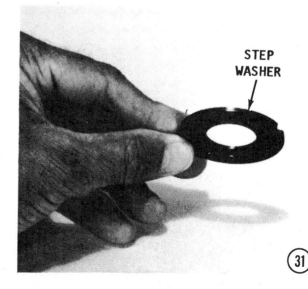

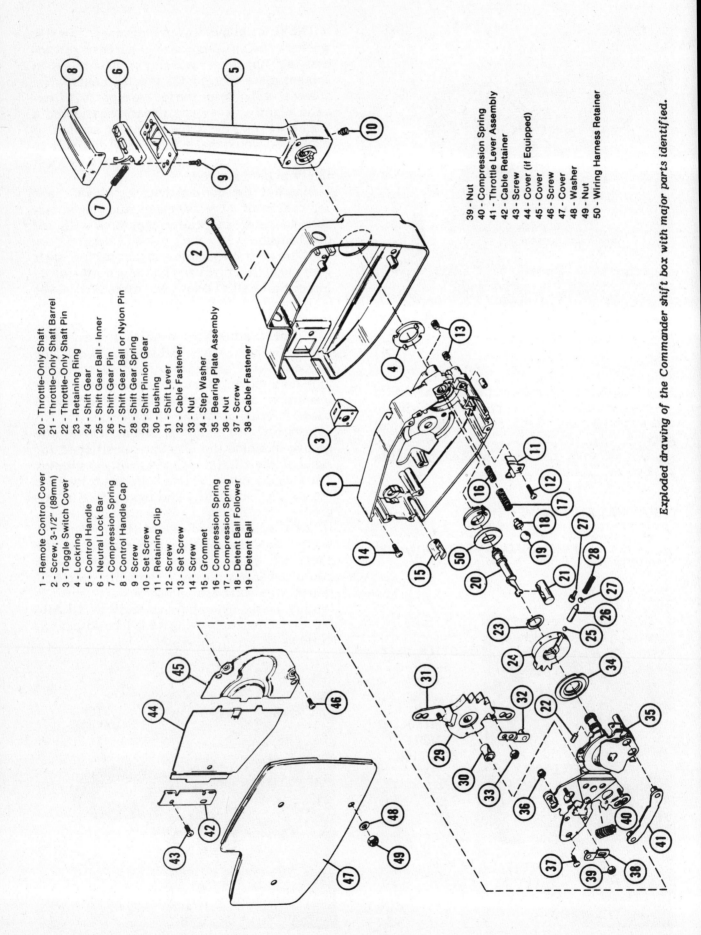

Exploded drawing of the Commander shift box with major parts identified.

ASSEMBLING SHIFT BOX

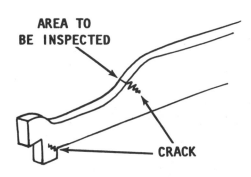

The throttle-only shaft should be inspected for wear along the ramp, as indicated.

Inspect the "throttle only" shaft for wear along the ramp. In early model units, this shaft was made of plastic. Later models have a shaft of stainless steel. Check for excessive wear or cracks on the ramp portion of the shaft, as indicated in the accompanying illustration. Also check the lower "stop" tab to be sure it has not broken away.

SPECIAL WORDS

Good shop practice dictates a thin coat of Multipurpose Lubricant be applied to all moving parts as a precaution against the "enemy" moisture. Of course the lubricant will help to ensure continued satisfactory operation of the mechanism.

ASSEMBLING AND INSTALLATION COMMANDER CONTROL SHIFT BOX

FIRST, THESE WORDS

The Commander control shift box, like others, has a number of small parts that MUST be assembled in only one order -- the proper order. Therefore, the work should not be "rushed" or attempted if the person assembling the unit is "under pressure". Work slowly, exercise patience, read ahead before performing the task, and follow the steps closely.

1- Place the step washer over the control shaft and ensure the steps of the washer seat onto the base of the bearing plate.

2- Rotate the control shaft until the "throttle only" shaft pin hole is aligned centrally between the neutral detent notch and the control handle friction pad. Lower the "throttle only" shaft into the barrel of the control shaft with the wide slot in the "throttle only" shaft aligned with the line drawn on the accompanying illustration. Secure the shaft in this position with the "throttle only" shaft pin.

SPECIAL WORDS

When the pin is properly installed, it should protrude slightly in line with the plastic bushing, as shown in the accompanying illustration.

Make an attempt to gently pull the "throttle only" shaft out of the control shaft. The attempt should fail, if the shaft and pin are properly installed.

3- Place the shift gear over the control shaft, and check to be sure the "throttle only" shaft pin clears the gear.

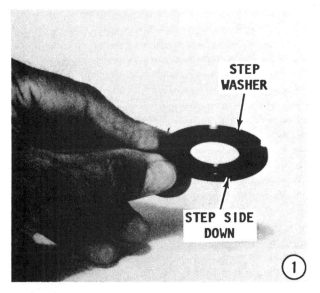

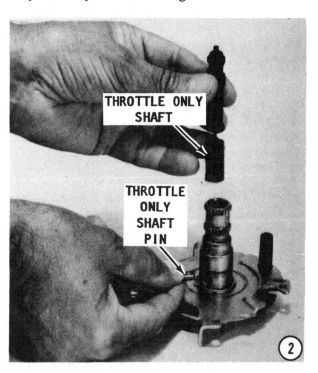

10-12 REMOTE CONTROLS

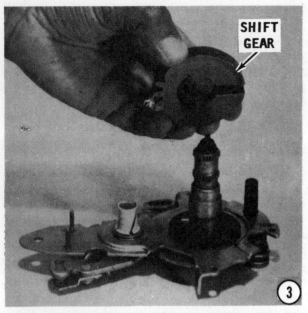

4- Install the retaining ring over the control shaft with a pair of Circlip pliers. Check to be sure the ring snaps into place within the groove.

Non-Power Trim/Tilt Units Only

5- Slide the wiring harness retainer and the trim harness bushing over the control shaft. The trim harness bushing is placed "stepped side" UP and the notched side toward the forward side of the control housing.

Power Trim/Tilt Units Only

6- Insert the trim harness bushing into the recess of the remote control housing and

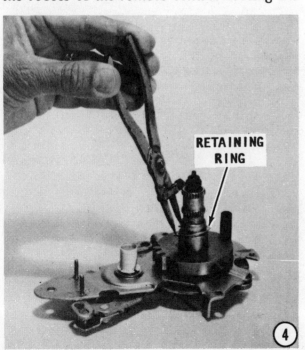

carefully coil the wires, as shown in the accompanying illustration. Ensure the black line on the trim harness is positioned at the exact point shown for correct installation. The purpose of the coil is to allow slack in the wiring harness when the control handle is shifted through a full cycle. The bushing and wires move with the handle.

All Units

7- Position the bushing, shift pinion gear and shift lever onto the pin on the bearing plate, with the shift gear indexing with the shift pinion gear.

8- Install the two compression springs,

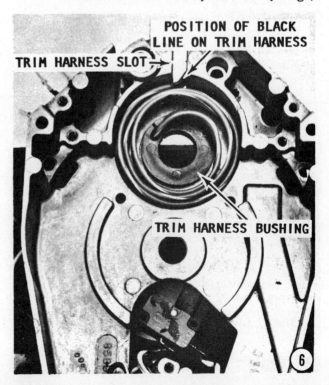

ASSEMBLING SHIFT BOX 10-13

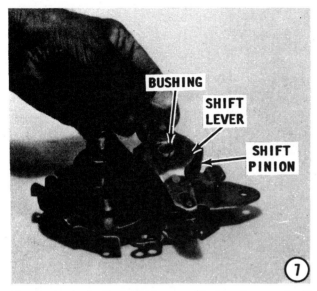

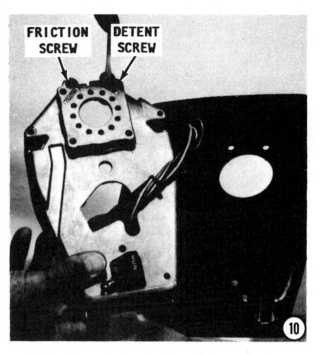

the detent ball follower and the detent ball into their recess in the control module housing.

9- If the friction sleeve is not a part of the friction pad, then place the control handle friction sleeve into its recess in the control module housing.

SPECIAL WORDS

In Step 16 of the disassembling procedures, instructions were given to count the number of turns required to remove the detent adjustment screw and the control handle friction screw. The number of turns is now necessary for ease in performing the next step.

10- Thread the detent adjustment screw and the control handle friction screw the exact number of turns as recorded during Step 16 of the disassembling procedures. A fine adjustment may be necessary after the unit is completely assembled.

11- Place the compression spring (if removed) in position on the shift lever and shift pinion gear assembly against the bearing plate, as shown. Use a rubber band to secure the shift pinion gear to the bearing plate. Lower the complete bearing plate assembly into the control module housing.

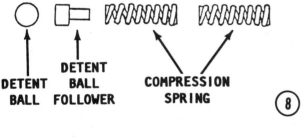

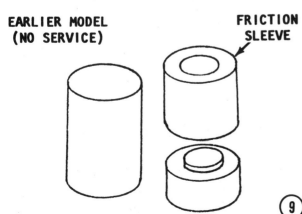

10-14 REMOTE CONTROLS

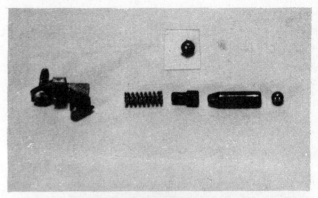

Arrangement of parts, cleaned and ready for installation into the control module recess.

12- Secure the bearing plate assembly to the control module housing with the three Phillips head screws, remove the rubber band.

13- Insert the gear shift ball (inner) into the recess of the shift gear and hole in the "throttle only" shaft barrel. Now, insert the shift gear pin into the recess with the rounded end of the pin away from the control shaft. Insert the nylon pin or shift gear ball (outer) into the same recess. Next, insert the gear shift spring.

14- Hold these small parts in place and at the same time secure them with the retaining clip and the Phillip head screw. (On power trim/tilt units, this retaining clip also secures the trim wire to the control module.)

15- Insert the neutral safety microswitch into the recess of the plate assembly and secure it with the two locknuts.

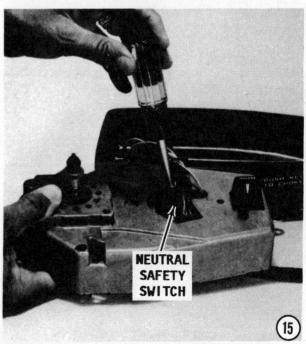

ASSEMBLING SHIFT BOX 10-15

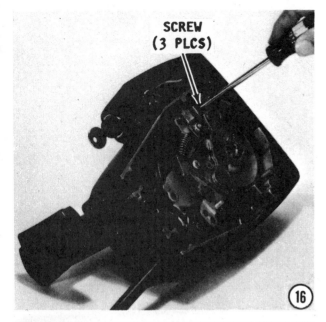

16

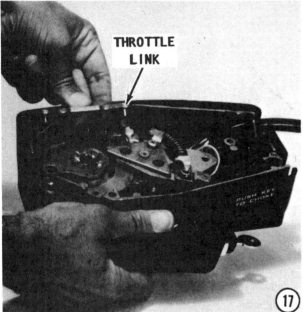

17

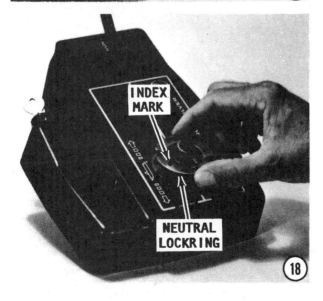

18

16- Secure the control module to the plastic control housing case with the three Phillips head screws. Two are located on either side of the bearing plate and the third in the recess where the throttle cable enters the control housing.

17- Temporarily install the control handle onto the control shaft. Shift the unit into forward detent **ONLY**, not full forward, to align the holes for installation of the throttle link. After the holes are aligned, remove the handle. Install the throttle link.

18- Again, temporarily install the control handle onto the control shaft. This time shift the unit into the **NEUTRAL** position, and then remove the handle. Place the neutral lockring over the control shaft, with the index mark directly beneath the small boot on the front face of the cover.

19- Install the control onto the splines of the control shaft. **TAKE CARE** not to cut, pinch, or damage the trim wires on the power trim/tilt unit.

CRITICAL WORDS

When positioning the control handle, ensure the trim wire bushing is aligned with its locating pin against the corresponding slot in the control handle. On a Power Trim/Tilt unit only: if this bushing is **NOT** installed correctly, it will not move with the control handle as it is designed to move -- when shifted. This may pinch or cut the trim wires, causing serious problems.

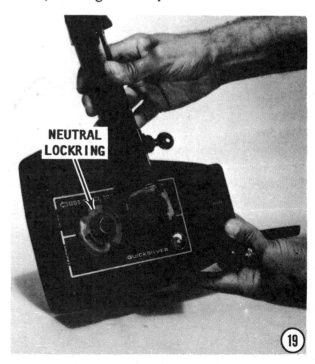

19

10-16 REMOTE CONTROLS

On a Non-Power Trim/Tilt unit, misplacement of this bushing (it is possible to install this bushing upside down) will not allow the control handle to seat properly against the lockring and housing. This situation will lead to the Allen screw at the base of the control handle to be incorrectly tightened to seat against the splines on the control shaft, instead of gripping the smooth portion of the shaft. Subsequently the control handle will feel "sloppy" and could cause the neutral lock to be ineffective.

WARNING
IF THIS HANDLE IS NOT SEATED PROPERLY, A SLIGHT PRESSURE ON THE HANDLE COULD THROW THE LOWER UNIT INTO GEAR, CAUSING SERIOUS INJURY TO CREW, PASSENGERS, AND THE BOAT.

20- Push the "throttle only" button in place on the control shaft.

21- Ensure the control handle has seated properly, and then tighten the set screw at the base of the handle to a torque value of 70 in. lbs (7.9Nm).

SAFETY WORDS
FAILURE to tighten the set screw to the required torque value, could allow the handle to disengage with a loss of throttle and shift control. An extremely DANGEROUS condition.

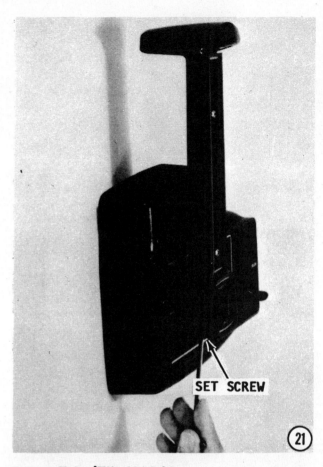

Power Trim/Tilt Models
or
Non-Power Models
If Handle Cover Was Removed

22- Slide the hooked end of the neutral lock rod into the slot in the neutral lock release. Route the trim wires in the control handle in their original locations. Connect them with the wires remaining in the handle and secure the connections with the wire retainer. Install the handle cover and tighten the two Phillips head screws.

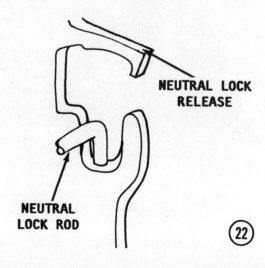

ASSEMBLING SHIFT BOX 10-17

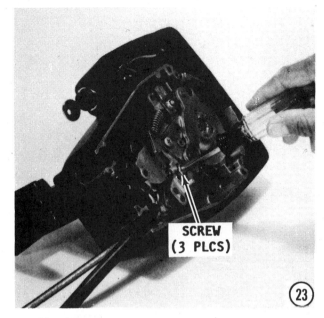

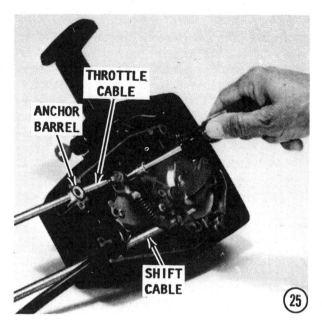

23- Move the wiring harness clear of the barrel recess. Thread the shift cable anchor barrel to the end of the threads, away from the cable converter, and place it into the recess. Hook the pin on the end of the cable fastener through the outer hole in the shift lever. Depress the **NEUTRAL** lock bar on the control handle and shift the handle into the **REVERSE** position. **TAKE CARE** to ensure the cable fastener will clear the neutral safety micro-switch. The access hole is now aligned with the locknut.

STOP

Check to be sure the pin on the cable fastener is all the way through the cable end and the shift lever. A pin partially engaging the cable and the shift lever may cause the cable fastener to **BEND** when the nut is tightened.

Tighten the locknut with a 3/8" deep socket to a torque value of 20 to 25 in. lbs (2.26 to 2.82 Nm). Position the wiring harness over the installed shift cable.

24- Install the grommet into the throttle cable recess.

25- Thread the throttle cable anchor barrel to the end of the threads, away from the cable connector, and then place it into the recess over the grommet. Hook the pin on the end of the cable fastener through the outer hole in the shift lever.

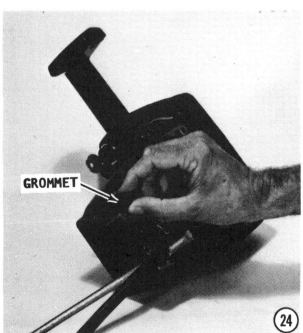

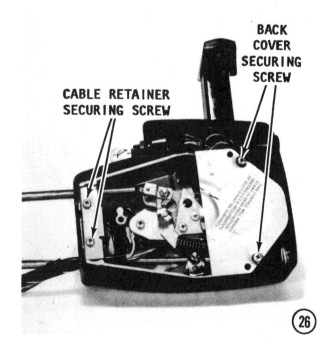

10-18 REMOTE CONTROLS

STOP AGAIN

Check to be sure the pin on the cable fastener is all the way through the cable end and the throttle lever. A pin partially engaging the cable and throttle lever may cause the cable fastener to **BEND** when the nut is tightened.

Tighten the locknut to a torque value of 20 to 25 in. lbs (2.26 to 2.82 Nm).

26- Position the control module back cover in place and secure it with the two Phillips-head screws. Tighten the screws to a torque value of 60 in. lbs (6.78 Nm). Install the cable retainer plate over the two cables and secure it in place with the two Phillip-head screws.

27- Place the plastic access cover over the control housing.

28- Position the control housing in place on the mounting panel and secure it with the three long (3-1/2") bolts, flat washers, and locknuts. One is located next to the **RUN** button (the ignition safety stop switch). The second is beneath the control handle on the power portion of the plastic case. The third bolt goes in behind the control handle when the handle is in the **NEUTRAL** position.

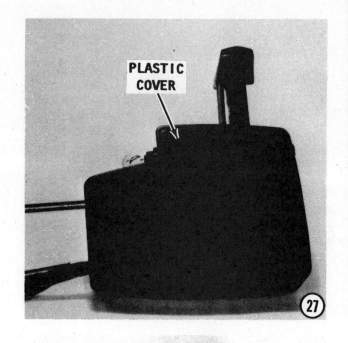

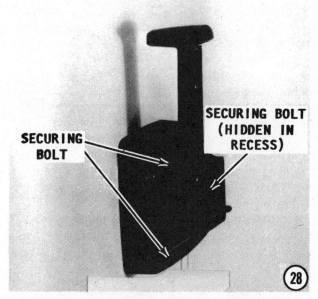

Commander shift box ready for installation.

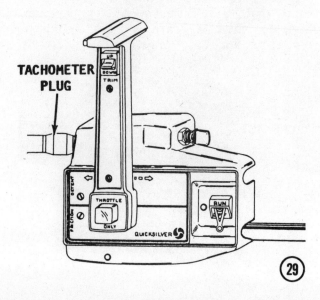

ASSEMBLING SHIFT BOX 10-19

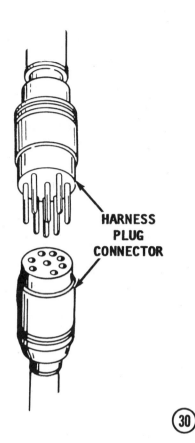

Therefore, in order to install this bolt, shift the handle into the **FORWARD** or the **REVERSE** position, and then install the bolt. After the bolt is secure, shift the handle back to the **NEUTRAL** position for the next few steps.

29- Connect the tachometer wiring plug to the forward end of the control housing.

GOOD WORDS

Clean the prongs of the connector with crocus cloth to ensure the best connection possible. Exercise care while cleaning to prevent bending the prongs.

30- Connect the remote control wiring harness plug from the outboard trim/tilt motor and pump assembly.

31- Install the high-tension leads to their respective spark plugs.

32- Route the wiring harness alongside the boat and fasten with the "Sta-Straps". Check to be sure the wiring will not be pinched or chafe on any moving part and will not come in contact with water in the bilge. Route the shift and throttle cables the best possible way to make large bends and as few as possible. Secure the cables approximately every three feet (one meter).

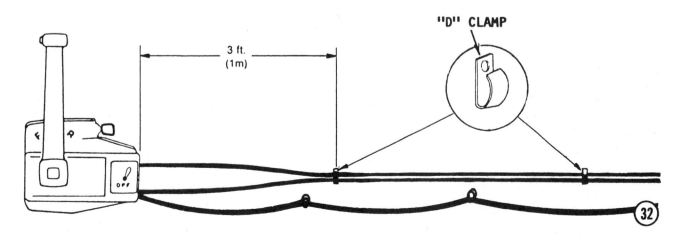

10-20 REMOTE CONTROLS

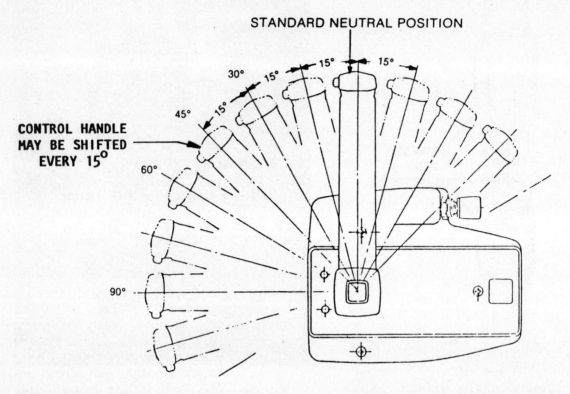

The neutral position of the remote control handle may be changd to any one of a number of convenient angles to meet the owner's preference. The change is accomplished by shifting the handle one spline on the shaft at a time. Each spline equals 15° of arc, as shown. The procedures on Page 10-15 explain the positioning in detail.

11
HAND REWIND STARTER

11-1 INTRODUCTION AND CHAPTER COVERAGE

Six different hand rewind starters are used on the powerheads covered in this manual. For simplicity, these six units have been identified as "A" thru "F".

The "A" hand rewind starter is installed on the single cylinder, 2hp models only and is covered in Section 11-2.

The "B" hand rewind starter is installed on the air-cooled, single cylinder 3.5hp and 5hp models and is covered in Section 11-3.

The "C" hand rewind starter is installed on the water-cooled single cylinder 4hp, and 5hp models; 2-cylinder 8hp (since 1980); 9.9hp; 15hp (from 1980); and 20hp, 25hp, 28hp, 30hp, and 40hp models. This starter is covered in Section 11-4.
Because the "C" is used on ten outboard models, at press time, slight variations in appearance and design may be noted. However, the procedures outlined in Section 11-4 are valid for all "C" hand starters.

The "D" hand rewind starter is installed on 2-cylinder 8hp models 1977-1979, and is covered in Section 11-5.

The "E" hand rewind starter is installed on 2-cylinder 15hp models 1977-1979, and is covered in Section 11-6.

The "F" hand rewind starter is installed on all 2 cylinder 48hp models, and is covered in Section 11-7.

All outboard units covered in this manual over 48hp are equipped with an electric cranking system.

11-2 SERVICING TYPE "A" HAND REWIND STARTER

The "A" hand rewind starter is installed on the single cylinder, 2hp model.

REMOVAL AND DISASSEMBLING

Preliminary Tasks
1- Remove the screw on half of the spark plug cover. Remove four more screws securing half of the cowling. Separate the cowling half from which the screws were removed. Remove the the four screws securing the other half of the cowling and then remove it from the engine. The spark plug cover will remain attached to one of the cowling halves.

SPECIAL WORDS
Observe the different length screws used to secure the cowling halves to the powerhead. Remember their location as an aid during assembling.

11-2 HAND REWIND STARTER

BOLT (3 PLACES)

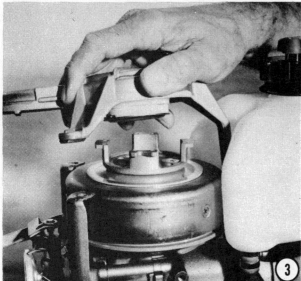

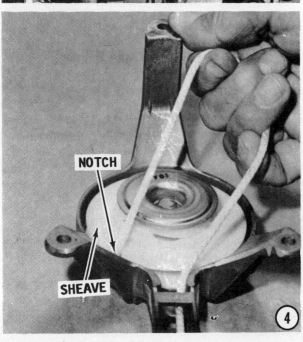

NOTCH
SHEAVE

2- Remove the three mounting bolts securing the legs of the hand rewind starter to the powerhead.

SPECIAL WORDS

A "no start-in-gear" protection device is not incorporated on single-cylinder powerheads equipped with the Type "A" hand starter.

3- Lift the hand rewind starter free of the powerhead.

4- Rotate the starter sheave to align the notch in the sheave with the starter handle. With the sheave in this position, pull the rope out a little and hook it into the notch. **CAREFULLY** allow the sheave to unwind **CLOCKWISE** until the spring has lost all its tension. Control the rotation with the rope secured in the notch to prevent the sheave from "free wheeling".

5- Invert the hand starter and remove the sheave retainer bolt with the proper size socket.

6- Lift the starter housing shaft free of the drive pawl. The drive pawl spring will remain attached to the starter housing shaft. Observe how the drive pawl spring was indexed over the peg on the drive pawl.

7- Using a small screwdriver, hold the back of the return spring to allow the drive pawl to be removed.

8- Carefully note, and **MARK** the recess holding the return spring before removing the spring.

CRITICAL WORDS

The recess **MUST** be marked to ensure proper installation. There are two recesses

BOLT

SERVICE TYPE "A"   11-3

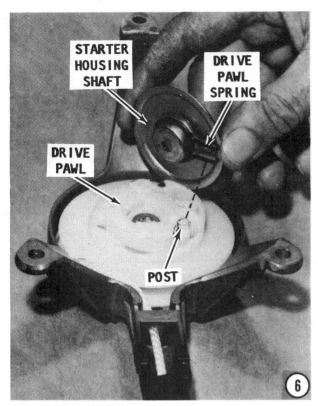

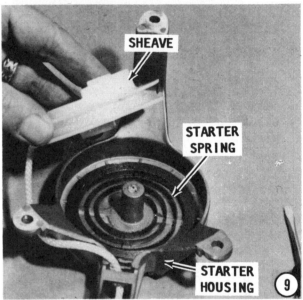

which appear to be identical, but actually are mirror images of each other.

After marking the recess, lift out the spring.

WARNING

THE REWIND SPRING IS A POTENTIAL HAZARD. The spring is under tremendous tension when it is wound -- a real **"TIGER"** in a cage! If the spring should accidentally be released, severe personal injury could result from being struck by the spring with force. Therefore, the following steps **MUST**

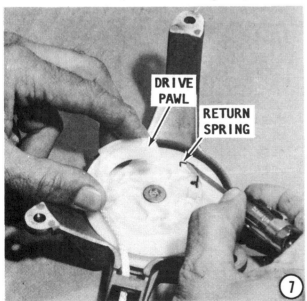

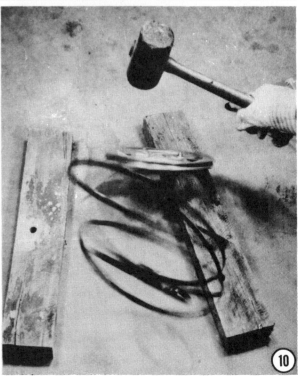

11-4 HAND REWIND STARTER

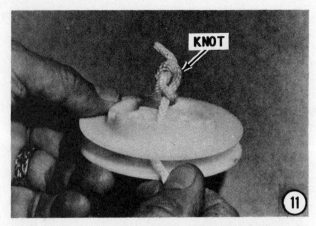

be performed with care to prevent personal injury to self and others in the area.

DO NOT attempt to remove the spring unless it is unfit for service and a new spring is to be installed.

9- **CAREFULLY** lift the sheave free of the starter housing, leaving the rewind spring still wound tightly inside the housing.

10- Obtain two pieces of wood, a short 2" x 4" (5cm x 10cm) will work fine. Place the two pieces of wood approximately 8" (20 cm) apart on the floor. Stand the housing on its legs between the two pieces of wood with the spring side facing **DOWN**.

AUTHORS' WORD

The accompanying illustration shows the spring being released from the same type,

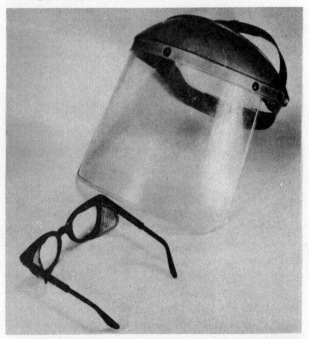

Protect your eyes with a face mask or safety glasses while working with the rewind spring, especially a used one. The spring is a real "tiger in a cage", almost 13' (4 m), of spring steel wound into less than 4" (about 10cm).

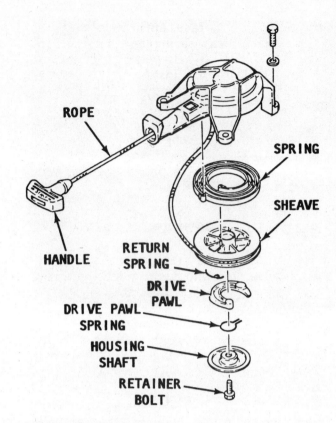

Exploded drawing of hand rewind starter "A", with major parts identified.

but different model rewind unit. **THEREFORE**, the principle is exactly the same. The procedure outlined in this step may be followed with safety.

Stand behind the wood, keeping away from the openings as the spring unwinds with considerable force and the housing will "jump" off the floor. Tap the housing a moderate blow with a soft mallet. The spring will fall and unwind almost instantly and with much **FORCE**.

11- If the rope is to be replaced, first push the rope back through the opening in the sheave, and then untie the knot and pull the rope free.

CLEANING AND INSPECTING

Wash all parts except the rope and the handle in solvent, and then blow them dry with compressed air.

Remove any trace of corrosion and wipe all metal parts with an oil dampened cloth.

Inspect the rope. Replace the rope if it appears to be weak or frayed. If the rope is frayed, check the holes through which the rope passes for rough edges or burrs. Remove the rough edges or burrs with a file and polish the surface until it is smooth. Inspect the starter spring end hooks. Replace the spring if it is weak, corroded or cracked. Inspect the inside surface of the sheave rewind recess for grooves or roughness. Grooves may cause erratic rewinding of the starter rope.

Coat the entire length of the used rewind spring (a new spring will be coated with lubricant from the package), with low-temperature lubricant.

ASSEMBLING AND INSTALLATION TYPE "A" HAND REWIND STARTER

The authors, the manufacturers, and almost anyone else who has handled the spring from this type rewind starter **STRONGLY** recommend a pair of safety goggles or a face shield be worn while the spring is being installed. As the work progresses a **"TIGER"** is being forced into a cage -- over 14' (4.3 m) of spring steel wound into about 4" (10.2cm) circumference. If the spring is accidentally released, it will lash out with tremendous ferocity and very likely could cause personal injury to the installer or other persons nearby.

SPECIAL WORDS

The rewind starter may be assembled with a new rewind spring or a used one. Procedures for assembling are **NOT** the same because the new spring will arrive held in a steel hoop already wound, lubricated, and ready for installation. The used spring must be manually wound into its recess.

NEW REWIND SPRING INSTALLATION

The situation may arise when it is only necessary to replace a broken spring. The following few procedures outline the tasks required to replace the spring. A new spring is already properly wound and will arrive in a special hoop. This hoop is designed to be used as an aid to installing the new spring.

New Spring Installation

Hook the outer end of the spring onto the insert in the starter housing, then place the spring into the housing. Seat the spring and then **CAREFULLY** remove the steel hoop.

Used Spring Installation

A used spring naturally will not be wound. Therefore, special instructions are necessary for installation.

SAFETY WORDS

Wear a good pair of gloves while winding and installing the spring. The spring will develop tension and the edges of the spring steel are extremely sharp. The gloves will prevent cuts to your palms and fingers.

Apply a light coating of water resistant anti-seize lubricant to the inside surface of the starter housing.

1- Wind the old spring loosely **CLOCKWISE** in one hand, as shown.

2- Insert the hook on the end of the spring into the notch of the recess. Feed the spring around the inner edge of the recess and at the same time rotate the housing **COUNTERCLOCKWISE.** The spring will be slippery with lubrication. Work slowly and with definite movements to prevent losing control of the spring. Proceed **WITH GREAT CARE.** Guide the spring into place.

3- Thread one end of the rope through the sheave and tie a figure "8" knot, as shown. Thread the other end of the rope

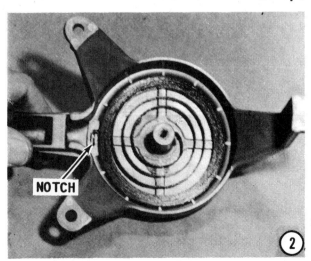

11-6 HAND REWIND STARTER

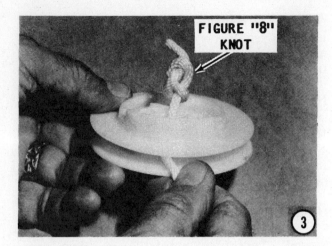

through the starter handle and again tie a figure "8" knot in the end.

4- Wind the rope 2-1/2 turns **COUNTERCLOCKWISE** around the sheave. Position the rope at the notch on the outer edge. Lower the sheave into the starter housing.

5- Insert a small screwdriver or an awl into the access window of the sheave and push the inner end of the spring into the recess under the sheave. This is not an easy task and may not be accomplished on the first try. If too much trouble is encountered, take a break, have a cup of coffee, cup of tea, whatever, and then try again. With patience, much patience, it can be done.

6- Congratulations on performing Step 5 successfully! Now, insert the return spring into the recess of the sheave which was previously marked during Step 8 of removal. If the recess was not marked during removal, rotate the sheave until both recesses are

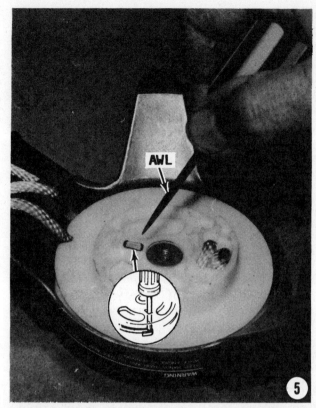

on the right and insert the spring into the upper one.

7- Hold back the return spring with a small screwdriver and at the same time install the drive pawl.

8- Place the starter housing shaft, with the drive pawl spring attached, over the drive pawl. Check to be sure the spring is centered over the post of the drive pawl.

9- Install and tighten the center bolt to a torque value of 5.8 ft lbs (8Nm).

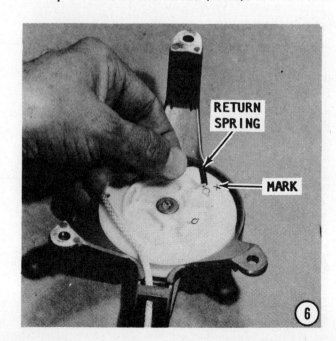

SERVICE TYPE "A"   11-7

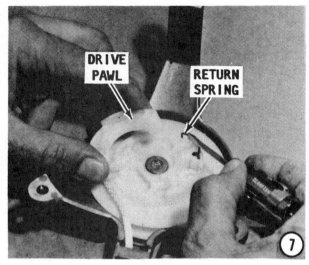

10- Pick up the slack in the rope, and then hold the rope firmly in the notch of the sheave. While the rope is being firmly held in the notch, wind the sheave **COUNTER-CLOCKWISE** until it can be wound no further. Now, ease the sheave just a little at a time until the notch in the sheave aligns with the starter handle. **SLOWLY** release the sheave and allow the rope to feed around the sheave as it unwinds, pulling in the slack rope.

Check the action of the rewind starter before proceeding with the installation. If all the rope is not taken in around the sheave as the spring unwinds, repeat this step, and then check again.

11- Position the rewind starter on top of the powerhead with the legs in place on the three bracket arms.

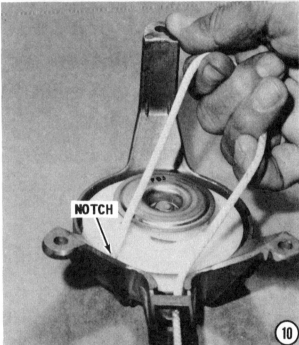

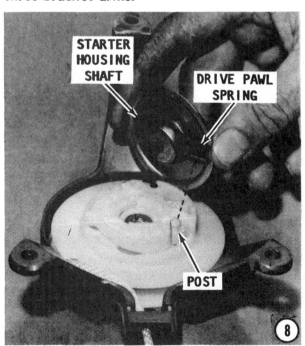

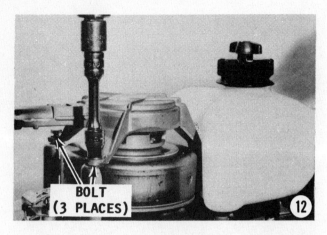

12- Secure the three starter legs with the attaching bolts. Tighten the bolts to a torque value of 5.9 ft lbs (8Nm).

13- Install the two halves of the cowling around the powerhead.

Secure the cowling with the attaching screws. Eight screws hold the cowling halves in place plus one more for the spark plug cover.

SPECIAL WORDS

As noted during disassembling, the screws are different lengths. Ensure the proper size is used in the correct location.

11-3 SERVICING TYPE "B" HAND REWIND STARTER

The "B" hand rewind starter is installed on air-cooled, single cylinder 3.5hp and 5hp models.

Refer to the accompanying exploded drawing while performing the following procedures.

Model 3.5hp

Untie the knot at the starter handle and allow the rope to slowly wind into the starter relieving spring tension. Remove the four bolts securing the starter case cover. Remove the single remaining bolt and washer securing the hand rewind starter to the powerhead. Remove the starter from the powerhead and place it upside down on a suitable work surface.

Model 5hp

Remove the three securing bolts and washers securing the starter to the powerhead. Remove the starter from the powerhead and place it upside down on a suitable work surface. Untie the knot at the starter handle and allow the rope to slowly wind into the starter relieving spring tension.

All models

Press downward on the drive plate against spring pressure, and then using a slotted screwdriver, pry the circlip from the center post of the starter. Remove the thrust washer from the post.

To prevent disturbing the rewind spring encased in the sheave, **CAREFULLY** and **SLOWLY** lift off the drive plate, the two springs, the three pawls, the thrust washer and finally the lubricator.

Rotate the sheave approximately one and a half turns **CLOCKWISE** until a "click" is heard. The "click" sound will indicate the hook on the starter housing has disengaged itself from the starter spring end. **CAREFULLY** separate the sheave, with spring, from the starter housing.

SPECIAL WORDS

If the only work to be performed on the hand rewind starter is to replace the rope, it is best **NOT** to disturb the spring wound inside the sheave.

Untie the knot at the end of the starter rope and remove it from the sheave.

SAFETY WORDS

Wear a good pair of heavy gloves and safety glasses while performing the following tasks.

WARNING

THE REWIND SPRING IS A POTENTIAL HAZARD. The spring is under tremendous tension when it is wound -- a real **"TIGER"** in a cage! If the spring should accidentally be released, severe personal injury could result from being struck by the spring with force. Therefore, the following steps **MUST**

SERVICE TYPE "B"   11-9

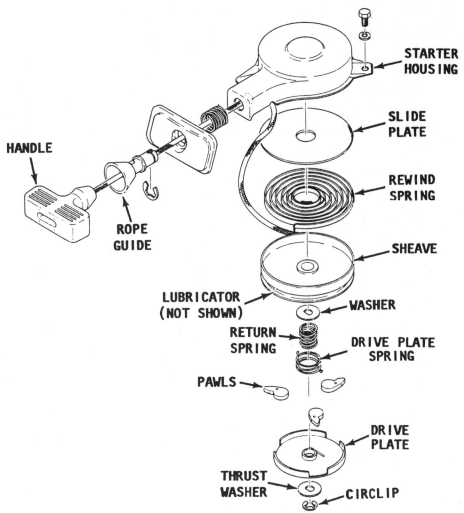

Exploded drawing of hand rewind starter "B" installed on the 3.5hp and 5hp air-cooled powerhead. Major parts are identified.

be performed with care to prevent personal injury to self and others in the area.

DO NOT attempt to remove the spring unless it is unfit for service and a new spring is to be installed.

AUTHORS' WORD

The accompanying illustration shows the spring being released from the same type but different model sheave. **THEREFORE,** the principle is exactly the same. The procedure outlined in the next step may be followed with safety.

Obtain two pieces of wood, short pieces of 2" x 4" (5cm x 10cm) will work fine. Place the two pieces of wood approximately 8" (20 cm) apart on the floor. Center the sheave on top of the wood with the spring side facing **DOWN**. Check to be sure the wood is not touching the spring.

Stand behind the wood, keeping away from the openings because the spring un-

winds with considerable force. Tap the sheave with a soft mallet. The spring retainer plate will drop down releasing the spring. The spring will fall and unwind almost instantly and with **FORCE**.

CLEANING AND INSPECTING

Wash all parts except the rope and the handle in solvent, and then blow them dry with compressed air.

Remove any trace of corrosion and wipe all metal parts with an oil dampened cloth.

Inspect the rope. Replace the rope if it appears to be weak or frayed. If the rope is frayed, check the holes through which the rope passes for rough edges or burrs. Remove the rough edges or burrs with a file and polish the surface until it is smooth. Inspect the starter spring end hooks. Replace the spring if it is weak, corroded or cracked. Inspect the inside surface of the

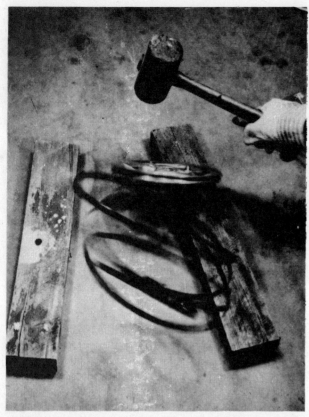

One method of releasing the rewind spring from the sheave or starter housing, as explained in the text.

sheave rewind recess for grooves or roughness. Grooves may cause erratic rewinding of the starter rope.

Coat the entire length of the used rewind spring (a new spring will be coated with lubricant from the package), with low-temperature lubricant.

ASSEMBLING AND INSTALLATION TYPE "B" HAND REWIND STARTER

Wear a good pair of gloves while winding and installing the spring. The spring will develop tension and the edges of the spring steel are extremely sharp. The gloves will prevent cuts to your palms and fingers.

New Spring Installation

Apply a light coating of water resistant anti-seize lubricant to the inside surface of the starter housing.

A new spring will be wound and held in a steel hoop. Hook the outer end of the new spring onto the starter sheave post, and then place the spring inside the sheave. **CAREFULLY** remove the steel hoop. The spring should unwind slightly and seat itself in the housing.

Old Spring Installation

SAFETY WORDS

The authors, the manufacturers, and almost anyone else who has handled the spring from this type rewind starter **STRONGLY** recommend a pair of safety goggles or a face shield and heavy gloves be worn while the spring is being installed. As the work progresses a **"TIGER"** is being forced into a cage -- over 14' (4.3 m) of spring steel wound into about 4" (10.2cm) circumference. If the spring is accidentally released, it will lash out with tremendous ferocity and very likely could cause personal injury to the installer or other persons nearby.

Apply a light coating of water resistant anti-seize lubricant to the inside surface of the starter sheave. Wind the old spring **COUNTERCLOCKWISE** loosely in one hand.

Hook the outer end of the spring onto the starter sheave post. Rotate the sheave **COUNTERCLOCKWISE** and at the same feed the spring into the housing **CLOCKWISE**. Continue working the spring into the sheave until the entire length has been confined. From this point handle the sheave with **EXTREME CAUTION** not to allow the spring to escape.

All Models

Insert one end of the rope through the hole in the starter sheave. Tie a figure "8" knot in the end of the rope leaving about one inch (2.5cm) beyond the knot.

Hold the sheave with the spring facing upward. Wind the rope around the perimeter of the sheave in a clockwise direction until only 16" (40mm) remain. Lower the sheave into the starter housing and at the same time position the free end of the rope into the long slot in the starter housing.

Hold the housing still and rotate the sheave about four or five turns **COUNTERCLOCKWISE.**

Hold the sheave against the housing to prevent the spring from unwinding while performing the following task.

Tie a temporary knot in the rope at the point at which it emerges from the housing to keep tension on the spring.

Install the three pawls into the starter sheave. Place the lubricator and thrust washer over the center post. Install the long narrow spring and then install the short wide spring. Hook one end of the second spring into the hole in the sheave. Install the drive plate and hook the other end of the second spring into the right side of the slot in the drive plate.

Rotate the plate 40° **CLOCKWISE**. Continue to hold the sheave against the housing to prevent the spring from unwinding and at the same time, install another thrust washer over the center post and snap the circlip into the goove.

Model 3.5hp

Secure the starter to the powerhead with the single attaching bolt and washer. Place the starter case cover over the starter and install the remaining three bolts and washers. Tighten the bolts to a torque value of 5.8 ft lbs (8Nm). Untie the temporary knot and feed the end of the rope through the rope guide and handle. Tie a figure "8" knot in the rope as close to the end as practical. Pull the knot back into the handle recess, and then install the seal in the handle to hide the rope knot.

Model 5hp

Untie the temporary knot and feed the end of the rope through the rope guide and handle. Tie a figure "8" knot in the rope as close to the end as practical. Pull the knot back into the handle recess, and then install the seal in the handle to hide the rope knot.

Position the starter in place on the powerhead. Secure the housing to the powerhead with the three attaching bolts and tighten them to a torque value of 5.8 ft lbs (8Nm).

11-4 SERVICING TYPE "C" HAND REWIND STARTER

The "C" hand rewind starter is installed on the water cooled single cylinder 4hp, and 5hp models, and 2-cylinder 8hp (from 1980), 9.9hp, 15hp (from 1980), 20hp, 25hp, 28hp, 30hp, and 40hp models.

Because the "C" is used on ten outboard models, at press time, slight variations in appearance and design may be noted. However, the procedures outlined in this chapter are valid for all "C" hand starters.

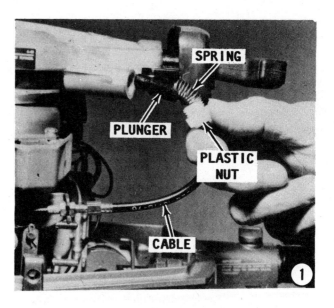

REMOVAL AND DISASSEMBLING

1- Unscrew the plastic nut and pull the shift interlock cable free of the starter housing. Remove the plunger and spring from the cable end, because they are easily lost.

2- Remove the three bolts securing the starter legs to the powerhead, and then remove the starter.

3- Place the starter assembly upside down on a suitable work surface.

4- Pry the circlip from the pawl post using a narrow slotted screwdriver.

5- Lift the pawl, with the spring attached free of the sheave.

6- Rotate the sheave to align the slot in the sheave with the starter handle, as

11-12 HAND REWIND STARTER

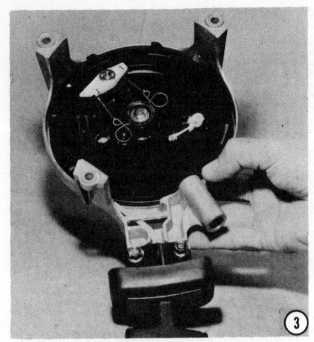

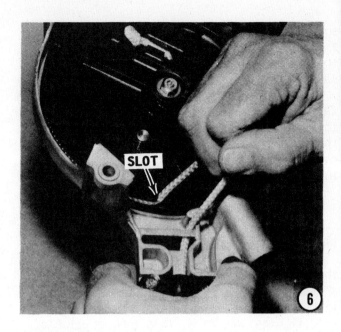

shown. Lift out a portion of rope, feed the rope into the slot and with a **CONTROLLED MOTION** allow the sheave to rotate **CLOCKWISE** until the tension on the rewind spring is completely released. **DO NOT** allow the sheave to spin without control.

7- Pry the seal from the handle and push out the knot in the end of the rope. Untie the knot and pull the handle free of the rope.

8- Remove the bolt and washer from the center of the sheave.

9- Remove the sheave bushing and starter housing shaft from the sheave.

SPECIAL WORDS

If the only work to be performed on the hand rewind starter is to replace the rope, it

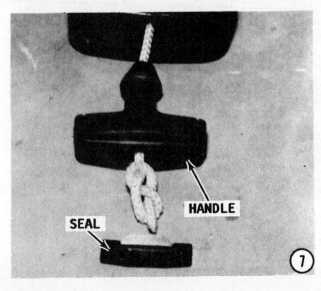

is best **NOT** to disturb the sheave and spring beneath the sheave.

Hold the sheave against the starter housing to prevent the spring from disengaging from the sheave and **CAREFULLY** rotate the sheave to allow the rope hole to align with the starter handle. Pull the knotted end out of the sheave until all of the rope is free.

If either the sheave or the starter rewind spring is to be replaced the rope may be left in place until the sheave is removed from the starter housing.

SAFETY WORDS

Wear a good pair of heavy gloves and safety glasses while performing the following tasks.

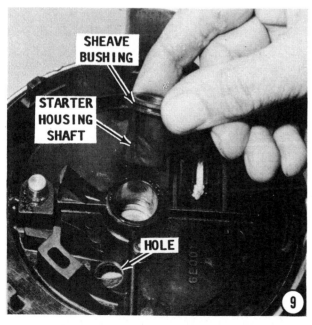

WARNING

THE REWIND SPRING IS A POTENTIAL HAZARD. The spring is under tremendous tension when it is wound -- a real **"TIGER"** in a cage! If the spring should accidentally be released, severe personal injury could result from being struck by the spring with force. Therefore, the following steps **MUST** be performed with care to prevent personal injury to self and others in the area.

DO NOT attempt to remove the spring unless it is unfit for service and a new spring is to be installed.

Insert a screwdriver into the hole in the sheave, push down on the section of spring visible through the hole. At the same time gently lift up on the sheave and hold the spring down to confine it in the housing and prevent it from escaping uncontrolled. If the rope has not been removed from the sheave, remove it at this time.

AUTHORS' WORD

The accompanying illustration shows the spring being released from the same type but different model rewind spring. **THEREFORE,** the principle is exactly the same. The procedure outlined in the next step may be followed with safety.

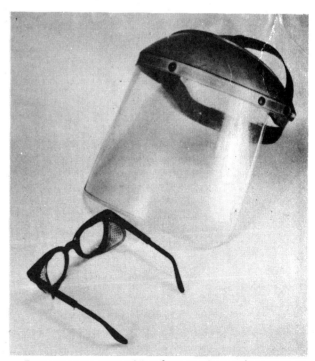

Protect your eyes with a face mask or safety glasses while working with the rewind spring, especially a used one. The spring is a "tiger in a cage", almost 13 feet (4 meters), of spring steel wound and confined into a space less than 4 inches (about 10 cm), in diameter.

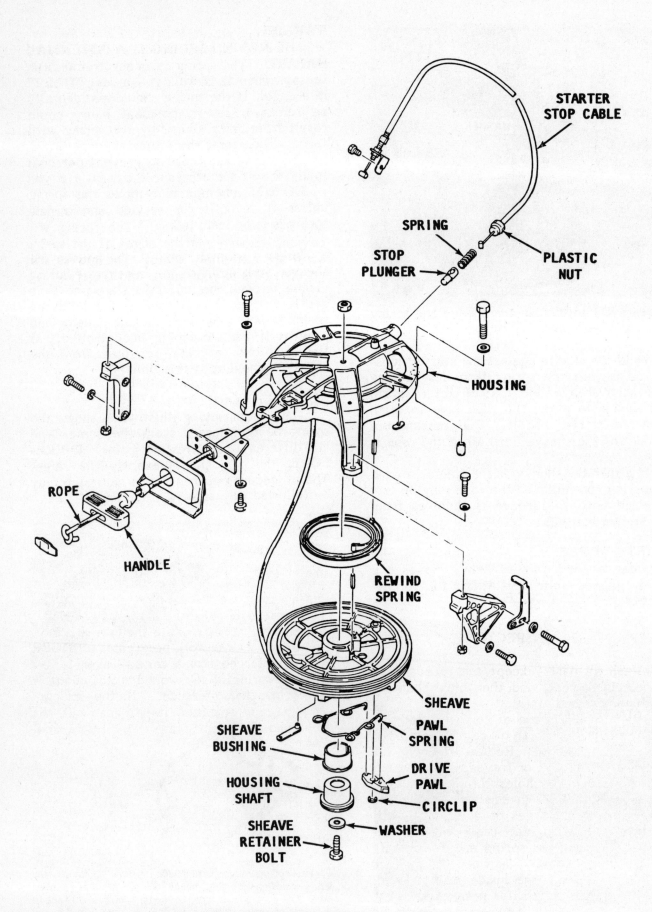

Exploded drawing of hand rewind starter "C", with major parts identified.

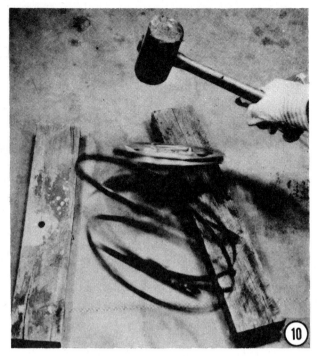

10- Obtain two pieces of wood, a short 2" x 4" (5cm x 10cm) will work fine. Place the two pieces of wood approximately 8" (20 cm) apart on the floor. Center the housing on top of the wood with the spring side facing **DOWN**. Check to be sure the wood is not touching the spring.

Stand behind the wood, keeping away from the openings as the spring unwinds with considerable force. Tap the sheave with a soft mallet. The spring retainer plate will drop down releasing the spring. The spring will fall and unwind almost instantly and with **FORCE**.

CLEANING AND INSPECTING

Wash all parts except the rope and the handle in solvent, and then blow them dry with compressed air.

Remove any trace of corrosion and wipe all metal parts with an oil dampened cloth.

Inspect the rope. Replace the rope if it appears to be weak or frayed. If the rope is frayed, check the holes through which the rope passes for rough edges or burrs. Remove the rough edges or burrs with a file and polish the surface until it is smooth. Inspect the starter spring end hooks. Replace the spring if it is weak, corroded or cracked. Inspect the inside surface of the sheave rewind recess for grooves or roughness. Grooves may cause erratic rewinding of the starter rope.

Coat the entire length of the used rewind spring (a new spring will be coated with lubricant from the package), with low-temperature lubricant.

ASSEMBLING AND INSTALLATION TYPE "C" HAND REWIND STARTER

GOOD WORDS

If the rewind spring or the sheave was not removed, proceed directly to Step 6.

Wear a good pair of gloves while winding and installing the spring. The spring will develop tension and the edges of the spring steel are extremely sharp. The gloves will prevent cuts to your palms and fingers.

New Spring Installation

1- Apply a light coating of water resistant anti-seize lubricant to the inside surface of the starter housing.

A new spring will be wound and held in a steel hoop. Hook the outer end of the new spring onto the starter housing post, and then place the spring inside the housing. **CAREFULLY** remove the steel hoop. The spring should unwind slightly and seat itself in the housing.

Old Spring Installation

SAFETY WORDS

The authors, the manufacturers, and almost anyone else who has handled the spring from this type rewind starter **STRONGLY** recommend a pair of safety goggles or a face shield be worn while the spring is being installed. As the work progresses a **"TIGER"** is being forced into a cage -- over 14' (4.3 m) of spring steel wound into about 4" (10.2cm) circumference. If the spring is accidentally released, it will lash out with tremendous ferocity and very likely could

11-16 HAND REWIND STARTER

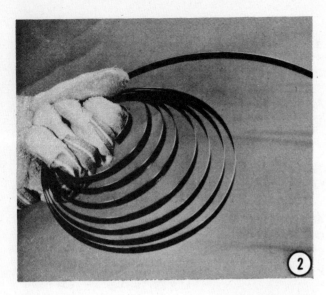

cause personal injury to the installer or other persons nearby.

2- Apply a light coating of water resistant anti-seize lubricant to the inside surface of the starter housing. Wind the old spring **CLOCKWISE** loosely in one hand, as shown.

3- Hook the outer end of the spring onto the starter housing post. Rotate the sheave **CLOCKWISE** and at the same feed the spring into the housing **COUNTERCLOCKWISE**. Continue working the spring into the housing until the entire length has been confined.

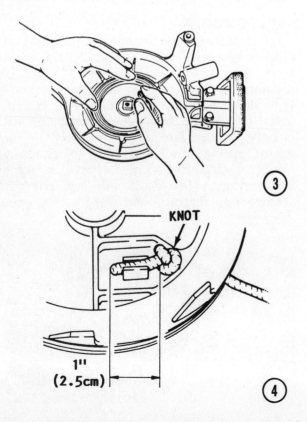

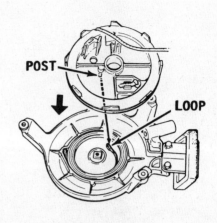

4- Insert one end of the rope through the hole in the starter sheave. Tie a figure "8" knot in the end of the rope leaving about one inch (2.5cm) beyond the knot. Tuck the end of the rope beyond the knot into the groove next to the knot.

5- Wind the rope in a **CLOCKWISE** direction 1-1/2 turns around the sheave, ending at the slot in the sheave. Lower the sheave into the starter housing. At the same time, use a small screwdriver through the hole to guide the inner loop of the spring onto the post on the underneath side of the sheave.

Units With Spring and Sheave Undisturbed

6- Align the hole in the edge of the sheave with the starter handle. Thread the rope through the hole and up through the top side. Tie a figure "8" knot in the end which was just brought through, leaving about one inch (25cm). Tuck the short free end into the groove next to the hole.

WITHOUT ROTATING THE SHEAVE, feed the rope between the sheave and the

SERVICE TYPE "C"   11-17

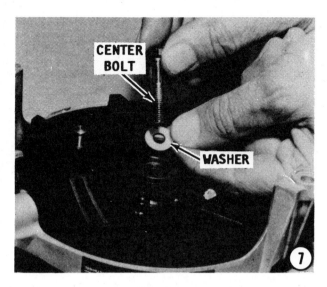

edge of the starter housing **CLOCKWISE**. Push the rope into place with a narrow screwdriver. Continue feeding and tucking the rope for 1-1/2 turns, ending with the rope at the slot of the sheave.

All Units

Slide the sheave bushing into the starter housing shaft. Insert the shaft and bushing into the center of the sheave.

7- Coat the threads of the center bolt with Loctite. Install the washer and bolt. Tighten the bolt to a torque value of 5.8 ft lbs (8Nm).

8- Thread the rope through the starter handle housing and through the handle. Tie a figure "8" knot in the rope as close to the end as practical. Pull the knot back into the handle recess, and then install the seal in the handle to hide the rope knot.

9- Lift up a portion of rope, and then hook it into the slot of the sheave. Hold the

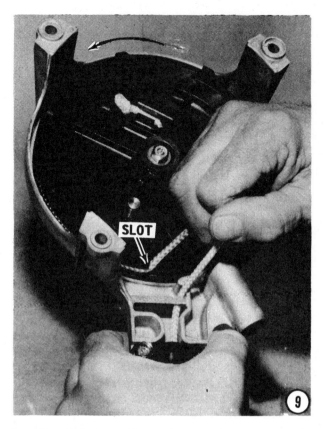

handle tightly and at the same time rotate the sheave **COUNTERCLOCKWISE** until the spring beneath is wound tight. This will take about three complete turns of the sheave. Slowly release the tension on the sheave and allow it to rewind **CLOCKWISE** while the rope is taken up as it feeds around the sheave.

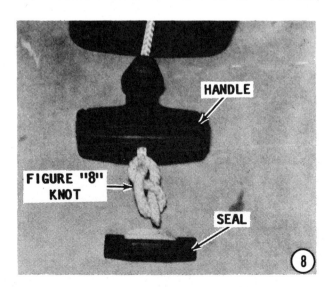

HAND REWIND STARTER

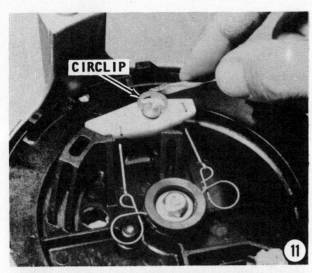

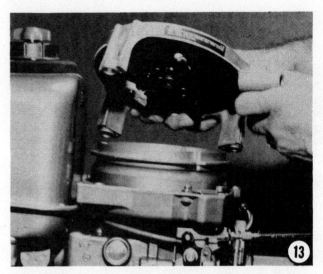

10- With the bevelled end of the pawl facing to the left, hook each end of the pawl spring into the two small holes in the pawl, from the underneath side of the pawl, and with the pattern of the spring, as shown. The short ends of the spring will then be on the upper surface of the pawl. Move the spring up against the center of the sheave shaft, and then slide the center of the pawl onto the pawl post, as indicated in the accompanying illustration.

11- Snap the circlip into place over the pawl post to secure the pawl in place.

12- Check the action of the rewind starter before further installation work proceeds. Pull out the starter rope with the handle, then allow the spring to slowly re-

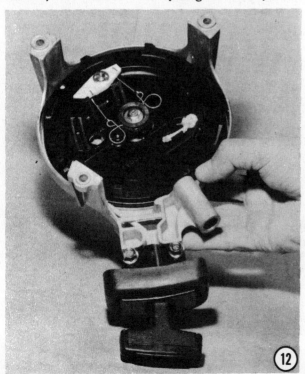

wind the rope. The starter should rewind smoothly and take up all the rope to lightly seat the handle against the starter housing.

13- Position the rewind starter in place on the powerhead. Apply Loctite to the threads of the three attaching bolts. Secure the starter legs to the powerhead with the bolts, and tighten them to a torque value of 5.8 ft lbs (8Nm).

14- Slip the starter stop cable end through the spring and then into the recess of the plunger. Hold these parts together and slide them into the starter housing. Tighten the plastic nut snugly.

SPECIAL WORDS

If the cable adjustment at the other end was undisturbed, the no-start-in-gear protection system should perform satisfactorily. When the unit is **NOT** in **NEUTRAL**, the plunger should push out to lock the sheave and prevent it from rotating. This means an attempt to pull on the rope with the lower unit in any gear except **NEUTRAL** should **FAIL**.

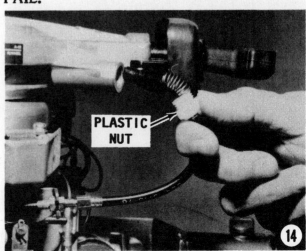

SERVICE TYPE "D"   11-19

Adjustment

If the no-start-in-gear protection system fails to function properly, first remove the rewind hand starter from the powerhead. Make an adjustment on the length of the cable at the two locknuts at either side of the bracket to bring the plunger flush with the inner surface of the starter housing. Install the starter on the powerhead and again check the no-start-in-gear system. If the system still fails to function correctly, replace the cable.

11-5 SERVICING TYPE "D" HAND REWIND STARTER

The "D" hand rewind starter is installed on 2-cylinder 8hp models from 1977 to 1979.

SPECIAL WORDS

Refer to the exploded diagram on Page 11-21 while performing the following service procedures.

REMOVAL AND DISASSEMBLING

Remove the cowling. Grasp a portion of the rope between the sheave and the adjacent pulley and pull out approximately a foot of rope. Tie a temporary knot in this portion of rope and then release the rope to allow the temporary knot to rest against the starter sheave. This action will prevent the rope from being completely lost inside the sheave when the handle is removed.

Pry the insert from the starter handle. Pull out the knot and untie it. Remove the insert and handle from the rope end. Feed the rope free of the rope guide and from around the two pulleys.

Remove the two starter mounting bolts, but not the bolt at the center of the starter. Lift the starter from the powerhead and place it on a suitable work surface. Hold the sheave and the housing together to prevent uncontrolled rewind while untying the temporary knot. Allow the sheave to rotate **SLOWLY** until the rewind spring is completely relaxed.

WARNING

THE REWIND SPRING IS A POTENTIAL HAZARD. The spring is under tremendous tension when it is wound -- a real **"TIGER"** in a cage! If the spring should accidentally be released, severe personal injury could result from being struck by the spring with force. Therefore, the following steps **MUST** be performed with care to prevent personal injury to self and others in the area.

DO NOT attempt to remove the spring unless it is unfit for service and a new spring is to be installed.

While performing the following tasks **TAKE CARE** not to lift the sheave in any way. The inner end of the rewind spring is indexed into the underside of the sheave. If the sheave is disturbed, the spring end will disengage and will "fly out" of the housing.

Remove the center bolt and washer. Lift off the following parts and keep them in order on the workbench: the mounting bracket, the wavy washer (with arms facing down), a thrust washer, a tall spacer, the toothed pulley, another thrust washer, the drive plate, the drive plate spring, a short spacer and finally a plain washer. Lift the two pawls from the sheave.

SPECIAL WORDS

If the only work to be performed on the hand rewind starter is to replace the rope, it is best **NOT** to disturb the spring encased inside the housing. The rope can be replaced at this time.

Hold the sheave against the starter housing to prevent the spring from disengaging from the sheave and **CAREFULLY** rotate the sheave to allow the rope hole to align with the starter handle. Pull the knotted end out of the sheave until all of the rope is free.

If the only work to be performed is to replace the rope, proceed directly to the paragraph headed "Rope Installation Only", on Page 11-22.

If the spring is to be replaced **CAREFULLY** "wiggle" the sheave to make certain the end of the rewing spring is not engaged, before lifting the sheave clear of the starter housing.

AUTHORS' WORD

The accompanying illustration shows the spring being released from the same type but different model rewind system. The rewind spring is housed inside the housing, not the sheave. **THEREFORE**, the principle remains the same even though the spring

HAND REWIND STARTER

One method of releasing the rewind spring from the sheave or starter housing, as explained in the text.

location is different. The procedure outlined in the next step may be followed with safety.

SAFETY WORDS

Wear a good pair of heavy gloves and safety glasses while performing the following tasks.

Obtain two pieces of wood, short pieces of 2" x 4" (5cm x 10cm) will work fine. Place the two pieces of wood approximately 8" (20 cm) apart on the floor. Center the housing on top of the wood with the spring side facing **DOWN**. Check to be sure the wood is not touching the spring.

Stand behind the wood, keeping away from the openings as the spring unwinds with considerable force. Tap the housing with a soft mallet. The spring will fall and unwind almost instantly and with **FORCE**.

CLEANING AND INSPECTING

Wash all parts, except the rope and the handle in solvent, and then blow them dry with compressed air.

Remove any trace of corrosion and wipe all metal parts with an oil dampened cloth.

Inspect the rope. Replace the rope if it appears to be weak or frayed. If the rope is frayed, check the holes through which the rope passes for rough edges or burrs. Remove the rough edges or burrs with a file and polish the surface until it is smooth. Inspect the starter spring end hooks. Replace the spring if it is weak, corroded or cracked. Inspect the inside surface of the sheave rewind recess for grooves or roughness. Grooves may cause erratic rewinding of the starter rope.

Coat the entire length of the used rewind spring (a new spring will be coated with lubricant from the package), with low-temperature lubricant.

Measure the free length of the drive plate spring. It must be between 5 7/64 - 5 7/32" (12.96 - 13.16cm). Replace the spring if the length is not within specifications.

Inspect the outer ends of the two pawls. The end making contact with the slots in the underside of the toothed pulley must be pointed for proper engagement. If the end has been rounded off, the starter action is severly impeded. If the pawls are worn or deformed, they must be replaced as a pair.

ASSEMBLING AND INSTALLATION TYPE "D" HAND REWIND STARTER

GOOD WORDS

If the rewind spring or the sheave was not removed, proceed directly to the paragraph headed "Rope Installation".

Wear a good pair of gloves while winding and installing the spring. The spring will develop tension and the edges of the spring steel are extremely sharp. The gloves will prevent cuts to your palms and fingers.

New Spring Installation

Apply a light coating of water resistant anti-seize lubricant to the inside surface of the starter housing.

A new spring will be wound and held in a steel hoop. Hook the outer end of the new spring onto the starter housing post, and then place the spring inside the housing. **CAREFULLY** remove the steel hoop. The spring should unwind slightly and seat itself in the housing.

SERVICE TYPE "D" 11-21

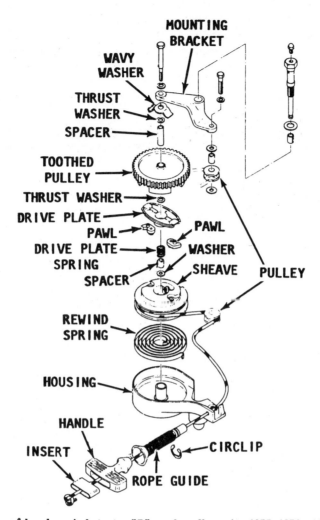

Exploded drawing of hand rewind starter "D" used on 8hp units 1977-1979. Major parts identified.

Old Spring Installation

SAFETY WORDS

The authors, the manufacturers, and almost anyone else who has handled the spring from this type rewind starter **STRONGLY** recommend a pair of safety goggles or a face shield and a pair of heavy gloves be worn while the spring is being installed. As the work progresses a **"TIGER"** is being forced into a cage. Actually, over 14' (4.3m) of spring steel wound into about 4" (10.2cm) circumference. If the spring is accidentally released, it will lash out with tremendous ferocity and very likely could cause personal injury to the installer or other persons nearby.

Apply a light coating of water resistant anti-seize lubricant to the inside surface of the starter housing. Wind the old spring **CLOCKWISE** loosely in one hand (see illustration top of Page 11-16).

Hook the outer end of the spring onto the starter housing post. Rotate the sheave **CLOCKWISE** and at the same feed the spring into the housing **COUNTERCLOCKWISE.** Continue working the spring into the housing until the entire length has been confined.

Insert one end of the rope through the hole in the starter sheave. Tie a figure "8" knot in the end of the rope leaving about one inch (2.5cm) beyond the knot. Wrap the rope around the sheave six times **CLOCKWISE** and hold the rope against the notch cut in the side of the sheave.

Lower the sheave into the starter housing and "wiggle" the sheave back and forth until the post on the underside of the sheave has engaged with the inner end of the rewind spring.

Proceed to the paragraph headed "The Work Continues".

11-22   HAND REWIND STARTER

Rope Installation Only

CAREFULLY rotate the sheave inside the housing COUNTERCLOCKWISE through six turns until the rope hole in the side of the sheave aligns with the housing arm. Then hold the sheave and the housing together to prevent rotation.

Guide one end of the rope through the hole from the housing arm into the sheave. Tie a figure "8" knot in the end of the rope leaving about one inch (2.5cm) beyond the knot. Slowly release pressure on the sheave and allow the sheave to rotate CLOCKWISE drawing in the rope.

The Work Continues

Install the two pawls into their respective cutouts in the sheave face. The pointed ends of the pawls must point CLOCKWISE. Place the following parts in order over the center hole of the sheave. All the parts will be secured together with the center bolt: the plain washer, a short spacer, the drive plate spring, the drive plate (indexing over the two installed pawls), a thrust washer, the toothed pulley (indexing with the drive plate), the long spacer, another thrust washer, the wavy washer (with the two arms facing down toward the starter housing), the mounting bracket and finally the washer and center bolt. Install the center bolt just hand tight at this time. Installation adjustments may have to be made before this bolt is finally tightened.

Mount the assembled starter onto the powerhead and secure it with the two attaching bolts. Tighten the bolts to a torque value of 3.5 ft lbs (4.75Nm).

Feed the rope around the two pulleys, through the rope guide, handle, and insert.

Tie a figure "8" knot in the rope as close to the end as practical. Pull the knot and insert into the handle recess.

Rotate the sheave CLOCKWISE until all slack in the rope is taken up and the starter handle rests lightly against the rope guide. Tighten the center bolt to a torque value of 3.5 ft lbs (4.75Nm).

11-6 SERVICING TYPE "E" HAND REWIND STARTER

The "E" hand rewind starter is installed on 2-cylinder 15hp models from 1977 to 1979.

SPECIAL WORDS

Refer to the exploded diagram on the next page while performing the following service tasks.

REMOVAL AND DISASSEMBLING

Remove the cowling from around the powerhead. Remove the three bolts and washers securing the three legs of the starter to the powerhead. Place the starter upside down on a suitable work surface.

Using a thin blade type screwdriver, pry the circlip from the center post and then lift off the thrust washer. Release the three return spring ends from the drive plate and remove the following parts and place them on the work bench in order as an assist in later installation: the drive plate, the drive plate spring, the cupped thrust washer, three pawls, and three return springs.

Grasp the portion of rope inside the starter housing, between the sheave and the handle support. Pull out approximately one foot of rope. Tie a temporary knot in this portion of rope and then release the rope to allow the temporary knot to rest against the starter sheave. This action will prevent the rope from being completely lost inside the sheave when the handle is removed.

Pry the insert from the starter handle. Pull out the knot and untie it. Remove the insert and handle from the rope end. Feed the rope free from the rope guide inside the starter housing.

Hold the sheave and the housing together to prevent uncontrolled rewind, and then untie the temporary knot. Allow the sheave to rotate SLOWLY until the rewind spring is completely relaxed.

WARNING

THE REWIND SPRING IS A POTENTIAL HAZARD. The spring is under tremendous tension when it is wound -- a real **"TIGER"** in a cage! If the spring should accidentally be released, severe personal injury could result from being struck by the spring with force. Therefore, the following steps **MUST** be performed with care to prevent personal injury to self and others in the area.

DO NOT attempt to remove the spring unless it is unfit for service and a new spring is to be installed.

SERVICE TYPE "E" 11-23

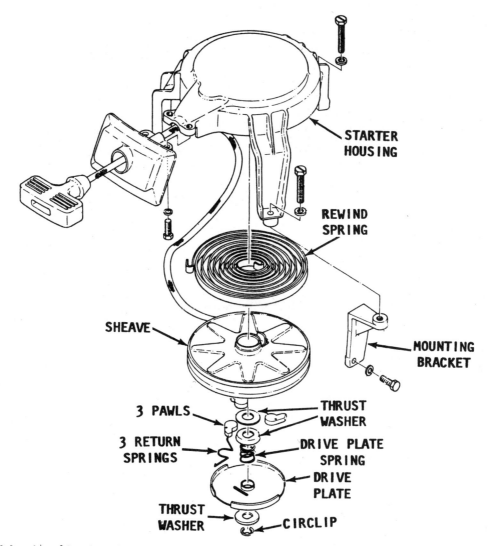

Exploded drawing of hand rewind starter "E" installed on 15hp units 1977-1979. Major parts are identified.

While performing the following tasks **TAKE CARE** not to lift the sheave in any way. The inner end of the rewind spring is indexed into the underside of the sheave. If the sheave is disturbed, the spring end will disengage and will "fly out" of the housing.

SPECIAL WORDS

If the only work to be performed on the hand rewind starter is to replace the rope, it is best **NOT** to disturb the spring encased inside the housing. The rope can be replaced at this time.

Hold the sheave against the starter housing to prevent the spring from disengaging from the sheave and **CAREFULLY** rotate the sheave to allow the rope hole to align with the starter handle. Unhook the rope loop from the post on the sheave and pull the rope free.

If the only work to be performed is to replace the rope, proceed directly to the paragraph headed "Rope Installation Only", on Page 11-25.

If the spring is to be replaced **CAREFULLY** "wiggle" the sheave to make certain the end of the rewind spring is not engaged, before lifting the sheave clear of the starter housing.

AUTHORS' WORD

The accompanying illustration shows the spring being released from the same type but different model rewind spring. The rewind spring is housed inside the sheave, not the housing. **THEREFORE**, the principle remains the same even though the spring location is different. The procedure outlined in the next step may be followed with safety.

SAFETY WORDS

Wear a good pair of heavy gloves and safety glasses while performing the following tasks.

Obtain two pieces of wood, short 2" x 4" (5cm x 10cm) pieces will work fine. Place the two pieces of wood approximately 8" (20 cm) apart on the floor. Center the housing on top of the wood with the spring side facing **DOWN**. Check to be sure the wood is not touching the spring.

Stand behind the wood, keeping away from the openings as the spring unwinds with considerable force. Tap the housing with a soft mallet. The spring will fall and unwind almost instantly and with **FORCE**.

CLEANING AND INSPECTING

Wash all parts except the rope and the handle in solvent, and then blow them dry with compressed air.

Remove any trace of corrosion and wipe all metal parts with an oil dampened cloth.

Inspect the rope. Replace the rope if it appears to be weak or frayed. If the rope is frayed, check the holes through which the rope passes for rough edges or burrs. Remove the rough edges or burrs with a file and polish the surface until it is smooth. Inspect the starter spring end hooks. Replace the spring if it is weak, corroded or cracked. Inspect the inside surface of the sheave rewind recess for grooves or roughness. Grooves may cause erratic rewinding of the starter rope.

Coat the entire length of the used rewind spring (a new spring will be coated with lubricant from the package), with low-temperature lubricant.

Inspect the outer ends of the three pawls. The end making contact with the slots cut out in the drive plate must be pointed for proper engagement. If the end has been rounded off, the starter action is severly impeded. If the pawls are worn or deformed, they must all be replaced.

ASSEMBLING AND INSTALLATION TYPE "E" HAND REWIND STARTER

GOOD WORDS

If the rewind spring or the sheave was not removed, proceed directly to the paragraph headed "Rope Installation".

Wear a good pair of gloves while winding and installing the spring. The spring will develop tension and the edges of the spring steel are extremely sharp. The gloves will prevent cuts to your palms and fingers.

New Spring Installation

Apply a light coating of water resistant anti-seize lubricant to the inside surface of the starter housing.

A new spring will be wound and held in a steel hoop. Hook the outer end of the new spring onto the starter housing post, and then place the spring inside the housing. **CAREFULLY** remove the steel hoop. The spring should unwind slightly and seat itself in the housing.

Old Spring Installation

SAFETY WORDS

The authors, the manufacturers, and almost anyone else who has handled the spring from this type rewind starter **STRONGLY** recommend a pair of safety goggles or a face shield and a pair of heavy gloves be worn while the spring is being installed. As the work progresses a **"TIGER"** is being forced into a cage -- over 14' (4.3m) of

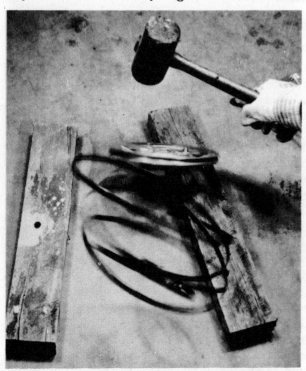

One method of releasing the rewind spring from the sheave or starter housing, as explained in the text.

spring steel wound into about 4" (10.2cm) circumference. If the spring is accidentally released, it will lash out with tremendous ferocity and very likely could cause personal injury to the installer or other persons nearby.

Apply a light coating of water resistant anti-seize lubricant to the inside surface of the starter housing. Wind the old spring **CLOCKWISE** loosely in one hand (see pix top of Page 11-16).

Hook the outer end of the spring onto the starter housing post. Rotate the sheave **CLOCKWISE** and at the same time feed the spring into the housing **COUNTERCLOCKWISE**. Continue working the spring into the housing until the entire length has been confined.

Insert one end of the rope through the hole in the starter sheave. Tie a figure "8" knot in the end of the rope about one inch from the end. Wrap the rope around the sheave through two revolutions **COUNTERCLOCKWISE** and hold the rope aginst the notch cut in the side of the sheave.

Lower the sheave into the starter housing and "wiggle" the sheave back and forth until it is felt the post on the underside of the sheave has engaged with the inner end of the rewind spring.

Proceed to the paragraph headed "The Work Continues".

Rope Installation Only

CAREFULLY rotate the sheave inside the housing **CLOCKWISE** through two turns until the rope hole in the side of the sheave aligns with the housing arm. Now, hold the sheave and the housing together to prevent rotation.

Guide one end of the rope through the rope guide into the sheave. Tie a figure "8" knot in the end of the rope about one inch from the end.

Slowly release pressure on the sheave and allow the sheave to rotate **COUNTERCLOCKWISE**, drawing in the rope.

The Work Continues

Position the rope at the notch on the sheave. Holding the starter housing steady, use the rope in the notch to rotate the sheave through five more revolutions **COUNTERCLOCKWISE. DO NOT LET GO** of the housing and sheave while performing the following tasks.

Feed the rope through the handle and insert.

Tie a figure "8" knot in the rope as close to the end as practical. Pull the knot and insert back into the handle recess.

Install the following parts in the order given: the three pawls into the recesses of the sheave, the three return springs, the flat thrust washer, the cupped thrust washer (with cup facing upward), the drive plate spring and finally the drive plate. Hook the return spring ends over the drive plate. Rotate the plate approximately one half turn **COUNTERCLOCKWISE**. Hold the plate against spring pressure while installing the thrust washer and the circlip over the center post. The sheave and starter housing may now be released. The rope will be pulled in to allow the starter handle to rest lightly against the rope guide. If the rope remains slack, remove the items just installed and rotate the sheave through six or seven revolutions instead of five and repeat the subsequent tasks.

Mount the assembled starter onto the powerhead and secure it with the three attaching bolts. Tighten the bolts to a torque value of 5.8 ft lbs (8Nm).

11-7 SERVICING TYPE "F" HAND REWIND STARTER

The **"F"** hand rewind starter is installed on the 2-cylinder 48hp model.

SPECIAL WORDS

In addition to the illustrations supporting the numbered steps, the exploded diagram on Page 11-28 may prove helpful while performing service work.

Shift Interlock System

This starter is equipped with a shift interlock system to prevent sheave rotation when the outboard is in **FORWARD** or **REVERSE** gear.

When the outboard is in **NEUTRAL**, a projection on the cam guide aligns with a stopper and pushes the stopper upward to clear the block-type "risers" on the upper surface of the sheave. This action allows the sheave to rotate.

When the outboard is **FORWARD** or **REVERSE** gear, the shift interlock cable actuates the cam guide. The projection on the cam guide no longer aligns with the stopper and the stopper drops down through gravity

HAND REWIND STARTER

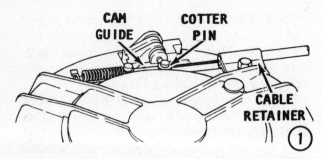

and a torsion spring, between two adjacent "risers" on the sheave. The end of the stopper thus blocks rotation of the sheave.

REMOVAL AND DISASSEMBLING

1- Remove the cowling from around the powerhead. Remove the cotter pin and disconnect the interlock shift cable from the cam guide. Remove the cable retainer to release the cable from the starter housing.

2- Remove the bolts and washers securing the rope guide and remove the bolts and washers securing the three legs of the starter to the powerhead. Place the starter upside down on a suitable work surface.

Grasp the portion of the rope inside the starter housing, between the sheave and the handle support. Pull out approximately one foot of rope. Tie a temporary knot in this portion of the rope and then release the rope to allow the temporary knot to rest against the starter sheave. This action will prevent the rope from being completely lost inside the sheave when the handle is removed.

Pry the insert free of the starter handle. Pull out the knot and untie it. Remove the insert and handle from the rope end. Feed the rope free of the rope guide inside the starter housing.

Hold the sheave and the housing together to prevent uncontrolled rewind and at the same time, untie the temporary knot. Allow the sheave to rotate **SLOWLY** until the rewind spring is completely relaxed.

3- Remove the center bolt and lockwasher. Remove the following parts and place them on the work bench in order as an assist in later installation: the thrust washer, the drive plate, the drive pawl, the return spring, and the drive plate spring.

CAREFULLY pry the large circlip from the center housing boss. Lift off the large flat washer.

WARNING

THE REWIND SPRING IS A POTENTIAL HAZARD. The spring is under tremendous tension when it is wound -- a real **"TIGER"** in a cage! If the spring should accidentally be released, severe personal injury could result from being struck by the spring with force. Therefore, the following steps **MUST** be performed with care to prevent personal injury to self and others in the area.

DO NOT attempt to remove the spring unless it is unfit for service and a new spring is to be installed.

While performing the following tasks **TAKE CARE** not to lift the sheave in any way. The inner end of the rewind spring is indexed into the underside of the sheave. If the sheave is disturbed, the spring end will disengage and will "fly out" of the housing.

SPECIAL WORDS

If the only work to be performed on the hand rewind starter is to replace the rope, it is best **NOT** to disturb the spring encased inside the housing. The rope can be replaced at this time.

4- Hold the sheave against the starter housing to prevent the spring from disengaging from the sheave and **CAREFULLY** rotate the sheave to allow the rope hole to

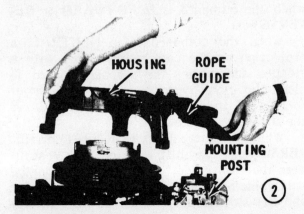

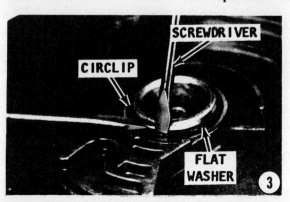

SERVICE TYPE "F" 11-27

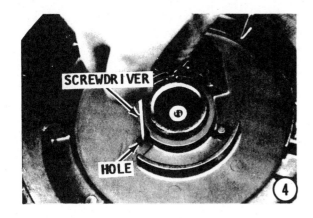

align with the starter handle. Pull the knotted end of the rope out of the sheave until the rope is free.

If the only work to be performed is to replace the rope, proceed directly to the paragraph headed "Rope Installation Only", on Page 11-29.

If the spring is to be replaced insert a small slotted head screwdriver into the hole in the sheave provided for this purpose. Hold down the inner spring loop and disengage the loop from the underside of the sheave.

"Wiggle" the sheave to make certain the end of the rewind spring is not engaged, before lifting the sheave clear of the starter housing.

5- Remove the large bushing.

AUTHORS' WORD

The accompanying illustration shows the spring being released from the same type but different model rewind spring. The rewind spring is housed inside the sheave, not the housing. **THEREFORE,** the principle remains the same even though the spring location is different. The procedure outlined in the next step may be followed with safety.

SAFETY WORDS

Wear a good pair of heavy gloves and safety glasses while performing the following tasks.

6- Obtain two pieces of wood, short 2" x 4" (5cm x 10cm) pieces will work fine. Place the two pieces of wood approximately 8" (20 cm) apart on the floor. Center the housing on top of the wood with the spring side facing **DOWN**. Check to be sure the wood is not touching the spring.

Stand behind the wood, keeping away from the openings as the spring unwinds with considerable force. Tap the housing with a soft mallet. The spring will fall and unwind almost instantly and with **FORCE**.

CLEANING AND INSPECTING

Wash all parts except the rope and the handle in solvent, and then blow them dry with compressed air.

Remove any trace of corrosion and wipe all metal parts with an oil dampened cloth.

Inspect the rope. Replace the rope if it appears to be weak or frayed. If the rope is frayed, check the holes through which the rope passes for rough edges or burrs. Remove the rough edges or burrs with a file and polish the surface until it is smooth.

11-28 HAND REWIND STARTER

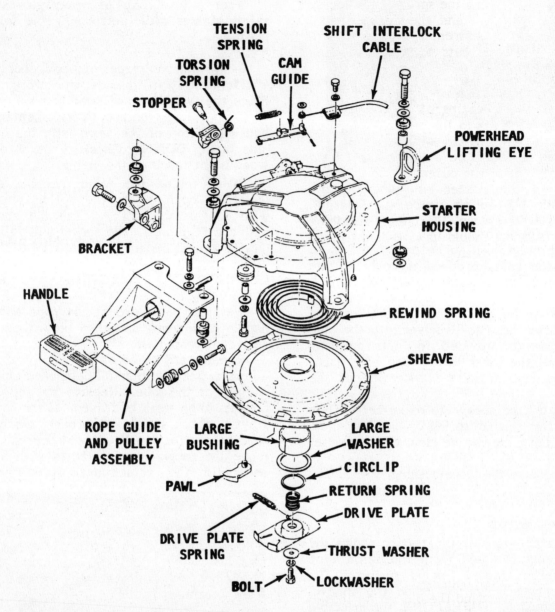

Exploded drawing of hand rewind starter "F" installed on 48hp units 1977-1979. Major parts are identified.

Inspect the starter spring end hooks. Replace the spring if it is weak, corroded or cracked. Inspect the inside surface of the sheave rewind recess for grooves or roughness. Grooves may cause erratic rewinding of the starter rope.

Coat the entire length of the used rewind spring (a new spring will be coated with lubricant from the package), with low-temperature lubricant.

Inspect the outer end of the pawl. The end making contact with the slots cut out in the drive plate must be pointed for proper engagement. If the end has been rounded off, the starter action is severly impeded. If the pawl is worn or deformed, it must be replaced.

ASSEMBLING AND INSTALLATION TYPE "F" HAND REWIND STARTER

GOOD WORDS

If the rewind spring or the sheave was not removed, proceed directly to the paragraph headed "Rope Installation".

1- Wear a good pair of gloves while winding and installing the spring. The spring will develop tension and the edges of the spring steel are extremely sharp. The gloves will prevent cuts to the palms and fingers.

New Spring Installation

Apply a light coating of water resistant anti-seize lubricant to the inside surface of the starter housing.

A new spring will be wound and held in a steel hoop. Hook the outer end of the new spring onto the starter housing post, and then place the spring inside the housing. **CAREFULLY** remove the steel retaining hoop. The spring should unwind slightly and seat itself in the housing.

Old Spring Installation

SAFETY WORDS

The authors, the manufacturers, and almost anyone else who has handled the spring from this type rewind starter **STRONGLY** recommend a pair of safety goggles or a face shield be worn while the spring is being installed. As the work progresses a **"TIGER"** is being forced into a cage -- over 14' (4.3 m) of spring steel wound into about 4" (10.2cm) circumference. If the spring is accidentally released, it will lash out with tremendous ferocity and very likely could cause personal injury to the installer or other persons nearby.

1- Apply a light coating of water resistant anti-seize lubricant to the inside surface of the starter housing. Wear a pair of heavy gloves and wind the old spring loosely **CLOCKWISE** in one hand, as shown.

Hook the outer end of the spring onto the starter housing post. Rotate the sheave **CLOCKWISE** and at the same time, feed the

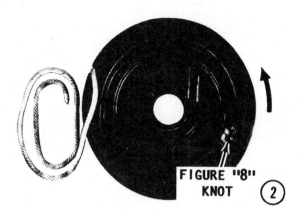

spring into the housing **COUNTERCLOCKWISE**. Continue working the spring into the housing until the entire length has been confined. Install the large bushing over the center boss.

2- Insert one end of the rope through the hole in the starter sheave. Tie a figure "8" knot in the rope end, leaving approximately one inch (2 to 3cm) beyond the knot.

Wrap the rope around the sheave through two revolutions **COUNTERCLOCKWISE** and hold the rope against the notch cut in the side of the sheave.

3- Lower the sheave into the starter housing and "wiggle" the sheave back and forth until the post on the underside of the sheave has engaged with the inner end of the rewind spring. Insert a small slotted head screwdriver into the hole in the sheave to help move the spring end into position, if necessary.

Install the bushing, the large flat washer and the large circlip over the center boss.

Proceed to the paragraph headed "The Work Continues".

Rope Installation Only

CAREFULLY rotate the sheave inside the housing **CLOCKWISE** through two turns until the rope hole in the side of the sheave

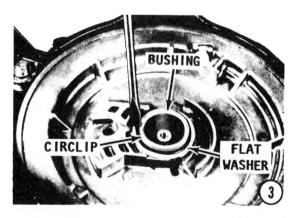

aligns with the housing arm. Now, hold the sheave and the housing together to prevent rotation.

Guide one end of the rope through the rope guide into the sheave. Tie a figure "8" knot in the rope end, leaving approximately one inch (2 to 3cm) beyond the knot.

Slowly release pressure on the sheave and allow the sheave to rotate **COUNTERCLOCKWISE** drawing in the rope.

The Work Continues

4- Position the rope at the notch on the sheave. Holding the starter housing steady, use the rope in the notch to rotate the sheave through two more revolutions **COUNTERCLOCKWISE. DO NOT LET GO** of the housing and sheave while performing the following tasks.

Feed the rope through the handle and insert and at the same time, guide it around the roller inside the starter housing.

5- Tie a figure "8" knot in the rope as close to the end as practical. Pull the knot and insert back into the handle recess.

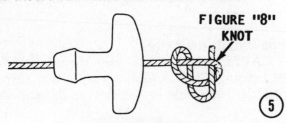

6- Install the pawl, indexing the pawl post into the hole in the sheave.

Install the drive plate spring, the return spring and the drive plate over the sheave. Next, install the thrust washer and secure all the components with the lockwasher and bolt. Tighten the bolt securely.

The sheave and starter housing may now be released. The rope will be pulled in to allow the starter handle to rest lightly against the rope guide. If the rope remains slack, remove the items just installed and rotate the sheave through three or four revolutions instead of two and repeat the subsequent tasks.

Mount the assembled starter onto the powerhead and secure it with the three attaching bolts. Tighten the bolts to a torque value of 5.8 ft lbs (8Nm).

Shift Interlock Installation and Adjustment

7- Hook the end of the shift interlock cable onto the post of the cam guide. Secure the end with the cotter pin. Install the cable stay to the starter housing.

SPECIAL WORDS

If the cable adjustment at the other end was undisturbed, the no-start-in-gear protection system should perform satisfactorily. When the unit is **NOT** in **NEUTRAL**, the stopper should drop down to lock the sheave and prevent it from rotating. This means an attempt to pull on the rope with the lower unit in any gear except **NEUTRAL** should **FAIL**.

If the no-start-in-gear protection system fails to function properly, adjust the position of the cable to align the mark on the stopper with the mark on the cam guide and again check the no-start-in-gear system. If the system still fails to function correctly, replace the cable.

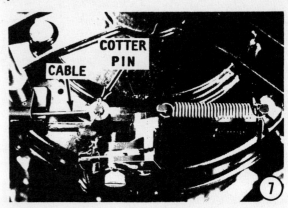

APPENDIX

METRIC CONVERSION CHART

LINEAR
inches	X 25.4	= millimetres (mm)
feet	X 0.3048	= metres (m)
yards	X 0.9144	= metres (m)
miles	X 1.6093	= kilometres (km)
inches	X 2.54	= centimetres (cm)

AREA
inches^2	X 645.16	= millimetres^2 (mm^2)
inches^2	X 6.452	= centimetres^2 (cm^2)
feet^2	X 0.0929	= metres^2 (m^2)
yards^2	X 0.8361	= metres^2 (m^2)
acres	X 0.4047	= hectares (10^4 m^2) (ha)
miles^2	X 2.590	= kilometres^2 (km^2)

VOLUME
inches^3	X 16387	= millimetres^3 (mm^3)
inches^3	X 16.387	= centimetres^3 (cm^3)
inches^3	X 0.01639	= litres (l)
quarts	X 0.94635	= litres (l)
gallons	X 3.7854	= litres (l)
feet^3	X 28.317	= litres (l)
feet^3	X 0.02832	= metres^3 (m^3)
fluid oz	X 29.60	= millilitres (ml)
yards^3	X 0.7646	= metres^3 (m^3)

MASS
ounces (av)	X 28.35	= grams (g)
pounds (av)	X 0.4536	= kilograms (kg)
tons (2000 lb)	X 907.18	= kilograms (kg)
tons (2000 lb)	X 0.90718	= metric tons (t)

FORCE
ounces - f (av)	X 0.278	= newtons (N)
pounds - f (av)	X 4.448	= newtons (N)
kilograms - f	X 9.807	= newtons (N)

ACCELERATION
feet/sec^2	X 0.3048	= metres/sec^2 (m/S^2)
inches/sec^2	X 0.0254	= metres/sec^2 (m/s^2)

ENERGY OR WORK (watt-second - joule - newton-metre)
foot-pounds	X 1.3558	= joules (j)
calories	X 4.187	= joules (j)
Btu	X 1055	= joules (j)
watt-hours	X 3500	= joules (j)
kilowatt - hrs	X 3.600	= megajoules (MJ)

FUEL ECONOMY AND FUEL CONSUMPTION
miles/gal	X 0.42514	= kilometres/litre (km/l)

Note:
235.2/(mi/gal) = litres/100km
235.2/(litres/100 km) = mi/gal

LIGHT
footcandles	X 10.76	= lumens/metre^2 (lm/m^2)

PRESSURE OR STRESS (newton/sq metre - pascal)
inches HG (60 F)	X 3.377	= kilopascals (kPa)
pounds/sq in	X 6.895	= kilopascals (kPa)
inches H$_2$O (60° F)	X 0.2488	= kilopascals (kPa)
bars	X 100	= kilopascals (kPa)
pounds/sq ft	X 47.88	= pascals (Pa)

POWER
horsepower	X 0.746	= kilowatts (kW)
ft-lbf/min	X 0.0226	= watts (W)

TORQUE
pound-inches	X 0.11299	= newton-metres (N·m)
pound-feet	X 1.3558	= newton-metres (N·m)

VELOCITY
miles/hour	X 1.6093	= kilometres/hour (km/h)
feet/sec	X 0.3048	= metres/sec (m/s)
kilometres/hr	X 0.27778	= metres/sec (m/s)
miles/hour	X 0.4470	= metres/sec (m/s)

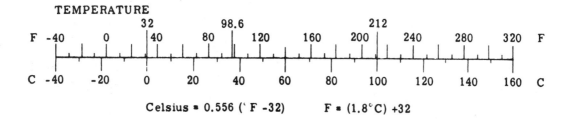

TEMPERATURE

Celsius = 0.556 (°F -32)   F = (1.8°C) +32

ENGINE SPECIFICATIONS AND

MODEL	CYL	CU IN DISPL.	CU CM DISPL.	BORE INCHES	STROKE INCHES	SPARK PLUG CHAMP	SPARK PLUG NGK	SPARK PLUG GAP IN.
				1977				
2	1	2.62	43	1.535	1.417	L86	B-6HS	0.024
3.5	1	3.84	63	1.772	1.575	L82	B-7HS	0.024
5	1	5.61	92	1.969	1.850	L82	B-7HS	0.024
8	2	10.01	165	1.969	1.654	L86	B-6HS	0.024
15	2	15.01	246	2.205	1.969	L82	B-7HS	0.024
20	2	23.91	392	2.520	2.402	L86	B-6HS	0.024
28	2	26.23	430	2.638	2.402	L82	B-7HS	0.024
48	2	46.37	760	3.228	2.835	L82	B-7HS	0.024
60	2	46.37	760	3.228	2.835	L78	B-8HS	0.025
				1978				
2	1	2.62	43	1.535	1.417	L86	B-6HS	0.024
3.5	1	3.84	63	1.772	1.575	L82	B-7HS	0.024
5	1	5.61	92	1.969	1.850	L82	B-7HS	0.024
8	2	10.01	165	1.969	1.654	L86	B-6HS	0.024
9.9	2	15.01	246	2.205	1.969	L82	B-7HS	0.024
15	2	15.01	246	2.205	1.969	L82	B-7HS	0.024
20	2	23.91	392	2.520	2.402	L86	B-6HS	0.024
28	2	26.23	430	2.638	2.402	L82	B-7HS	0.024
40	2	36.13	592	2.953	2.638	L78	B-8HS	0.025
48	2	46.37	760	3.228	2.835	L82	B-7HS	0.024
60	2	46.37	760	3.228	2.835	L78	B-8HS	0.025
				1979				
2	1	2.62	43	1.535	1.417	L86	B-6HS	0.024
3.5	1	3.84	63	1.772	1.575	L82	B-7HS	0.024
5	1	5.61	92	1.969	1.850	L82	B-7HS	0.024
8	2	10.01	165	1.969	1.654	L86	B-6HS	0.024
9.9	2	15.01	246	2.205	1.969	L82	B-7HS	0.024
15	2	15.01	246	2.205	1.969	L82	B-7HS	0.024
20	2	23.91	392	2.520	2.402	L86	B-6HS	0.024
28	2	26.23	430	2.638	2.402	L82	B-7HS	0.024
40	2	36.13	592	2.953	2.638	L78	B-8HS	0.025
48	2	46.37	760	3.228	2.835	L82	B-7HS	0.024
60	2	46.37	760	3.228	2.835	L78	B-8HS	0.025

TUNE-UP ADJUSTMENTS

CARB TYPE	PRIMARY PICKUP TIMING	TIMING AT IDLE	MAXIMUM SPARK ADV AT WOT	IDLE PRM	OPERATING RANGE RPM	POINT GAP IN.
1977						
A	--	--	18° BTDC	1200	4000-5000	0.014
B	--	--	17° BTDC	1150	4000-5000	0.014
D	--	--	18° BTDC	1200	4500-5500	0.014
E	--	0.039 BTDC	25° BTDC	1200	4500-5500	0.014
H	--	--	22° BTDC	1150	4500-5500	0.014
J	--	6° ATDC	20° BTDC	700	4500-5500	0.014
J	--	2° BTDC	24° BTDC	800	4500-5500	0.014
L	4° BTDC	4° ATDC	20° BTDC	800	4500-5500	--
L	4° BTDC	5° ATDC	6° BTDC	800	5000-5800	--
1978						
A	--	--	18° BTDC	1200	4000-5000	0.014
B	--	--	17° BTDC	1150	4000-5000	0.014
D	--	--	18° BTDC	1200	4500-5500	0.014
E	--	0.039 BTDC	25° BTDC	1200	4500-5500	0.014
F or G	--	5° BTDC	25° BTDC	650	4500-5500	0.014
H	--	--	22° BTDC	1150	4500-5500	0.014
J	--	6° ATDC	20° BTDC	700	4500-5500	0.014
J	--	2° BTDC	24° BTDC	800	4500-5500	0.014
K	--	4° BTDC	26° BTDC	900	4500-5500	0.014
L	4° BTDC	4° ATDC	20° BTDC	800	4500-5500	--
L	4° BTDC	5° ATDC	26° BTDC	800	5000-5800	--
1979						
A	--	--	18° BTDC	1200	4000-5000	0.014
B	--	--	17° BTDC	1150	4000-5000	0.014
D	--	--	18° BTDC	1200	4500-5500	0.014
E	--	0.039 BTDC	25° BTDC	1200	4500-5500	0.014
F or G	--	5° BTDC	25° BTDC	650	4500-5500	0.014
H	--	--	22° BTDC	1150	4500-5500	0.014
J	--	6° ATDC	20° BTDC	700	4500-5500	0.014
J	--	2° BTDC	24° BTDC	800	4500-5500	0.014
K	--	4° BTDC	26° BTDC	900	4500-5500	0.014
L	4° BTDC	4° ATDC	20° BTDC	800	4500-5500	--
L	4° BTDC	5° ATDC	26° BTDC	800	5000-5800	--

ENGINE SPECIFICATIONS AND

MODEL	CYL	CU IN DISPL.	CU CM DISPL.	BORE INCHES	STROKE INCHES	SPARK PLUG CHAMP	SPARK PLUG NGK	SPARK PLUG GAP IN.
				1980				
2	1	2.62	43	1.535	1.417	L86	B-6HS	0.024
3.5	1	3.84	63	1.772	1.575	L82	B-7HS	0.024
5	1	5.61	92	1.969	1.850	L82	B-7HS	0.024
8	2	10.01	165	1.969	1.654	L86	B-6HS	0.024
9.9	2	15.01	246	2.205	1.969	L82	B-7HS	0.024
15	2	15.01	246	2.205	1.969	L82	B-7HS	0.024
20	2	26.23	430	2.638	2.402	L82	B-7HS	0.024
25	2	26.23	430	2.638	2.402	L82	B-7HS	0.024
30	2	30.30	496	2.835	2.402	L82	B-7HS	0.024
40	2	36.13	592	2.953	2.638	L78	B-8HS	0.025
60	2	46.37	760	3.228	2.835	L78	B-8HS	0.025
				1981				
2	1	2.62	43	1.535	1.417	L86	B-6HS	0.024
3.5	1	3.84	63	1.772	1.575	L82	B-7HS	0.024
5	1	5.61	92	1.969	1.850	L82	B-7HS	0.024
8	2	10.01	165	1.969	1.654	L86	B-6HS	0.024
9.9	2	15.01	246	2.205	1.969	L82	B-7HS	0.024
15	2	15.01	246	2.205	1.969	L82	B-7HS	0.024
20	2	26.23	430	2.638	2.402	L82	B-7HS	0.024
25	2	26.23	430	2.638	2.402	L82	B-7HS	0.024
30	2	30.30	496	2.835	2.402	L82	B-7HS	0.024
40	2	36.13	592	2.953	2.638	L78	B-8HS	0.025
60	2	46.37	760	3.228	2.835	L78	B-8HS	0.025
				1982				
2	1	2.62	43	1.535	1.417	L86	B-6HS	0.024
4	1	5.06	83	1.969	1.654	L82	B-7HS	0.024
5	1	6.28	103	2.126	1.772	L82	B-7HS	0.024
8	2	10.01	165	1.969	1.654	L86	B-6HS	0.024
9.9	2	15.01	246	2.205	1.969	L82	B-7HS	0.024
15	2	15.01	246	2.205	1.969	L82	B-7HS	0.024
20	2	26.23	430	2.638	2.402	L82	B-7HS	0.024
25	2	26.23	430	2.638	2.402	L82	B-7HS	0.024
30	2	30.30	496	2.835	2.402	L82	B-7HS	0.024
40	2	36.13	592	2.953	2.638	L78	B-8HS	0.025
60	2	46.37	760	3.228	2.835	L78	B-8HS	0.025

TUNE-UP ADJUSTMENTS

CARB TYPE	PRIMARY PICKUP TIMING	TIMING AT IDLE	MAXIMUM SPARK ADV AT WOT	IDLE RPM	OPERATING RANGE RPM	POINT GAP IN.
1980						
A	--	--	18° BTDC	1200	4000-5000	0.014
B	--	--	17° BTDC	1150	4000-5000	0.014
C	--	--	28° BTDC	1000	4500-5500	--
F or G	--	4° BTDC	25° BTDC	1200	4500-5500	--
F or G	--	5° BTDC	25° BTDC	650	4500-5500	--
F or G	--	5° BTDC	22° BTDC	1150	4500-5500	--
J	2° ATDC	6° ATDC	21° BTDC	700	4500-5500	--
J	--	2° BTDC	24° BTDC	700	4500-5500	--
J	--	TDC	25° BTDC	900	4500-5500	--
K	--	4° BTDC	26° BTDC	900	4500-5500	--
L	4° BTDC	5° ATDC	22° BTDC	800	5000-5800	--
1981						
A	--	--	18° BTDC	1200	4000-5000	0.014
B	--	--	17° BTDC	1150	4000-5000	0.014
C	--	--	28° BTDC	1000	4500-5500	--
F or G	--	4° BTDC	25° BTDC	1200	4500-5500	--
F or G	--	5° BTDC	25° BTDC	650	4500-5500	--
F or G	--	5° BTDC	22° BTDC	1150	4500-5500	--
J	2° ATDC	6° ATDC	21° BTDC	700	4500-5500	--
J	--	2° BTDC	24° BTDC	700	4500-5500	--
J	--	TDC	25° BTDC	900	4500-5500	--
K	--	4° BTDC	26° BTDC	900	4500-5500	--
L	4° BTDC	5° ATDC	22° BTDC	800	5000-5800	--
1982						
A	--	--	18° BTDC	1200	4000-5000	0.014
C	--	5° BTDC	28° BTDC	1000	4500-5500	--
C	--	5° BTDC	28° BTDC	1000	4500-5500	--
F or G	--	4° BTDC	25° BTDC	1200	4500-5500	--
F or G	--	5° BTDC	25° BTDC	650	4500-5500	--
F or G	--	5° BTDC	22° BTDC	1150	4500-5500	--
J	2° ATDC	6° ATDC	21° BTDC	700	4500-5500	--
J	--	2° BTDC	24° BTDC	700	4500-5500	--
J	--	TDC	25° BTDC	900	4500-5500	--
K	--	4° BTDC	26° BTDC	900	4500-5500	--
L	4° BTDC	5° ATDC	22° BTDC	800	5000-5800	--

ENGINE SPECIFICATIONS AND

MODEL	CYL	CU IN DISPL.	CU CM DISPL.	BORE INCHES	STROKE INCHES	SPARK PLUG CHAMP	SPARK PLUG NGK	SPARK PLUG GAP IN.
				1983				
2	1	2.62	43	1.535	1.417	L86	B-6HS	0.024
4	1	5.06	83	1.969	1.654	L82	B-7HS	0.024
5	1	6.28	103	2.126	1.772	L82	B-7HS	0.024
8	2	10.01	165	1.969	1.654	L86	B-6HS	0.024
9.9	2	15.01	246	2.205	1.969	L82	B-7HS	0.024
15	2	15.01	246	2.205	1.969	L82	B-7HS	0.024
20	2	26.23	430	2.638	2.402	L82	B-7HS	0.024
25	2	26.23	430	2.638	2.402	L82	B-7HS	0.024
30	2	30.30	496	2.835	2.402	L82	B-7HS	0.024
40	2	36.13	592	2.953	2.638	L78	B-8HS	0.025
60	2	46.37	760	3.228	2.835	L78	B-8HS	0.025
				1984				
2	1	2.62	43	1.535	1.417	L86	B-6HS	0.024
4	1	5.06	83	1.969	1.654	L82	B-7HS	0.024
5	1	6.28	103	2.126	1.772	L82	B-7HS	0.024
8	2	10.01	165	1.969	1.654	L86	B-6HS	0.024
9.9	2	15.01	246	2.205	1.969	L82	B-7HS	0.024
15	2	15.01	246	2.205	1.969	L82	B-7HS	0.024
20	2	26.23	430	2.638	2.402	L82	B-7HS	0.024
25	2	26.23	430	2.638	2.402	L82	B-7HS	0.024
30	2	30.30	496	2.835	2.402	L82	B-7HS	0.024
40	2	36.13	592	2.953	2.638	L78	B-8HS	0.025
				1985				
2	1	2.62	43	1.535	1.417	L86	B-6HS	0.024
4	1	5.06	83	1.969	1.654	L82	B-7HS	0.024
5	1	6.28	103	2.126	1.772	L82	B-7HS	0.024
8	2	10.01	165	1.969	1.654	L86	B-6HS	0.024
9.9	2	15.01	246	2.205	1.969	L82	B-7HS	0.024
15	2	15.01	246	2.205	1.969	L82	B-7HS	0.024
20	2	26.20	430	2.638	2.402	L82	B-7HS	0.024
25	2	26.20	430	2.638	2.402	L82	B-7HS	0.024
30	2	30.30	496	2.835	2.402	L82	B-7HS	0.024
40	2	36.13	592	2.953	2.638	L78	B-8HS	0.025
55	2	46.37	760	3.228	2.835	L78	B-8HS	0.025

TUNE-UP ADJUSTMENTS

CARB TYPE	PRIMARY PICKUP TIMING	TIMING AT IDLE	MAXIMUM SPARK ADV AT WOT	IDLE RPM	OPERATING RANGE RPM	POINT GAP IN.
1983						
A	--	--	18° BTDC	1200	4000-5000	0.014
C	--	5° BTDC	28° BTDC	1000	4500-5500	--
C	--	5° BTDC	28° BTDC	1000	4500-5500	--
F or G	--	4° BTDC	25° BTDC	1200	4500-5500	--
F or G	--	5° BTDC	25° BTDC	650	4500-5500	--
F or G	--	5° BTDC	22° BTDC	1150	4500-5500	--
J	2° ATDC	6° ATDC	21° BTDC	700	4500-5500	--
J	--	2° BTDC	24° BTDC	700	4500-5500	--
J	--	TDC	25° BTDC	900	4500-5500	--
K	--	4° BTDC	26° BTDC	900	4500-5500	--
L	4° BTDC	5° ATDC	22° BTDC	800	5000-5800	--
1984						
A	--	--	18° BTDC	1200	4000-5000	0.014
C	--	5° BTDC	28° BTDC	1000	4500-5500	--
C	--	5° BTDC	28° BTDC	1000	4500-5500	--
F or G	--	4° BTDC	25° BTDC	1200	4500-5500	--
F or G	--	5° BTDC	25° BTDC	650	4500-5500	--
F or G	--	5° BTDC	22° BTDC	1150	4500-5500	--
J	2° ATDC	6° ATDC	21° BTDC	700	4500-5500	--
J	--	2° BTDC	24° BTDC	700	4500-5500	--
J	--	TDC	25° BTDC	900	4500-5500	--
K	--	4° BTDC	26° BTDC	900	4500-5500	--
1985						
A	--	--	18° BTDC	1200	4000-5000	0.014
C	--	5° BTDC	28° BTDC	1000	4500-5500	--
C	--	5° BTDC	28° BTDC	1000	4500-5500	--
F or G	--	4° BTDC	25° BTDC	1200	4500-5500	--
F or G	--	5° BTDC	25° BTDC	650	4500-5500	--
F or G	--	5° BTDC	22° BTDC	1150	4500-5500	--
J	2° ATDC	6° ATDC	21° BTDC	700	4500-5500	--
J	--	2° BTDC	24° BTDC	700	4500-5500	--
J	--	TDC	25° BTDC	900	4500-5500	--
K	--	4° BTDC	26° BTDC	900	4500-5500	--
L	TDC	2° ATDC	26° BTDC	800	4500-5500	--

ENGINE SPECIFICATIONS AND

MODEL	CYL	CU IN DISPL.	CU CM DISPL.	BORE INCHES	STROKE INCHES	SPARK PLUG CHAMP	SPARK PLUG NGK	SPARK PLUG GAP IN.
				1986				
2	1	2.62	43	1.535	1.417	L86	B-6HS	0.024
4	1	5.06	83	1.969	1.654	L82	B-7HS	0.024
5	1	6.28	103	2.126	1.772	L82	B-7HS	0.024
8	2	10.01	165	1.969	1.654	L86	B-6HS	0.024
9.9	2	12.76	210	2.126	1.772	L82	B-7HS	0.024
15	2	15.01	246	2.205	1.969	L82	B-7HS	0.024
20	2	24.40	400	2.562	2.375	L86	B-6HS	0.024
25	2	24.40	400	2.562	2.375	L86	B-6HS	0.024
30	2	30.30	496	2.835	2.402	L82	B-7HS	0.024
40	2	36.13	592	2.953	2.638	L78	B-8HS	0.025
55	2	46.37	760	3.228	2.835	L78	B-8HS	0.025
				1987				
2	1	2.62	43	1.535	1.417	L86	B-6HS	0.024
4	1	5.06	83	1.969	1.654	L82	B-7HS	0.024
5	1	6.28	103	2.126	1.772	L82	B-7HS	0.024
8	2	10.01	165	1.969	1.654	L86	B-6HS	0.024
9.9	2	12.76	210	2.126	1.772	L82	B-7HS	0.024
15	2	15.01	246	2.205	1.969	L82	B-7HS	0.024
20	2	24.40	400	2.562	2.375	L86	B-6HS	0.024
25	2	24.40	400	2.562	2.375	L86	B-6HS	0.024
30	2	30.30	496	2.835	2.402	L82	B-7HS	0.024
40	2	36.13	592	2.953	2.638	L78	B-8HS	0.025
				1988 & 1989				
2	1	2.62	43	1.535	1.417	L86	B-6HS	0.024
4	1	5.06	83	1.969	1.654	L82	B-7HS	0.024
5	1	6.28	103	2.126	1.772	L82	B-7HS	0.024
8	2	12.76	210	2.126	1.772	L82	B-7HS	0.024
9.9	2	12.76	210	2.126	1.772	L82	B-7HS	0.024
15 (1988)	2	15.01	246	2.205	1.969	L82	B-7HS	0.024
15 (1989)	2	15.98	262	2.362	1.811	L82	B-7HS	0.024
20	2	24.40	400	2.562	2.375	L86	B-6HS	0.024
25	2	24.40	400	2.562	2.375	L86	B-6HS	0.024
30	2	30.30	496	2.835	2.402	L82	B-7HS	0.024
40	2	36.13	592	2.953	2.638	L78	B-8HS	0.025

TUNE-UP ADJUSTMENTS

CARB TYPE	PRIMARY PICKUP TIMING	TIMING AT IDLE	MAXIMUM SPARK ADV AT WOT	IDLE РRM	OPERATING RANGE RPM	POINT GAP IN.
1986						
A	--	--	18° BTDC	1200	4000-5000	0.014
C	--	5° BTDC	28° BTDC	1000	4500-5500	--
C	--	5° BTDC	28° BTDC	1000	4500-5500	--
F or G	--	4° BTDC	25° BTDC	1200	4500-5500	--
F or G	--	5° BTDC	25° BTDC	650	4500-5500	--
F or G	--	5° BTDC	22° BTDC	1150	4500-5500	--
J	2° ATDC	6° ATDC	21° BTDC	700	4500-5500	--
J	--	2° BTDC	24° BTDC	700	4500-5500	--
J	--	TDC	25° BTDC	900	4500-5500	--
K	--	4° BTDC	26° BTDC	900	4500-5500	--
L	TDC	2° ATDC	26° BTDC	800	4500-5500	--
1987						
A	--	--	18° BTDC	1200	4000-5000	0.014
C	--	5° BTDC	28° BTDC	1000	4500-5500	--
C	--	5° BTDC	28° BTDC	1000	4500-5500	--
F or G	--	4° BTDC	25° BTDC	1200	4500-5500	--
F or G	--	5° BTDC	25° BTDC	650	4500-5500	--
F or G	--	5° BTDC	22° BTDC	1150	4500-5500	--
J	2° ATDC	6° ATDC	21° BTDC	700	4500-5500	--
J	--	2° BTDC	24° BTDC	700	4500-5500	--
J	--	TDC	25° BTDC	900	4500-5500	--
K	--	4° BTDC	26° BTDC	900	4500-5500	--
1988 & 1989						
A	--	--	18° BTDC	1200	4000-5000	0.014
C	--	--	28° BTDC	1000	4500-5500	--
C	--	--	28° BTDC	1000	4500-5500	--
F or G	--	4° BTDC	25° BTDC	1200	4500-5500	--
F or G	--	5° BTDC	25° BTDC	650	4500-5500	--
F or G	--	5° BTDC	22° BTDC	1150	4500-5500	--
F or G	--	5° BTDC	22° BTDC	1150	4500-5500	--
J	2° ATDC	6° ATDC	21° BTDC	700	4500-5500	--
J	--	2° BTDC	24° BTDC	700	4500-5500	--
J	--	TDC	25° BTDC	900	4500-5500	--
K	--	4° BTDC	26° BTDC	900	4500-5500	--

A-10 APPENDIX

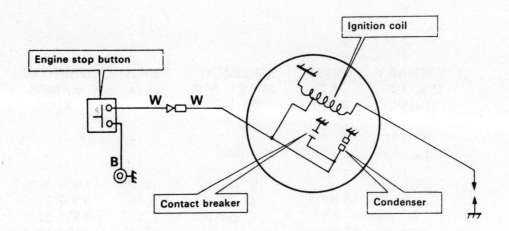

Simplified wiring diagram for the 2hp powerhead.

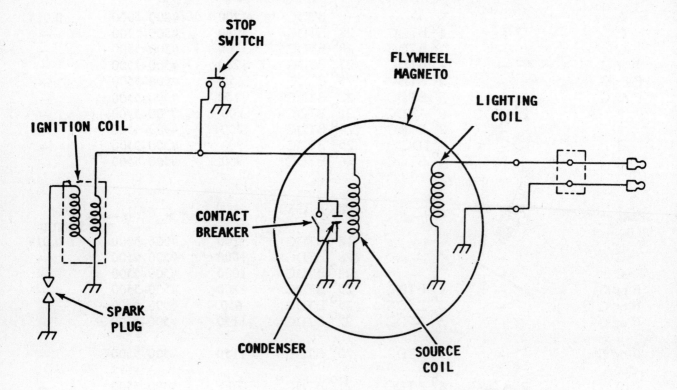

Wire identification for the 3.5hp and 5hp Air-Cooled Powerheads with Breaker Point Type Ignition.

WIRING DIAGRAMS   A-11

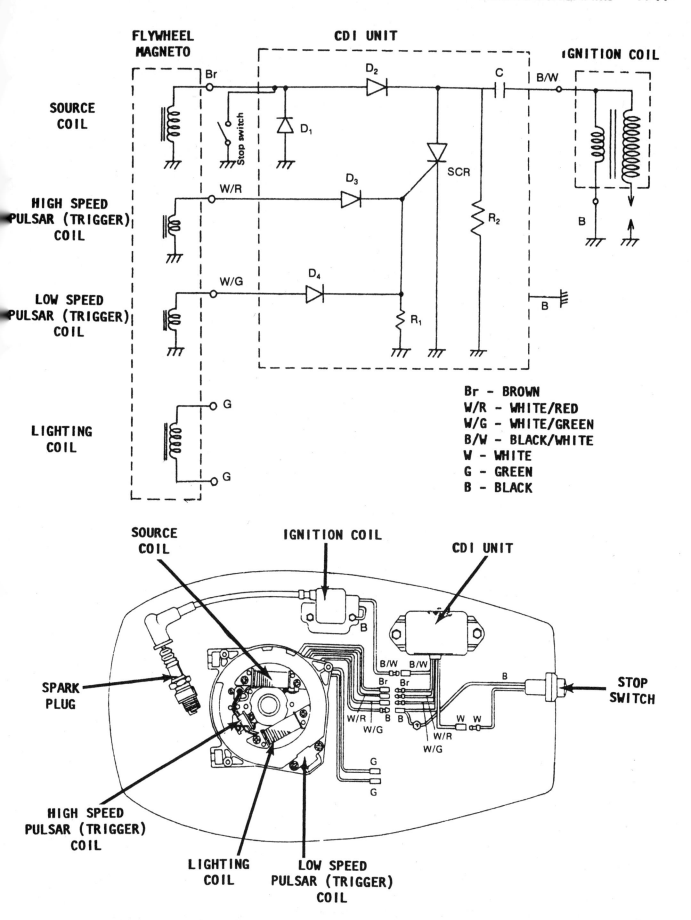

Wire identification for the 4hp and 5hp Water-Cooled Powerheads with CDI Type Ignition.

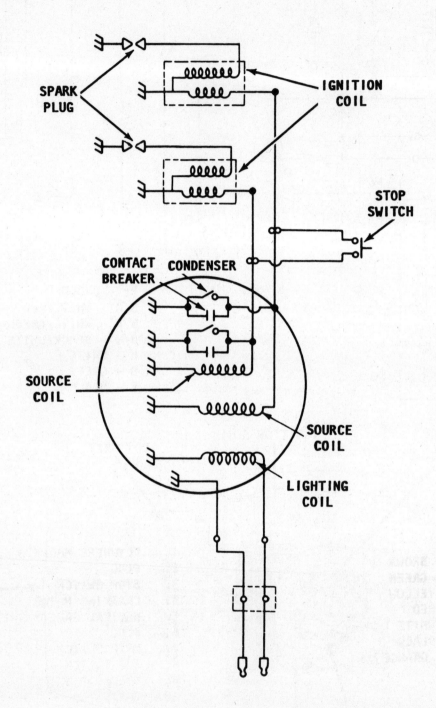

Wire identification for the Ignition System for the 8hp, 15hp, 20hp and 28hp Powerheads with Breaker Point Type Ignition. A separate wiring diagram for auxiliary wiring on Models 8hp, 15hp, and 20hp equipped with electric cranking motors is presented on the following page. A wiring diagram for Model 28hp is presented on Page A-16.

WIRING DIAGRAMS   A-13

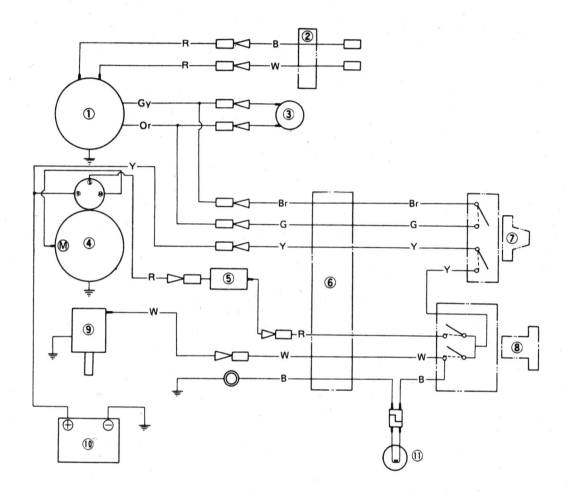

Br - BROWN
Gy - GREEN
Y - YELLOW
R - RED
W - WHITE
B - BLACK
Or - ORANGE

1. FLYWHEEL MAGNETO
2. PLUG
3. STOP SWITCH
4. CRANKING MOTOR
5. NEUTRAL SAFETY SWITCH
6. PLUG
7. MAIN SWITCH
8. SUB-SWITCH
9. SOLENOID SWITCH
10. BATTERY
11. PILOT LAMP

Wire identification for the Electrical System for the 8hp, 15hp, 20hp and 28hp Powerheads with Breaker Point Type Ignition. A separate wiring diagram for the Ignition System is presented on the previous page.

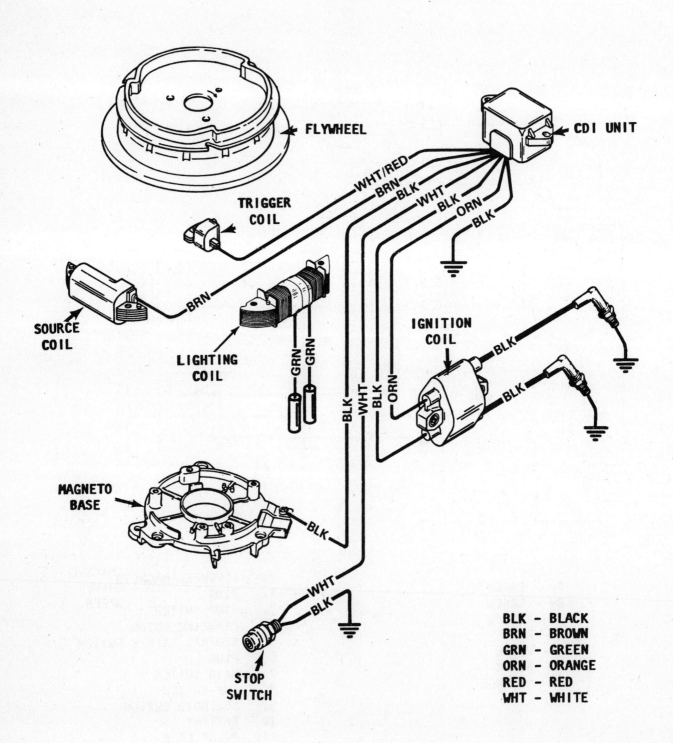

Wire identification for the 8hp, 9.9hp, and 15hp Powerheads with CDI Type Ignition.

WIRING DIAGRAMS   A-15

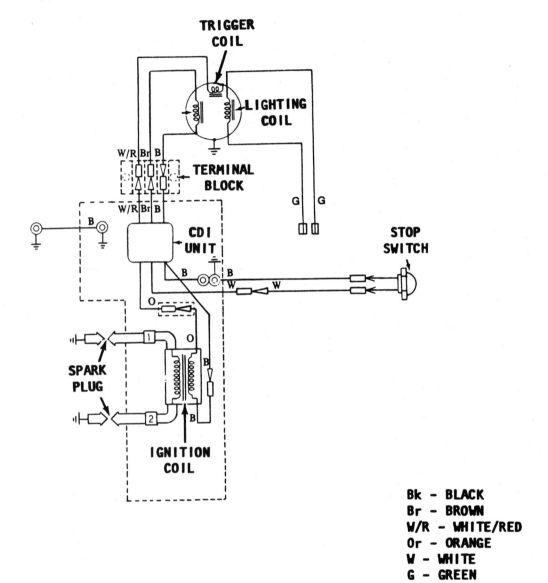

Wire identification for the 20hp, 25hp, and 30hp Powerheads with CDI Type Ignition.

A-16 APPENDIX

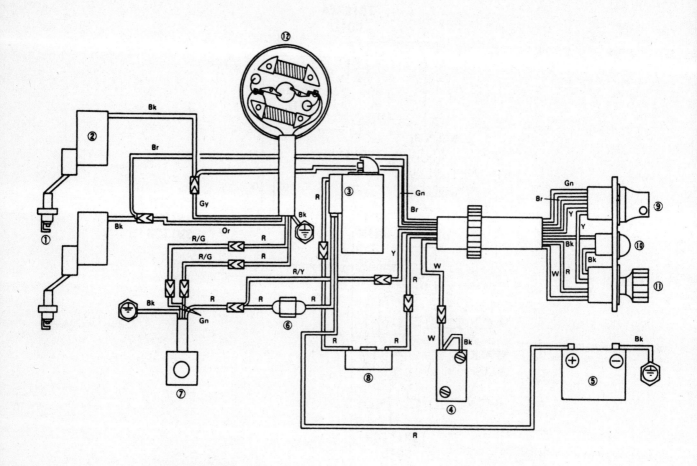

1. SPARK PLUG
2. IGNITION COIL
3. CRANKING MOTOR
4. SOLENOID SWITCH
5. BATTERY
6. FUSE
7. RECTIFIER
8. NEUTRAL SAFETY SWITCH
9. MAIN SWITCH
10. PILOT LAMP
11. STOP SWITCH
12. FLYWHEEL MAGNETO BASE

Bk - BLACK
Br - BROWN
Gy - GRAY
Gn - GREEN
Bl - BLUE
Y - YELLOW
R - RED
R/Y - RED/YELLOW
R/G - RED/GREEN
W - WHITE
Or - ORANGE

Wire identification for the Electrical System for the 28hp Powerhead with Breaker Point Type Ignition. A separate wiring diagram for the Ignition System is presented on Page A-12.

WIRING DIAGRAMS A-17

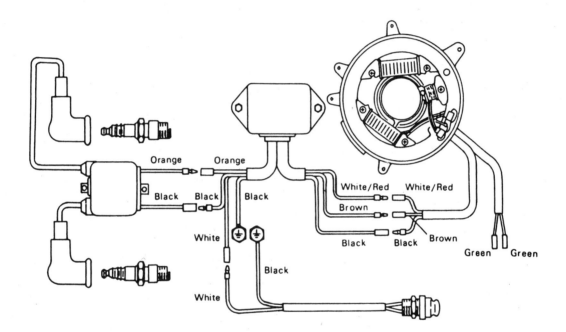

Wire identification for the 40hp Powerhead with CDI Type Ignition.

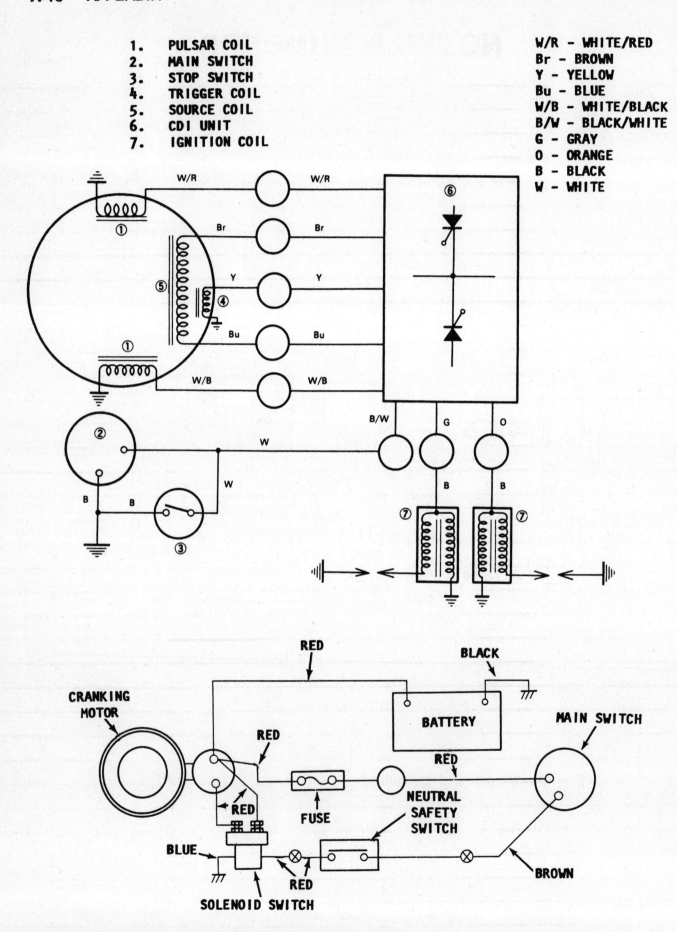

Wire identification for the 48hp, 55hp and 60hp Powerheads with CDI Type Ignition.

NOTES & NUMBERS

NOTES & NUMBERS

NOTES & NUMBERS

NOTES & NUMBERS

NOTES & NUMBERS

Other Seloc Marine Manuals

New titles are constantly being produced and the updating work on existing manuals never ceases.
All manuals contain complete detailed instructions, specifications, and wiring diagrams.